Communications and Control Engineering

Springer
London
Berlin
Heidelberg
New York
Barcelona
Budapest
Hong Kong
Milan
Paris
Santa Clara
Singapore
Tokyo

Romeo Ortega, Antonio Loría, Per Johan Nicklasson
and Hebertt Sira-Ramírez

Passivity-based Control of Euler-Lagrange Systems

Mechanical, Electrical and Electromechanical Applications

Springer

Romeo Ortega, PhD
Laboratoire des Signaux et Systèmes, UMR CNRS 014, Gif-sur-Yvette 91192, France
Antonio Loría, PhD
Electrical and Computer Engineering Department, University of California,
Santa Barbara, CA 93106-9560, USA
Per Johan Nicklasson, Dr Ing
SINTEF Electronics and Cybernetics, N-7034 Trondheim, Norway
Hebertt Sira-Ramírez, PhD
Departamento Sistemas de Control, Escuela de Ingenieria de Sistemas,
Universidad de los Andes, Merida, Venezuela

Series Editors
B.W. Dickinson • A. Fettweis • J.L. Massey • J.W. Modestino
E.D. Sontag • M. Thoma

ISBN 1-85233-016-3 Springer-Verlag London Berlin Heidelberg

British Library Cataloguing in Publication Data
Passivity-based control of Euler-Lagrange systems :
   mechanical, electrical and electromechanical applications
   - (Communications and control engineering)
   1.Control theory   2.Language equations
   I.Ortega, Romeo
   629.8'312
ISBN 1852330163

Library of Congress Cataloging-in-Publication Data
A catalog record for this book is available from the Library of Congress

Typesetting: Camera ready by authors
Printed and bound at the Athenæum Press Ltd., Gateshead, Tyne & Wear
69/3830-543210 Printed on acid-free paper

To Amparo with all my love,
Romeo.

To Lena with $r = 1 - \sin\theta$,
to Mari my sister with my deepest admiration,
Toño.

To my parents,
Per Johan.

To José Humberto Ocariz E. with respect and affection,
to María Elena and María Gabriela with all my love,
Hebertt.

Preface

By its own definition the final purpose of control is to *control* something. In fact, the foundational developments of Huygens, Maxwell, Routh, Minorsky, Nyquist and Black (to name a few) were motivated by real–world applications. In the hands of mathematicians such as Wiener, Bellman, Lefschetz, Kalman and Pontryagin (again, to name just a few) control theory developed in the 1950s and 1960s as a branch of applied mathematics, independent of its potential application to engineering problems. Some tenuous arguments were typically invoked to provide some practical motivation to the research on this so–called mathematical control theory. For instance, the study of the triple (A, B, C) was rationalized as the study of the linearization of an arbitrary nonlinear system –an argument that had a grain of truth. By the end of the 1980s a fairly complete body of knowledge for general linear systems –including powerful techniques of controller synthesis– had been completed. Some spectacular applications of this theory to practical situations that fitted the linear systems paradigm were reported.

The attempt to mimic the developments of linear systems theory in the general nonlinear case enticed many researchers. Extensions to a fairly general class of nonlinear systems of the basic concepts of controllability, observability, and realizability were crowned with great success. The controller synthesis problem proved to be, however, much more elusive. Despite some significant progress, to date, general techniques for stabilization of nonlinear systems are available only for special classes of nonlinear systems. This is, of course, due to the daunting complexity of the behaviour of nonlinear dynamic systems which puts a serious question mark on the interest of aiming at a monolithic synthesis theory. On the other hand, new technological developments had created engineering problems where certain well–defined nonlinear effects had to be taken into account. Unfortunately, the theory developed for general nonlinear systems could not successfully deal with them, basically because the "admissible structures" were determined by analytical considerations, which do not necessarily match the physical constraints. It became apparent that to solve these new problems, the "find an application for my theory" approach had to be abandoned, and a new theory tailored for the application had to be worked out.

The material reported in this book is an attempt in this direction. Namely, we start from a well–defined class of systems to be controlled and try to develop a theory best suited for them. As the title suggests, the class we consider covers a very broad spectrum (Euler–Lagrange systems with mechanical, electrical and electromechanical applications) however, detailed analysis is presented only for robots, AC machines and power converters. We have found that this set of applications is sufficiently general —it has at least kept us busy for the last 10 years!

Different considerations and techniques are used to solve the various problems however, in all cases we strongly rely on the information provided by the variational modeling and, in particular, concentrate our attention on the energy and dissipation functions that define the dynamics of the system. A second unifying thread to all the applications is the fundamental concept of passivity. Finally, a recurrent theme throughout our work is the notion of interconnection that appears, either in the form of a feedback decomposition instrumental for the developments, or as a framework for focusing on the relevant parts of a model.

An important feature of the proposed controller design approach is that it is based on the input–output property of passivity, hence it will typically not require the measurement of the full state to achieve the control objectives. Consequently, throughout the book we give particular emphasis to (more realistic, but far more challenging) output–feedback strategies.

The book is organized in the following way. In Chapter **1** we present first a brief introduction that explains the background of the book and elaborates upon its three keywords: Euler–Lagrange (EL) systems, passivity and applications. The notion of passivity–based control (PBC) is explained in detail also in this chapter, underscoring its conceptual advantages. The main background material pertaining to EL systems is introduced in Chapter **2**. In particular we mathematically describe the class of systems that we study throughout the book, exhibit some fundamental input–output and Lyapunov stability properties, as well as some basic features of their interconnection. We also give in this section the models of some examples of physical systems that will be considered in the book.

The remaining of the book is divided in three parts devoted to mechanical, electrical and electromechanical systems, respectively. The first part addresses a class of mechanical systems, of which a prototypical example are the robot manipulators, but it is not restricted to them. For instance, we consider also applications to simple models of marine vessels and rotational translational actuators. The results concerning mechanical systems are organized into set point regulation (Chapter **3**), trajectory tracking (Chapter **4**) and adaptive disturbance attenuation, with application to friction compensation (Chapter **5**). The theoretical results are illustrated with realistic simulation results. In this part, as well as in other sections of the book, we carry out comparative studies of the performance obtained by PBC with those achievable with other schemes. In particular, for robots with flexible joints, we compare in Chapter **4** PBC with schemes based on backstepping and cascaded systems.

The second part of the book is dedicated to electrical systems, in particular DC-

DC power converters. In Chapter **6** the EL model is derived and the control relevant properties are presented. We present both, a switched model that describes the exact behaviour of the system with a switching input, and an approximate model for the pulse width modulator controlled converters. While in the first model we have to deal with a hybrid system (with inputs 0 or 1), in the latter model, which is valid for sufficiently high sampling frequencies, the control input is the duty ratio which is a continuous function ranging in the interval $[0, 1]$. Besides the standard EL modeling, we also present a rather novel, and apparently more natural, Hamiltonian model that follows as a particular case of the extended Hamiltonian models proposed by Maschke and van der Schaft.

Chapter **7** is devoted to control of DC-DC power converters. We present, of course, PBC for the average models. To deal with hybrid models we also introduce the concept of PBC with sliding modes. We show that combining this two strategies we can reduce the energy consumption, a well–known important drawback of sliding mode control. Adaptive versions of these schemes, that estimate on–line the load resistance are also derived. An exhaustive experimental study, where various linear and nonlinear schemes are compared, is also presented.

In the third part of the book we consider electromechanical systems. To handle this more challenging problem we introduce a feedback decomposition of the system into passive subsystems. This decomposition naturally suggests a nested–loop controller structure, whose basic idea is presented in a motivating levitated system example in Chapter **8**. This simple example helps us also to clearly exhibit the connections between PBC, backstepping and feedback linearization. In Chapters **9–11** we carry out a detailed study of nonlinear control of AC motors. The torque tracking problem is first solved for the generalized machine model in Chapter **9**. As an off–spin of our analysis we obtain a systems invertibility interpretation of the well-known condition of Blondel–Park transformability of the machine.

The next two chapters, **10** and **11**, are devoted to voltage–fed and current–fed induction machines, respectively. For the voltage–fed case we present, besides the nested–loop scheme, a PBC with total energy shaping. Connections with the industry standard field oriented control and feedback linearization are thoroughly discussed. These connections are further explored for current–fed machines in Chapter **11**. First, we establish the fundamental result that, for this class of machines, PBC exactly reduces to field oriented control. Then, we prove theoretically and experimentally that PBC outperforms feedback linearization control. The robustness of PBC, as well as some simple tuning rules are also given. Finally, motivated by practical considerations, a globally stable discrete–time version of PBC is derived. Both chapters contain extensive experimental evidence.

At last, in Chapter **12** we study the problem of electromechanical systems with nonlinear mechanical dynamics. The motivating example for this study is the control of robots with AC drives, for which we give a complete theoretical answer. The chapter clearly illustrates how PBC, as applied to Euler–Lagrange models, yields a modular design which effectively exploits the features of the interconnections. In a simulation study we compare our PBC with a backstepping design showing, once again, the superiority of PBC.

Background material on passivity, variational modeling and vector calculus are included in Appendices **A**, **B** and **C**, respectively.

The book is primarily aimed at graduate students and researchers in control theory who are interested in engineering applications. It contains, however, new theoretical results whose interest goes beyond the specific applications, therefore it might be useful also to more theoretically oriented readers. The book is written with the conviction that to deal with modern engineering applications, control has to reevaluate its role as a component of an interdisciplinary endeavor. A lot of emphasis is consequently given to modeling aspects, analysis of current engineering practice and experimental work. For these reasons it may be also of interest for students and researchers, as well as practitioning engineers, involved in more practical aspects of robotics, power electronics and motor control. For this audience the book may provide a source to enhance their theoretical understanding of some well–known concepts and to establish bridges with modern control theoretic concepts.

We have adopted the format of theorem–proof–remark, which may give the erroneous impression that it is a "theoretical" book, this is done only for ease of presentation. Although most of the results in this book are new, they are presented at a level accessible to audiences with a standard undergraduate background in control theory and a basic understanding of nonlinear systems theory. In order to favour the "readability" of our book we have moved some of the most "technical" proofs to Appendix **D**.

The material contained in the book summarizes the experience of the authors on control engineering applications over the last 10 years. It builds upon the PhD theses of the second and fourth author as well as collaborative research among all of us, and with several other researchers. Numerous colleagues and collaborators contributed directly and indirectly, and in various ways to this book.

The first author is particularly indebted to his former PhD students: G. Espinosa and R. Kelly triggered his interest in the areas of electrical machines and robotics, respectively, we have since kept an intensive and very productive research collaboration; G. Escobar, K. Kim and D. Taoutaou carried out some of the experimental work on converters and electrical machines. He has also enjoyed a long scientific collaboration with L. Praly who always provided insightful remarks and motivation to his work. Many useful scientific exchanges have been carried out over the years with H. Nijmeijer, M. Spong and A. J. van der Schaft, while Henk and Arjan motivated him to improve his theoretical background, Mark always found the threshold necessary to make a robot turn. He would like to thank all his co–authors from whom he learned the importance of collaborative work. Finally, he wants to express his deep gratitude to the french CNRS, which provides his researchers with working conditions unparalleled by any other institution in the world.

The second author wishes to acknowledge specially the collaboration with his former undergraduate-school teacher R. Kelly who earlier introduced him to robot control and Lyapunov theory. The enthusiasm of Rafael on these topics increased the motivation of the second author to pursue a doctoral degree in the field. During his doctoral research period he was also enriched with the advice and collaboration of H. Nijmeijer and L. Praly. The author wishes to express as well his deepest gratitude to his fiancee and collaborator E. Panteley for her fundamental moral support in this project and for helping with the figures of Chapter **6**. Last but not least, the work of the second author has been sponsored by the institutions he has been affiliated to in the past 5 years, in chronological order: CONACyT, Mexico; University of Twente, The Netherlands; University of Trondheim, Norway; and University of California at Santa Barbara, USA.

The third author wants to thank Research Director Peter Singstad, SINTEF Electronics and Cybernetics, Automatic Control, for supporting parts of this project financially.

The fourth author is indebted to his colleagues and students of the Control Systems Department of the Universidad de Los Andes (ULA) in Mérida (Venezuela) for the continuous support over the years in many academic endeavors. Special thanks and recognition are due to his former student, Dr. Orestes Llanes-Santiago, for his creative enthusiasm and hard work in the area of switched power converters. Visits to R. Ortega, since 1995, have been generously funded by the Programme de Cooperation Postgradué (PCP), by the National Council for Scientific Research of Venezuela (CONICIT), as well as by the Centre National de la Recherche Scientifique (CNRS) of France. Thanks are due to Professor Marisol Delgado, of the Universidad Simón Bolívar, who has acted as a highly efficient PCP Coordinator in Venezuela. Over the years, the author has benefited from countless motivational discussions with his friend Professor Michel Fliess of the Laboratoire des Signaux et Systèmes (CNRS), France. His experience and vision has been decisively helpful in many of the author's research undertakings.

R. Ortega, A. Loría, P. J. Nicklasson, H. Sira-Ramírez,
May 1998.

Contents

List of Figures

List of Tables

Notations, acronyms, and numbering

Most commonly used mathematical symbols.

$\mathbb{R}$	Field of real numbers.
$\mathbb{R}^n$	Linear space of real vectors of dimension n.
$\mathbb{R}^{n \times m}$	Ring of matrices with n rows and m columns and elements in $\mathbb{R}$.
$\mathbb{R}_{\geq 0}$	Field of nonnegative real numbers.
I_n	The identity matrix of dimension n.
t	Time, $t \in \mathbb{R}_{\geq 0}$.
$\mathcal{L}_2^n$	Space of n-dimensional square integrable functions.
$\mathcal{L}_{2e}^n$	Extended space of n-dimensional square integrable functions.
$\langle \cdot \vert \cdot \rangle$	Inner product in $\mathcal{L}_2^n$.
$\langle \cdot \vert \cdot \rangle_T$	Truncated inner product.
$\Vert \cdot \Vert_{2T}$	The truncated $\mathcal{L}_2$ norm.
$\rightarrow$	Mapping from a domain into a range. Also "tends to".
$\mapsto$	Mapping of two elements into their image.
$\triangleq$	"defined as".
$\frac{dz}{d\xi}$	Derivative of $z = f(\xi)$.
$\frac{d}{dt}(\cdot) = (\dot{\cdot})$	Total time derivative.
$p \triangleq \frac{d}{dt}$	Differentiation operator.
$\frac{\partial}{\partial \xi}$	Differentiation operator with respect to ξ.
$\underline{n}$	The set of integers $[1, \dots, n]$.

Matrices.

A usual convention undertaken in this book (though not exhaustively) is to use UPPERCAPS for matrices and lowercaps for vectors and scalars.

$\Vert x \Vert$	The Euclidean norm of $x \in \mathbb{R}^n$.
$\Vert A \Vert$	with $A \in \mathbb{R}^{n \times m}$: Induced 2-norm.
$\vert a \vert$	Absolute value of the scalar a.

K_p	Frequently used (e.g. Part I) to denote a "proportional" gain.
K_d	Frequently used (e.g. Part I) to denote a "derivative" gain.
K_i	Frequently used (e.g. Part I) to denote an "integral gain".
diag{} or diag()	Diagonal operator: transforms a vector into a diagonal matrix.
$(\cdot)^{-1}$	Inverse operator.
A_{ij} or a_{ij}	ij-th element of the matrix A.
k_m	Positive constant corresponding to a square matrix K, such that $k_m \|x\|^2 \leq x^\top K x$ for all $x \in \mathbb{R}^n$. Correspondingly k_{p_m} for K_p.
k_M	Positive constant corresponding to a square matrix K, such that $k_M \|x\|^2 \geq x^\top K x$ for all $x \in \mathbb{R}^n$. Correspondingly k_{p_M} for K_p.
$\underline{\lambda}(K)$	Smallest eigenvalue of K.
$\overline{\lambda}(K)$	Largest eigenvalue of K.
$(\cdot)^\top$	Transpose operator.

Euler-Lagrange systems.

q	Vector of generalized positions.
q_*	Denotes the desired reference value for q imposed by the designer, hence an *external* signal.
q_d	Denotes a "desired" value of certain *internal* signals produced by the passivity-based-control design.
$\dot{q}$	Vector of generalized velocities.
$\tilde{()}$	Denotes an error between two quantities, typically a state variable and its desired value. Notice that the desired value can be denoted using $_*$ or $_d$ depending on the nature of the reference.
ϑ	Output of of the "dirty derivatives" filter.
$\hat{\theta}$	Denotes the estimate of a parameter (or vector of parameters) θ.
$\tilde{\theta}$	The estimation error $\hat{\theta} - \theta$.
$\mathcal{T}(q, \dot{q})$	Kinetic energy.
$\mathcal{V}(q)$	Potential energy.
$\mathcal{F}(\dot{q})$	Rayleigh dissipation function.
$\mathcal{M}$	Inputs matrix.
$\mathcal{H}$	Energy or storage function.
$D(q), \mathcal{D}(q)$	Inertia matrix.
D_m	Moment of inertia.
$D_e(q_m)$	Inductance matrix.
$C(q, \dot{q}), \mathcal{C}(q, \dot{q})$	Coriolis and centrifugal forces matrix.
$g(q)$	Potential forces vector.
Q	External generalized forces.
R_m	Mechanical viscous damping constant.
u, u_p	Control inputs.
$()_e, (\cdot)_m$	Denote electrical and mechanical variables respectively.

q_p	Generalized positions vector of an EL *plant*.
$\mathcal{T}_p(q_p, \dot{q}_p)$	Kinetic energy of a plant.
$\mathcal{V}_p(q_p)$	Potential energy of a plant.
$\mathcal{F}_p(\dot{q}_p)$	Rayleigh dissipation function of a plant.
$\mathcal{M}_p$	Matrix which maps the inputs to an EL plant's coordinates.
q_c	Generalized positions of an EL *controller*.
$\mathcal{T}_c(q_c, \dot{q}_c)$	Controller's kinetic energy.
$V_c(q_c, q_p)$	Controller's potential energy.
$\mathcal{F}_c(\dot{q}_c)$	Controller's Rayleigh dissipation function.
λ	Flux linkage vector.
$\dot{q}_e$	Current vector.
$\mathbf{e}^{(\cdot)}$	Matrix exponential function.
$\mathcal{J}$	The matrix $\begin{bmatrix} 0 & -1 \\ 1 & 0 \end{bmatrix}$.
$\mathbf{e}^{\mathcal{J}(\cdot)}$	The 2×2 rotation matrix $\begin{bmatrix} \cos(\cdot) & -\sin(\cdot) \\ \sin(\cdot) & \cos(\cdot) \end{bmatrix}$.

Frequent acronyms

AC	Alternate current.
BP	Blondel–Park.
DC	Direct current.
EL	Euler-Lagrange.
FET	Field effect transistor.
FLC	Feedback linearization control.
FOC	Field-oriented-control.
GUAS	Global uniform asymptotic stability.
GAS	Global asymptotic stability.
IBC	Integrator backstepping control.
ISP	Input Strictly Passive.
KYP	Kalman-Yakubovich-Popov.
LC	Inductor-Capacitor.
LQG	Linear quadratic Gaussian.
OBFL	Observer–based feedback linearizing.
OSP	Output Strictly Passive.
PBC	Passivity-based control.
PD	Proportional derivative.
PID	Proportional integral derivative.
PI^{2}D	Proportional double-integral derivative.
PWM	Pulse-width modulation.
RLC	Resistor-Inductor-Capacitor.
SM	Sliding mode.
TORA	Translational-rotational actuator.

Numbering of chapters, sections, equations, theorems, propositions, properties, assumptions, examples, definitions, facts, corollaries, lemmas and remarks

This book contains three parts labeled I-III and 12 chapters labeled **1-12**. The chapters are structured in sections labeled in each chapter as 1, 2, etc. The subsection numbers are relative to the section which they belong to that is, 1.1, 1.2, etc. The sub-subsections are labeled A., B., C., etc. Below sub-subsections we use the numbering A.1, A.2., etc. Only sections and subsections are listed in the table of contents.

When citing a section or subsection belonging to a different chapter, we use the **bold** font for the chapter number. Thus, the reader should not be confused for instance, between Section 2 of Chapter **1**, cited **1**.2 in Chapter **3**, from Subsection 2 of Section 1, cited in Chapter **3**, as 1.2.

The equation numbers are relative to the corresponding chapter, for instance, equation (3.56), is the 56th equation of Chapter **3**. Theorems, propositions, lemmas, definitions, examples, etc succeed each other in the numbering, which is relative to the chapter number. For instance, Remark 3.2 follows after Proposition 3.1 in the third chapter.

Properties and assumptions are labeled as **P#** and **A#** respectively where the number # is relative to the chapter. For instance **P1.1** corresponds to Property 1.1 which is defined in Chapter **1**.

Figure and table counters are also relative to the chapter number.

Chapter 1

Introduction

"A control theorist's first instinct in the face of a new problem is to find a way to use the tools he knows, rather that a commitment to understand the underlying phenomenon. This is not the failure of individuals but the failure of our profession to foster the development of experimental control science. In a way, we have become the prisoners of our rich inheritance and past successes".

Y. C. Ho (1982).

The final objective of the research reported in this book is to contribute to the development of a system theoretic framework for control of nonlinear systems that incorporates at a fundamental level the systems physical structure and provides solutions to practical engineering problems. This is, of course, a very ambitious and somehow imprecise objective. To help delineate what we really want to accomplish we underscore the three major keywords of our work: *Euler-Lagrange* (EL) systems, *passivity* and *applications*. The first keyword mathematically defines the class of systems that we study, the second one the main physical property that we focus on, while the last one is our final objective. In this chapter we will develop upon this three keywords to explain the background and the contents of the book, and to motivate our approach.

1 From control engineering to mathematical control theory and back

The early developments in control theory (from the 1930s to the early 1960s) were motivated by technological problems ranging from feedback amplifier design to space applications. Actually, the fundamental notion of feedback was first introduced by

Black in his celebrated feedback amplifier. Experiments were usually carried out to establish insightful frequency response models, which via an intrinsic averaging process, captured the effect of uncertainties and in this way led to robust designs. The mathematical framework that was suited for modeling the plant, to analyze it, and to synthesize controllers was based on input–output ideas with Fourier and Laplace transforms being the key tools. This viewpoint took a major shift with the introduction of state–space ideas, where the mapping between inputs and outputs takes place via the transformation of the internal state of the system. The natural mathematical framework in the state–space approach became then the differential equations.

The mathematical advantages of the state–space formalism were rapidly exploited in the 1960s and 1970s resulting in the introduction of many new fundamental concepts, notably those of controllability, observability and optimality, and the development of powerful techniques for controller synthesis (identification, adaptive control, digital control, among others). The theoretical research expanded very rapidly, with the control field attracting many of the best students who indulged themselves in beautiful mathematical problems. Unfortunately, with this paradigm shift control drifted away from its design vocation and became more mathematically driven than driven by its original aim of providing the tools to cope with nonlinearity and uncertainty in practical engineering problems.[1]

The research in the 1970s and early 1980s seemed to be motivated more by trendy academic problems, mathematical abstraction, and an overemphasis on certain given methodologies than by engineering needs. In particular, mathematical modeling, a central ingredient for successful control design, was essentially relegated. Also, the interest in control experiments faded out in view of the conventional wisdom that it was scientifically pointless to build an experiment to test a mathematical statement that was self–consistent and provably correct.

By the mid–1980s the widening gap between theory and practice did not pass unnoticed to funding agencies and industries which, via economic pressure, forced the control community to reevaluate its research directions and put more emphasis on practical applications. This trend has not reduced the importance of control theory, since new technological developments have created difficult engineering challenges that require the use of sophisticated theoretical concepts. It has, however, underscored the role of control as a component of an interdisciplinary approach needed to solve the new practical problems.[2] The control specialist has to interact then with people from other engineering fields. The establishment of such communication is, of course, a long process and it is difficult to decide where to start and which route to take.

---

[1] A brief respite to this scenario resulted from the revival of robustness issues in the early 1980s. In particular, the elegant formulation and solution of what is called the $\mathcal{H}_\infty$ problem and the research on structured uncertainties.

[2] A successful story that illustrates the advantages of synergetic collaboration is the exciting new field of mechatronics.

2   A route towards applications

The work reported in this book, which summarizes our experience in control engineering applications over the last 10 years, has roughly proceeded along the following path:

- Since the main skills of control engineers concern the study of dynamical systems, the first step is to give back to *modeling* the central role that it deserves and focus on structural aspects of some specific systems –typically nonlinear– that can be exploited for control system design.

- The next step is then to try to *mathematically formalize current engineering practice*, which is usually developed from physical considerations and practical experience, as opposed to theoretical analysis. This is a fundamental step, whose importance can hardly be overestimated. It provides a solid system–theoretic foundation to existing control strategies which enhances their understanding allowing us to estimate their achievable performance and paves the way for subsequent improvements.

- Based on our understanding of the model, the mathematical rationalization of current engineering practice and new analytical considerations, the third step is to *propose new controllers*. In some fortunate –and some times serendipitous– cases these may be viewed as "upgrades" of the existing schemes, facilitating in this way the transfer of this knowledge to practitioners.

- The final, and *sine qua non* step is to *test* the theoretical developments in realistic simulations or experiments and confront the achievable performance with standard controllers. Even though in all steps the interaction with the specialist of the application field is essential, it is perhaps in the latter one that it is particularly crucial.

The book describes two instances in which the payoff for abiding to these guidelines was particularly rewarding. The first one is in *robotics*, to which the first part of the book is devoted. First, a careful study of the robot dynamics, and in particular the establishment of its passivity, provided the theoretical underpinnings for the industry standard PD control. Later developments, exploiting suitable passive subsystems decompositions, proved that non-measurable velocities could be replaced by its approximate derivative –again, a standard engineering practice– without affecting its stability properties. Simulation results showed later that a control input beyond the physical saturation constraints was sometimes required to achieve fast motions. Hence, it became necessary to study the effect of saturations, which turned out to be possible with a simple and natural addition to our design toolbox. As a result of further analysis it was possible to prove that adding a *double* integral action to the PD controller would not just remove the need for velocity measurements, but also enhance its robustness making it insensitive to uncertainty in the payload. The overall result of our research was to provide a nice and natural upgrade, with guaranteed stability properties, of the classical PID control with saturated inputs. The mathematical machinery we had set up for the stability analysis could now be used for tuning and performance improvement.

The second case study concerns *AC electric machines*, (described in Part III of the book), which in spite of their complex dynamics are shown to define passive maps. An insightful decomposition of the machine dynamics into passive subsystems opened the gate for a rigorous theoretical analysis, in particular its proof of global asymptotic stability, of the industry standard field–oriented control. A first upgrade to this controller was then proposed for the case when, due to performance considerations, the model of the machine has to be refined to include the stator dynamics. The modularity of our control design was exploited at an even higher level when we had to take into account a nonlinear model of the load to enhance performance. This was the case of controlling fast moving robots with AC drives. We proved that we could cascade the control law for the machine with an outer–loop controller, designed upon the same passivity principles, for the robot dynamics. Extensive experimental evidence proved later the validity of our theoretical considerations. Furthermore, as an outcome of our stability analysis, we provided some simple tuning rules to commission and retune either the field–oriented controller or our upgraded version.

There are, of course, alternative routes towards control engineering applications to the one delineated above. For instance, instead of our model–based starting point other research groups have developed non–model–based techniques. This approach is motivated by the massive availability of data –either from extensive simulations or from experiments–, and centers around the use of powerful "data–fitters" for modeling and control purposes, the best well–known example being neural nets and fuzzy sets. Even though at this stage it is not clear how the reported analytical results uses properties particular to neural nets, it is indeed the case that unveiling the complicated mechanisms of operation of neural nets requires a deep understanding of nonlinear dynamics, the natural realm of application of control theory. Although this approach does not seem to enhance a synergetic collaboration with specialist of other areas, it is fair to say that, in some sense, it fosters the development of an experimental control science, and certainly has a profitable role to play in a control practitioner's toolbox.

3 Why Euler–Lagrange systems?

We are interested in this book on controller design for dynamical systems where the nonlinear components cannot be neglected. For instance, motion control of robots actuated by AC drives where besides the intrinsic nonlinear operation of the AC motors we have that a simple double integrator model of the mechanics cannot capture the behaviour of the robot under fast motions. It is widely recognized, at least as far back as Poincaré, that nonlinear systems can exhibit extremely complex dynamic behaviour. This renders futile the quest of a monolithic control theory applicable for all systems, we must therefore specialize the class of systems under consideration in one way or another.[3]

We have proposed above to single out from the outset a class of physical systems and concentrate our efforts upon them. This should be contrasted with the research on, what we have called, mathematical control theory, where one takes off from a class of systems whose solution is known, e.g. LTI systems, and start to enlarge the class including some special nonlinearities (or structures) for which the available analysis and design tools can be suitably extended. A typical example of this scenario is the absolute stability problem consisting of an LTI system in feedback with a static nonlinearity. One way to study this paradigm is to add to the standard quadratic Lyapunov function of the LTI system an integral term that takes care of the presence of the nonlinearity, leading to a Lur'e–Postnikov Lyapunov functional.

The final aim of this type of research is to build up tools and procedures to treat a more or less general paradigm, the final aim typically being $\dot{x} = f(x, u)$, $y = h(x, u)$. Needless to say that this fundamental research can provide some stepping stones in our route towards applications, it is particularly useful to define achievable performances. However, a fundamental obstacle for the application of the resulting techniques to physical systems is that it is hard, if not impossible, to incorporate *a posteriori* the natural structures imposed by the systems physical character. This stems from the fact that the design techniques are applicable only to systems with particular structures which are defined only by mathematical considerations, hence there is no apparent reason why a physical system belongs to the admissible class.

It is sometimes possible to "force" the system into the required form, via changes of coordinates or nonlinearity cancelations, it is in any case somehow distressing that to control a physical system we have to start by "destroying" its structure. The extreme case where the physical structure of the system is neglected is, of course, feedback linearization, where the aim of the control is to render the system linear in closed loop. Sliding mode control lies somewhere in between in so far as once a relative degree one minimum phase output is constructed the remaining dynamics can be swamped via high–gain (relay) feedback.

It is the authors' belief that to develop a practically meaningful nonlinear control theory we should start by considering a practically meaningful class of systems, whose physical structure should be taken into account, from the outset, in the design procedure. It is in the definition of this class that EL systems enter into the picture. The most important reason for singling out the study of EL systems is that

---

[3]In this respect it is interesting to mention the concept of flatness [83], which identifies a class of systems for which trajectory planning becomes trivial. The stabilization itself is then carried–out along the reference trajectory with a lower hierarchy standard control loop.

they describe the behaviour of a large class of contemporary engineering systems, specially some which are intractable with linear control tools. EL systems are the outcome of a powerful modeling technique –the variational method– whose starting point is the definition of the energy functions in terms of sets of generalized variables (typically positions and charges for mechanical and electrical systems, respectively), which leads to the definition of the Lagrangian function. The equations of motion are then derived invoking well–known principles of analytical dynamics, in particular the fundamental Hamilton's principle, which roughly speaking states that the system moves along trajectories that minimize the integral of the Lagrangian.[4]

The variational modeling method is one of the most powerful techniques of dynamics. As we will show throughout the book this method is particularly suited for our purposes for the following reasons:

- Given that the modeling problem is formulated in terms of energy quantities the variational method allows us to treat, without the need of any special bookkeeping, systems of "mixed nature", which often appear in engineering applications, –e.g., having both electrical and mechanical components.

- It automatically provides us with the storage and dissipation functions of the system. These are the cornerstones of the design technique that we advocate in this book which is based on energy dissipation.

- Since the modeling is based on some kind of network representation and energy flow, it is compatible with one of the important viewpoints of systems theory that complicated systems are best thought of as being *interconnections* of simpler subsystems, each one of them being characterized by its energy functions. This aggregation procedure underscores the role of the interconnections between the subsystems helping us to think in terms of the structure of the system and to realize that sometimes the pattern of the interconnections is more important than the detailed behaviour of the components.

- We will show that the EL structure is preserved under feedback interconnection. That is, the interconnection of two EL systems is still an EL system. (As we will explain below, this invariance property is also enjoyed by passive systems, which is the second key component of our approach.)

Before closing this subsection a word on notation is in order. As explained in Section **2**.1 throughout the book we use the name EL systems because we choose to represent their dynamics with the EL equations of motion. Our choice stems from the fact that in this representation it is easier to reveal some structural properties instrumental for our controller design. They can alternatively be represented with

---

[4]Putting it in another way: The possible trajectories of a dynamical system are precisely those that minimize a suitable action integral. In the words of Legendre, *"Ours is the best of all worlds."*

Hamilton's equation and called Hamiltonian systems. A more generic, but in other respects somehow confusing name encountered in the literature is mechanical systems.

4 On the role of interconnection

The features above will be particularly relevant for our research where interconnection is a recurrent theme. At a conceptual level, we follow Willems' behavioral perspective and think of control as the establishment of a suitable interconnection of the system with its environment. For instance, we will show in Section **3**.2.3 that the outcome of our design procedure, which relies upon energy dissipation considerations, as applied to the set point regulation of a flexible pendulum is the interconnection of the system with a virtual pendulum as depicted in Fig. 1.1. That is, the controller consists of a rigid pendulum (of unitary mass) without gravity forces that we attach to a plane (with inclination δ) with a spring (of stiffness coefficient K_1) and a damper (with damping coefficient R_c). Then, we attach a spring (of stiffness coefficient K_2) between this virtual pendulum and the first link angle in such a way that the virtual pendulum "pulls" the actual pendulum to the desired equilibrium (q_*). The position of the plane is selected so that in steady–state the springs store the excess of potential energy, the damper is added to ensure that energy is dissipated and asymptotic convergence is henceforth achieved. To take into account saturation in the input torque, which was mentioned in Section 2, we simply make the controller springs nonlinear.

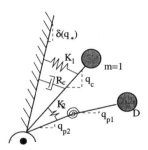

Figure 1.1: Elastic pendulum regulated with a PBC.

Cela va sans dire that this controller has an obvious physical interpretation, a fact whose importance can hardly be overestimated. Its relevance pertains not just to aesthetical considerations, but more importantly to its impact on the controller tuning stage. Commissioning of a nonlinear controller is a difficult task, which has unfortunately been overlooked in most theoretical developments. A final, but not less important, advantage of a physically interpretable design is that it clarifies the role of the sensors and actuators. For instance, in the present example we assumed available the measurement of the actuated side of the link, it is easy to modify the design if we add to it (or replace it by) measurement of the unactuated side.[5]

Interconnection appears also in the book when, in the quest to achieve the desired performance, we propose to iterate at the modeling and control levels. More precisely, a first design is made on the basis of a simple model (say the purely rigid mechanical dynamics); if performance is below par we refine the model (for instance, include flexible modes or the electrical dynamics of the actuators) and cascade the new controller with the previous one.

To be able to carry out this interlaced modeling–control procedure we effectively exploit the aggregation property of EL models described above. We also require that the controller design should be based upon a property that remains *invariant* upon interconnection, it is at this stage that we invoke the fundamental physical property of passivity.

5   Why passivity?

Passive systems are a class of dynamical systems in which the energy exchanged with the environment plays a central role. In passive systems the rate at which the energy flows into the system is not less than the increase in storage. In other words, a passive system cannot store more energy than is supplied to it from the outside, with the difference being the dissipated energy. See Appendix A for a basic circuit–theoretic introduction to passivity, and the formal definition,[6] as well as various properties of passive systems that will be used throughout the book. Let us simply point out here that in the LTI case passive systems are minimum phase and have relative degree not greater than one, i.e. their frequency response satisfies $\mathbb{R}_e\{H(j\omega)\} \geq 0$. The groundbreaking result of [46] furthermore shows that a nonlinear system can be rendered passive via state–feedback if and only if (the nonlinear analog of) these two conditions hold.

It is clear from the energy interpretation of passivity given above that it is intimately related with the physics of the system, and in particular its stability properties. For instance, viewing a feedback interconnection as a process of energy exchange, it is not surprising to learn that passivity is *invariant* under negative feedback interconnection. In other words, the feedback interconnection of two passive systems is still passive, see Theorem A.3. If the overall energy balance is positive, in the sense that the energy generated by one subsystem is dissipated by the other one, the closed loop will be furthermore stable as stated in Proposition A.10. As an immediate corollary

---

[5]The interplay of actuators, sensors and dissipation has traditionally being a basic building block for structural control, active vibration suppression and control of large space structures.

[6]It should be mentioned here that, even though our definition is given in terms of inputs and outputs, invoking the behavioral framework of Willems it is possible to define this concept without reference to inputs and outputs.

we have that passive systems are "easy to control", for instance, with a simple constant gain, which can actually be made arbitrarily large –recall the relative degree and minimum phase properties of passive systems mentioned above. This property together with the clean characterization of passifiable systems reported in [46] explains the interest of passivity as the basic building block for control of nonlinear systems.

One final, but not less important property of passivity is that it is a property which is independent of the notion of the *state*, hence as we will see throughout the book, state–feedback (which we know is unrealistic in most applications) will not be a prerequisite to achieve the control objectives.

Summarizing: Passivity (and the closely related property of finite gain stability) provide a natural generalization (to the nonlinear time–varying case) of the fact that stability of an LTI feedback system depends on the amount of gain and phase shift injected into the loop. Furthermore, and perhaps more importantly, the measures of signal amplification (the operator gain) and signal shift (its passivity) can be in some instances associated to physical quantities. These fundamental properties motivates us to take passivity as the basic building block of our constructions.

A more precise evaluation of the concepts above can be made for EL systems, namely:

- We prove that EL systems define passive maps with storage function their total energy. This is the most fundamental property for our work since it identifies the output which is "easy" to control and provides a storage function that will typically motivate the desired closed–loop storage function.

- In most problems we need a stronger so–called "skew–symmetry" property of EL systems that essentially says that the dynamics of EL systems can be treated as a set of linear double integrators, provided a suitable factorization of the (workless) forces is used. This factorization is essential in all our developments.

- Invoking the property of invariance under feedback interconnection of the EL structure we will, in some control tasks, view the action of the controller as an *EL–structure–preserving interconnection*. In this way the properties of the closed–loop are still captured by the energy and dissipation functions that characterize an EL system. Furthermore, these are simply the sum of the corresponding functions of the plant and the controller.

- Under some reasonable assumptions we prove that EL systems can be decomposed into the feedback interconnection of two passive subsystems. This decomposition is instrumental for the interlaced modeling–control procedure described above.

All these properties are used in the sequel to design what we call passivity–based control (PBC) schemes.

Before closing this section it is convenient to clarify that the fact that EL systems are passive, (or can even be "treated as double integrators") does not make the problem of control of EL system trivial. First of all, the output with respect to which the system is passive will in general not be the signal that we want to control. For instance, in mechanical systems it is velocity, and we are often interested in position control. In electrical machines it will be current, which is nonlinearly related to torque, which is the output we want to control.

Secondly, imposing a desired behaviour to the passive output will only be a preliminary step of the control design, we have in addition to define the shape of this signal that will ensure the output we want to control behaves as desired, e.g., the currents that will generate the desired torque. In this second step of the design we are confronted with the complexity of the nonlinear dynamics which can be daunting even restricted to EL systems. It suffices to say that, as pointed out in Section 1 of Chapter **11** and thoroughly discussed in [73], the dynamics of a *simplified* model of the induction machine coincides (up to a linear drift term) with the nonholonomic integrator of Brockett, which has enticed control theory researchers for many years.

6   What is passivity–based control?

The term PBC was first introduced in [212] to define a controller methodology whose aim is to render the closed–loop passive. This objective seemed very natural within the context of adaptive control of robot manipulators, since as shown in that paper the robot dynamics defines a passive map, and it had been known since the early work of Landau [149] that parameter estimators are also passive. (See Section 7.2 for a brief history of passivity and stabilization of nonlinear systems.)

The PBC approach presented in this book may be viewed as an extension of the by now well–known energy–shaping plus damping injection technique introduced to solve state–feedback set point regulation problems in fully actuated robotic systems by Takegaki and Arimoto in [261]. For this particular problem we can concentrate our attention on the potential energy and the dissipation functions and proceed along two basic stages.[7] First an *energy shaping* stage where we modify the potential energy of the system in such a way that the "new" potential energy function has a global and unique minimum in the desired equilibrium. (It is clear that, similarly to the choice of a Lyapunov function, no systematic procedure exists to select the desired potential energy. The basic difference is that inspiration comes now –at least for EL systems– from physical considerations.) Second, a *damping injection* stage where we now modify the dissipation function to ensure asymptotic stability. In the case

---

[7]See the pendulum example of Chapter **3** for a detailed discussion of these points.

considered in [261] the assumption of full actuation allows us to assign any arbitrary potential energy function, while the availability of the full–state trivializes the task of damping injection. The controller resulting from this technique is a simple PD law.

Viewed from the broader PBC perspective pursued in this book the energy shaping stage accomplishes a *passivation* objective with a desired storage function that consists of the original kinetic energy and the new desired potential energy. The damping injection reinforces this property to *output strict passivity*. Finally, Lyapunov stability follows from the input–output stability of the output strictly passive map provided some dissipation propagation (i.e., detectability) conditions are met.

This basic (energy shaping plus damping injection) methodology is extended in this book in several directions: first, we consider *output* feedback and *underactuated* systems; second, we address also *tracking* problems; finally, all these –more realistic cases– are considered for a broader class of EL systems which contains, besides mechanical, electrical and electromechanical systems. In carrying out these extensions several additional fundamental difficulties –which require nontrivial modifications– appear:

- When the state is not available for measurement the damping must be added via a dynamic extension.

- In underactuated cases the systems potential energy cannot be cancelled, but it has to be now dominated. If furthermore we have to modify the kinetic energy, the controller will first be defined implicitly and an additional "inversion" step will be needed to obtain an explicit realization.

- In all cases, besides regulation of mechanical systems, the kinetic energy plays a role in the control task, it must therefore also be "shaped". In terms of passivation this is tantamount to saying that the desired storage function that we want to assign to the closed–loop cannot be simply chosen as the sum of the systems kinetic energy and a new potential energy as in [261]. We will typically select the desired storage function from the consideration of an error dynamics where, thanks to a suitable factorization of the system (workless) forces, we obtain linearity with respect to the error signal.

To solve all these new challenging problems we often make appeal to feedback interconnection decompositions and strongly rely on the invariance properties of EL systems and passivity discussed above.

7   Some historical remarks

7.1   Euler–Lagrange systems and nonlinear dynamics

Nonlinear dynamics in the guise of planetary motions, has some claim to be the most ancient of scientific problems. The celebrated names of Newton, Kepler, Cartan, Hamilton and Dirac (to mention a few) are attached to the development of classical mechanics as a primary motivation for the study of dynamical systems and their properties. As usual for many other important developments in nonlinear control theory, credit goes to Brockett for the introduction of controlled Hamiltonian systems with inputs and outputs.[8] Controllability and observability theories were later developed by van der Schaft. Parallels with the linear systems theory were obtained in realization theory, where the "existence and uniqueness" of realizations has been proved conclusively for suitably "smooth" systems by Crouch and co–workers. The special nature of mechanical systems for control purposes was first exploited in robotics, whose history is reviewed in Remark 2.9 and Section 1.3 of Chapter **3**. The study of mechanical control systems has also provided some interesting case studies. For instance, the spacecraft attitude control problem is an excellent test bed for controllability of nonlinear control systems. Another more recent example is the class of mechanical control systems, whose underlying plant is characterized as "nonholonomic", which clearly exhibited limits on the extent to which intuition about linear systems can be transfered to the nonlinear domain. It also illustrated the practical effect of Lie brackets in the synthesis of controllers for nonlinear systems and the use of flatness for trajectory planning.

7.2   Passivity and feedback stabilization

The concept of passivity has played a prominent role in many areas of systems theory for many years now. See, for instance, the introduction to Chapter 3 of [286] for an authoritative answer to the question *"Why are passive (positive) operators important?"* One of the early connections between passivity and stability is due to Youla [291] who proved that a passive network in closed–loop with a resistive element is $\mathcal{L}_2$ stable, meaning by this that finite energy inputs will be mapped into finite energy outputs. There are also many scholar books that cover the subject of passivity, or more generally input–output theory, and contain detailed descriptions of its history. We recommend in particular the seminal books [286] and [69] for encyclopedic coverage. Viewed from a more recent perspective we have [272] and [237]. See also [203] for a brief history of this topic.

Interestingly, one fundamental early connection between passivity and stability

---

[8]See [197] for a more detailed account on controlled mechanical systems, including the references cited below.

in nonlinear systems was obtained via optimal control in 1973. This connection has a surprising, and somehow little well known history which is described in [203], and we repeat here for ease of reference. In [188] the authors solved an inverse optimal regulator problem for nonlinear systems and establish as a corollary that optimal systems define passive maps. This is the nonlinear extension of the celebrated Kalman's inverse optimal control paper.

A far reaching implication of this result is that passivity provides a criterion for deciding the optimality or otherwise of a feedback loop. An offspin of this fundamental result is the nonlinear Kalman–Yakubovich–Popov (KYP) lemma of [103], which has triggered so much interest in the recent years. To the best of the authors' knowledge [188] is the first paper where the important concepts of stabilization, existence of Lyapunov functions and optimality are shown to be closely connected via passivity. An early, and quite modest, attempt to explore these connections in an adaptive stabilization problem was reported in 1988 in [202]. The recent book [237] further investigates applications of inverse optimality and passivity for stabilization of nonlinear systems.

In recent years passivity, and more specifically feedback passivation, has been used to reformulate, in an elegant and unifying manner, the fundamental problem of feedback stabilization of nonlinear systems in [46]. The history of this result (which once again is retraced in [203]) is intertwined with the history of backstepping, a widely popular stabilization technique for nonlinear systems that we also discuss in this book. We refer the reader to Chapter 3 of [142] for a detailed description of the history of backstepping and to the authoritative survey paper [61] for an overview on nonlinear stabilization. For the sake of completeness we give here some elements missing in [142]. As pointed out in [187] backstepping has also roots in the work on adaptive control of Feuer and Morse in 1978. The term "integrator backstepping" had not been coined at that time but all the elements of the technique, including the nonlinear damping, may be found there. In the last section of this important paper it is explained how the prototypical problem, whose solution triggered the re–emergence of backstepping in nonlinear stabilization problems, can be solved with the techniques advanced by Feuer and Morse.

One final apostille in this brief historical review is that, to the best of our knowledge, the first attempt to use feedback passivation for stabilization was made in [225] and [210], where the work of [188] and the nonlinear KYP lemma of [103] are used as design tools for adaptive stabilization of non-feedback linearizable, but passifiable, nonlinear systems.

Chapter 2

Euler-Lagrange systems

"All happy families [linear systems] are alike, every unhappy family [nonlinear system] is unhappy [nonlinear] in its own way"

L. Tolstoi

It has been argued in the Introduction that a good starting point to develop a practically meaningful nonlinear control theory is to specialize the class of systems under consideration. The main reason being, of course, that the vast array of nonlinear systems renders futile the quest of a monolithic theory applicable for all systems. In particular, it defies the approach of mimicking the, by now fairly complete, linear theory. Specializing the systems, on the other hand, introduces additional constraints and structure, which may enable otherwise intractable problems to be answered. In this chapter we describe the class of systems that we will consider throughout the book and which we call Euler–Lagrange (EL) systems. The most important reason for singling out the study of EL systems is that they capture a large class of contemporary engineering problems, specially some which are intractable with linear control tools. Finally, by restricting ourselves to systems with physical constraints we believe we can contribute to reverse the tide of "find a plant for my controller" which still permeates most of the research on control of general nonlinear systems.

What is an EL system? To answer this question we will borrow inspiration from the definition of adaptive control quoted in the seminal book of Åström and Wittenmark [13] (i.e., "An adaptive system is a system that has been designed with an adaptive viewpoint"). Hence we will say: An EL system is a system whose motion is described by the EL equations. The logical question which arises next is "What are the EL equations?" From a purely mathematical viewpoint they are a set of nonlinear ordinary differential equations with a certain specific structure. A far more interesting question is "Where do they come from?" In contrast to the first two questions, the answer to the latter is far from simple and involves principles of minimization, calculus of variations and other tools from analytical dynamics. For the purposes of this book the EL equations are important because they are the outcome of a powerful modeling

15

technique –the variational method– which describes the behaviour of a large class of physical systems. There are several excellent textbooks on variational modeling both for mechanical [91, 97] and electromechanical systems [179, 285]. We refer the interested reader to these books to prove further. For the sake of self-containment we have summarized in Appendix **B** some of the relevant elements of this fundamental theory.

We thus start this chapter by describing the EL equations and introducing the notation used throughout the remaining parts. EL systems models are, roughly speaking, obtained from the minimization of an energy function. It is therefore expected that they enjoy some energy dissipation properties, in particular that they define passive maps, that one can profitably use for the controller design. These, as well as some other interconnection features and Lyapunov stability properties of EL systems are also reviewed in this chapter.

1   The Euler–Lagrange equations

In modeling physical systems with lumped parameters two basic approaches have been typically used: derivation of the equations of motion using forces laws or application of variational principles to selected energy functions. For simple systems having only elements of the "same nature" the first approach is usually sufficient. For instance, for purely mechanical or electrical systems, Newton's second law and Kirchhoff's laws respectively, yield the desired equations. This method can still be applied for systems having "mixed natures", e.g., having both electrical and mechanical portions. In this case, the forces of interaction can be obtained invoking the method of arbitrary displacements and conservation of energy. This approach requires a lot of insight, and much bookkeeping must be done in complicated problems. In order to develop in a systematic manner the equilibrium equations a more general formulation is required, this is the aim of the variational approach. (See Appendix **B**).

The *common link* between the different subsystems is that all of them transform energy. Therefore, it seems natural to formulate the modeling problem in terms of energy quantities. The starting point of the variational approach to modeling is the definition of the energy functions in terms of sets of generalized variables (typically positions and charges for mechanical and electrical systems, respectively), this procedure leads to the introduction of the Lagrangian function. The equations of motion are then derived invoking well–known principles of analytical dynamics, in particular the fundamental Hamilton principle, which roughly speaking states that the system moves along trajectories that minimize the integral of the Lagrangian. The variational modeling method is one of the most powerful techniques of dynamics. As we will show throughout the book this method is particularly suited for PBC since it underscores the role of the interconnections between the subsystems and provides us with the storage and dissipation functions, which are the cornerstones of the PBC design technique.

As a first step towards the development of PBC, we have reviewed in Appendix **B** some of the relevant elements of analytical dynamics, and highlighted the role of energy and Rayleigh dissipation functions in the model development. Hamilton's principle and its application for the derivation of the equations of motion for dynamical systems are also presented there.

We have discussed in Appendix **B** that the resulting dynamics can be described by the EL or the Hamiltonian equations of motion. In this book we favor the Lagrangian representation. Although, for the purposes of the material contained in this book, there is no essential difference between the two approaches, we believe that in the EL equations it is easier to reveal some structural aspects of the workless forces (e.g., Coriolis terms in robots manipulators), which is somehow obscured in the Hamiltonian model. As we will see later a suitable factorization of the workless forces is an essential step for the application of the PBC methodology in most tasks, including trajectory tracking of mechanical systems and regulation or tracking for electrical and electromechanical systems. See also [285] for further motivation of this choice.

We have shown in Appendix **B** that an n degrees of freedom dynamical system with generalized coordinates $q \in \mathbb{R}^n$ and external forces $Q \in \mathbb{R}^n$, is described by the EL equations

$$\frac{d}{dt}\left(\frac{\partial \mathcal{L}}{\partial \dot{q}}(q, \dot{q})\right) - \frac{\partial \mathcal{L}}{\partial q}(q, \dot{q}) = Q, \tag{2.1}$$

where

$$\mathcal{L}(q, \dot{q}) \triangleq \mathcal{T}(q, \dot{q}) - \mathcal{V}(q) \tag{2.2}$$

is the Lagrangian function, $\mathcal{T}(q, \dot{q})$ is the kinetic energy (or co–energy) function which we assume to be of the form

$$\mathcal{T}(q, \dot{q}) = \frac{1}{2}\dot{q}^\top D(q)\dot{q}, \tag{2.3}$$

where $D(q) \in \mathbb{R}^{n \times n}$ is the generalized inertia matrix that satisfies $D(q) = D^\top(q) > 0$, and $\mathcal{V}(q)$ is the potential function which is assumed to be bounded from below that is, there exists a $c \in \mathbb{R}$ such that $\mathcal{V}(q) \geq c$ for all $q \in \mathbb{R}^n$.

Throughout the book we will consider (unless otherwise specified) three types of external forces: the action of controls, dissipation and the interaction of the system with its environment. We assume controls enter linearly[1] as $\mathcal{M}u \in \mathbb{R}^n$, where $\mathcal{M} \in \mathbb{R}^{n \times n_u}$ is a constant matrix and $u \in \mathbb{R}^{n_u}$ is the control vector. Dissipative forces

---

[1] In Part II of the book we will study a class of electrical systems where the control does not enter additively.

are of the form $-\frac{\partial \mathcal{F}}{\partial \dot{q}}(\dot{q})$, where $\mathcal{F}(\dot{q})$ is the Rayleigh dissipation function which by definition satisfies

$$\dot{q}^{\top} \frac{\partial \mathcal{F}}{\partial \dot{q}}(\dot{q}) \geq 0 \qquad (2.4)$$

In summary we have the external forces

$$Q = -\frac{\partial \mathcal{F}}{\partial \dot{q}}(\dot{q}) + Q_{\zeta} + \mathcal{M}u \qquad (2.5)$$

where Q_{ζ} is an external signal that models the effect of disturbances.

As explained in the introduction, in PBC the control objective is achieved by imposing to the closed–loop dynamics a certain passivity property, which in its turn reduces to assigning some desired storage and dissipation functions. It comes natural then to define EL systems in the following, admittedly redundant, way

Definition 2.1 (EL equations and EL parameters.) *The EL equations of motion*

$$\frac{d}{dt} \left(\frac{\partial \mathcal{L}}{\partial \dot{q}}(q, \dot{q}) \right) - \frac{\partial \mathcal{L}}{\partial q}(q, \dot{q}) + \frac{\partial \mathcal{F}}{\partial \dot{q}}(\dot{q}) = \mathcal{M}u + Q_{\zeta}, \qquad (2.6)$$

with (2.2), (2.3), (2.4) define an EL system which is characterized by its EL parameters :

$$\{\mathcal{T}(q, \dot{q}), \mathcal{V}(q), \mathcal{F}(\dot{q}), \mathcal{M}, Q_{\zeta}\}.$$

The use of the EL parameters as defined by the quintuple above captures a fairly general notation which we use throughout this book. However, when clear from the context, we may use the more compact notation $\{\mathcal{T}(q, \dot{q}), \mathcal{V}(q), \mathcal{F}(\dot{q}), \mathcal{M}\}$ for systems for which $Q_{\zeta} \equiv 0$ or $\{\mathcal{T}(q, \dot{q}), \mathcal{V}(q), \mathcal{F}(\dot{q})\}$ in the case when only dissipative forces affect the EL system.

The matrix $\mathcal{M}$ is a full column rank matrix relating the external inputs to the generalized coordinates. We find thus convenient to distinguish two classes of EL systems according to the structure of this matrix:

Definition 2.2 (Underactuated EL systems.) *An EL system is fully-actuated if it has equal number of degrees of freedom than available control inputs (that is if $n = n_u$, e.g. if $\mathcal{M} = I_n$). Otherwise, if $n_u < n$ we say that the system is underactuated. In the latter case, q can be partitioned into non-actuated $\mathcal{M}^{\perp}q$ and actuated components $\mathcal{M}q$, where $\mathcal{M}^{\perp}$ denotes the perpendicular complement of $\mathcal{M}$.*

A second classification that we find convenient to introduce at this point, involves the presence of damping. We can thus distinguish two classes of systems:

Definition 2.3 (Underdamped and fully-damped systems.) *The EL system (2.6) is said to be* fully-damped[2] *if the Rayleigh dissipation function satisfies*

$$\dot{q}^\top \frac{\partial \mathcal{F}}{\partial \dot{q}}(\dot{q}) \geq \sum_{i=1}^{n} \alpha_i \dot{q}_i^2$$

with $\alpha_i > 0$ *for all* $i \in \underline{n} \triangleq \{1, \cdots, n\}$. *It is, on the other hand,* underdamped *if* $\exists i \in \underline{n}$ *such that* $\alpha_i = 0$.

Remark 2.4 (Generality of the model) In most of the practical cases we will assume that the Rayleigh dissipation function is quadratic (this models, e.g., linear friction or constant resistances) as

$$\mathcal{F}(\dot{q}) \triangleq \frac{1}{2}\dot{q}^\top R \dot{q}$$

with $R = R^\top \geq 0$ and diagonal. Fully damped and underdamped EL systems correspond to R being positive definite or only positive semi–definite, respectively. Also, to simplify the presentation we will assume in the sequel a special structure of $\mathcal{M}$. It will become clear later that these assumptions are not essential for the basic developments.

2 Input–output properties

As discussed in the introduction, the input–output approach to systems analysis provides a natural generalization (to the nonlinear time–varying case) of the fact that stability of an LTI feedback system depends on the amount of gain and phase shift injected into the loop. Furthermore, and perhaps more importantly, the measures of signal amplification (the operator gain) and signal shift (its passivity) can be in some instances associated to physical quantities. The input–output approach is also consistent with one of the important viewpoints of control theory that complicated systems are best thought of as being interconnections of simpler subsystems. This aggregation procedure has two important implications. On one hand, it help us to think in terms of the structure of the system and to realize that more often than not the pattern of the interconnections is more important than the detailed behaviour of the components, which can be characterized by an input–output property, e.g. its passivity. On the other hand, it yields a design–oriented methodology since it allows us to isolate the controller as a "free" subsystem. As we will show in the book both features of the input–output approach are clearly illustrated when applied to EL systems.

---

[2]This class of systems is sometimes called *pervasively* damped in analytical mechanics and structural vibration.

In this important section we establish the main tools that will be used for our PBC design. They concern not just the passivity of EL systems, but more importantly their *interconnection* properties. In the present book we effectively exploit these properties of EL systems for the design of PBC in three different ways:

- We prove that EL systems define passive maps. This is the most fundamental property for PBC since it identifies the output which is "easy" to control and provides a storage function that will typically motivate the desired closed–loop storage function. Actually, we will prove a stronger so–called "skew–symmetry" property that essentially says that the dynamics of EL systems can be treated as a set of linear double integrators. The proof of these properties has an interesting intermingled story which is briefly reviewed in Remark 2.9 and in Subsection **3**.1.3.

- To simplify the controller design we introduce a decomposition of the EL system dynamics –into the feedback interconnection of two passive subsystems. We give conditions under which this decomposition is possible, these include some cases of practical interest, in particular in electromechanical systems. This property was first established in [76] and, as shown in Chapter **9**, is instrumental for the solution of the global tracking problem of electrical machines.

- We view the action of the controller as an *EL–structure–preserving intercon- nection*. In this way the properties of the closed–loop are still captured by the energy and dissipation functions, which are simply the sum of the corresponding functions of the plant and the controller. This property, which was first reported in [208], will be exploited in Chapter **3** to design PBC, which are themselves EL systems, for output feedback regulation of mechanical systems.

2.1   Passivity of EL systems

Passive systems are a class of dynamical systems in which the energy exchanged with the environment plays a central role. As we have seen in Appendix **A** in passive systems the rate at which the energy flows into the system is not less than the increase in storage. In other words, a passive system cannot store more energy than it is supplied to it from the outside, with the difference being the dissipated energy. The proposition below proves that EL systems define passive maps.

Proposition 2.5 *The EL system (2.6) with $Q_\zeta \equiv 0$ defines a passive operator $\Sigma :$ $u \mapsto \mathcal{M}^\top \dot{q}$ with storage function the systems total energy[3] $\mathcal{H}(q, \dot{q})$. That is,*

$$\langle u \mid \mathcal{M}^\top \dot{q} \rangle_T \geq \mathcal{H}[q(T), \dot{q}(T)] - \mathcal{H}[q(0), \dot{q}(0)] \tag{2.7}$$

---

[3]Recall from Appendix **A** that a storage function need only be positive semidefinite and bounded from below, thus not necessarily zero for zero argument.

for all $T \geq 0$ and all $u \in \mathcal{L}_{2e}^m$. *Further, this property is strengthened to* output strict passivity *(OSP) if the system is fully damped. In this case*

$$\langle u \mid \mathcal{M}^{\top}\dot{q}\rangle_T \;\geq\; \alpha\|\mathcal{M}^{\top}\dot{q}\|_{2T}^2 + \mathcal{H}[q(T),\dot{q}(T)] - \mathcal{H}[q(0),\dot{q}(0)] \tag{2.8}$$

for some $\alpha > 0$ and all $u \in \mathcal{L}_{2e}^m$. $\qquad\square$

Proof. The property can be established taking the time derivative of the Lagrangian function $\mathcal{L}(q,\dot{q})$, where for simplicity in the notation we drop the arguments,

$$\frac{d\mathcal{L}}{dt} = \left(\frac{\partial\mathcal{L}}{\partial q}\right)^{\top}\frac{dq}{dt} + \left(\frac{\partial\mathcal{L}}{\partial\dot{q}}\right)^{\top}\frac{d\dot{q}}{dt} \tag{2.9}$$

and using the EL equations (2.1) to write

$$\frac{\partial\mathcal{L}}{\partial q} = \frac{d}{dt}\left(\frac{\partial\mathcal{L}}{\partial\dot{q}}\right) - Q,$$

so (2.9) can be rewritten as

$$\frac{d\mathcal{L}}{dt} = \left(\frac{\partial\mathcal{L}}{\partial\dot{q}}\right)^{\top}\frac{d\dot{q}}{dt} + \frac{d}{dt}\left(\frac{\partial\mathcal{L}}{\partial\dot{q}}\right)^{\top}\dot{q} - \dot{q}^{\top}Q$$

then, reordering the terms above and using (2.5), it follows that

$$\frac{d}{dt}\left[\left(\frac{\partial\mathcal{L}}{\partial\dot{q}}\right)^{\top}\dot{q} - \mathcal{L}\right] = \dot{q}^{\top}\left(\mathcal{M}u - \frac{\partial\mathcal{F}}{\partial\dot{q}}\right), \tag{2.10}$$

Now, notice that the term in parenthesis on the left hand side coincides with the systems total energy, which we denote by $\mathcal{H}(q,\dot{q})$, that is

$$\left(\frac{\partial\mathcal{L}(q,\dot{q})}{\partial\dot{q}}\right)^{\top}\dot{q} - \mathcal{L}(q,\dot{q}) = \mathcal{T}(q,\dot{q}) + \mathcal{V}(q) \triangleq \mathcal{H}(q,\dot{q})$$

Integrating (2.10) from 0 to T we establish the key *energy balance* equation

$$\underbrace{\mathcal{H}[q(T),\dot{q}(T)] - \mathcal{H}[q(0),\dot{q}(0)]}_{\text{stored energy}} + \underbrace{\int_0^T \dot{q}^{\top}\frac{\partial\mathcal{F}(\dot{q})}{\partial\dot{q}}ds}_{\text{dissipated}} = \underbrace{\int_0^T \dot{q}^{\top}\mathcal{M}uds}_{\text{supplied}}. \tag{2.11}$$

Now, observe that, since $\mathcal{V}(q)$ is bounded from below by c, and $\mathcal{T}(q,\dot{q}) \geq 0$ we have that $\mathcal{H}(q,\dot{q}) \geq c$. Finally, the Rayleigh dissipation function satisfies (2.4) hence (2.7) follows.

If the system is fully damped it follows immediately from Definition 2.3 and (2.11) that (2.8) holds with $\alpha \triangleq \frac{\min_i\{\alpha_i\}}{\|\mathcal{M}\|^2}$.

$\blacksquare$

Remark 2.6 The energy balance equation (2.11) reveals several interesting properties of EL systems. First, if we set $u = 0$ we see that energy is not increasing, hence the trivial equilibrium of the unforced system is *stable* in the sense of Lyapunov. Needless to say that these considerations constituted the starting point of Lyapunov's original work! Second, stability is also preserved if we now fix the output $\mathcal{M}\dot{q}$ to zero, hence reflecting the fact that the system is *minimum phase*.[4] Thirdly, we see that damping can be easily added –along the actuated channels– if $\dot{q}$ is measurable. Notice, however, that the operator $u \mapsto \mathcal{M}^\top \dot{q}$ may be output strictly passive even if energy is *not dissipated* "in all directions". Namely, it is enough to ensure $\dot{q}^\top \frac{\partial \mathcal{F}(\dot{q})}{\partial \dot{q}} \geq \alpha \|\mathcal{M}^\top \dot{q}\|^2$. This property will be used in Chapter **3** to solve output feedback stabilization problems.

2.2 Passivity of the error dynamics

The passivity property described above is sufficient to solve regulation tasks in mechanical systems, where the PBC only needs to modify the potential energy and the dissipation function. However, to study tracking problems or treat electrical or electromechanical systems, we need a stronger property. The main reason, that will be explained in detail later, is that in these cases a desired behaviour should be imposed, not only on q, but on $\dot{q}$ as well, which in its turn translates into the need for modifying the kinetic energy.

Leaving aside the problem of regulation in mechanical systems (which is basically the only case that has been documented in the literature, and we will only very briefly review it here) throughout the book we will see that a first step in PBC may be to achieve a closed–loop dynamics of the form

$$D(q)\dot{s} + [C(q, \dot{q}) + K_d(q, \dot{q})]s = 0 \qquad (2.12)$$

where s denotes an *error signal* that we want to drive to zero, $K_d(q, \dot{q}) = K_d^\top(q, \dot{q}) > 0$ is a damping injection matrix, and $C(q, \dot{q})$ is a matrix, univocally defined by $D(q)$, which satisfies

$$\dot{D}(q) = C(q, \dot{q}) + C^\top(q, \dot{q}). \qquad (2.13)$$

Notice that (2.13) is equivalent to the skew–symmetry property[5]

$$z^\top[\dot{D}(q) - 2C(q, \dot{q})]z = 0, \quad \forall\, z \in \mathbb{R}^n$$

The motivation for aiming at (2.12) stems from the following key lemma from [212].

Lemma 2.7 *The differential equation*

$$D(q)\dot{s} + [C(q, \dot{q}) + K_d(q, \dot{q})]s = \Psi$$

---

[4] Actually stability and inverse stability hold true for general passive systems.

[5] As explained in [212] no matter how $C(q, \dot{q})$ is defined in $C(q, \dot{q})\dot{q}$ it is *always* true that $\dot{q}^\top[\dot{D}(q) - 2C(q, \dot{q})]\dot{q} = 0$. This, however, does not imply (2.13).

where $D(q)$ and $K_d(q, \dot{q})$ are positive definite and $C(q, \dot{q})$ satisfies (2.13) defines an output strictly passive operator $\Sigma_d : \Psi \mapsto s$. Consequently, if $\Psi \equiv 0$ we have $s \in \mathcal{L}_2$.
□

Proof. The proof follows using the storage function

$$\mathcal{H}_d = \frac{1}{2}s^\top D(q)s \geq 0. \tag{2.14}$$

By differentiating (2.14) with respect to time and using the skew-symmetry of $\dot{D}(q) - 2C(q, \dot{q})$ we obtain that $\dot{\mathcal{H}}_d \leq -K_d(q, \dot{q})s^2 + \Psi^\top s$. The OSP property follows by integrating on both sides of this inequality from 0 to T. The second part of the proof follows from the fact that OSP systems are $\mathcal{L}_2$-stable (see Corollary A.5). Henceforth, with $\Psi \equiv 0$ we have $s \in \mathcal{L}_2$. ∎

The skew–symmetry property (2.13) is essential for the lemma above. The lemma below, first published in [212] (but implicit in the work of [133]), establishes this fact.

Lemma 2.8 *Let the ik–th entry of the matrix $C(q, \dot{q})$ (called in the robotics literature the "Coriolis and centrifugal forces" matrix) be given by*

$$C_{ik}(q, \dot{q}) = \sum_{j}^{n} c_{ijk}(q)\dot{q}_j.$$

where

$$c_{ijk}(q) \overset{\triangle}{=} \frac{1}{2}\left(\frac{\partial d_{ik}(q)}{\partial q_j} + \frac{\partial d_{jk}(q)}{\partial q_i} - \frac{\partial d_{ij}(q)}{\partial q_k} \right)$$

are the so called Christoffel symbols of the first kind. Then, (2.13) holds. □

Using the special factorization described above the EL equations (2.6) can be written in the equivalent form

$$D(q)\ddot{q} + C(q, \dot{q})\dot{q} + g(q) + \frac{\partial \mathcal{F}(\dot{q})}{\partial \dot{q}} = \mathcal{M}u + Q_\zeta \tag{2.15}$$

where

$$g(q) \overset{\triangle}{=} \frac{\partial \mathcal{V}}{\partial q}(q)$$

which is called in robotics the gravity forces.

Remark 2.9 The passivity property presented in Proposition 2.5 and the "skew–symmetry" property, although clearly closely related, are fundamentally different. Attaching the basic principle of passivity to the dynamics is a far reaching systems theoretic concept that goes beyond the mathematical convenience of a "suitable factorization". The establishment of passivity is at the core of the whole PBC methodology, which started with its application in robot control. To the best of our knowledge, the property was first explicitly expressed *in terms of passivity* in the conference version of [124] in 1988, for the case when the potential energy is absent. The complete proof with the potential energy term appeared first in [212]. The qualifier "PBC" was coined in the latter paper, and has since enjoyed widespread popularity. On the other hand, the "skew–symmetry" property was first obtained by [133]. Both properties have played a fundamental role in the development of energy shaping plus damping injection controller designs, which have a parallel interesting history. We refer the reader to Subsection **3**.1.3 for further historical precisions.

2.3   Other properties and assumptions

In this book we focus our attention on those systems for which the following properties and assumptions hold

P 2.1 The system (2.15) can be parameterized as

$$D(q)\ddot{q} + C(q,\dot{q})\dot{q} + g(q) + \frac{\partial \mathcal{F}(\dot{q})}{\partial \dot{q}} = \Phi(q,\dot{q},\ddot{q})^{\top}\theta \tag{2.16}$$

where $\theta \in \mathbb{R}^p$ is a vector of constant parameters and $\Phi(q,\dot{q},\ddot{q}) \in \mathbb{R}^{q \times n}$ is called *regressor matrix*.

A 2.1 The matrix $D(q)$ is symmetric positive definite and there exist some positive constants d_m and d_M such that

$$d_m I < D(q) < d_M I \tag{2.17}$$

A 2.2 There exists some positive constants k_g and k_v such that

$$k_g \geq \sup_{q \in \mathbb{R}^n} \left\| \frac{\partial^2 \mathcal{V}(q)}{\partial q^2} \right\| \tag{2.18}$$

$$k_v \geq \sup_{q \in \mathbb{R}^n} \left\| \frac{\partial \mathcal{V}(q)}{\partial q} \right\| \tag{2.19}$$

P 2.2 The matrix $C(x,y)$ is bounded in x and linear in y, then for all $z \in \mathbb{R}^n$

$$C(x,y)z = C(x,z)y \tag{2.20}$$
$$\|C(x,y)\| \leq k_c\|y\|, \quad k_c > 0. \tag{2.21}$$

As a matter of fact, (2.20) is a direct consequence of the definition of $C(q,\dot{q})$. Also, inequality (2.21) follows using (2.17) and the definition of Christoffel symbols.

2.4 Passive subsystems decomposition

In some practical cases the generalized inertia matrix is *block diagonal*, expressing the fact that the angular part of the kinetic energy of each sub–block is due only to its own rotation. See Section 4.3 for a discussion of this point in robots with flexible joints, and Chapter **9** for its application in electrical machines. For this class of systems we have the following nice property.

Proposition 2.10 *Assume the Lagrangian (2.2) can be decomposed in the form*

$$\mathcal{L}(q, \dot{q}) = \mathcal{L}_e(q_e, \dot{q}_e, q_m) + \mathcal{L}_m(q_m, \dot{q}_m)$$

where $q \stackrel{\triangle}{=} [q_e^\top, q_m^\top]^\top$ *with* $q_e \in \mathbb{R}^{n_e}$ *and* $q_m \in \mathbb{R}^{n_m}$. *Then, the EL system (2.1) can be represented as the negative feedback interconnection of two passive subsystems (as shown in Fig. 2.1)*

$$\Sigma_e \quad : \quad \begin{bmatrix} Q_e \\ -\dot{q}_m \end{bmatrix} \mapsto \begin{bmatrix} \dot{q}_e \\ \tau \end{bmatrix}$$

$$\Sigma_m \quad : \quad (\tau + Q_m) \mapsto \dot{q}_m$$

with storage functions $\mathcal{L}_e(q_e, \dot{q}_e, q_m)$ *and* $\mathcal{L}_m(q_m, \dot{q}_m)$, *respectively, where*

$$\tau \stackrel{\triangle}{=} \frac{\partial \mathcal{L}_e}{\partial q_m}(q_e, \dot{q}_e, q_m)$$

is the subsystems coupling signal, and $Q \stackrel{\triangle}{=} [Q_e^\top, Q_m^\top]^\top$ *with* $Q_e \in \mathbb{R}^{n_e}$, $Q_m \in \mathbb{R}^{n_m}$.
□

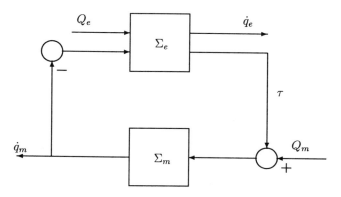

Figure 2.1: Feedback decomposition of an EL system.

Proof. Splitting (2.2) into its components we get

$$\frac{d}{dt}\left(\frac{\partial \mathcal{L}_e}{\partial \dot{q}_e}\right) - \frac{\partial \mathcal{L}_e}{\partial q_e} = Q_e \tag{2.22}$$

$$\frac{d}{dt}\left(\frac{\partial \mathcal{L}_m}{\partial \dot{q}_m}\right) - \frac{\partial \mathcal{L}_m}{\partial q_m} = Q_m + \tau \tag{2.23}$$

Evaluating the total time derivative of $\mathcal{L}_e$ results in

$$\dot{\mathcal{L}}_e = \left(\frac{\partial \mathcal{L}_e}{\partial q_e}\right)^\top \dot{q}_e + \left(\frac{\partial \mathcal{L}_e}{\partial \dot{q}_e}\right)^\top \ddot{q}_e + \left(\frac{\partial \mathcal{L}_e}{\partial q_m}\right)^\top \dot{q}_m \tag{2.24}$$

Noting that

$$\left(\frac{\partial \mathcal{L}_e}{\partial \dot{q}_e}\right)^\top \ddot{q}_e = \frac{d}{dt}\left[\left(\frac{\partial \mathcal{L}_e}{\partial \dot{q}_e}\right)^\top \dot{q}_e\right] - \frac{d}{dt}\left(\frac{\partial \mathcal{L}_e}{\partial \dot{q}_e}\right)^\top \dot{q}_e$$

inserting this into (2.24), using (2.22) and rearranging the terms, it follows that

$$\frac{d}{dt}\mathcal{H}_e = Q_e^\top \dot{q}_e - \dot{q}_m^\top \tau$$

where, similarly to the proof of Proposition 2.5, we have defined

$$\mathcal{H}_e(q_e, \dot{q}_e, q_m) \triangleq \left(\frac{\partial \mathcal{L}_e}{\partial \dot{q}_e}\right)^\top \dot{q}_e - \mathcal{L}_e$$

which is the total energy of the subsystem Σ_e. Passivity of Σ_e follows, as done above, integrating from 0 to T.

A similar procedure can be used to establish the passivity of Σ_m, using the energy function $\mathcal{H}_m = \frac{\partial \mathcal{L}_m^\top}{\partial \dot{q}_m}\dot{q}_m - \mathcal{L}_m$. ∎

2.5   An EL structure-preserving interconnection

We have explained in the Introduction that the aim of PBC is to achieve a desired passive map in closed–loop. Now, we have seen in Proposition 2.5 that EL systems define passive operators, while in Proposition A.10 of Appendix **A** we mentioned that passivity is invariant under feedback interconnection. Putting all these pieces together motivates us to look for our PBC among the class of EL systems. Interestingly enough, if we restrict ourselves to this class we can define a controller interconnection that not only preserves the EL structure (and hence closed–loop passivity), but furthermore – and probably more significantly– the new storage and energy dissipation functions are obtained by *adding up* the corresponding functions of the plant and the controller. As we will see later, this essentially trivializes the energy shaping and damping injection stages of the passivity–based design in regulation tasks for mechanical systems.

Proposition 2.11 (Interconnected EL systems.) *Consider two EL systems* Σ_p : $\{\mathcal{T}_p(q_p, \dot{q}_p), \mathcal{V}_p(q_p), \mathcal{F}_p(\dot{q}_p), \mathcal{M}_p\}$ *and* $\Sigma_c \{\mathcal{T}_c(q_c, \dot{q}_c), \mathcal{V}_c(q_c, q_p), \mathcal{F}_c(\dot{q}_c)\}$ *with generalized coordinates* $q_p \in \mathbb{R}^{n_p}$ *and* $q_c \in \mathbb{R}^{n_c}$, *respectively, (notice that the potential energy of* Σ_c *depends on* q_p*). Interconnect the systems via*

$$\mathcal{M}_p u = -\frac{\partial \mathcal{V}_c(q_c, q_p)}{\partial q_p}$$

where u *is the input of the subsystem* Σ_p*. See Fig. 2.2. Under these conditions, the closed–loop system is an EL system* $\Sigma : \{\mathcal{T}(q, \dot{q}), \mathcal{V}(q), \mathcal{F}(\dot{q})\}$*, with generalized coordinates* $q \stackrel{\triangle}{=} [q_p^\top, q_c^\top]^\top$ *and EL parameters*

$$\mathcal{T}(q, \dot{q}) = \mathcal{T}_c(q_c, \dot{q}_c) + \mathcal{T}_p(q_p, \dot{q}_p) \qquad \mathcal{V}(q) = \mathcal{V}_c(q_c, q_p) + \mathcal{V}_p(q_p),$$

$$\mathcal{F} = \mathcal{F}_c(\dot{q}_c) + \mathcal{F}_p(\dot{q}_p).$$

$\square$

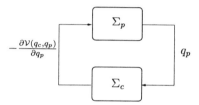

Figure 2.2: Feedback interconnection of two EL systems.

3 Lyapunov stability properties

In this section, we stress other properties of EL systems which are related to their stability in the sense of Lyapunov. For the sake of clarity, we separate the results for fully-damped and underdamped systems.

3.1 Fully-damped systems

The proposition below is a well-known result that establishes conditions for *internal stability* of fully damped EL systems. It was stated by Joseph L. La Grange (c. 1788) and later proved by Dirichlet.[6] The proof is typically established invoking La Salle–Krasovskii's lemma. To underscore the role of passivity and detectability we give

---

[6]See [180] for an interesting historical review on the important question of stability of mechanical systems including several converse versions of this theorem.

below a proof that combines the strict passivity property established in Proposition 2.5 with Proposition A.9[7].

Proposition 2.12 (GAS with full damping.) *The equilibria of a fully damped unforced EL system (i.e., with $u = Q_\zeta = 0$) are $(q, \dot{q}) = (\bar{q}, 0)$, where $\bar{q}$ is the solution of*

$$\frac{\partial \mathcal{V}(q)}{\partial q} = 0. \tag{2.25}$$

The equilibrium is stable if $\bar{q}$ is a strict local minimum of the potential energy function $\mathcal{V}(q)$. Furthermore, if $\mathcal{V}(q)$ is proper (for instance if $\mathcal{V}(q)$ satisfies the conditions of Lemma C.7) and the minimum is unique, then this equilibrium is GAS. □

Proof.   The existence of the equilibrium follows immediately writing the EL equations with $u = Q_\zeta = 0$

$$D(q)\ddot{q} + N(q, \dot{q}) + \frac{\partial \mathcal{V}(q)}{\partial q} + \frac{\partial \mathcal{F}(\dot{q})}{\partial \dot{q}} = 0 \tag{2.26}$$

where in order to simplify the notation we defined

$$N(q, \dot{q}) \triangleq \frac{d}{dt}\{D(q)\} - \frac{\partial \mathcal{T}(q, \dot{q})}{\partial q} \tag{2.27}$$

which satisfies $N(q, 0) = 0$, hence noting that $\frac{\partial \mathcal{F}(\dot{q})}{\partial \dot{q}}\big|_{\dot{q}=0} = 0$ it follows that the equilibria $\bar{q}$ are the solutions of (2.25).

We will now verify the conditions of (ii) of Proposition A.9. We have shown already in Proposition 2.5 that, for fully damped EL systems, the map $u \mapsto \dot{q}$ is OSP with storage function the total energy. In view of the assumption on the potential energy, the total energy is bounded from below, and we can simply add a constant to make it positive definite. To prove stability it only remains to show that the EL system is zero–state detectable –actually, it will be shown to be zero–state observable. This follows immediately setting $\dot{q} = 0$ in (2.26) and noting that this implies that $\frac{\partial \mathcal{V}(q)}{\partial q} = 0$. Finally, GAS results from the assumptions of uniqueness of the equilibrium and that $\mathcal{V}(q)$ is proper. ■

3.2   Underdamped systems

In the proposition below we show that *global* asymptotic stability of a unique equilibrium point can still be ensured even when the system is not fully damped provided the inertia matrix $D(q)$ has a certain block diagonal structure, and the dissipation is

---

[7]An obvious change of coordinates is needed to shift the equilibrium to the origin.

suitably propagated. Even though other, more general, versions of this theorem can be easily proved, we give here the one that will be needed in the sequel. It is interesting to remark that, as far as we know, the first paper which established sufficient conditions for asymptotic stability of underdamped EL systems [222], was published more than 35 years ago.

Our motivation to present this result stems from its application for PBC with EL systems which will be presented in Chapter **3**. For this case it is natural to partition q as

$$q_c \triangleq [0 \ I_{n_c}]q, \quad q_p \triangleq [I_{n_p} \ 0]q, \quad n = n_p + n_c, \tag{2.28}$$

to distinguish the *damped* and *undamped* coordinates. Our motivation to use the subindices $(\cdot)_c$ and $(\cdot)_p$, which suggest controller and plant, respectively, will become clear in Chapter **3**.

Proposition 2.13 (GAS with partial damping)

Consider an unforced underdamped EL system with the coordinate partition (2.28). The equilibrium $(\dot{q}, q) = (0, \bar{q})$ is GAS if the potential energy function is proper and has a global and unique minimum at $q = \bar{q}$, and if

(i) $D(q) \triangleq \begin{bmatrix} D_p(q_p) & 0 \\ 0 & D_c(q_c) \end{bmatrix}$, *where $D_p(q_p) \in \mathbb{R}^{n_p \times n_p}$ and $D_c(q_c) \in \mathbb{R}^{n_c \times n_c}$;*

(ii) $\dot{q}^\top \frac{\partial \mathcal{F}(\dot{q})}{\partial \dot{q}} \geq \alpha \|\dot{q}_c\|^2$ *for some $\alpha > 0$;*

(iii) For each q_c, the function $\frac{\partial \mathcal{V}(q)}{\partial q_c} = 0$ has only isolated zeros in q_p. □

Proof. First, proceeding as is Proposition 2.5, it clearly follows that the map $u \mapsto \dot{q}_c$ is OSP.

Now, as in Proposition 2.12, we write the EL equations (2.6) with $u = Q_\zeta = 0$ by exploiting the block diagonal structure of $D(q)$, in the form

$$D_p(q_p)\ddot{q}_p + N_p(q_p, \dot{q}_p) + \frac{\partial \mathcal{V}(q)}{\partial q_p} = 0 \tag{2.29}$$

$$D_c(q_c)\ddot{q}_c + N_c(q_c, \dot{q}_c) + \frac{\partial \mathcal{F}(\dot{q})}{\partial \dot{q}_c} + \frac{\partial \mathcal{V}(q)}{\partial q_c} = 0 \tag{2.30}$$

where $N_c(q_c, \dot{q}_c)$, $N_p(q_p, \dot{q}_p)$ are suitably defined vectors of the form (2.27). The equilibria are determined by the critical points of the potential energy function $\mathcal{V}(q)$, that is, the solutions of $\frac{\partial \mathcal{V}(q)}{\partial q} = 0$.

The stability proof is carried out, again, using Proposition A.9. Hence, we need to verify only zero–state detectability with respect to the output $\dot{q}_c$. To this end, we set $\dot{q}_c = 0$. From $q_c = const$ in (2.30) and (iii) it follows that q_p is also constant. Since $q = \bar{q}$ is the unique global equilibrium of (2.29), (2.30), zero–state detectability follows. ■

Remark 2.14 It is interesting to notice that if we rewrite (2.29), (2.30) as a lossless system

$$D_p(q_p)\ddot{q}_p + N_p(q_p, \dot{q}_p) + \frac{\partial \mathcal{V}(q)}{\partial q_p} = 0$$

$$D_c(q_c)\ddot{q}_c + N_c(q_c, \dot{q}_c) + \frac{\partial \mathcal{V}(q)}{\partial q_c} = u$$

in closed–loop with the dissipation term

$$y = \frac{\partial \mathcal{F}(\dot{q})}{\partial \dot{q}_c}$$

$$u = -y$$

we can prove Proposition 2.13 invoking zero–state observability properties and the following interesting result:

Proposition 2.15 (Corollary 3.4 of [47].) *Consider a lossless system with a positive definite and proper storage function* $\mathcal{H}$*. The output feedback law*

$$u = -ky, \quad k > 0,$$

renders the origin of the closed loop system globally asymptotically stable if and only if the system is zero-state observable.   □

4   Examples

In Appendix **B** we have shown that the Lagrangian formalism provides a very convenient way of setting up the equations of motion. One simply has to write $\mathcal{T}$ and $\mathcal{V}$ in generalized coordinates and use (2.6) to derive the equations of motion. In this section we give some examples of dynamical systems whose equations of motion can be derived, using the EL equations, from their corresponding EL parameters.

4.1   A rotational/translational proof mass actuator

A translational oscillator with an attached eccentric rotational proof mass actuator (TORA) is shown in Fig. 2.3. It consists of a cart of mass M connected by a linear spring with stiffness k to a fixed wall. This system model was recently proposed as a benchmark problem for nonlinear system design. See [41] for further details.

The cart has only one-dimensional motion parallel to the spring axis. The proof mass actuator attached to the cart has mass m and moment of inertia I around its center of mass. The latter is located at a distance l from its rotational axis. The gravitational forces are neglected because the motion occurs in an horizontal plane. The control torque applied to the proof mass is denoted by u.

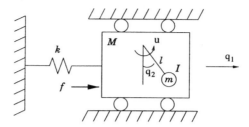

Figure 2.3: Rotational/translational proof mass actuator.

Let q_1 be the translational position of the cart and q_2 the angular position of the proof mass, where $q_2 = 0$ is perpendicular to the motion of the cart, and $q_2 = 90°$ is aligned with the positive q_1 direction. The generalized coordinates are then $q = [q_1, q_2] \in \mathbb{R}^2$, the kinetic and potential energy functions are defined as

$$\mathcal{T}(q_2, \dot{q}) = \frac{1}{2}\dot{q}^\top D(q_2)\dot{q} = \dot{q}^\top \begin{bmatrix} M+m & -ml\cos(q_2) \\ -ml\cos(q_2) & I+ml^2 \end{bmatrix} \dot{q}$$

$$\mathcal{V}(q_1) = \frac{1}{2}kq_1^2$$

Applying the EL equations (2.6) with $Q_\zeta = 0$ yields the model

$$D(q_2)\begin{bmatrix} \ddot{q}_1 \\ \ddot{q}_2 \end{bmatrix} + \begin{bmatrix} -ml\dot{q}_2^2\sin(q_2)\dot{q}_1 \\ 0 \end{bmatrix} + \begin{bmatrix} kq_1 \\ 0 \end{bmatrix} = \begin{bmatrix} 0 \\ u \end{bmatrix}$$

which can be written in compact notation as

$$D(q_2)\ddot{q} + C(q_2, \dot{q}_2)\dot{q} + g(q_1) = \mathcal{M}u \qquad (2.31)$$

where $\mathcal{M} = [0, 1]^\top$, $g(q_1) \triangleq [kq_1,\ 0]^\top$. The system is clearly underactuated, with actuated coordinate q_2, and undamped. Further, it defines a passive operator $u \mapsto \dot{q}_2$.

Notice that, via the definition of the matrix

$$C(q_2, \dot{q}_2) = \begin{bmatrix} 0 & -ml\dot{q}_2\sin(q_2) \\ 0 & 0 \end{bmatrix}$$

we have introduced the factorization of the vector

$$\begin{bmatrix} -ml\dot{q}_2\sin(q_2)\dot{q}_1 \\ 0 \end{bmatrix} = C(q_2, \dot{q}_2)\dot{q}$$

discussed in Proposition 2.12. As pointed out there this factorization is uniquely defined by the inertia matrix $D(q_2)$ and constitutes a *fundamental* step, which reveals the workless forces of the system, for all our further developments.

4.2   Levitated ball

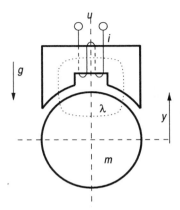

Figure 2.4: Ball in a vertical magnetic field.

Consider the system of Fig. 2.4, consisting of an iron ball in a vertical magnetic field created by a single electromagnet. We denote by q_m the position of the ball measured with respect to the nominal position, with the q_m-axis oriented upwards, and with $\dot{q}_e$ the electric current in the inductance. To apply the variational modeling technique, we define first the generalized coordinates, which in this case are $q = [q_e, q_m]^\top$. If we assume linearity of the magnetic circuit and neglect all fringing fields[8] we obtain the magnetic field co–energy and the mechanical kinetic energy as

$$\mathcal{T}_e(q_m, \dot{q}_e) = \frac{1}{2}L(q_m)\dot{q}_e^2, \quad \mathcal{T}_m(\dot{q}_m) = \frac{1}{2}m\dot{q}_m^2$$

where, $L(q_m)$ is the inductance and $m > 0$ the mass of the ball. A suitable approximation for the former, in the open domain $q_m \in (-\infty, c_2)$ where $c_2 > 0$ is the nominal air gap, is given by

$$L(q_m) = \frac{c_1}{c_2 - q_m}$$

where c_1 is some positive constant that depends on the number of coil turns, air permeability and the cross–sectional area of the electromagnet. To simplify the presentation in the sequel we will assume $c_1 = c_2 = 1$.

We must also derive the potential energy which is given as $\mathcal{V}(q_m) = mg(1 - q_m)$. The Rayleigh dissipation function is $\mathcal{F}(\dot{q}_e) = \frac{1}{2}R_e\dot{q}_e^2$, where $R_e > 0$ is the electrical resistance. The control u is the input voltage, therefore $\mathcal{M} = [1,\ 0]^\top$.

---

[8]See Examples 3.1 and 3.5 of [179] for a detailed derivation of this model.

The resulting Lagrangian is

$$\mathcal{L}(q_m, \dot{q}_e, \dot{q}_m) = \mathcal{T}_e(q_m, \dot{q}_e) + \mathcal{T}_m(\dot{q}_m) - \mathcal{V}(q_m) = \frac{1}{2}\dot{q}^\top D(q_m)\dot{q} - \mathcal{V}(q_m)$$

where we have introduced the generalized inertia matrix

$$D(q_m) \stackrel{\triangle}{=} \left[\begin{array}{cc} L(q_m) & 0 \\ 0 & m \end{array} \right]$$

Applying the EL equations (2.6) with $Q_\zeta = 0$ yields

$$\left\{ \begin{array}{rcl} \frac{1}{(1-q_m)}\ddot{q}_e + \frac{1}{(1-q_m)^2}\dot{q}_m\dot{q}_e + R_e\dot{q}_e & = & u \\[2mm] m\ddot{q}_m - \frac{1}{2}\frac{1}{(1-q_m)^2}\dot{q}_e^2 - mg & = & 0 \end{array} \right.$$

These equations look messy, and as we have discussed above to reveal its fundamental structural features it is convenient to write them in compact form

$$D(q_m)\ddot{q} + C(\dot{q}_e, q_m, \dot{q}_m)\dot{q} + R\dot{q} + G = \mathcal{M}u$$

where we have defined

$$C(\dot{q}_e, q_m, \dot{q}_m) \stackrel{\triangle}{=} \frac{1/2}{(1 - q_m)^2} \left[\begin{array}{cc} \dot{q}_m & \dot{q}_e \\ -\dot{q}_e & 0 \end{array} \right], \quad R \stackrel{\triangle}{=} \left[\begin{array}{cc} R_e & 0 \\ 0 & 0 \end{array} \right], \quad G \stackrel{\triangle}{=} \left[\begin{array}{c} 0 \\ -mg \end{array} \right]$$

This system is also underactuated, with actuated coordinate q_e, and underdamped with q_e the damped coordinate. It defines a passive operator $u \mapsto \dot{q}_e$, as expected if we view the system as a two port with a terminal voltage u and an input current $\dot{q}_e$. We will see in the following chapter that for the PBC design it is sometimes more convenient to use other passivity properties.

Notice we have again introduced a particular factorization of the vector

$$\frac{1}{(1 - q_m)^2} \left[\begin{array}{c} \dot{q}_e\dot{q}_m \\ -\frac{1}{2}\dot{q}_e^2 \end{array} \right] = C(\dot{q}_e, q_m, \dot{q}_m)\dot{q}$$

Admittedly, it is difficult at this point to convince the reader of the advantage of using the variational approach to model this extremely simple system which can be easily obtained from basic physical laws as

$$\left\{ \begin{array}{rcl} \dot{\lambda} & = & -R_e(1 - q_m)\lambda + u \\ m\ddot{q}_m & = & \frac{1}{2}\lambda^2 - mg \end{array} \right.$$

where $\lambda \stackrel{\triangle}{=} L(q_m)\dot{q}_e$ is the flux in the inductance. The advantage will become apparent when we will come to the controller design.

4.3   Flexible joints robots

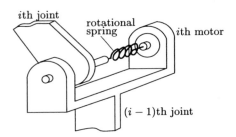

Figure 2.5: Ideal model of a flexible joint.

Some robotic manipulators have flexible joints, for instance, manipulators where harmonic drives, elastic bands, or motors with long shafts are used. The joint flexibility phenomenon can be modeled by a rotational spring, see e.g. [168, 253] and references therein, as it is illustrated in Fig. 2.5.

Flexible joint robots are underactuated EL systems with generalized coordinates $q \triangleq [q_1^\top, q_2^\top]^\top$, $q_1, q_2 \in \mathbb{R}^{\frac{n}{2}}$ being the link and motor shaft angles respectively. The control variables are the torques at the shafts, thus $m = \frac{n}{2}$ and $\mathcal{M} \triangleq [0 \mid I_m]^\top$.

The *kinetic and potential energies* of a flexible joint robot are given by –see [266]–

$$\mathcal{T}(q_1, \dot{q}) \triangleq \frac{1}{2}\dot{q}^\top \mathcal{D}(q_1)\dot{q}, \ \mathcal{V}(q) \triangleq \frac{1}{2}q^\top \mathcal{K}q + \mathcal{V}_g(q_1) \tag{2.32}$$

where

$$\mathcal{K} \triangleq \begin{bmatrix} K & -K \\ -K & K \end{bmatrix}, \ \mathcal{D}(q_1) \triangleq \begin{bmatrix} D_{11}(q_1) & D_{12}(q_1) \\ D_{12}^\top(q_1) & J \end{bmatrix} \tag{2.33}$$

with $D_{12}(q_1)$ of the form

$$D_{12}(q_1) = \begin{bmatrix} 0 & d_{12}(q_{1,1}) & d_{13}(q_{1,1}, q_{1,2}) & \cdots & d_{1m}(q_{1,1}, q_{1,m-1}) \\ 0 & 0 & d_{23}(q_{1,2}) & \cdots & d_{2m}(q_{1,2}, q_{1,m-1}) \\ \vdots & \vdots & \vdots & \ddots & \vdots \\ 0 & 0 & 0 & \cdots & 0 \end{bmatrix} \tag{2.34}$$

and $q_{1,j}$, for $j \in [1..m]$, is the j-th component of the vector q_1. Matrix $\mathcal{D}(q_1) = \mathcal{D}^\top(q_1) > 0$ is the robot inertia matrix, $J \in \mathbb{R}^{m \times m}$ is a diagonal matrix of actuator inertias reflected to the link side, K is a diagonal matrix containing the joint stiffness coefficients, and $\mathcal{V}_g(q_1)$ is the potential energy due to the gravitational forces.

Assuming no internal damping, that is, $\mathcal{F}(\dot{q}) = 0$ we obtain the *dynamic equations* of the flexible joint robot:

$$\mathcal{D}(q_1)\ddot{q} + \mathcal{C}(q_1, \dot{q})\dot{q} + \mathcal{G}(q_1) + \mathcal{K}q = \mathcal{M}u \tag{2.35}$$

where $\mathcal{G}(q_1) \triangleq [g_{p1}^\top(q_1), 0]^\top = \frac{\partial V_g(q_1)}{\partial q_1}$. Once again we have factored the second left hand term in a suitable manner defining the Coriolis matrix:

$$\mathcal{C}(q_1, \dot{q}) \triangleq \begin{bmatrix} C_{11}(q_1, \dot{q}_1) + C'_{11}(q_1, \dot{q}_2) & C_{12}(q_1, \dot{q}_1) \\ C_{21}(q_1, \dot{q}_1) & 0 \end{bmatrix}, \tag{2.36}$$

$$(C_{21})_{i,j}(q_1, \dot{q}) \triangleq \frac{1}{2}\left[\dot{q}_1^\top \frac{\partial(D_{12})_{j,i}}{\partial q_1} + \frac{\partial(D_{12})^i}{\partial \dot{q}_1}\dot{q}_1\right] \tag{2.37}$$

where $(\cdot)_{i,j}$, $(\cdot)^i$ denote the (i, j)-th term and i-th row of a matrix respectively. When the angular part of the kinetic energy of each rotor can be considered due only to its own rotation then we obtain the simplified model of [253]

$$\begin{cases} D_l(q_1)\ddot{q}_1 + C(q_1, \dot{q}_1)\dot{q}_1 + g(q_1) = K(q_2 - q_1) \\ J\ddot{q}_2 + K(q_2 - q_1) = u. \end{cases} \tag{2.38}$$

In the case where flexibility is negligible ($K \to \infty$) it is shown in [253] that the model (2.38) reduces to the well–known rigid robot model

$$D(q_1)\ddot{q}_1 + C(q_1, \dot{q}_1)\dot{q}_1 + g(q_1) = u \tag{2.39}$$

where $D(q_1) \triangleq D_l(q_1) + J$.

4.4 The Duffing system

Consider a mechanical system with generalized coordinates $q \in \mathbb{R}$ and EL parameters

$$\begin{aligned} \mathcal{T}(\dot{q}) &= \frac{1}{2}\dot{q}^2 \\ \mathcal{V}(q) &= \frac{1}{2}p_2q^2 + \frac{1}{4}p_4 \\ \mathcal{F}(\dot{q}) &= \frac{1}{2}p_1\dot{q}^2 \\ \mathcal{M} &= 1. \end{aligned}$$

where p_1 is a nonnegative constant and p_2, p_3 are real. The direct evaluation of the Lagrangian equations using the EL parameters defined above yields the Duffing equation

$$\ddot{q} + p_1\dot{q} + p_2q + p_3q^3 = u \tag{2.40}$$

where u is external force. If we set $u = Q \cos \omega t$ to the right hand side of (2.40) we obtain the well known periodically-forced Duffing equation which borrows the name from his creator, who used it in 1918 to study the dynamics of a pendulum moving in a viscous medium.

Notice that the Duffing equation is similar to a common mass-spring-damper system except from the term $p_3 q^3$. As a matter of fact, the Duffing equation also models the motion of a mass-spring-damper system where the spring induces a restoring force which obeys Hooks law ($F = -kq$) only for small displacements. For large displacements however, the restoring force is given by the expression $F = -\kappa q^3$, such spring is called "hardening" since after a certain limit a small displacement induces a large restoring force.

4.5   A marine surface vessel

One of the simplest models of a marine surface vessel is the Lagrangian type model

$$M\dot{\nu} + R\nu = u + J^\top(\eta)b \qquad (2.41)$$
$$\dot{\eta} = J\nu \qquad (2.42)$$

where $M \in \mathbb{R}^{3 \times 3}$ is the ship's mass (positive definite) matrix, D is a damping constant matrix, not necessarily positive definite but bounded, η is the vector of position and orientation of the vessel with respect to a fixed frame and ν is the vector of velocities referred to a mobile frame attached to the ship. The term b represents the influence of external forces acting on the ship due to environmental disturbances.

The Jacobian J is an orthogonal rotation matrix, hence full rank. This property is fundamental since we can rewrite the equations (2.41)-(2.42) in the familiar form (2.26). Let us write with an abuse of notation $\dot{J}^\top \triangleq \frac{d}{dt}(J^\top)$. Derivate (2.42) once with respect to time to obtain $\dot{\nu} = \dot{J}^\top \dot{\eta} + J^\top \ddot{\eta}$. Then substituting this in (2.41) and premultiplying on both sides of the equality by $J(\eta)$ we may write:

$$D(q)\ddot{q} + C(q, \dot{q})\dot{q} + \bar{R}(q)\dot{q} + g(q) = u \qquad (2.43)$$

where we have defined $q \triangleq \eta$

$$\begin{aligned}
D(q) &\triangleq J(q)MJ^\top(q) \\
C(q, \dot{q}) &\triangleq J(q)M\dot{J}^\top(q, \dot{q}) \\
g(q) &\triangleq b \\
u &\triangleq J(q)\tau
\end{aligned}$$

Further, since $J(q)$ is orthogonal then (2.43) possesses the properties described in Section 2.3.

It is interesting to remark that even though there is no gravitational energy as in the case of a manipulator, one can think of the bias b as derived from of a conservative potential, in other words, the energy of the weather perturbations.

Finally, we point out that the model (2.41)-(2.42) is most appropriate for *setpoint* regulation purposes. However, for trajectory tracking control additional dynamics must be taken into account, such as the added mass of the water as the ship moves on the surface. For further detail on this and other *Lagrangian* models of ships we refer the reader to [86].

5   Concluding remarks

In this chapter we have defined EL systems, which is the class that we will consider throughout the book. We have also recalled some important properties of EL systems that will be used in the sequel in our control design. These properties can be summarized as follows:

- EL systems are characterized by their EL parameters: Kinetic energy, Potential energy, Rayleigh dissipation function, Inputs matrix and Disturbance signal.

- EL systems define passive operators with storage function the total energy.

- The EL dynamics can be parameterized so as to generate a linear error dynamics which is output strictly passive (OSP).

- The feedback interconnection of two EL systems yields an EL system. The EL parameters of the resulting system are simply the addition of those of both subsystems.

- Under suitable conditions, EL systems can be decomposed as a feedback interconnection of passive subsystems.

- The stable equilibria of an EL system correspond to the minima of its potential energy function.

- EL systems are asymptotically stable if they have a suitable damping.

Part I

Mechanical Systems

Chapter 3

Set-point regulation

In the previous chapter we underlined several fundamental properties of EL systems. In particular we saw that the equilibria of an EL plant are determined by the critical points of its potential energy function, moreover the equilibrium is unique and globally stable if this function has a global and unique minimum. We also saw that this equilibrium is asymptotically stable if suitable damping is present in the system. These two fundamental properties motivated Takegaki and Arimoto in [261] to formulate the problem of *set point regulation* of robots in two steps, first an *energy shaping* stage where we modify the potential energy of the system in such a way that the "new" potential energy function has a global and unique minimum in the desired equilibrium. Second, a *damping injection* stage where we now modify the Rayleigh dissipation function. This seminal contribution contained the first clear exposition of the use of energy functions in robotics. (See Subsection 1.3 for a brief review of the literature). It generated a lot of interest in the robotics community since it rigorously established that computationally simple control laws, derived from *energy considerations*, could accomplish rather sophisticated tasks.

Of course, the key property that underlies the success of this procedure is the *passivity* of the EL system (Proposition 2.5.) Viewed from the broader PBC perspective pursued in this book the energy shaping stage accomplishes a *passivation* objective with a storage function that contains the desired potential energy. The damping injection reinforces this property to *output strict passivity*. Finally, Lyapunov stability follows from the input–output $\mathcal{L}_2$ stability of the output strictly passive map provided some detectability (i.e., dissipation propagation) conditions are met. This two stage procedure will be used throughout the book with the fundamental differences that in tasks, other than regulation of mechanical systems, we must also shape the kinetic energy of the system and an additional "system inversion" step will also be required.

In this chapter we study the problem of *set-point regulation* of mechanical systems described by the EL equations (2.6). We start by recalling the fundamental result of [261] for *fully-actuated* systems with *full state* measurement. To underscore the main steps of this technique we carefully explain the basic steps of energy shaping plus damping injection in the classical pendulum example. For the application of this methodology there are two structural obstacles related with the actuators and sensors of the EL system. On one hand, if the system is *not fully-actuated* it is clear from (2.6) that we cannot assign an arbitrary potential energy function. On the other hand, it follows from the energy balance equation (and it has been pointed out already in Remark 2.6) that damping can be easily injected if the generalized *velocities* are available for measurement, but as shown in this chapter a dynamic extension is needed otherwise. Cost considerations and the fact that velocity measurements are often contaminated with noise are two clear motivations to look for regulators without velocity feedback. Also, for some applications the assumption of full actuation is not realistic. For instance, in robot manipulators with fast motions the existence of joint flexibilities significantly affects the dynamic behaviour.

The main contribution of this chapter is the development of a class of globally stable output feedback PBC for underactuated mechanical systems. To solve this problem we exploit two additional properties of EL systems established in the previous chapter. First, the fact (pointed out in Proposition 2.13) that, for asymptotic stability, the damping needs not be pervasive (i.e., the EL system need not be fully-damped), it suffices that it propagates through the whole system coordinates. Second, the key property that the feedback interconnection of two EL systems yields an EL system with *added* EL parameters (Proposition 2.11). Since the latter property essentially trivializes the design we are motivated to choose the controllers as *EL systems*, that is, the dynamics of our controllers is described by the EL equations.

This contribution is further extended to the case of constrained inputs, we then identify a subclass of EL systems which can be controlled by EL controllers which yield control inputs satisfying an *a priori* imposed bound.

Finally, we explore the setpoint control problem of EL systems with uncertain potential energy knowledge. We give particular importance to PID control and show how this popular control law may be interpreted as an interconnection of passive systems. In the case of unmeasurable velocities we add a second integral action to the PID, yielding a PI^{2}D controller, which we also present from a passivity perspective.

We illustrate the results with the example of robots with flexible joints and the TORA system and conclude the chapter with some numerical simulations.

1   State feedback control of fully-actuated systems

1.1   A basic result: The PD controller

For the sake of clarity let us recall here the seminal result of [261]. It concerns the global asymptotic stabilization via energy shaping plus damping injection of the equilibrium $[q, \dot{q}]^\top = [q_*, 0]^\top$, with q_* a constant vector, for the EL system (2.15).

Proposition 3.1 *Consider an n–degrees of freedom fully-actuated EL system with no internal damping nor external forces described by (2.15), which for convenience of presentation we repeat below with $Q_\zeta \equiv 0$*

$$D(q)\ddot{q} + C(q, \dot{q})\dot{q} + g(q) = u$$

where $q \in \mathbb{R}^n$ and $u \in \mathbb{R}^n$ is the vector of control inputs, and $D(q), C(q, \dot{q})$ and $g(q)$ are defined in Chapter 2, hence satisfy the properties of Section 2.2.3[1]

Let the state–feedback control law be given as

$$u = -\frac{\partial \mathcal{V}_c}{\partial q}(q) - \frac{\partial \mathcal{F}_c}{\partial \dot{q}}(\dot{q})$$

and let us assume the following.

A 3.1 *The function $\mathcal{V}_c(q)$ is such that the potential energy of the closed–loop system*

$$\mathcal{V}_d(q) \triangleq \mathcal{V}(q) + \mathcal{V}_c(q)$$

has a unique global minimum at $q = q_*$ (a constant) and is radially unbounded (with respect to $q - q_*$);

A 3.2 *The dissipation function $\mathcal{F}_c(\dot{q})$ satisfies*

$$\frac{\partial \mathcal{F}_c}{\partial \dot{q}}(0) = 0 \quad \text{and} \quad \dot{q}^\top \frac{\partial \mathcal{F}_c}{\partial \dot{q}}(\dot{q}) > 0, \ \forall \dot{q} \neq 0$$

Under these conditions, the equilibrium $[q, \dot{q}]^\top = [q_*, 0]^\top$ is globally asymptotically stable. □

Proof. The proof follows *verbatim* from Proposition 2.12 considering as storage function the total energy of the closed–loop

$$\mathcal{H}_d(q, \dot{q}) = \mathcal{T}(q, \dot{q}) + \mathcal{V}_d(q) - \mathcal{V}_d(q_*)$$

where the last term is included to enforce $V_d(q_*, 0) = 0$. ■

Remark 3.2 If $\mathcal{V}_d(q)$ has a unique *local* minimum at $q = q_*$ we can conclude only local asymptotic stability.

Remark 3.3 The state feedback controller of Proposition 3.1 has been extended by Tomei to the case of underactuated mechanical systems (in particular robots with flexible joints) in the important paper [266]. The result in that paper is established invoking Lyapunov arguments. In [208], where we presented a solution to the output feedback problem, we give a passivity interpretation to Tomei's controller.

---

[1]In particular, the EL parameters are $\{\mathcal{T}(q, \dot{q}), \mathcal{V}(q), 0, I_n\}$, where as explained after Definition 2.1, when clear from the context, an EL parameter which is equal to zero will be simply omitted from the list, in this case $Q_\zeta = 0$.

1.2   An introductory example

We illustrate here the application of the proposition above with the simple pendulum shown in Fig. 3.1. Although this system has been exhaustively studied in many nonlinear systems texts, we believe it is still an ideal example to put forth the basic principles of PBC. We will therefore go through it in some detail.

The total (kinetic + potential) energy of the simple pendulum is

$$\mathcal{H} = \underbrace{\frac{1}{2}ml^2\dot{q}^2}_{\mathcal{T}(q,\dot{q})} + \underbrace{mgl(1 - \cos(q))}_{\mathcal{V}(q)}$$

where $q \in \mathbb{R}$ and g is the gravity acceleration. We assume torque as the control input u, hence in the absence of friction the EL parameters of such system are $\{\mathcal{T}(q,\dot{q}), \mathcal{V}(q), 0, 1\}$. Using the EL equations we can easily derive the dynamics

$$ml^2\ddot{q} + g(q) = u \tag{3.1}$$

where $g(q)$, which we call the gravitational force, is the force derived from the potential energy, that is,

$$g(q) \overset{\triangle}{=} \frac{\partial \mathcal{V}(q)}{\partial q} = mgl \sin(q)$$

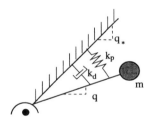

Figure 3.1: Simple pendulum.

Now let us calculate the equilibria of the unforced system (3.1) (i.e. with $u \equiv 0$). As discussed in the previous chapter, and is clear from (3.1), the positions of equilibria correspond to the critical points of the potential energy function, that is, the solutions of the equation:

$$\frac{\partial \mathcal{V}(q)}{\partial q} = 0 \iff mgl \sin(q) = 0,$$

Hence the equilibria are $[q, \dot{q}]^\top = [i\pi, 0]$, $i = \cdots, -1, 0, 1, \cdots$. Next, taking the second partial derivative of $\mathcal{V}(q)$ with respect to q yields

$$\frac{\partial^2 \mathcal{V}(q)}{\partial q^2} = mgl \cos(q),$$

which is positive for $q = 0$ and negative for $q = \pi$. It is clear then that the origin corresponds to a minimum of the potential energy function, hence recalling Proposition 2.12, we conclude that $[q, \dot{q}]^\top = [0, 0]$ is a stable equilibrium. On the other hand, $q = \pi$ is a local maximum, and it can be shown that this is an unstable equilibrium.

Our design problem is to stabilize the pendulum at a constant equilibrium $[q, \dot{q}]^\top = [q_*, 0]$. As suggested by the proposition above we will seek to modify the potential energy and the Rayleigh dissipation function of the system, leaving untouched the kinetic energy, since it plays no role on the stability properties of the equilibrium. That is we want the closed–loop system to be an EL system with EL parameters $\{\mathcal{T}(q, \dot{q}), \mathcal{V}_d(q), \mathcal{F}_d(\dot{q})\}$.

Since we know that a minimum of the potential energy corresponds to a stable equilibrium point, the "new" potential energy function should have a global and unique minimum at the desired position. A natural candidate is then

$$\mathcal{V}_d(q) = \frac{1}{2} k_p \tilde{q}^2 \tag{3.2}$$

where $k_p > 0$ and $\tilde{q} \overset{\triangle}{=} q - q_*$. It is trivial to see that this function satisfies condition **A3.1** of Proposition 3.1. To make this stable equilibrium attractive we choose the desired Rayleigh dissipation function $\mathcal{F}_d(\dot{q}) = \frac{1}{2} k_d \dot{q}^2$, $k_d > 0$, which clearly verifies **A3.2** of the proposition. These choices lead to the control

$$
\begin{aligned}
u &= \underbrace{\frac{\partial}{\partial q} \left(\mathcal{V}(q) - \mathcal{V}_d(q) \right)}_{u_{\mathrm{ES}}} - \underbrace{\frac{\partial \mathcal{F}_c}{\partial \dot{q}}(\dot{q})}_{u_{\mathrm{DI}}} \\
&= g(q) - k_p \tilde{q} - k_d \dot{q}
\end{aligned}
\tag{3.3}
$$

Using Proposition 2.12 global asymptotic stability follows. Remark that the control law consists of two terms taking care of the energy shaping and the damping injection, respectively. This nice nested–loop structure of the control will be encountered in all subsequent PBC.

The PD control law above is one of the simplest one can obtain, however it has the drawback that besides the computational charge that it represents to compute on line the term $g(q)$, it is widely believed that *dominating* instead of *cancelling* the nonlinear term $g(q)$ enhances the robustness of the system vis-a-vis parametric uncertainties[2].

Thus, we consider the desired potential energy function [261]

$$\mathcal{V}_d(q) = \mathcal{V}(q) + \frac{1}{2} k_p [q - \delta(q_*)]^2 \tag{3.4}$$

---

[2] Although this is very hard to prove in general, interested readers are referred to [132] and Section **11.4** where, for an induction motor example, this claim is theoretically proven and experimentally illustrated.

where $\delta(q_*)$ is a *constant* chosen to assign a global and unique minimum at $q = q_*$ to $\mathcal{V}_d(q)$. Its computation then proceeds by evaluating

$$\frac{\partial \mathcal{V}_d}{\partial q}(q) = g(q) + k_p[q - \delta(q_*)],$$

upon setting the latter to zero at $q = q_*$ we get

$$\delta(q_*) = q_* - \frac{1}{k_p}g(q_*).$$

To ensure that this critical point is a global and unique minimum we calculate

$$\frac{\partial^2 \mathcal{V}_d}{\partial q^2}(q) = \frac{\partial g}{\partial q}(q) + k_p.$$

We now invoke Assumption **A2.2** of Chapter **2** and choose $k_p > k_g$ where k_g is defined by (2.18) so that $\frac{\partial^2 \mathcal{V}_d}{\partial q^2}(q) \geq \varepsilon > 0$ for all $q \in \mathbb{R}^n$.

The energy shaping part of the control law is given by

$$u_{\text{ES}} = \frac{\partial}{\partial q}\left(\mathcal{V}(q) - \mathcal{V}_d(q)\right) = -k_p \tilde{q} + g(q_*) \tag{3.5}$$

and adding the same damping as above we find the well known PD plus precompensated gravity controller:

$$u = -k_p \tilde{q} - k_d \dot{q} + g(q_*). \tag{3.6}$$

As before, the closed loop system (3.1), (3.6) is a fully-damped Euler-Lagrange system with EL parameters $\{\mathcal{T}(q, \dot{q}), \mathcal{V}_d(q), \mathcal{F}_d(\dot{q})\}$, therefore, by virtue of Proposition 2.12 the equilibrium point $q = q_*$ is globally asymptotically stable.

Remark 3.4 This pendulum example also illustrates an interesting connection between the energy-shaping condition of Proposition 2.12 and the zero-state detectability condition of 2.15. Notice that the closed loop (3.5) with (3.1) is a lossless system. By defining the output $y = \dot{q}$ it is also zero-state detectable if $k_p \geq k_g$ (energy shaping). Hence the feedback $-k_d \dot{q}$, $k_d > 0$ globally asymptotically stabilizes the closed loop system (3.5), (3.1).

1.3   Physical interpretation and literature review

The two PBCs above have clear *physical interpretations* depicted in Figs. 3.1 and Fig. 3.2. The proportional gain k_p can be regarded as the stiffness constant of a linear spring connecting the pendulum to a virtual line. In the first controller (3.3) gravity forces are cancelled therefore the spring will not store energy at the equilibrium, and we can set the virtual line at the angle q_*. On the other hand, for the PBC (3.6) the spring must store the energy due to the gravitational forces and the plane should be at an angle $\delta(q_*)$. The damping injection term in both cases represents a viscous damper with gain k_d which introduces viscous friction to attain asymptotic stability. We can hardly overestimate the importance of having a direct physical interpretation of the PBC action. Its relevance pertains not just to aesthetical considerations, but more importantly to its impact on the controller tuning stage. Commissioning a nonlinear controller is a difficult task, which has unfortunately been overlooked in most theoretical developments; it is rendered quite transparent to the practicing engineer by PBC. At a more fundamental level:

• PBC underscores the role of control as the establishment of suitable *interconnections* of the system with its environment.

This is perhaps the main asset of PBC, which explains its great success in applications.

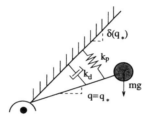

Figure 3.2: Physical interpretation of a PD plus gravity compensation controller.

It is interesting to note that, simultaneously and independently of the work of Arimoto and co–workers in robot control, Jonckeere suggested in [113] also the utilization of energy shaping plus damping injection ideas for controller design of a broader class of EL systems which included electrical and electromechanical systems.[3] It should be noted that while Arimoto and co–workers presented their developments with the Hamiltonian formulation of the dynamics, Jonckeere used instead the EL formalism. Subsequent to the publication of [261] and [113], independent derivations of the same result were reported by van der Schaft [272] and Koditschek [133]. As pointed out in Remark 2.9 in the latter work the important "skew–symmetry" property of Proposition 2.8 was first reported. We refer the reader to [135] for an interesting review of this circle of ideas that spans, in a *tour de force*, from the times of Lagrange and Lord Kelvin to the late 80s.

We conclude this section by singling out three main drawbacks of the utilization of the PD controller of Proposition 3.1 which stymie its utilization in some applications.

---

[3]This very interesting paper, which is the outgrowth of Jonckeere's 1975 PhD thesis in Toulouse, passed relatively unnoticed but strongly influenced the work of the first author.

These motivate all the further developments of the present chapter.

1. Measurement of the generalized velocities $\dot{q}$, and full actuation are required to add the necessary damping.

2. No amplitude constraints are imposed on the control input.

3. The potential energy function $\mathcal{V}(q)$ is supposed to be exactly known.

Each of these drawbacks being of indisputable practical importance, we devote to them the next three sections, respectively.

2   Output feedback stabilization of underactuated systems

In order to put the material of this section in perspective we first briefly review the literature.

2.1   Literature review

Speed measurement increases cost and imposes constraints on the achievable bandwidth because of the presence of noise. This has motivated the researchers to look for regulators which avoid velocity measurements. Linear control laws that obviate the need of this signal preserving GAS have been recently (and independently) proposed in [3, 20, 42, 123]. In the first three papers it is shown that velocity can be replaced by approximate differentiation. In [3] a linear PBC that shapes both the kinetic and the potential energy, and adds damping injection is presented. It must be remarked that in this paper the solution was given also for underactuated EL systems, e.g., flexible-joint robots. Thus extending, to the output feedback case, the controller of [266]. Finding a common feature to all of these controllers and extending the results to the more general frame of underactuated EL systems, motivated us to look for a new methodology for output feedback regulation of underactuated EL systems. This research culminated in the definition of the *EL controllers* in [208]. Other research efforts related with the material of this section may be found in [17, 146]. In particular, in [17], some connections between PBC and control based on output injection ideas are explored.

2.2   Problem formulation

We consider in this section the underactuated EL plants with no internal damping which is the worst case scenario because, as it will become clear later, damping in the plant helps us to relax the conditions for stabilization. That is, we consider plants with EL parameters

$$\{\mathcal{T}_p(q_p, \dot{q}_p), \mathcal{V}_p(q_p), 0, \mathcal{M}_p\}$$

Since the PBC of this section will be dynamical (and furthermore also an EL system) we have added the subindex $_p$ for "plant". The dynamic model has the form (2.15) which for convenience of presentation we write below

$$D_p(q_p)\ddot{q}_p + C(q_p, \dot{q}_p)\dot{q}_p + g(q_p) = \mathcal{M}_p u \tag{3.7}$$

where $q_p \in \mathbb{R}^n$, $u \in \mathbb{R}^m$. For ease of presentation and without loss of generality we will assume that $\mathcal{M}_p = [0, I_m]^\top$.

The problem we study in this section is formulated as follows:

Definition 3.5 (Output feedback global stabilization problem) *Consider the EL system (3.7) where q_p is partitioned as $q_p = [q_{p_1}^\top,\ q_{p_2}^\top]^\top$, $q_{p_2} = \mathcal{M}_p q_p$. Assume that the measurable outputs are q_{p_2} and the regulated outputs are q_{p_1} with constant desired value q_{p1*}. Then, design a controller $q_{p_2} \mapsto u_p$ that makes the closed loop system GAS at an equilibrium point $\bar{q} = [\bar{q}_p^\top,\ \bar{q}_c^\top]^\top$ such that such that $q_{p1*} = \mathcal{M}_p^\perp \bar{q} = [I_{n_{p1}},\ 0]\bar{q}$.*

2.3 Euler–Lagrange controllers

Motivated by the energy shaping plus damping injection technique, and the properties of EL systems described in the previous chapter we will define here a class of controllers which, preserving the EL structure, suitably modifies the potential energy and dissipation properties of the EL plant. Towards this end, we invoke the EL structure preserving interconnection property of Proposition 2.11, and propose to consider *EL controllers* with generalized coordinates $q_c \in \mathbb{R}^{n_c}$ and EL parameters $\{\mathcal{T}_c(q_c, \dot{q}_c), \mathcal{V}_c(q_c, q_p), \mathcal{F}_c(\dot{q}_c)\}$. We remark that since we are dealing here with a regulation and not a tracking problem there are no external inputs to the controller, which explains our choice of 0 as the "input matrix" $\mathcal{M}_c$. Thus, the controller dynamics is given by

$$D_c(q_c)\ddot{q}_c + N_c(q_c, \dot{q}_c) + \frac{\partial \mathcal{V}_c(q_c, q_p)}{\partial q_c} + \frac{\partial \mathcal{F}_c(\dot{q}_c)}{\partial \dot{q}_c} = 0. \tag{3.8}$$

Notice that the potential energy of the controller depends on the measurable output q_{p_2}, therefore q_{p_2} enters into the controller via the term $\frac{\partial \mathcal{V}_c(q_c, q_{p_2})}{\partial q_c}$. On the other hand, following Proposition 2.11, the *feedback interconnection* between plant and controller is established by

$$\mathcal{M}_p u = -\frac{\partial \mathcal{V}_c(q_c, q_p)}{\partial q_{p_2}}. \tag{3.9}$$

In this way, it follows from Proposition 2.11 that the closed-loop system is still an EL system with EL parameters $\{\mathcal{T}(q, \dot{q}), \mathcal{V}(q), \mathcal{F}(\dot{q})\}$ where

$$\mathcal{T}(q, \dot{q}) \triangleq \mathcal{T}_p(q_p, \dot{q}_p) + \mathcal{T}_c(q_c, \dot{q}_c), \ \mathcal{V}(q) \triangleq \mathcal{V}_p(q_p) + \mathcal{V}_c(q_c, q_p), \ \mathcal{F}(\dot{q}) \triangleq \mathcal{F}_c(\dot{q}_c)$$

and $q = [q_p{}^\top, \ q_c^\top]^\top$. The resulting feedback system is depicted in Fig. 3.3, where $\Sigma_p : u_p \mapsto q_{p_2}$ is an operator defined by the dynamic equations (3.7) and operator $\Sigma_c : q_{p_2} \mapsto u_p$ is defined by (3.8), (3.9).

The following remarks are in order: 1) The interconnection constraint (3.9) imposes some clear limitations on the achievable desired potential energy functions. It is clear that the constraints are removed for fully-actuated systems; 2) From the new definition of $\mathcal{F}(\dot{q})$ we see that the dynamic extension we just introduced injects damping to the system through the controller dynamics. As it will be seen later this damping has to be suitably propagated for asymptotic stability.

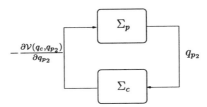

Figure 3.3: EL closed loop system.

We will apply now Proposition 2.13 to establish conditions for solvability of the problem using EL controllers. These conditions are summarized in the proposition below.

Proposition 3.6 (Output feedback stabilization) *An EL controller (3.8), (3.9) with EL parameters*

$$\{\mathcal{T}_c(q_c, \dot{q}_c), \mathcal{V}_c(q_c, q_{p2}), \mathcal{F}_c(\dot{q}_c)\}$$

solves the global output feedback stabilization problem above if

(i) (Energy shaping)
 $\mathcal{V}(q)$ is proper and has a global and unique minimum at $q = \bar{q}$, where $\bar{q}$ is such that $q_{p1*} = [I_{n_{p1}} \mid 0]\bar{q}$.

(ii) (Damping injection)
 $\mathcal{F}_c(\dot{q}_c)$ satisfies

$$\dot{q}_c^\top \frac{\partial \mathcal{F}_c(\dot{q}_c)}{\partial \dot{q}_c} \geq \alpha \|\dot{q}_c\|^2$$

 for some $\alpha > 0$

(iii) (Dissipation propagation)

> For each trajectory such that $q_c \equiv const$ and $\frac{\partial \mathcal{V}_c(q_c, q_{p2})}{\partial q_c} = 0$, we have that $q_p \equiv const$.

$\square$

Proof. The proof follows from Proposition 2.13 observing that the closed loop system is an underdamped EL system with damped coordinates q_c and undamped coordinates q_p. Notice that condition *(iii)* above implies that $\dot{q}_p \neq 0$ and $\frac{\partial \mathcal{V}_c(q_c, q_p)}{\partial q_c} = 0$ cannot happen simultaneously, hence this condition implies *(iii)* of Proposition 2.13.
∎

Remark 3.7 From Proposition 3.6 it is clear that the kinetic energy of the controller plays no role on the stabilization task, it may however affect the transient performance. In particular, the result applies even in the case when $\mathcal{T}_c(q_c, \dot{q}_c) = 0$. Furthermore, the conditions on the Rayleigh dissipation function of Proposition 3.6 are satisfied with $\mathcal{F}_c(\dot{q}_c) = \frac{1}{2}\dot{q}_c^\top R_c \dot{q}_c$, where $R_c = R_c^\top > 0$. Thus, with this choice of $\mathcal{F}_c(\dot{q}_c)$ and setting $\mathcal{T}_c(q_c, \dot{q}_c) = 0$ we get controllers with dynamics

$$\dot{q}_c = -R_c^{-1} \frac{\partial \mathcal{V}_c(q_c, q_{p2})}{\partial q_c}$$
$$\mathcal{M}_p u = -\frac{\partial \mathcal{V}_c(q_c, q_p)}{\partial q_{p_2}}$$

With an obvious abuse of terminology, and with the purpose of enlarging the class of controllers (and actually to obtain simpler solutions), we will also call these controllers EL controllers. As it will become clear later this scheme corresponds to the *approximate differentiation* filter, widely used in practical applications.

Remark 3.8 It is interesting to underline that for some underactuated systems one can instead reshape the *kinetic* energy for *regulation* purposes, as shown in [27] where the authors introduced the related "controlled Lagrangians" methodology. This alternative framework allows in principle, to include mechanic systems with nonholonomic constraints[4].

2.4 Examples

We have proven above that for underactuated EL systems global stabilization with PBC controllers is still possible with *only output feedback* provided a *dissipation propagation* condition is satisfied. In this section we apply this general result to two physical systems: the TORA and the flexible-joint robots, whose dynamical models have been derived in Sections 4.1 and 4.3 of Chapter 2, respectively.

---

[4]See the URL http://www.cds.caltech.edu/~ marsden/ for an extensive bibliography.

A   The TORA

For ease of reference we recall here from Section **2**.4.3 the EL parameters of the TORA system

$$T_p(q_p, \dot{q}_p) = \frac{1}{2}\dot{q}_p^\top D_p(q_p)\dot{q}_p = \frac{1}{2}\dot{q}_p^\top \begin{bmatrix} M+m & -ml\cos(q_{p2}) \\ -ml\cos(q_{p2}) & I+ml^2 \end{bmatrix} \dot{q}_p$$

$$V_p(q_p) = \frac{1}{2}kq_{p1}^2, \quad \mathcal{F}_p(\dot{q}_p) = 0, \quad \mathcal{M}_p = \begin{bmatrix} 0 \\ 1 \end{bmatrix}.$$

This is a very simple EL system that can be globally asymptotically stabilized at any given constant equilibrium point, in particular at zero, using Proposition 3.1. This leads to a proportional plus derivative (PD) controller

$$u_p = -k_p q_{p2} - k_d \dot{q}_{p2}$$

with $k_p, k_d > 0$. The proof of GAS is easily established checking conditions **A3.1** and **A3.2** of this proposition. Notice thet the total energy of the closed–loop is

$$\mathcal{H}(q_p, \dot{q}_p) = T_p(q_p, \dot{q}_p) + V_p(q_p) + \frac{1}{2}k_p q_{p2}^2.$$

The more practically interesting problem of making the zero equilibrium GAS assuming that *only* q_{p2} is available for measurement can also be easily solved by invoking Proposition 2.13. To this end, we propose an EL controller with EL parameters

$$T_c(q_c, \dot{q}_c) = 0, \quad \mathcal{F}_c(\dot{q}_c) = \frac{k_d}{2ab}\dot{q}_c^2$$

$$V_c(q_c, q_{p2}) = \frac{k_p}{2}q_{p2}^2 + \frac{k_d}{2b}(q_c + bq_{p2})^2$$

where $a, b, k_p, k_d > 0$. Notice that this corresponds to the "degenerate" case of EL system with zero inertia discussed in Remark 3.7, and a standard quadratic damping. The first term of $V_c(q_c, q_{p2})$ corresponds to the potential energy stored by a spring connecting the pendulum to a vertical plane. We will see below that the second term of $V_c(q_c, q_{p2})$ corresponds to a "spring action" between $\dot{q}_c$ and a zero reference. Actually, we will prove that this PBC corresponds to the PD controller above with $\dot{q}_{p2}$ replaced by its dirty derivative.

Following Proposition 2.13 this defines our EL controller dynamics as

$$\dot{q}_c = -a(q_c + bq_{p_2}) \tag{3.10}$$

$$u = -k_p q_{p_2} - k_d(q_c + bq_{p_2}) \tag{3.11}$$

We will now verify the conditions of the proposition. First, the energy–shaping condition is satisfied for all $k_p, k_d > 0$ because the total potential energy is a quadratic form

$$V(q) \triangleq V_p(q_p) + V_c(q_c) = \frac{1}{2}q^\top \begin{bmatrix} k & 0 & 0 \\ 0 & k_p + k_d & \frac{k_d}{b} \\ 0 & k_d & \frac{k_d}{b} \end{bmatrix} q$$

which is positive definite. Second, the damping injection is clearly satisfied with $\alpha = k_d$. Finally, we check the damping propagation condition (iii). From (3.10) and $q_c = const$ we have that $q_{p_2} = const$. This, replaced in (3.11) implies that $u = const$, which in its turn implies that $\ddot{q}_{p1} = const$, from the first dynamic equation of the TORA. We conclude the proof noting that the second dynamic equation of the TORA yields $q_{p_1} = const$ as desired.

Let us now prove that the controller above is a PD controller with approximate differentiation. Let us introduce the notation

$$\vartheta \stackrel{\triangle}{=} (q_c + bq_{p_2})$$

Hence,

$$u = -k_p q_{p_2} - k_d \vartheta$$

and ϑ satisfies

$$\dot{\vartheta} = -a\vartheta + b\dot{q}_{p_2}$$

thus the control can be written as

$$u = -k_p q_{p_2} - k_d \left\{ \frac{bp}{p+a} \right\} q_{p_2}$$

proving the claim.

The simplicity of this controller should be contrasted with the derivations reported in [111] (see also [237]). In the latter paper the first step is to make a coordinate change that transforms the system into the cascaded structure required by the backstepping technique. Unfortunately, since this coordinate transformation destroys the physical structure of the system, the controller design is not transparent –though in some sense systematic.

In Subsection 3.4.A, where we will further assume the input is subject to a *saturation constraint*, we will present some simulation results comparing the two controllers.

B Flexible-joint robots

We will derive here a class of EL PBC solving the output feedback global stabilization problem for flexible-joint robots. This provides a unified framework to compare different schemes via analysis of their energy dissipation properties. Further, we show that as particular cases of this class we can obtain the (apparently unrelated) controllers of [125], [20] and [3], and extend to the output feedback case the controller of [266].

We use in this section, the dynamic model (2.35), whose EL parameters are given by (2.32). We recall the reader that in our notation, q_{p_1} stands for the vector of link positions and q_{p_2} for the vector of motor shafts angles. The control variables are the torques at the shafts. We are interested in the set-point control of the link angles to a constant desired value q_{p1*}, and we assume that only the motor shaft angles, q_{p2} are available for measurement[5]. We also recall that the map $u \mapsto \dot{q}_{p2}$ is passive, as we have shown in the previous section.

B.1 EL controller with a virtual rigid-joints robot

In [3] an EL controller for a flexible-joint robot was designed using the following physically motivated approach. First, we design a rigid robot without gravity forces that we attach to a hyperplane with a PD control (*à la* Takegaki–Arimoto). Then, we attach some springs between this virtual robot and the motor shaft angles in such a way that the virtual robot "pulls" the actual robot to the desired equilibrium. The position of the hyperplane is selected so that in steady–state the springs store the potential energy required to place the robot links at the desired angle. This reasoning led us to consider an EL controller with EL parameters

$$\mathcal{T}_c(q_c, \dot{q}_c) = \frac{1}{2}\|\dot{q}_c\|^2, \quad \mathcal{F}_c(\dot{q}_c) = \frac{1}{2}\dot{q}_c^\top R_c \dot{q}_c, \tag{3.12}$$

$$\mathcal{V}_c(q_c, q_{p2}) = \frac{1}{2}\{(q_c - q_{p2})^\top K_2(q_c - q_{p2}) + (q_c - \delta(q_{p1*}))^\top K_1(q_c - \delta(q_{p1*}))\}. \tag{3.13}$$

That is, the virtual rigid robot is fully-damped $R_c = R_c^\top > 0$, has unitary inertia matrix, is attached to a fixed hyperplane at an angle $\delta(q_{p1*})$ via a spring of stiffness coefficient $K_1 = K_1^\top > 0$ (as in the pendulum example above). Finally, it is connected to the motor shaft angles through some additional springs of stiffness coefficient $K_2 = K_2^\top > 0$. See Fig. 1.1.

Using the EL equations (2.6), we easily derive the controller dynamics

$$\ddot{q}_c + R_c \dot{q}_c + K_1(q_c - \delta) + K_2(q_c - q_{p2}) = 0. \tag{3.14}$$

while the feedback interconnection is given, according to (3.9), by

$$u = -\frac{\partial \mathcal{V}_c(q_c, q_{p2})}{\partial q_{p2}} = K_2(q_c - q_{p2}). \tag{3.15}$$

It is interesting to note that the control signal of this PBC does not have the structure $u = u_{\mathrm{ES}} + u_{\mathrm{DI}}$, this is because we are also shaping the kinetic energy.

Now, adding up the EL parameters of the robot and the controller we get

$$\mathcal{T}(q, \dot{q}) = \frac{1}{2}\dot{q}_p^\top D(q_p)\dot{q}_p + \frac{1}{2}\|\dot{q}_c\|^2, \quad \mathcal{F}(\dot{q}) = \frac{1}{2}\dot{q}_c^\top R_c \dot{q}_c$$

$$\mathcal{V}(q) = \frac{1}{2}q_p{}^\top \mathcal{K}_p q_p + \mathcal{V}_g(q_{p1}) + \frac{1}{2}\{(q_c - q_{p2})^\top K_2(q_c - q_{p2}) + (q_c - \delta)^\top K_1(q_c - \delta)\}$$

---

[5]See [146] and [17] for globally stabilizing schemes measuring q_{p1} instead.

We proceed to verify now conditions (i) and (iii) of Proposition 3.6, the condition (ii) being trivially satisfied with $\alpha = \underline{\lambda}(R_c) > 0$.

(i) (Energy Shaping)

First, we need to insure that our EL controller (3.14), (3.15) shapes the closed loop system's potential energy to make it have a global and unique minimum at the desired equilibrium point. This step involves the definition of $\delta(q_{p1*})$ and some restrictions on the various springs. To this end, we first calculate $\frac{\partial V(q)}{\partial q}$ and set it equal to zero at an equilibrium containing q_{p1*}

$$\begin{bmatrix} K & -K & 0 \\ -K & K+K_2 & -K_2 \\ 0 & -K_2 & K_1+K_2 \end{bmatrix} \begin{bmatrix} q_{p1*} \\ \bar{q}_{p2} \\ \bar{q}_c \end{bmatrix} + \begin{bmatrix} g_{p1}(q_{p1*}) \\ 0 \\ 0 \end{bmatrix} = \begin{bmatrix} 0 \\ 0 \\ K_1\delta(q_{p1*}) \end{bmatrix}. \quad (3.16)$$

We see that by defining the constant

$$\delta(q_{p1*}) \triangleq q_{p1*} + (K^{-1} + K_1^{-1} + K_2^{-1})g_{p1}(q_{p1*})$$

we assure that (3.16) has a solution of the required form

$$\bar{q} = \begin{bmatrix} \bar{q}_{p1} \\ \bar{q}_{p2} \\ \bar{q}_c \end{bmatrix} = \begin{bmatrix} q_{p1*} \\ q_{p1*} + K^{-1}g_{p1}(q_{p1*}) \\ q_{p1*} + (K^{-1} + K_2^{-1})g_{p1}(q_{p1*}) \end{bmatrix}. \quad (3.17)$$

Now, to enforce $V(q)$ to have a global and unique minimum, that is

$$\frac{\partial^2 V(q)}{\partial q^2} = \begin{bmatrix} K + \frac{\partial g_{p1}(q_{p1})}{\partial q_{p1}} & -K & 0 \\ -K & K+K_2 & -K_2 \\ 0 & -K_2 & K_1+K_2 \end{bmatrix} \geq I_n\varepsilon > 0$$

which happens to hold if $\underline{\lambda}(K_a) > k_g$ where

$$K_a \triangleq \begin{bmatrix} K & -K & 0 \\ -K & K+K_2 & -K_2 \\ 0 & -K_2 & K_1+K_2 \end{bmatrix},$$

and k_g is defined by (2.18). The authors of [3] observed that K_a accepts the congruence transformation

$$\begin{bmatrix} I & 0 & 0 \\ I & I & 0 \\ I & I & I \end{bmatrix} K_a \begin{bmatrix} I & I & I \\ 0 & I & I \\ 0 & 0 & I \end{bmatrix} = \begin{bmatrix} K & 0 & 0 \\ 0 & K_1 & 0 \\ 0 & 0 & K_2 \end{bmatrix},$$

therefore $\underline{\lambda}(K_a) > k_g$ if and only if

$$\text{block-diag}\{K, K_1, K_2\} > k_g \begin{bmatrix} I & 0 & 0 \\ I & I & 0 \\ I & I & I \end{bmatrix} \begin{bmatrix} I & I & I \\ 0 & I & I \\ 0 & 0 & I \end{bmatrix} = k_g \begin{bmatrix} I & I & I \\ I & 2I & 2I \\ I & 2I & 3I \end{bmatrix}.$$

On the other hand, it can be shown that there exists a permutation matrix $P \in \mathbb{R}^{3n \times 3n}$ such that

$$P \begin{bmatrix} I & I & I \\ I & 2I & 2I \\ I & 2I & 3I \end{bmatrix} P^{\top} = \text{block-diag}\{E\}$$

where

$$E \triangleq \begin{bmatrix} 1 & 1 & 1 \\ 1 & 2 & 2 \\ 1 & 2 & 3 \end{bmatrix}.$$

Thus, recalling that permutation matrices are orthogonal, the required condition on K_1, K_2, K reduces to

$$K_1, K_2, K > I_n \bar{\lambda}(E) k_g.$$

(*iii*) (*Dissipation propagation*)
Notice that

$$\frac{\partial \mathcal{F}_c(\dot{q}_c)}{\partial \dot{q}_c} + \frac{\partial \mathcal{V}_c(q_c, q_{p2})}{\partial q_c} = R_c \dot{q}_c + K_1(q_c - \delta) + K_2(q_c - q_{p2}) \quad (3.18)$$

then setting $q_c = const$ and equating the right hand side of (3.18) to zero, it follows that $q_{p_2} = const$. The proof is completed as done by [266] observing the upper triangular structure of $D_{12}(q_{p_1})$ (see eq. 2.34) in order to conclude that q_{p_1} is also constant: Since q_{p_2} is constant then we can write the last n differential equations of (2.35) using (2.33) and (2.36) as

$$D_{12}^{\top}(q_{p_1})\ddot{q}_{p_1} + C_{21}(q_{p_1}, \dot{q}_{p_1})\dot{q}_{p_1} - Kq_{p_1} = -Kq_{p_2} - K_2(q_c - q_{p_2}) + g_{p_1}(q_{p1d}) = \text{constant.} \quad (3.19)$$

Considering (2.37) and (2.34), the first equation of (3.19) becomes $q_{p_{1,1}} = \text{constant}$. Substituting the latter into the second equation of (3.19) we get $q_{p_{1,2}} = \text{constant}$. Proceeding in the same way till the nth differential equation of (3.19), we can conclude that q_{p_1} is constant. Since we proved that $q = \bar{q}$ is the only equilibrium point, the control goal has been achieved.

B.2 EL controller with approximate differentiation

The PBC that we derive in this section, corresponds to the PD of [266] where velocity is replaced by its approximate differentiation (or "dirty derivative"). It constitutes an extension, to flexible-joint robots, of the controller reported in [123], and was originally proposed in [125]. To understand the rationale of the controller we recall that the flexible-joint robot defines a passive map $u \mapsto \dot{q}_{p_2}$ with storage function the total energy $\mathcal{H}_p = \mathcal{T}_p(q_p, \dot{q}_p) + \mathcal{V}_p(q_p)$, that is, it satisfies

$$\dot{\mathcal{H}}_p = \dot{q}_{p_2}^{\top} u$$

Let us close a first loop with a proportional energy–shaping term

$$u = u_{ES} + u_{DI} = -K_p[q_{p_2} - \delta(q_{p_1*})] + u_{DI}$$

where $\delta(q_{p_1*})$ is a constant term that plays the same role as in the previous controller and K_p is a diagonal positive definite matrix. Clearly, we have passivity of the map $u_{DI} \mapsto \dot{q}_{p_2}$ with the new storage function

$$\mathcal{H}_p + \frac{1}{2}[q_{p_2} - \delta(q_{p_1*})]^\top K_p[q_{p_2} - \delta(q_{p_1*})]$$

Now, we propose to add damping by feeding back the *dirty derivative* of q_{p_2}, that is

$$u_{DI} = -K_d \text{diag}\left\{\frac{b_i p}{p + a_i}\right\} q_{p_2} = -K_d(pI + A)^{-1}B\dot{q}_{p_2}$$

where $K_d \triangleq \text{diag}\{k_{di}\} > 0$ and $A \triangleq \text{diag}\{a_i\} > 0$. The transfer matrix $K_d(pI+A)^{-1}B$ is strictly positive real, hence output strictly passive with storage function

$$\mathcal{H}_F = \frac{1}{2}\vartheta^\top K_d B^{-1}\vartheta.$$

The overall closed–loop consists of the passive map $u_{DI} \mapsto \dot{q}_{p_2}$ in feedback with an output strictly passive LTI system $K_d(pI+A)^{-1}B$, hence it is still passive. The proof of GAS can be completed using passivity and detectability arguments, or taking the sum of the storage functions as a Lyapunov function candidate and invoking LaSalle. See [125] for such a proof.

Let us prove now that this controller belongs to the EL class defined in Section 2.3. Towards this end it is convenient to define the generalized coordinates as

$$q_c \triangleq \vartheta - Bq_{p_2}.$$

In terms of these coordinates we can rewrite the controller above as

$$\begin{aligned}
\dot{q}_c &= -A(q_c + K_d q_{p_2}) \\
u &= -K_p(q_{p_2} - \delta(q_{p_1*})) - K_d(q_c + Bq_{p_2})
\end{aligned}$$

which corresponds to an EL system with EL parameters

$$\mathcal{T}_c(q_c, \dot{q}_c) = 0, \quad \mathcal{F}_c(\dot{q}_c) = \frac{1}{2}\dot{q}_c^\top K_d B^{-1} A^{-1}\dot{q}_c$$

$$\mathcal{V}_c(q_c, q_{p2}) = \frac{1}{2}(q_{p_2} - \delta(q_{p_1*}))^\top K_p(q_{p_2} - \delta(q_{p_1*})) + \frac{1}{2}(q_c + Bq_{p_2})^\top K_d B^{-1}(q_c + Bq_{p_2})$$

Notice that, as in the TORA example, this corresponds to the "degenerate" case of EL system with zero inertia and a standard quadratic damping. The first term of

$\mathcal{V}_c(q_c, q_{p2})$ corresponds again to the potential energy stored by a spring connecting the motor axes to the hyperplane at angle $\delta(q_{p1*})$. The second term of $\mathcal{V}_c(q_c, q_{p2})$ corresponds to a "spring action" between $\dot{q}_c$ and a zero reference.

The EL parameters of the closed loop are

$$\mathcal{T}(q, \dot{q}) = \frac{1}{2}\dot{q}_p^\top D(q_p)\dot{q}_p, \quad \mathcal{F}(\dot{q}) = \frac{1}{2}\dot{q}_c^\top K_d B^{-1} A^{-1}\dot{q}_c, \tag{3.20}$$

$$\mathcal{V}(q) = \frac{1}{2}\{(q_c + Bq_{p2})^\top K_d B^{-1}(q_c + Bq_{p2}) + (q_{p2} - \delta)^\top K_p(q_{p2} - \delta)\} +$$

$$+ \frac{1}{2}q_p^\top K_p q_p + \mathcal{V}_g(q_{p1}) \tag{3.21}$$

Now, we determine the values of K_p, K_d and δ such that the conditions (i) and (iii) of Proposition 3.6 hold.

(i) (*Energy shaping*)
To verify this condition first notice that setting $\frac{\partial \mathcal{V}(q)}{\partial q}(\bar{q}) = 0$ yields

$$\begin{bmatrix} K & -K & 0 \\ -K & K + K_p + K_d & K_d B^{-1} \\ 0 & K_d & K_d B^{-1} \end{bmatrix} \begin{bmatrix} \bar{q}_{p1} \\ \bar{q}_{p2} \\ \bar{q}_c \end{bmatrix} + \begin{bmatrix} g(\bar{q}_{p1}) \\ 0 \\ 0 \end{bmatrix} = \begin{bmatrix} 0 \\ K_p\delta \\ 0 \end{bmatrix}$$

which has a (unique) solution of the required form $\bar{q} = [q_{p1*}^\top, *, *]^\top$ with

$$\delta = q_{p1*} + (K^{-1} + K_p^{-1})g_{p1}(q_{p1*}).$$

The second part of this condition is met if

$$\frac{\partial^2 \mathcal{V}(q)}{\partial q^2} = \begin{bmatrix} K + \frac{\partial g_{p1}(q_{p1})}{\partial q_{p1}} & -K & 0 \\ -K & K + K_p + K_d & K_d B^{-1} \\ 0 & K_d & K_d B^{-1} \end{bmatrix} \geq I_n\varepsilon > 0, \quad \forall\, q \in \mathbb{R}^n.$$

To satisfy this requirement we partition the two diagonal sub-blocks of the Hessian matrix as

$$Q_1 \triangleq \begin{bmatrix} K + \frac{\partial g_{p1}(q_{p1})}{\partial q_{p1}} & -K \\ -K & K + \frac{1}{2}K_p \end{bmatrix}, \quad Q_2 \triangleq \begin{bmatrix} \frac{1}{2}K_p + K_d & K_d B^{-1} \\ K_d & K_d B^{-1} \end{bmatrix}$$

and look for conditions that ensure that both are bounded from below by some matrix $I_{n_p}\varepsilon$, $\varepsilon > 0$ for all $q \in \mathbb{R}^n$. The submatrix Q_1 is positive definite if $\underline{\lambda}(K_a) > k_g$ where we have redefined

$$K_a \triangleq \begin{bmatrix} K & -K \\ -K & K + \frac{1}{2}K_p \end{bmatrix}$$

then we can proceed as before observing that K_a satisfies the congruence transformation [3]

$$\begin{bmatrix} I & 0 \\ I & I \end{bmatrix} K_a \begin{bmatrix} I & I \\ 0 & I \end{bmatrix} = \begin{bmatrix} K & 0 \\ 0 & K_p \end{bmatrix},$$

on the other hand we can prove that there exists a permutation matrix $P \in \mathbb{R}^{n_p \times n_p}$ such that

$$P \begin{bmatrix} I & I \\ I & 2I \end{bmatrix} P^\top = \text{block-diag}\{E\}$$

where

$$E \triangleq \begin{bmatrix} 1 & 1 \\ 1 & 2 \end{bmatrix}.$$

Thus, the condition on K, K_p reduces to $K_p, K > \underline{\lambda}(E)k_g I_{m_p}$. Following the same procedure, it can be also shown that the submatrix Q_2 is positive definite for all $K_p, K_d > 0$.

(*iii*) (*Dissipation propagation*)
Finally, this condition is verified as follows: we set the right hand side of

$$\frac{\partial V_c(q_c, q_{p_2})}{\partial q_c} = K_d(q_c + Bq_{p_2})$$

to zero and consider $q_c \equiv const$, we then get that $q_{p_2} \equiv const$. The proof is completed as for the previous EL controller exploiting the special triangular structure of the robot inertia matrix in order to conclude that also q_{p_1} is constant. ∎

Remark 3.9 It is important to remark that the PBC discussed in Subsection B.1 modifies the kinetic energy of the EL system. The dynamic extension is then of the order of the EL system itself. On the other hand, the PBC in Subsection B.2, leaves the kinetic energy unchanged, and requires an n-th order dynamic extension only to inject the damping.

B.3 Simulation results

In this section we illustrate through simulations the performance of both EL controllers presented above. We have used the two degrees of freedom simplified model of [20], whose EL parameters are (with zero payload)

$$\mathcal{T}(q_p, \dot{q}_p) \triangleq \begin{bmatrix} \dot{q}_{p_1} \\ \dot{q}_{p_2} \end{bmatrix}^\top \begin{bmatrix} 1.02\cos(q_{p_2}) + 8.77 & 0.76 + 0.51\cos(q_{p_2}) \\ 0.76 + 0.51\cos(q_{p_2}) & 0.62 \end{bmatrix} \begin{bmatrix} \dot{q}_{p_1} \\ \dot{q}_{p_2} \end{bmatrix} \tag{3.22}$$

$$\mathcal{V}_p(q_p) \triangleq 9.81(7.6\sin(q_{p_1}) + 0.63\cos(q_{p_1} + q_{p_2})) \tag{3.23}$$

$$\mathcal{F}_p(\dot{q}_p) = 0. \tag{3.24}$$

Moreover the bounds mentioned in properties **P2.1 – P2.2** for this particular case are (assuming a payload of $m_l \leq 2$ kg)

$$d_m = 0.45, \quad d_M = 9.96, \quad k_c = 1.53, \quad k_g = 81.2, \quad k_v = 80.7, \tag{3.25}$$

and we have considered a joint stiffness of $K = 3500I_2$.

The constant reference to be followed is $q_{p1*} = [\frac{\pi}{4}, \frac{\pi}{4}]^\top$ with zero initial conditions. Figure 3.4 shows the result of the controller described in [125]; in this case we have set the gains to $K_1 = \text{diag}([5000, 6000])$, $K_2 = \text{diag}([4000, 5000])$, $a_i = 30$ and $b_i = 10$. In Figure 3.5 we illustrate the response to the same reference using the controller of [3]. We have used the same model as before and the controller gains were set in such a way to have a transient approximately similar in time to that of the first controller. These values are $K_1 = K_2 = \text{diag}([8500, 8500])$, $R_c = \text{diag}([3500, 3500])$. Gains can be tuned to have a smoother but slower transient response.

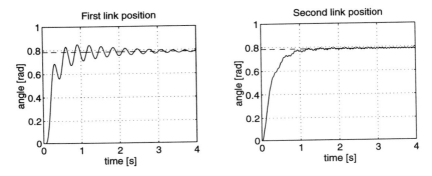

Figure 3.4: EL Controller of 2.4.B.2.

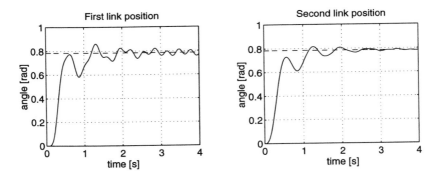

Figure 3.5: EL Controller of 2.4.B.1.

In this brief section we have limited ourselves to illustrate the behaviour of the controllers contained in Subsection 2.4.B, further simulation results comparing the performance of [266] and [125] can be found in the latter reference.

3 Bounded output feedback regulation

In the previous section we characterized a class of Euler-Lagrange systems that can be globally asymptotically stabilized via nonlinear dynamic *output* feedback. The controller design, relies on the fact that the storage function for feedback interconnected systems is the sum of the corresponding storage functions. This basic property is expressed in terms of the physically appealing principles of shaping the potential energy and injection of the required damping. In particular, we considered the case where the controller is also an EL system; in this way the closed loop is still an EL system with total energy and dissipation function the sum of the corresponding plant and controller total energies and dissipation functions.

Motivated by practical problem of windup present in numerous applications, we will consider in this section the setpoint control problem with *amplitude constrained* inputs. More particularly, we focus our attention on saturated *set-point* control of a particular class of fully-actuated EL systems.

3.1 Literature review

One of the first contributions in the robotics literature is due to [49] who proposed a saturated PD plus gravity compensation like controller, to deal with stick-slip friction effects. Later, [126] extended this result using precompensated gravity, however both results use velocity measurements. Some recent extensions of these works which only use position measurements are given in [44] and [162]. Burkov [44] proposed a PD-like controller which uses exact gravity compensation, while we introduced a subclass of EL controllers (which contains that of [44]), thus extending this methodology to the case of fully-actuated EL systems under input constraints.

3.2 Problem formulation

In this section we assume that **A2.2** holds, then under this condition we deal with the

Definition 3.10 (Set-point control under input constraints) *For the system (2.39):*

$$D(q_p)\ddot{q}_p + C(q_p, \dot{q}_p)\dot{q}_p + g(q_p) = u_p, \tag{3.26}$$

assume that only generalized position measurements, q_p, are available and that the system's inputs are constrained to

$$|u_{p_i}| \leq u_{p_i}^{\max} \qquad \forall i \in \underline{n} \tag{3.27}$$

then, find an output feedback controller which renders the closed loop system globally asymptotically stable, that is, an output feedback controller such that

$$\lim_{t\to\infty} \|\tilde{q}_p(t)\| = \lim_{t\to\infty} \|q_p(t) - q_{p*}\| = 0, \tag{3.28}$$

where q_{p*} is the desired constant position.

Based on the results of the previous section we define in the sequel, a family of EL controllers which yield bounded control inputs. We repeat for convenience that the aim of the EL controllers is to shape the closed loop energy $\mathcal{V}(q)$ and to inject a partial damping. It is clear that the easiest way of shaping the potential energy of the closed loop, $\mathcal{V}(q)$, is to *cancel* the potential energy of the plant, $\mathcal{V}_p(q_p)$, and then to impose a "new" potential energy shape. However, to enhance the robustness of the controller, instead of cancelling the potential energy we aim at *dominating* it.

According to the EL controllers methodology, the control input is defined by

$$u_p = -\frac{\partial \mathcal{V}_c(q_c, q_p)}{\partial q_p},$$

from this, we can deduce that the input constraint established by (3.27), shall entail some growth restrictions on $\mathcal{V}_c(q)$. For instance, if we take $\mathcal{V}_c(q)$ quadratic, as we have done before, the control u_p will grow linearly with q_p, and we cannot expect to satisfy (3.27) for large $\|q_p\|$. The basic idea is then to choose a $\mathcal{V}_c(q)$ which grows *linearly* with respect to $\|q_p\|$ outside some ball, in this case we can expect to verify (3.27). As it will become more clear later, a suitable class of functions is of the form

$$\mathcal{V}_c(q_c, q_p) = \sum_{i=1}^{n} \left(\int_0^{f_i(q_p, q_c)} \text{sat}(x)dx \right) + \mathcal{V}_{cN}(q_c, q_p)$$

where $f_i : \mathbb{R} \times \mathbb{R} \to \mathbb{R}$, $i \leq \underline{n}$ and $\mathcal{V}_{cN}(q_c, q_p)$ are *linear* in q_p and $\text{sat}(x)$ is a saturation function defined below. Replacing this expression above we get

$$u_{p_i} = -\frac{\partial f_i(q_c, q_p)}{\partial q_{p_i}} \text{sat}(f_i(q_c, q_p)) - \frac{\partial \mathcal{V}_{cN}(q_c, q_p)}{\partial q_p}$$

which is a bounded function of q_p.

On the other hand, since $\mathcal{V}_c(q)$ should be designed in a way that it dominates $\mathcal{V}_p(q_p)$, it is also necessary to impose the same growth restrictions on $\mathcal{V}_p(q_p)$. Thus, we consider in this section, a subclass of fully-actuated EL systems whose potential energy function satisfies (2.19). Loosely speaking, this condition restricts the growth rate of $\mathcal{V}_p(q_p)$ to be of order $\mathcal{O}(\|q_p\|^2)$ for all q_p in some ball B_β and to $\mathcal{O}(\|q_p\|)$ outside B_β.

The following definition is in order.

Definition 3.11 (Saturation function.) *A saturation function* $\text{sat}(x) : \mathbb{R} \to \mathbb{R}$ *is a* C^2 *strictly increasing odd function that satisfies*

1. $\text{sat}(0) = 0$,

2. $|\,\text{sat}(x)| < 1$,

3. $\frac{\partial^2 \,\text{sat}(x)}{\partial x^2} \neq 0 \quad \forall\, x \neq 0 \in \mathbb{R}$.

Our motivation for considering saturation functions as defined above is that these functions satisfy the following properties.

P 3.1 $\int_0^{\tilde{q}_{p_i}} \text{sat}(x) dx \geq \frac{1}{2}\,\text{sat}(\tilde{q}_{p_i})\tilde{q}_{p_i}, \quad \tilde{q}_{p_i} \in \mathbb{R}$.

P 3.2 There exists some $\varepsilon > 0$ such that

$$\text{sat}(\tilde{q}_{p_i})\tilde{q}_{p_i} \geq \frac{\text{sat}(\varepsilon)}{\varepsilon}\tilde{q}_{p_i}^2 \quad |\tilde{q}_{p_i}| < \varepsilon, \tag{3.29}$$

$$\text{sat}(\tilde{q}_{p_i})\tilde{q}_{p_i} \geq \text{sat}(\varepsilon)|\tilde{q}_{p_i}| \quad |\tilde{q}_{p_i}| \geq \varepsilon. \tag{3.30}$$

For instance, we can take $\text{sat}(x) \overset{\triangle}{=} \tanh(\omega x)$, $\omega > 0$, as proposed in [49, 126].

3.3 Globally stabilizing saturated EL controllers

Since we are dealing with fully-actuated systems, the simplest way to "dominate" the plant's potential energy is to cancel $\mathcal{V}_p(q_p)$ and to impose a desired shape to the closed loop function. This however entails some potential robustness problems, hence we favour a solution that does not rely on this cancelation. Interestingly enough, if we use a controller that does not cancel the vector of potential forces, the growth rate restriction on $\mathcal{V}_p(q_p)$ mentioned above is imposed only at the desired position. The price paid, however, is that in this case we need to use "high" gains in $\mathcal{V}_c(q_c, q_p)$ to dominate $\mathcal{V}_p(q_p)$ and this translates into stiffer requirements on the input saturation bound, $u_{p_i}^{\max}$.

Proposition 3.12 (Control with cancelation of potential forces) *Assume that the systems potential energy verifies the strict inequality*

$$\sup_{q_p \in \mathbb{R}^n} \left| \left(\frac{\partial \mathcal{V}_p(q_p)}{\partial q_p} \right)_i \right| < u_{p_i}^{\max}, \quad i \in \underline{n} \tag{3.31}$$

with $(\cdot)_i$ *the* $i - th$ *component of the vector. Under these conditions, there exists a dynamic output feedback EL controller* $\Sigma_c : \{\mathcal{T}_c(q_c, \dot{q}_c), \mathcal{V}_c(q_c, q_p), \mathcal{F}_c(\dot{q}_c)\}$ *that insures the input constraint (3.27) holds, and makes*

$$(\dot{q}_p, q_p, \dot{q}_c, q_c) = (0, q_{p*}, 0, q_{cd}) \tag{3.32}$$

with q_{cd} *some constant, a GAS equilibrium point of the closed loop system.* □

• **A controller with cancelation of potential forces**: Consider the EL controller characterized by

$$
\begin{aligned}
\mathcal{T}_c(q_c, \dot{q}_c) &= 0, \quad \mathcal{F}_c(\dot{q}_c) = \frac{1}{2}\|\dot{q}_c\|^2 \\
\mathcal{V}_c(q_c, q_p) &= \mathcal{V}_{c_1}(q_c) + \mathcal{V}_{c_2}(q_c, q_p) - \mathcal{V}_p(q_p) \\
\mathcal{V}_{c_1}(q_c) &= \frac{1}{2}q_c^\top K_1 q_c \\
\mathcal{V}_{c_2}(q_c, q_p) &= \sum_{i=1}^{n} k_{2_i} \int_0^{(q_{c_i} - \tilde{q}_{p_i})} \mathrm{sat}(x_i)dx_i
\end{aligned}
$$

where $\tilde{q}_{p_i} \triangleq q_{p_i} - q_{p*_i}$, $k_{1_i}, k_{2_i} > 0$, $K_1 \triangleq \mathrm{diag}\{k_{1_i}\}$. Using Lagrange's equations we can derive the controller dynamics

$$
\dot{q}_{c_i} = -k_{1_i}q_{c_i} - k_{2_i}\,\mathrm{sat}(q_{c_i} - \tilde{q}_{p_i}) \tag{3.33}
$$

$$
u_{p_i} = k_{2_i}\,\mathrm{sat}(q_{c_i} - \tilde{q}_{p_i}) + \left(\frac{\partial \mathcal{V}_p(q_p)}{\partial q_p}\right)_i \tag{3.34}
$$

which corresponds to that proposed by Burkov in [44].

Proposition 3.13 (Control without cancelation of potential forces) *Assume that, at the desired reference, the gradient of the systems potential energy satisfies the inequality*

$$
\left|\left(\frac{\partial \mathcal{V}_p}{\partial q_p}(q_{p*})\right)_i\right| \le k_{g_i}^{\max}, \quad i \in \underline{n} \tag{3.35}
$$

with $k_{g_i}^{\max} < u_{p_i}^{\max}$, and let its Hessian satisfy (2.18). Under these conditions, there exists a globally asymptotically stabilizing EL controller that does not cancel the potential forces and insures the input constraints (3.27) provided $u_{p_i}^{\max}$ is sufficiently large. □

• **A controller without cancelation of potential forces**: In this case the EL parameters of the controller can be chosen as

$$
\mathcal{T}_c(q_c, \dot{q}_c) = 0, \quad \mathcal{F}_c(\dot{q}_c) = \frac{1}{2}\dot{q}_c^\top K_2 B^{-1} A^{-1} \dot{q}_c \tag{3.36}
$$

$$
\mathcal{V}_c(q_c, q_p) = \mathcal{V}_{c_2}(q_c, q_p) - \mathcal{V}_p(q_{p*}) + q_p^\top \frac{\partial \mathcal{V}_p}{\partial q_p}(q_{p*})
$$

$$
\mathcal{V}_{c_2}(q_c, q_p) = \sum_{i=1}^{n} \left(\frac{k_{2_i}}{b_i} \int_0^{(q_{c_i} + b_i q_{p_i})} \mathrm{sat}(x_i)dx_i + k_{3_i} \int_0^{\tilde{q}_{p_i}} \mathrm{sat}(x_i)dx_i \right)
$$

where $A \triangleq \text{diag}\{a_i\}$, $B \triangleq \text{diag}\{b_i\}$, $K_2 \triangleq \text{diag}\{k_{2_i}\} > 0$, and we select $k_{3_i} > 0$ *sufficiently large*. This choice yields the EL controller

$$\dot{q}_{c_i} = -a_i \, \text{sat}(q_{c_i} + b_i q_{p_i}) \tag{3.37}$$

$$u_{p_i} = -k_{2_i} \, \text{sat}(q_{c_i} + b_i q_{p_i}) - k_{3_i} \, \text{sat}(\tilde{q}_{p_i}) + \left(\frac{\partial \mathcal{V}_p}{\partial q_p}(q_{p*})\right)_i \tag{3.38}$$

A   Some remarks on saturated EL controllers

- The propositions above characterize, –in terms of the EL parameters $\mathcal{T}_c(q_c, \dot{q}_c)$, $\mathcal{F}_c(\dot{q}_c)$–, a *class* of output feedback GAS controllers for EL systems with *saturated inputs*. Thus, providing an extension, to the constrained input case, of the result presented in Section 2.3.

- A key feature of the controller given above is that, to enhance its robustness, we *avoid* explicit cancelations of the plant dynamics. As mentioned before, the price paid for this is the requirement that the plants potential energy grow not faster than linearly; also, higher gains have to be injected into the loop through k_{3_i} (see Appendix **D**). As seen from our proposition, this imposes an additional requirement of sufficiently large input constraints for stability. The condition on k_{3_i} stems from the fact that, to impose a desired minimum point to the closed loop potential energy, now we have to dominate (and not to cancel) the systems potential energy. In this respect controller (3.37)- (3.38) supersedes the result of [44] which relies on exact cancelation of $\mathcal{V}_p(q_p)$.

- As a corollary of our proposition, we obtain an extension to the *output* feedback case of the result in [126] where a *full state* feedback solution to the problem of global regulation of rigid-joints robots with saturated inputs was presented. It is also interesting to remark that if we write $\dot{q}_{c_i} = -a_i(q_{c_i} + b_i q_{p_i})$ instead of $\dot{q}_{c_i} = -a_i \, \text{sat}(q_{c_i} + b_i q_{p_i})$, in (3.37) we *exactly recover* the (approximate differentiation) output feedback GAS controller of [123]. Our proposition then shows that by simply including the saturations we can preserve GAS even under input constraints.

B   Proofs

The proofs of Propositions 3.12 and 3.13 are constructive. For this, we provide below the stability proofs of the above-proposed saturated EL controllers.

As pointed out before, the interconnection of two EL systems, yields an EL system with potential energy $\mathcal{V}(q) = \mathcal{V}_c(q_c, q_p) + \mathcal{V}_p(q_p)$, hence the proofs of both results are carried out by proving the conditions of Proposition 3.6. Notice that the difficulty lies in proving that $\mathcal{V}(q)$ has a global and unique minimum including $q_p = q_{p*}$. For this, we will use Lemma C.8.

B.1 Proof of Proposition 3.12

Notice first that the condition on the Rayleigh dissipation function of Proposition 3.6 is trivially satisfied, hence we go on proving that the potential energy is adequately shaped and that the damping suitably propagates from the controller coordinates q_c to the plant coordinates q_p.

(i) (Energy shaping)

The potential energy of the closed loop system (3.26), (3.27), (3.34) is given by

$$\mathcal{V}(q) = \frac{1}{2}q_c^\top K_1 q_c + \sum_{i=1}^{n} k_{2_i} \int_0^{(q_{c_i} - \tilde{q}_{p_i})} \text{sat}(x_i)dx_i.$$

Now we use Lemma C.8 to prove that $\mathcal{V}(q)$ has a global and unique minimum at the origin $(q_c - \bar{q}_c, \tilde{q}_p) = (0, 0)$. The positivity condition of Lemma C.8 follows from Definition 3.11 and Property **P3.1** while the second condition follows from equalizing $\frac{\partial \mathcal{V}(q)}{\partial q} = 0$:

$$\begin{bmatrix} K_1 q_c + K_2 \, \text{sat}(q_c - \tilde{q}_p) \\ -K_2 \, \text{sat}(q_c - \tilde{q}_p) \end{bmatrix} = \begin{bmatrix} 0 \\ 0 \end{bmatrix}. \tag{3.39}$$

Notice that, since $\text{sat}(x)$ is strictly increasing and vanishes only at $x = 0$, and K_1, K_2 are full rank, (3.39) is satisfied if and only if $(q_c, \tilde{q}_p) = (0, 0)$.

(iii) (Dissipation propagation)

This condition is easily verified by equalizing $\frac{\partial \mathcal{V}_c(q_c, q_p)}{\partial q_c} = 0$:

$$K_1 q_c + K_2 \, \text{sat}(q_c - \tilde{q}_p) = 0,$$

setting $q_c \equiv const$ and observing that effectively, $q_p \equiv const$.

The proof is completed applying the triangle inequality to (3.34), and using the fact that $|\,\text{sat}(x)| < 1$, to get the bound

$$|(u_p)_i| < k_{2_i} + k_{3_i} + \left| \left(\frac{\partial \mathcal{V}_p(q_p)}{\partial q_p} \right)_i \right|$$

thus, under assumption (3.31), we can always choose sufficiently small k_{2_i}, $k_{3_i} > 0$ such that (3.27) holds. ∎

B.2 Proof of Proposition 3.13

We provide in this section a stability proof for the controller (3.37), (3.38). As in the previous proof, we verify the conditions of Proposition 3.6.

We prove next that, if we take k_{2_i} sufficiently small and $\min_i \{k_{3i}\} > k_{3_i}^{\min}$, with $k_{3_i}^{\min}$ some suitably defined positive constant, then (3.32) is a GAS equilibrium point of the closed loop (3.26), (3.27), (3.38) provided that the gradient of the systems potential energy, evaluated *at the desired* reference, satisfy

$$u_{p_i}^{\max} > \left| \left(\frac{\partial \mathcal{V}_p}{\partial q_p}(q_{p*}) \right)_i \right| + k_{3_i}, \quad i \in \underline{n}. \tag{3.40}$$

We finally show that, if in particular we take $\text{sat}(x) = \tanh(x)$ then

$$k_{3_i}^{\min} \triangleq \frac{4 k_v}{\tanh\left(\frac{4 k_v}{k_g} \right)} \tag{3.41}$$

where k_v and k_g are given by (2.19) and (2.18) respectively.

(i) (Energy Shaping)

The closed loop potential energy is now

$$\mathcal{V}(q) = \sum_{i=1}^{n} \left(\frac{k_{2_i}}{b_i} \int_0^{(q_{c_i} + b_i q_{p_i})} \text{sat}(x_i) dx_i + k_{3_i} \int_0^{\tilde{q}_{p_i}} \text{sat}(x_i) dx_i \right)$$
$$+ \mathcal{V}_p(q_p) - \mathcal{V}_p(q_{p*}) - q_p^\top \frac{\partial \mathcal{V}_p}{\partial q_p}(q_{p*}). \tag{3.42}$$

Hereafter we show that, if there exists a k_v as defined by (2.19), then there exists $k_{3_i}^{\min} > 0$ such that $\mathcal{V}(q)$ has a global and unique minimum at the desired equilibrium for all $k_{3_i} \geq k_{3_i}^{\min}$.

Notice that the first right hand term of (3.42) is a nonnegative function of q_c, q_p which is zero at $q_c = -B^{-1} q_p$. Hence, to prove that (3.42) has a global and unique minimum at $(q_{p*}, \bar{q}_c)$ it suffices to show that the last three terms have a global and unique minimum at $q_p = q_{p*}$, or equivalently, that the function $f(\tilde{q}_p) : \mathbb{R}^n \to \mathbb{R}$

$$f(\tilde{q}_p) \triangleq \sum_{i=1}^{n} \left\{ k_{3i} \int_0^{\tilde{q}_{p_i}} \text{sat}(x_i) dx_i \right\} + \mathcal{V}_p(\tilde{q}_p + q_{p*}) - \mathcal{V}_p(q_{p*}) - \tilde{q}_p^\top \frac{\partial \mathcal{V}_p}{\partial q_p}(q_{p*}) \tag{3.43}$$

has a global and unique minimum at zero. The proof of the latter statement requires some lengthy but straightforward calculations which we include in Appendix **D**.

(iii) (Dissipation propagation)

The second condition is verified by equalizing $\frac{\partial \mathcal{V}(q)}{\partial q_c} = 0$:

$$K_2 B^{-1} \text{sat}(q_c + B\tilde{q}_p) = 0$$

and observing that it holds true only if $q_c = -B^{-1}\tilde{q}_p$, since K_2 is full rank, hence, $q_c \equiv$ const implies that $q_p \equiv$ const.

The proof is completed applying the triangle inequality and using (3.38) to get the bound

$$|(u_p)_i| < k_{2_i} + k_{3_i} + \left|\left(\frac{\partial \mathcal{V}_p}{\partial q_p}(q_{p*})\right)_i\right|$$

thus under assumption (3.35), we can always choose sufficiently small $k_{2_i}, k_{3i} > 0$ such that (3.27) be satisfied. ∎

3.4 Examples

A The TORA system.

We will consider now the TORA system, whose model was given in Section 4.1 of Chapter **2**, and for which we already designed in Subsection 2.4.A an output feedback EL PBC. The material in this section follows closely [72] to which we refer the readers for further details.

We will take the same approximate differentiation approach of Subsection 2.4.A, but to take into account the input saturation (3.27) we modify the EL parameters of the controller as proposed in (3.36). That is, we choose

$$\mathcal{V}_c(q_{p2}, q_c) = \frac{1}{b}\int_0^{(q_c+bq_{p2})} k_2 \, \text{sat}(s)ds + \int_0^{q_{p2}} k_3 \, \text{sat}(s)ds$$
$$\mathcal{F}_c(\dot{q}_c) = \frac{1}{2ab}\dot{q}_c^2$$

where a, b are positive constants, and $k_i > 0$, $i = 2, 3$. The controller dynamics is then given by

$$u_p = -k_2 \, \text{sat}(q_c + bq_{p2}) - k_3 \, \text{sat}(q_{p2})$$
$$\dot{q}_c = -ak_2 \, \text{sat}(q_c + bq_{p2})$$

The conditions of Proposition 3.13 can be easily verified, hence the trivial equilibrium is GAS if

$$k_1 + k_2 \leq u_p^{max},$$

Computer simulations have been carried out to show the performance of the proposed controller. We use the parameters shown in table 3.1 with the physical constraints $|q_{p1}| \leq 0.025$ m and $|u_p| \leq 0.100$ Nm given in [41].

Description	Parameter	Value	Units
Cart mass	M	1.3608	Kg
Arm mass	m	0.096	Kg
Arm eccentricity	l	0.0592	m
Arm inertia	I	0.0002175	Kg/m^2
Spring stiffness	k	186.3	N/m

Table 3.1: TORA parameters.

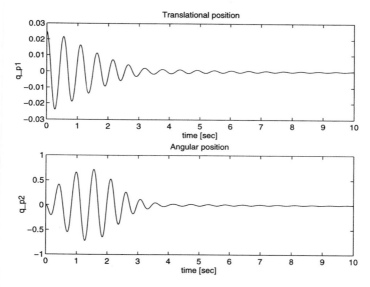

Figure 3.6: Transient behaviour for translational and angular positions.

All initial conditions are set to zero except the initial translational position which is set at its extreme value $q_{p1}(0) = 0.025$. We selected $\text{sat}(s) \triangleq \tanh(s)$ for the control law and after a few iterations in simulation to get the best transient behaviour we chose the following parameters: $a = 550$, $b = 4.5$, $k_1 = 0.035$, $k_2 = 0.018$. Notice that $k_1 + k_2$ is much smaller than the allowable bound 0.1. We observed, however, that performance was actually degraded for larger values of these gains because of the peaking phenomenon.

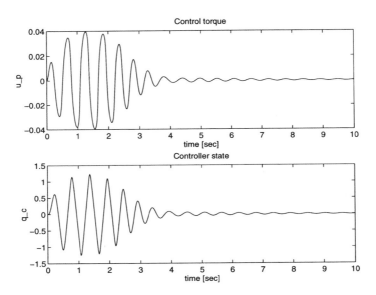

Figure 3.7: Applied control and controller state.

A typical response of the system in closed loop with the proposed controller is shown in Figs. 3.6 and 3.7. As we see the system exhibits good settling behaviour, it stabilizes around 3 sec., with a control effort, $|u_p| \leq 0.04$ Nm, well below its admissible upper bound.

It is interesting to compare our results with the ones obtained with the *full state feedback unsaturated* controllers reported in [111]. First of all, notice that our results pertain to the original system, while those given in [111] are carried out for its scaled version, Eq. (1) in [111] where the arbitrary value 0.1 is given for the scale coupling factor $\epsilon = ml/\sqrt{(I + ml^2)(M + m)}$. This factor equals 0.2 for the benchmark problem of [41], hence the plots are not directly comparable. Secondly, as pointed out in [111] the best results were obtained with the controller P3, which is a passifying controller consisting of a standard PD plus a nonlinear term that enforces the passivity property. Thus is very similar, at least in spirit, to the controller presented here, although our controller is saturated and uses only output feedback. It is quite clear from the figures of [111] that the behaviour of the backstepping–based controllers that did not exploit passivity properties, that is P1 and P2, is significantly inferior. Further comparisons of the two controllers may be found in [72].

In order to evaluate the robustness of the closed loop system with respect to the external disturbance f, we apply a pulse of amplitude 3 N, and duration 0.1 sec. to the system once it has reached its equilibrium point. See Figs. 3.8, and 3.9.

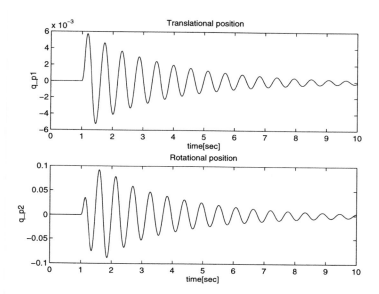

Figure 3.8: Translational and rotational responses to an external disturbance.

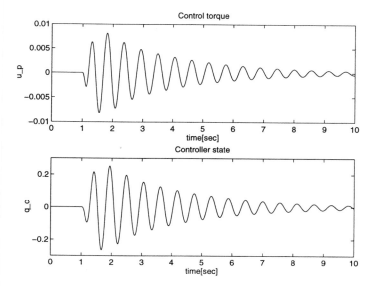

Figure 3.9: Applied control and controller state for an external disturbance.

To illustrate the global nature of our controller we present a simulation where we want to "unwind" the arm from an initial value of $q_{p2}(0) = 10\pi$ to the zero position, with all other initial conditions equal to zero.[6] In Figs. 3.10, and 3.11, we show the transient behaviour. Note from Fig. 3.11 that the controller actually saturates but global asymptotic stability is preserved as predicted by the theory.

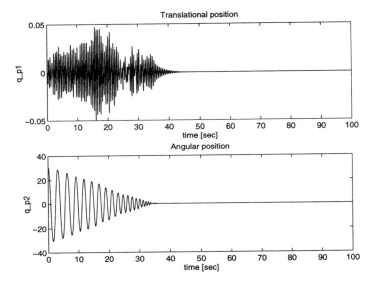

Figure 3.10: Translational and rotational responses for $q_{p2}(0) = 10\pi$.

---

[6]Notice that we are looking at the evolution of the system in Euclidean space, hence the points $q_{p2} = n\pi$, $n = \cdots, -1, 0, 1, \cdots$ are different. This is done just for the purposes of illustration.

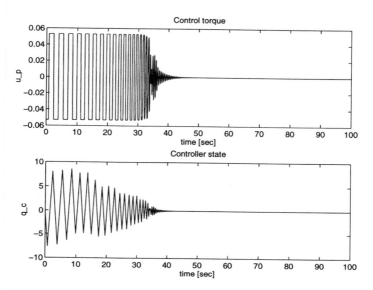

Figure 3.11: Applied control and controller state for $q_{p2}(0) = 10\pi$.

B Robot manipulators

The rigid-joint robot manipulators (2.39) is another system to which our results can be applied. Note that for manipulators with purely rotational joints, the potential energy function has only linear and trigonometric functions of the generalized positions while for a purely translational-joints manipulators, the potential energy function has linear functions only. A more general case is the combination of the two previous ones thus, the restriction (3.35) holds and our results can be directly applied.

Using SIMULINK™ of MATLAB™, we tested our algorithm in the two link robot arm of [19] –see also p. 59– with a desired reference $q_{p*} = [\pi/2, \ \pi/2]^\top$. We have imposed the input constraint $u_{p_i}^{\max} = 220$ Nm in (3.27). To meet the conditions of Proposition 3.13 we chose $A = \text{diag}\{100, \ 100\}$, $B = \text{diag}\{130, \ 130\}$ while the controller gains were set to $K_3 = \text{diag}\{180, \ 180\}$, $K_2 = \text{diag}\{125, \ 125\}$ according to (3.40).

Then, in order to evaluate the performance of our controller, we tested as well the one proposed in [123] with exactly the same gain values and starting from initial conditions $q_{p_0} = [\pi/4, \ \pi/4]^\top$, and in accordance with the previous discussions we set $q_{c_0} = [-32.5\pi, \ -32.5\pi]^\top$ in order to make $\vartheta_0 = 0$.

In Fig. 3.12 we show the transient of the first link position using the algorithm of [123], i.e. the *non* saturated controller and the control input signal yielded by this

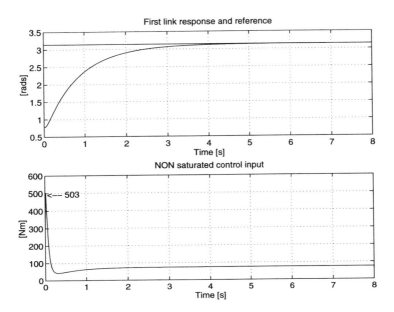

Figure 3.12: EL Controller of Subsection 2.4.B.2

controller. In Fig. 3.13 we show the response of the same link driven by the saturated controller of Proposition 3.13 and its control input.

On one hand, notice that the transient produced by the non saturated controller is much faster than the response using saturated controls. On the other hand, it must be remarked that the control input yielded by the linear controller fails to satisfy the input constraint; in particular the maximum absolute value of u_p is 503Nm for the first link. In contrast to this the saturated controller yields a control input with $|u_{p_i}|^{\max} = 211$ Nm.

Thus we verify what is not surprising: that there is a compromise between a fast and smooth transient and small control inputs.

Remark 3.14 (Stabilization of a marine vessel.) Another EL system to which our saturated EL controllers can be applied is the Lagrangian ship model discussed in **2.4.5**. The construction of the controller follows along the same lines as for the rigid-joints robot. It is interesting to note however that since in this case $g(q) = b$ which is *constant* then there is no difference between cancelling and compensating for the potential energy.

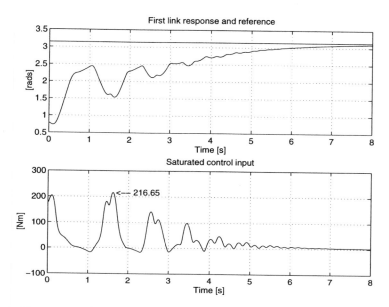

Figure 3.13: Saturated EL controller without cancelation of $g(q)$.

4 Set-point regulation under parameter uncertainty

As it was previously discussed, in our EL controllers approach it is it assumed that the potential forces are accurately known, which in practice rarely happens. In this section we explore the set-point control problem with uncertain knowledge of the potential energy. We make particular emphasis on PID control. For the position feedback case we propose the so-called PI^2D controller which, by means of a linear dynamic extension and a second integrator solves this problem without velocity measurements.

Notation: In this section we adopt the compact notation k_m and k_M to denote positive constants for a positive definite matrix $K \in \mathbb{R}^{n \times n}$, such that $k_m \|x\|^2 \leq x^\top K x \leq k_M \|x\|^2$, for all $x \in \mathbb{R}^n$.

Definition 3.15 (Set-point control problem under parameter uncertainty)
For the system (2.39):

$$D(q)\ddot{q} + C(q, \dot{q})\dot{q} + g(q) = u, \tag{3.44}$$

assume that the potential energy function $\mathcal{V}(q)$ is not known accurately but only the bounds k_g and k_v defined in (2.2) are available. Then, find a (output feedback) controller which renders the closed loop system asymptotically stable, that is, a (output *feedback) controller such that*

$$\lim_{t\to\infty} \|\tilde{q}_p(t)\| = \lim_{t\to\infty} \|q_p(t) - q_{p*}\| = 0,$$

where q_{p*} is the desired constant position.

4.1   Literature review

It is interesting to stress that the conception of the popular PID controller is apparently due to Nicholas Minorsky, and goes back to 1922. In the robotics literature, there is a huge body of research on the subject, among the first we can mention [9]. In [282] it is shown that the PID controller (3.47), (3.48) is *locally* asymptotically stable provided the gains K_p, K_i and K_d satisfy some complex relationships.

A tool commonly used in the proof of asymptotic stability are Lyapunov functions with cross terms of the form $\varepsilon \tilde{q}^\top D(q)\dot{q}$ or $\varepsilon \tilde{q}^\top \dot{q}$, $\varepsilon > 0$. This type of functions, which are also used throughout this section, have been widely used in the literature starting probably with [135] (see also [7, 122, 207, 284] and references therein). The motivation being usually to construct a Lyapunov function with a derivative containing more negative terms than $-\dot{q}^\top K_d \dot{q}$, for instance quadratic negative terms of the position error. This allows to prove GAS of the origin by invoking standard Lyapunov techniques.

Unfortunately the price paid for the extra terms in the Lyapunov function derivative is the apparition of cubic terms of the form $\varepsilon k_c \|\tilde{q}\| \|\dot{q}\|$ which can be dominated only locally. Interestingly enough, this technical problem can be overcome by normalizing the cross term. The use of normalized cross terms goes back at least to the work by Koditscheck [134].

In this section we will revise these techniques showing that the use of the cross terms leads in fact to the definition of passive maps with respect to outputs of the form $\varepsilon \tilde{q} + \dot{q}$. This is important in the sense that the relative degree of the system with respect to this output is one, and henceforth, properties as OSP can be claimed.

Unfortunately, all the above-mentioned approaches can be applied only to fully-actuated systems. As far as we know the only results where underactuation is considered are [2] and [45] for the particular case of flexible-joint robots. In the first reference, the author makes use of the global contraction mapping theorem to prove semi-global ultimate boundedness of the solutions. In the second reference Burkov proposes a common PID controller, the novelty is the use of singular perturbation techniques in order to prove that "there exists" a sufficiently small integral gain such that the closed loop system is globally asymptotically stable. Unfortunately, no specific bounds for the integral gain are given. Both approaches are based on the assumption that velocity measurements are available.

4.2 Adaptive control

We have shown above that an easy way of shaping the potential energy of the plant is by substituting it with another function of the generalized positions which has suitable properties, most typically we choose a quadratic potential. The latter leads to the PD controller (3.3) which *cancels* the natural gravitational forces vector from the dynamics. In the case when this vector is not accurately known a natural remedy is to try to parameterize this function in terms of the unknown parameters –usually those dependent on the payload– and use an adaptation law to estimate them. As shown in Eq. (2.16), the parameterization is *linear*, hence standard estimation techniques may be used. That is, we can write $g(q) \triangleq \Phi(q)\theta$, where $\Phi(q)$ contains some known functions and $\theta \in \mathbb{R}^q$ is the vector of unknown parameters. An adaptive version of the PD plus gravity cancelation controller is obtained as

$$u = -K_p\tilde{q} - K_d\dot{q} + \Phi(q)\hat{\theta} \qquad (3.45)$$

where $\hat{\theta}$ are the parameter estimates to be updated with some adaptation law. The dynamics (3.44) in closed loop with (3.45) are given by

$$D(q)\ddot{q} + C(q,\dot{q})\dot{q} + K_p\tilde{q} + K_d\dot{q} = \Phi(q)\tilde{\theta}$$

where $\tilde{\theta} = \hat{\theta} - \theta$ is the parameter error. It is easy to see that these dynamics define an OSP operator $\Phi(q)\tilde{\theta} \mapsto \dot{q}$ with storage function

$$\mathcal{H}(q,\dot{q}) = \mathcal{T}(q,\dot{q}) + \frac{1}{2}\tilde{q}^\top K_p\tilde{q}.$$

On the other hand, it is well known [13] that the standard gradient estimator

$$\dot{\hat{\theta}} = -\gamma\Phi^\top(q)\dot{q} \qquad (3.46)$$

with $\gamma > 0$, defines a passive operator $\dot{q} \mapsto -\Phi(q)\tilde{\theta}$ with storage function

$$\mathcal{H}_\theta = \frac{1}{2\gamma}\tilde{\theta}^\top\tilde{\theta}.$$

That is, the closed–loop system consists of the feedback interconnection of a passive and an OSP operators. Invoking Proposition A.10 of Appendix **A** we have the following.

Fact 3.16 The system (3.44) in closed loop with (3.45), (3.46) and external input $\zeta \in \mathcal{L}_{2e}^n$, i.e.,

$$D(q)\ddot{q} + C(q,\dot{q})\dot{q} + K_p\tilde{q} + K_d\dot{q} - \Phi(q)\tilde{\theta} = \zeta$$
$$\dot{\hat{\theta}} = -\gamma\Phi^\top(q)\dot{q}$$

defines an OSP (and in view of Proposition A.5 $\mathcal{L}_2$–stable) operator $\zeta \mapsto \dot{q}$ with storage function $\mathcal{H} + \mathcal{H}_\theta$.

Noting that the storage function is positive definite and invoking Proposition A.10 of Appendix **A** we could conclude from here asymptotic stability *if* we could verify zero-state detectability (with respect to the output $\dot{q}$). Unfortunately, this is not the case. To see this remark that $\dot{q} \equiv 0$ does not imply that $(\tilde{\theta}, \tilde{q}, \dot{\tilde{q}}) \to 0$.

To go around this problem we need to establish a new passivity property for the EL dynamics. To this end we need to add to the total energy function cross-terms to define the new storage function. This topic has been thoroughly studied (in the form of Lyapunov analysis) in [7, 122, 207, 284] and will be explained in the next section. In particular we have the following proposition.

Proposition 3.17 (Passivity of PD plus gravity compensation control) *The dynamics*

$$D(q)\ddot{q} + C(q, \dot{q})\dot{q} + K_p\tilde{q} + K_d\dot{q} = \Phi(q)\tilde{\theta}$$

defines, an OSP operator $\Sigma_1 : \Phi(q)\tilde{\theta} \mapsto (\dot{q} + \varepsilon\tilde{q})$, *where*

$$\varepsilon \triangleq \frac{1}{\gamma(1 + \|\tilde{q}\|^2)}$$

with positive definite storage function[7]

$$\mathcal{H}_N(q, \dot{q}) = \frac{1}{2}\dot{q}^\top D(q)\dot{q} + \frac{1}{2}\tilde{q}^\top K_p\tilde{q} + \varepsilon\tilde{q}^\top D\dot{q}$$

provided $\gamma > 0$ *is sufficiently large.* $\square$

This property motivates us to consider the estimation law

$$\dot{\tilde{\theta}} = -\Phi^\top(q)\left[\gamma\dot{q} + \frac{\tilde{q}}{1 + \|\tilde{q}\|^2}\right].$$

first proposed by Tomei in [265], which we know defines a passive operator $(\dot{q} + \varepsilon\tilde{q}) \mapsto -\Phi(q)\tilde{\theta}$, yielding the desired OSP for the closed–loop. Unfortunately, we still do not have the required detectability property, because $\tilde{\theta}$ has a manifold of equilibria. We need then Theorem 1.2 from [47] which states that for the system $\dot{x} = f(x)$, $y = h(x)$ the following implication holds

$$x(t) \in \mathcal{L}_\infty \quad \text{and} \quad y \in \mathcal{L}_2 \quad \Rightarrow \quad \lim_{t\to\infty} h(x(t)) = 0$$

We can apply this theorem to our system with the new "output" $\dot{\tilde{q}} + \varepsilon\tilde{q}$, which is square integrable in view of the OSP. Boundedness of the trajectories follows noting that the function $\mathcal{H}_N(q, \dot{q}) + \frac{1}{2\gamma}\tilde{\theta}^\top\tilde{\theta}$ is nonincreasing.

---

[7]Notice the present of the cross product of q and $\dot{q}$ in the third right hand term.

4.3 Linear PID control

For the sake of clarity, let us first review the passivity properties of the EL system formed by the closed loop of (3.44) (for simplicity let us assume $M = I$) with a PD plus gravity compensation controller as depicted in Fig. 3.14.

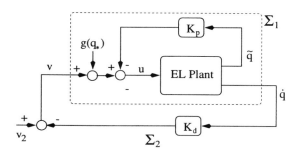

Figure 3.14: PD plus gravity compensation.

From Proposition 2.5 we know that the plant's total energy function $\mathcal{T}(q, \dot{q}) + \mathcal{V}(q)$ qualifies as a storage function for the supply rate $w(u, \dot{q}) = u^\top \dot{q}$. It is not difficult to see that this passivity property is conserved after the feedback $u = -K_p \tilde{q} + g(q_*) + v$ with v an external input. By setting $v = -K_d \dot{q} + v_2$ the passivity is strengthen to *output* strict passivity from the input v_2.

Interestingly enough, if the potential energy function is not known accurately the passivity properties of the closed loop are preserved. Let $v_2 = -g(q_*) + \hat{g}(q_*) + v_3$ then the actual control input will be $u = -K_p \tilde{q} + \hat{g}(q_*) + v_3$. A simple analysis shows that the closed loop system of (3.44) with this new control input defines an output strictly passive map $\Sigma : v_3 \mapsto \dot{q}$ for any positive definite K_d and storage function

$$\mathcal{H} = \mathcal{T}(q, \dot{q}) + U_g(q) + \frac{1}{2}(q - \hat{\delta})^\top K_p (q - \hat{\delta})$$

where in analogy with (3.4) $\hat{\delta}(q_*) = q_* + K_p^{-1}\hat{g}(q_*)$. It is important to remark however that even though the condition for $\mathcal{H}$ to be positive definite is still that K_p be sufficiently large (specifically $K_p \geq k_g I$) the global minimum of $\mathcal{H}$ is not necessarily $q = q_*$. In other words, even though the input output properties are unchanged, the system is no longer GAS at the desired equilibrium. That is, with respect to Fig. 3.2 this control law "pulls" the pendulum to a constant position different from the desired one q_*.

As it is well known, the steady state error resulting from the mismatch between $g(q_*)$ and $\hat{g}(q_*)$ can be compensated by means of a simple integrator. Unfortunately, the price paid in terms of passivity is that the property looses its global character, that is, the resulting system is only *locally* output strictly passive [221, 237]. This is summarized in the following proposition.

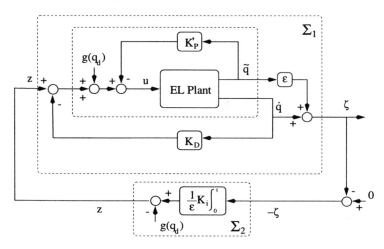

Figure 3.15: Passivity interpretation of PID Control.

Proposition 3.18 (PID is locally passive) *Consider the feedback interconnected system depicted in Fig. 3.15 which corresponds to the closed loop of the EL plant (3.44) and the PID controller*

$$u = -K_p\tilde{q} - K_d\dot{q} + \nu \qquad (3.47)$$
$$\dot{\nu} = -K_i\tilde{q} \qquad (3.48)$$

where $\tilde{q} \overset{\triangle}{=} q - q_*$, q_* is the desired link position which is assumed to be constant, K_p, K_d, K_i are positive definite diagonal matrices. Then for a sufficiently small constant $\varepsilon > 0$ the following is true:

1. *The map $\Sigma_1 : z \mapsto \zeta$ is locally output strictly passive.*

2. *The map $\Sigma_2 : -\zeta \mapsto z$ is passive*   □

Proof. Let us choose any positive definite diagonal matrix K_p' such that

$$K_p \overset{\triangle}{=} K_p' + \frac{1}{\varepsilon}K_i$$

where $\varepsilon > 0$ is a (small) constant to be determined, clearly K_p is also positive definite and diagonal for any $\varepsilon > 0$. Then the error equations (3.44), (3.47), (3.48) which correspond to the internal dynamics of the maps Σ_1 and Σ_2 can be written as

$$\Sigma_1 \; : \; D(q)\ddot{q} + C(q,\dot{q})\dot{q} + g(q) - g(q_*) + K'_p\tilde{q} + K_d\dot{q} = z \qquad (3.49)$$
$$\Sigma_2 \; : \; \dot{z} = -\varepsilon\tilde{q} + \dot{q} \qquad (3.50)$$

where we have defined $\tilde{\nu} \triangleq \nu - g(q_*)$ and $z = \tilde{\nu} - \frac{1}{\varepsilon}K_i\tilde{q}$. Consider then the storage function

$$\mathcal{H}_1(\tilde{q},\dot{q}) = \frac{1}{2}\dot{q}^{\top}D\dot{q} + U_g - U_{g_d} - \tilde{q}^{\top}g_d + \frac{1}{2}\tilde{q}^{\top}K'_p\tilde{q} + \varepsilon\tilde{q}^{\top}D\dot{q} \qquad (3.51)$$

where we have dropped the arguments and defined $U_{g_d} \triangleq U_g(q_*)$, $g_d \triangleq g(q_*)$ to simplify the notation. We find it convenient to this point to split the kinetic, and part of the potential energy terms as

$$\tilde{q}^{\top}K'_p\tilde{q} = (\lambda_1 + \lambda_2 + \lambda_3)\tilde{q}^{\top}K'_p\tilde{q}$$
$$\dot{q}^{\top}D(q)\dot{q} = (\lambda_1 + \lambda_2 + \lambda_3)\dot{q}^{\top}D(q)\dot{q}$$

with $1 > \lambda_i > 0$, $i = 1, 2, 3$. Then one can show that if

$$k'_{p_m} \geq \max\left\{\frac{k_g}{\lambda_1}, \frac{\varepsilon^2 d_M}{\lambda_1\lambda_2}\right\},$$

then function $\mathcal{H}_1(q,\dot{q})$ satisfies the lowerbound:

$$\mathcal{H}_1(\tilde{q},\dot{q}) \geq \frac{\lambda_3}{2}\tilde{q}^{\top}K'_p\tilde{q} + \frac{\lambda_2 + \lambda_3}{2}\dot{q}^{\top}D\dot{q}$$

hence it is positive definite and radially unbounded. The motivation for this partitioning of the energy terms will become more evident in the sequel. Next, using the well known bounds (2.21) and

we obtain that the time derivative of $\mathcal{H}_1(q,\dot{q})$ along the trajectories of (3.49), (3.50) is bounded by

$$\dot{\mathcal{H}}_1 \leq -\left(k_{d_m} - \frac{\varepsilon}{2}k_{d_M} - \varepsilon k_c \|\tilde{q}\| - \varepsilon d_M\right)\|\dot{q}\|^2 - \varepsilon\left(k'_{p_m} - k_g - \frac{1}{2}k_{d_M}\right)\|\tilde{q}\|^2 + z^{\top}\varsigma. \qquad (3.52)$$

Notice that $\dot{\mathcal{H}}_1$ with $z \equiv 0$ is negative semidefinite for instance if

$$k_{d_m} > \varepsilon(k_{d_M} + 2d_M)$$
$$k'_{p_m} > k_g + \frac{1}{2}k_{d_M}$$
$$\|\tilde{q}\| \leq \frac{k_{d_m}}{2\varepsilon k_c}. \qquad (3.53)$$

After completing some squares, one can show that there exist strictly positive constants γ_o and β such that

$$\dot{\mathcal{H}}_1 \leq -\gamma_o \|\zeta\|^2 + z^\top \zeta$$

thus completing the proof of 1. Notice that the existence of $\gamma_o > 0$ is conditioned to (3.53) thus the local character of the passivity property.

To complete the proof notice that the operator Σ_2 is an integrator, which is the simplest passive system one can think of. The property is established immediately by considering the storage function

$$\mathcal{H}_2 = \frac{1}{\varepsilon} z^\top K_i^{-1} z. \tag{3.54}$$

∎

We have established above the passivity properties of a simple PID in closed loop with a fully-actuated EL system. The convergence of the signals $\tilde{q}$ and $\dot{q}$ to the origin can be easily established using Lemma A.12 and observing that $\zeta \in \mathcal{L}_2^n$, $\zeta \to 0$ as $t \to \infty$ and $\varepsilon > 0$. Furthermore from (3.49), (3.50) we conclude that $\dot{z} \to 0$ and $z(t) \to 0$. From the definition of z we obtain the convergence of $\tilde{\nu}(t) \to 0$. Hence the system is zero-state detectable and GAS follows.

Alternatively one can take the Lyapunov function candidate $V = \mathcal{H}_1 + \mathcal{H}_2$ which is positive definite under the conditions established above and whose time derivative is negative semidefinite. *Local* asymptotic stability of the origin $\mathrm{col}(\tilde{q}, \dot{q}, \tilde{\nu}) = \mathrm{col}(0, 0, 0)$ follows invoking Krasovskii-LaSalle's invariance principle. Furthermore one can define a domain of attraction for the closed loop system (3.49), (3.50) as follows. Define the level set

$$B_\delta \stackrel{\triangle}{=} \left\{ x \in \mathbb{R}^{3n} : V(x) \leq \delta \right\}$$

where δ is the largest positive constant such that $\dot{V}(x) \leq 0$ for all $x \in B_\delta$.

4.4   Nonlinear PID control

As it is clear now from the proof of Proposition 3.18, what impedes claiming output strict passivity for a PID controller in the usual (global) sense, is the presence of the cubic term $\varepsilon k_c \|\tilde{q}\| \|\dot{q}\|^2$ in the time derivative $\dot{\mathcal{H}}_1$. This technical difficulty can be overcome by making some "smart" modifications to the PID control law, leading to the design of *nonlinear* PID controllers.

To the best of our knowledge, the first non-linear PID controllers that appeared in the literature are [122] and [7]. In this section we discuss these controllers from a passivation point of view.

A The normalized PID

In order to cope with the cubic term $\varepsilon k_c \|\tilde{q}\| \|\dot{q}\|^2$ in (3.52), and inspired upon the results of Tomei revised in Section 4.2, Kelly [122] proposed the "adaptive" PD controller

$$u = -K'_p \tilde{q} - K_d \dot{q} + \Phi(q_*)\hat{\theta} \tag{3.55}$$

together with the update law

$$\dot{\hat{\theta}} = \dot{\tilde{\theta}} = -\frac{1}{\gamma}\Phi(q_*)^\top \left[\dot{q} + \frac{\varepsilon_0 \tilde{q}}{1 + \|\tilde{q}\|} \right] \tag{3.56}$$

where $\varepsilon_0 > 0$ is a small constant. In the original contribution [122], Kelly proved that this "adaptive" controller in closed loop with a rigid-joint robot results in a globally convergent system. However, since the regressor vector $\Phi(q_*)$ is *constant* the update law (3.56), together with the control input (3.55) can be implemented as a *nonlinear* PID controller by integrating out the velocities vector from (3.56):

$$\hat{\theta}(t) = -\frac{1}{\gamma}\Phi(q_*)^\top \left[\tilde{q}(t) + \int_0^t \frac{\varepsilon_0 \tilde{q}(\tau)}{1 + \|\tilde{q}(\tau)\|} d\tau \right] + \hat{\theta}(0). \tag{3.57}$$

Notice that the choice $K_p = K'_p + K_i$, with $K_i = \frac{1}{\gamma}\Phi(q_*)\Phi(q_*)^\top$, yields the controller implementation

$$u = -K_p \tilde{q} - K_d \dot{q} + \nu \tag{3.58}$$
$$\dot{\nu} = -\varepsilon K_i \tilde{q}, \qquad \nu(0) = \nu_0 \in \mathbb{R}^n. \tag{3.59}$$

where we have redefined

$$\varepsilon \triangleq \frac{\varepsilon_0}{1 + \|\tilde{q}\|}. \tag{3.60}$$

Since controllers (3.55), (3.57) and (3.58), (3.59) are equivalent, following the steps of Kelly [122] one can prove *global* asymptotic stability of the closed loop system (3.44), (3.58)–(3.59).

Furthermore, following the same steps as in the proof of Propositions 3.18 and 3.17 one can show that the normalization introduced in the integrator helps in enlarging the domain where the passivity property holds, hence rendering the system $\Sigma_1 : z \mapsto \zeta$ where we redefined

$$z \triangleq -K_i \tilde{q} + \tilde{\nu}, \tag{3.61}$$

output strictly passive. This can be shown by evaluating the time derivative of the storage function $\mathcal{H}_1$ (using (3.60) and (3.61)) along the trajectories of Σ_1 whose dynamics is defined by (3.49). One obtains after some straightforward bounding

$$\dot{\mathcal{H}}_1 \leq -\left(k_{d_m} - \varepsilon \|\tilde{q}\| - \varepsilon_0 d_M - \frac{\varepsilon_0 k_{d_M}}{2} \right) \|\dot{q}\|^2 - \varepsilon(k'_{p_m} - k_g - \frac{\varepsilon_0 k_{d_M}}{2}) \|\tilde{q}\|^2 + z^\top \zeta$$

which is similar to (3.52) however, notice from (3.60) that $\varepsilon k_c \|\tilde{q}\| \|\dot{q}\|^2 \le k_c \|\dot{q}\|^2$, hence the condition (3.53) is no longer needed. The output strict passivity is then established following similar arguments as in the proof of Proposition 3.18.

It is worth remarking that even though the normalization of the cross term in the storage function and in the integral gain, allows to claim OSP notice that the number $-\varepsilon = -\frac{\varepsilon_0}{1+\|\tilde{q}\|}$ is multiplying the term $\tilde{q}$ in $\mathcal{H}_1$. While Lyapunov *global* asymptotic stability can be claimed using the Lyapunov function $V = \mathcal{H}_1 + \mathcal{H}_2$, it can be expected that the rate of convergence for large position errors, be small since $-\dot{V}$ has then a linear growth in $\|\tilde{q}\|$.

A second important remark is that the normalization just introduced can be thought of as the same effect of a saturation function as defined in Def. 3.11. This leads us to the second passive nonlinear PID controller.

B   The saturated PID controller

An alternative trick to achieve output strict passivity for Σ_1 is to saturate the proportional feedback term instead. Even though with a Lyapunov design motivation such idea was firstly presented as far as we know in [7] where the following nonlinear PID[8] was proposed:

$$
\begin{aligned}
u &= -K'_p \operatorname{sat}(\tilde{q}) - \frac{1}{\varepsilon} K_i \tilde{q} - K_d \dot{q} + \nu & (3.62)\\
\dot{\nu} &= -K_i \tilde{q}, \qquad \nu(0) = \nu_0 \in \mathbb{R}^n
\end{aligned}
$$

where in this case $\varepsilon > 0$ is a small *constant* and sat shall be considered component-wise. Arimoto proved in [7] that if $k_{p_m} > k_g$, and K_i is sufficiently small, the closed loop is passive and moreover it is *globally* asymptotically stable.

The key idea used in [7] is to dominate the cubic terms in the derivative of the storage function by means of the saturated proportional feedback in (3.62). More precisely, one can prove that the map $\Sigma_1 : z \mapsto \zeta$ with input $z = \tilde{\nu} - \frac{1}{\varepsilon} K_i \tilde{q}$, output $\zeta = \varepsilon \operatorname{sat}(\tilde{q}) + \dot{q}$ and internal dynamics

$$\Sigma_1 : D(q)\ddot{q} + C(q,\dot{q})\dot{q} + g(q) - g(q_*) + K'_p \operatorname{sat}(\tilde{q}) + K_d \dot{q} = z$$

is output strictly passive. This can be established by derivating the storage function

$$\mathcal{H}_{1a}(\tilde{q},\dot{q}) = \frac{1}{2}\dot{q}^\top D(q)\dot{q} + U_g(q) - U(q_*) - \tilde{q}^\top g(q_*) + \frac{1}{2}\tilde{q}^\top K'_p \tilde{q} + \varepsilon \operatorname{sat}(\tilde{q})^\top D(q)\dot{q}$$

(which is positive definite for $\varepsilon > 0$ sufficiently small) to obtain

$$
\begin{aligned}
\dot{\mathcal{H}}_{1a}(\tilde{q},\dot{q}) \le{}& -(k_{d_m} - \varepsilon k_c \|\operatorname{sat}(\tilde{q})\| - \varepsilon d_M - \frac{\varepsilon}{2}k_{d_M}) \|\dot{q}\|^2 \\
& -\varepsilon(k'_{p_m} - k_g - \frac{1}{2}k_{d_M})\tilde{q}^\top \operatorname{sat}(\tilde{q}) + z^\top \zeta & (3.63)
\end{aligned}
$$

---

[8]It is worth mentioning that in [7], Arimoto used a saturation function which is a particular case of sat considered here, however this point is not fundamental for the validity of the result.

then using $\| \operatorname{sat}(\tilde{q}) \| \leq 1$ and similar arguments as in the previous proofs.

Notice however, that as in the previous case one gets rid of the cubic terms but the price paid for enhancing thereby the passivity property is a slower convergence of the signal $\tilde{q}$ due to the linear growth of $\tilde{q}^\top \operatorname{sat}(\tilde{q})$ for large error values.

4.5 Output feedback regulation: The PI²D controller

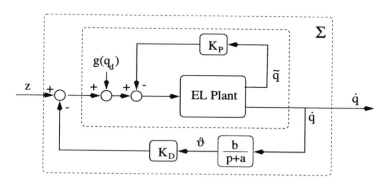

Figure 3.16: The EL controller of Kelly [125].

In this section we solve the last practical problem raised at the end of Section 1, that of position feedback setpoint control with uncertain gravity knowledge. Our contribution is the so called PI²D controller originally presented in [207]. In order to put the PI²D controller in perspective, let us briefly summarize the results we have presented so far concerning the set-point control problem for EL mechanical systems.

As we have seen, one can identify several solutions to both problems *separately* that is, on one hand the controllers which do not need measurement of generalized velocities need exact a priori knowledge of the potential energy, and on the other hand, the different approaches to regulation with uncertain potential energy knowledge needed the measurement of the generalized velocities.

To the best of our knowledge the first solutions to the problem of designing an asymptotically stable regulator that does not require the *exact knowledge* of $g(q_d)$ nor the *measurement of speed* appeared independently in [207] and [60]. The contribution of [207], the PI²D controller, is a semiglobally stable control law that solves this problem. The authors of [60] obtained a similar result for robot manipulators considering the dynamics of DC actuators.

We proceed now to show how the PI²D controller can be constructed from the interconnection of passive blocks we are already familiarized with. Our starting point is the EL controller of Kelly [125] revisited in Subsection 2.4.B.2. As we have pointed out, the closed loop system depicted in Fig. 3.16 constitutes a passive operator $\Sigma : z \mapsto \dot{q}$. On the other hand we know that the PID controller in closed loop with the EL plant (3.44) is locally output strictly passive.

To this point, we might wonder if there is a way to introduce the OSP block defined by the approximate differentiation filter, into the PID controller (3.47), (3.48) without destroying the passivity properties of the closed loop, as we have done with the PD plus gravity compensation. Fortunately the answer is affirmative as the following proposition shows.

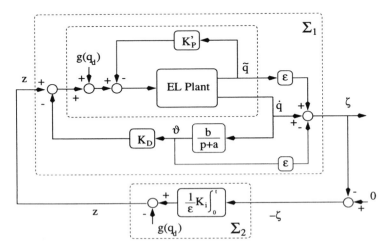

Figure 3.17: PI²D Controller, block diagram.

Proposition 3.19 (Passivity of the PI²D controller.) *For the system depicted in Fig. 3.17 defined by*

$$D(q)\ddot{q} + C(q, \dot{q})\dot{q} + g(q) + K'_p(q - \delta) + K_d\vartheta = z \qquad (3.64)$$

$$\dot{\vartheta} = -A\vartheta + B\dot{q} \qquad (3.65)$$

$$\dot{z} = -K_i(\tilde{q} + \frac{1}{\varepsilon}\dot{q} - \vartheta) \qquad (3.66)$$

where the first two equations represent the dynamics of Σ_1 and (3.66) is the integrator block, there exist some control gains K_p, K_d, K_i, A and B such that

(i) System Σ_1: (3.64), (3.65) defines a Locally Output Strictly Passive (LOSP) operator $z \mapsto \zeta$ where $\zeta \stackrel{\triangle}{=} \varepsilon[\tilde{q} - \vartheta]$ and storage function

$$\mathcal{H}_3(\tilde{q}, \dot{q}, \vartheta) \stackrel{\triangle}{=} \mathcal{H}_1(\tilde{q}, \dot{q}) + \frac{1}{2}\vartheta^\top K_d B^{-1}\vartheta - \varepsilon[\tilde{q} - \vartheta]^\top D(q)\dot{q}.$$

where $\mathcal{H}_1$ is defined by (3.51).

(ii) *System* Σ_2: *(3.66) defines a passive operator* $-\zeta \mapsto z$ *with storage function* $\mathcal{H}_2(z)$, *where* $\mathcal{H}_2$ *is defined by (3.54).*

□

Proof. For clarity of exposition we present below the main steps of the proof, for more detail we refer the reader to the Appendix **D**.

(i): By defining

$$z \triangleq -\frac{1}{\varepsilon} K_i \tilde{q} + \tilde{\nu}$$

we show in Appendix **D** that for all $(\tilde{q}, \vartheta, \dot{q})$ satisfying some suitable conditions, the time derivative of $\mathcal{H}_3(\tilde{q}, \dot{q}, \vartheta)$ along the trajectories of (3.64), (3.65) satisfies the bound

$$\dot{\mathcal{H}}_3(\tilde{q}, \dot{q}, \vartheta) \leq -\gamma_1 \|\dot{q}\|^2 - \gamma_2 \|\tilde{q}\|^2 - \gamma_3 \|\vartheta\|^2 + z^\top [\varepsilon(\tilde{q} - \vartheta) + \dot{q}] \qquad (3.67)$$

where the strictly positive constants γ_1, γ_2, γ_3 are defined in (D.7). Also notice that

$$\|\varepsilon[\tilde{q} - \vartheta] + \dot{q}\|^2 \leq \varepsilon^2(\|\tilde{q}\|^2 + 2\|\tilde{q}\|\|\vartheta\| + \|\vartheta\|^2) + 2\varepsilon\|\dot{q}\|(\|\tilde{q}\| + \|\vartheta\|) + \|\dot{q}\|^2$$

from this, we obtain after some straightforward calculations that

$$\gamma_1 \|\dot{q}\|^2 + \gamma_2 \|\tilde{q}\|^2 + \gamma_3 \|\vartheta\|^2 \geq \|\varepsilon[\tilde{q} - \vartheta] + \dot{q}\|^2, \qquad (3.68)$$

if $\gamma_1 \geq 2\varepsilon + 1$ and γ_2, $\gamma_3 \geq 2\varepsilon^2 + \varepsilon$. This clearly imposes new conditions on the controller gains:

$$\frac{8}{b_m d_m - 8} < \varepsilon \qquad (3.69)$$

$$\frac{k_{p_m} - 2}{4} > \varepsilon \qquad (3.70)$$

$$\frac{k_{d_m} a_m}{4 b_M} > 2\varepsilon^2 + \varepsilon. \qquad (3.71)$$

The condition imposed by inequality (3.69) is met for sufficiently large b_m while (3.70) holds for a sufficiently small ε and (3.71) holds for sufficiently large a_m. Integrating (3.67) from 0 to T, on both sides of the inequality and using (3.68) we get

$$\mathcal{H}_3(T) - \mathcal{H}_3(0) \leq -\|\varepsilon[\tilde{q} - \vartheta] + \dot{q}\|^2_{2T} + \langle z \mid [\varepsilon(\tilde{q} - \vartheta) + \dot{q}] \rangle_T$$

since $\mathcal{H}_3(\tilde{q}, \dot{q}, \vartheta)$ is positive definite for all $T \geq 0$,

$$\langle z \mid [\varepsilon(\tilde{q} - \vartheta) + \dot{q}] \rangle_T \geq \|\varepsilon[\tilde{q} - \vartheta] + \dot{q}\|^2_{2T} - \beta_2$$

hence completing the proof.

(i): Straightforward. (Σ_2 is a simple integrator).

■

Remark 3.20 Notice that inequality (3.67) only holds locally, that is if (D.10) is satisfied. This is what leads us to *local* output strict passivity.

We have thus shown how to construct a PI²D controller by interconnecting passive blocks. The following proposition which is the original contribution of [207], defines the PI²D controller.

Proposition 3.21 (SGAS of the PI²D controller.) *Consider the dynamic model (3.44) in closed loop with the PI²D control law*

$$u = -K_p \tilde{q} + \nu - K_d \vartheta \qquad (3.72)$$

$$\dot{\nu} = -K_i(\tilde{q} - \vartheta), \quad \nu(0) = \nu_0 \in \mathbb{R}^n \qquad (3.73)$$

$$\dot{q}_c = -A(q_c + Bq) \qquad (3.74)$$

$$\vartheta = q_c + Bq \qquad (3.75)$$

Let K_p, K_i, K_d, $A \triangleq \mathrm{diag}\{a_i\}$, $B \triangleq \mathrm{diag}\{b_i\}$ be positive definite diagonal matrices with $K_p \triangleq K_p' + \frac{1}{\varepsilon}K_i$ and

$$B > \frac{4d_M}{d_m}I \qquad (3.76)$$

$$K_p' > 4k_g$$

where k_g is defined by (2.18).

Under these conditions, we can always find a (sufficiently small) integral gain K_i such that the equilibrium $x \triangleq [\tilde{q}^\top,\ \dot{q}^\top,\ \vartheta^\top,\ \nu^\top - g(q_*)^\top]^\top = 0$ is asymptotically stable with a domain of attraction including

$$\{x \in \mathbb{R}^{4n} : \|x\| < c_3\}$$

where $\lim_{b_m \to \infty} c_3 = \infty$. In other words, given any (possibly arbitrarily large) initial condition $\|x(0)\|$, there exist controller gains that ensure $\lim_{t \to \infty} \|x(t)\| = 0$. □

Remark 3.22 Notice that control law (3.72) can be alternatively written as

$$u(t) = -K_p \tilde{q}(t) - K_i \int_0^t \tilde{q}(\tau)d\tau - K_d \mathrm{diag}\left\{\frac{b_i p}{p + a_i}\right\} q(t) - K_i \int_0^t \mathrm{diag}\left\{\frac{b_i p}{p + a_i}\right\} q(\tau)d\tau.$$

The first three right hand side terms constitute the proportional, integral and filtered derivative actions, while the presence of the fourth right hand term motivates the name PI²D. We have kept the notation D for the derivative term because in practical applications of PID regulators this is commonly implemented incorporating a filter.

Proposition 3.21 establishes the *semiglobal stability* of the PI²D controller, in the sense that the domain of attraction can be arbitrarily enlarged with a suitable choice of the gains, namely by increasing b_m. However, as shown in the proof (cf. Subsection A), the stability conditions impose an order relationship between K_i and B such that we must correspondingly decrease K_i.

A   Sketch of proof of Proposition 3.21

Once we have established the local output strict passivity of the map $\Sigma_1 : z \mapsto \zeta$ and accounting for the passivity of the integrator Σ_2, the next step to prove *internal* stability would be to prove zero-state detectability for the system Σ_1 however, in contrast to Section 4.3 one cannot conclude much about the behaviour of the state $(\tilde{q}, \dot{q}, \vartheta)$ from the fact that $\zeta \to 0$. Therefore we rely on Lyapunov theory.

Since we have established some passivity properties for the closed loop by using storage functions which are based on the total energy of the system, we can use now the Lyapunov function candidate $V = \mathcal{H}_3 + \mathcal{H}_2$ which is positive definite under the conditions established in the Appendix **D**. Further, the time derivative of V along the closed loop trajectories (3.64) - (3.66) is negative semidefinite as it can be appreciated from (3.67). Asymptotic stability of the equilibrium $(\tilde{q}, \dot{q}, \vartheta, z) = 0$ follows invoking Krasovskii-LaSalle's invariance principle, observing that the latter is the largest invariant set obtained in $\dot{V} \equiv 0$.

Furthermore as proven in Appendix **D** a domain of attraction is given by

$$\|x\| \leq c_3 \triangleq \frac{1}{2k_c} \left[\frac{1}{2} b_m d_m - d_M \right] \sqrt{\frac{\alpha_1}{\alpha_2}}.$$

where α_1, α_2 and b_m are positive constants such that $b_m I \geq B$ and

$$\alpha_1 \|x\|^2 \leq V(x) \leq \alpha_2 \|x\|^2.$$

A.1 Semiglobal asymptotic stability

To establish *semiglobal asymptotic stability* (SGAS) we must prove that, with a suitable choice of the controller gains, we can arbitrarily enlarge the domain of attraction. To this end, we propose to increase b_m and b_M at the same rate. The key question here is whether this can be done without violating the order relationships between B and ε imposed by the stability conditions (D.5), (D.8) and (D.11). The order relationship due to (D.5) is $\varepsilon(b_m) = \mathcal{O}(1/\sqrt{b_m})$, while that of (D.8) and (D.11) is $\varepsilon(b_m) = \mathcal{O}(1/b_m)$. The latter being implied by the former for ε sufficiently small.

On the other hand, for b_m sufficiently large we can always find $\varepsilon > 0$ so that $\alpha_1 = c_4/b_m$ and $\alpha_2 = c_5$, where c_4, c_5 are constants independent of B. Replacing this in (D.13) we get

$$\lim_{b_m \to \infty} c_3^2 = \lim_{b_m \to \infty} c_6 \sqrt{b_m}$$

where c_6 is also independent of B. This proves that there exists $\varepsilon > 0$ such that 1) the stability conditions, are satisfied, i.e., verifying $\varepsilon = \mathcal{O}(1/\sqrt{b_m})$; 2) the domain of attraction is arbitrarily enlarged, that is, $\lim_{b_m \to \infty} c_3 = \infty$.

The proof is completed choosing, for the given ε, $\bar{\lambda}(K_i) = \mathcal{O}(\varepsilon)$.   ∎

Remark 3.23 It is important to remark that when implementing the PI^{2}D controller, one should be careful with setting the initial conditions of the dirty derivatives filter (3.74)–(3.75): from the calculations above, the initial conditions $\vartheta(0)$ should be small enough to guarantee asymptotic stability. Also, in order to enlarge the domain of attraction, we proceed to increase B. Nevertheless, notice from (3.75) that the initial condition $\vartheta(0) = q_c(0) + Bq(0)$, hence the larger B is, the larger $\vartheta(0)$ may also be. A simple way to overcome this problem, is to set $q_c(0) = \vartheta(0) - Bq(0)$ for some fixed $\vartheta(0)$, $q(0)$ and B.

B   Simulation results

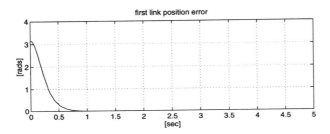

Figure 3.18: PI^{2}D control. First link position error.

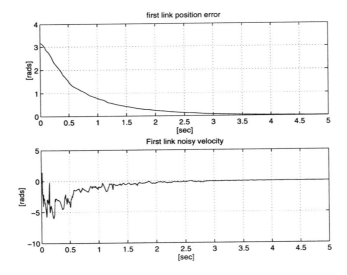

Figure 3.19: PID control under noisy velocity measurement.

In this section we illustrate in simulation the performance of our algorithm in the two-rigid-link robot manipulator model of [20], see also p. 59. For the sake of comparison we tested the performance of a simple PID controller with velocity measurements. We assumed that the measurements are affected by a random noise of 10% which obeys a uniform distribution. In both cases, the controller parameters were set to $K_P \triangleq \mathrm{diag}\{[400,\ 400])\}$, $K_I \triangleq \mathrm{diag}\{[0.004,\ 0.005]\}$, $K_D \triangleq \mathrm{diag}\{[300,\ 300]\}$, $A \triangleq \mathrm{diag}\{[3000,\ 3000]\}$ and $B \triangleq \mathrm{diag}\{[1000,\ 1000]\}$. We started from zero initial conditions to achieve the constant reference $q_* \triangleq [-\pi;\ -\pi]^\top$.

In Fig. 3.18 we show the first link response using our PI^{2}D controller. In Fig. 3.19 we show the first link response using a common PID controller as well as the noisy velocity measurements corresponding to the same link. As a criterion of comparison, we have evaluated the integral square position error, (ISE) i.e. $J = \int_0^5 (\tilde{q}_p) dt$. For the PI^{2}D controller, this calculation resulted in $J = 1.59$ while for the ordinary PID control, $J = 3.67$.

5   Concluding remarks

We have addressed in this chapter the set-point control problem of EL systems by output feedback. The controller, which we choose to be also an EL system, is designed using the energy shaping plus damping injection ideas of the passivity-based approach.

Our contribution to this problem is the proof that asymptotic stabilization *without* generalized velocity measurement is possible via the inclusion of a passive *dynamic extension* provided the system satisfies a *dissipation propagation* condition.

An important extension of these results includes the EL controllers which yield bounded inputs. We have identified a class of EL plants which can be regulated with saturated controllers depending on the growth rate of the potential energy function. In words, the key requirement is that this function does not grow faster than linearly for large position values.

We have illustrated this technique with the case studies of the TORA system and the flexible joints robots. For the TORA system we proved the superiority of our PBC with respect to the backstepping controllers of [111], both in performance and simplicity of implementation. For flexible joints robots we showed that several apparently unrelated controllers which appeared independently in the literature happen to belong to the EL class. We illustrated in simulations the performance of the two EL controllers [3, 125].

Finally we revised the setpoint control problem with uncertain gravity knowledge, in particular, we revisited the popular PID control from a passivity point of view. For the case of unmeasurable velocities we have presented the PI^{2}D controller, which relies on the passivity properties of the EL plant to ensure SGAS.

Chapter 4

Trajectory tracking control

In this chapter we extend the passivity-based method, developed for regulation in the previous chapter, to solve trajectory *tracking* problems. The first main modification that we have to make is that for tracking, besides reshaping the potential energy of the EL plant, we must also shape the "kinetic energy" function. Whereas modifying the potential energy function means to relocate the equilibria of the system, the modification of the "kinetic energy" function can, roughly speaking, be rationalized as imposing a *specific* pattern to the transformation of potential into kinetic energy. However, the quotes here are important because the storage function that we assign to the closed loop *is not* an energy function in the sense that it defines the equations of motion. With an obvious abuse of notation we will still refer to this step as energy shaping, but it is better understood as *passivation* with a desired *storage* function (see Appendix A). The damping injection step is added then to make the passivity *strict*. The passivation objective is achieved invoking the key passive error dynamics Lemma 2.7, which states that we can always factor the workless forces in such a way that, in terms of the error signals s, the EL system behaves like a linear passive system.

In Section **3**.2 we have introduced already a PBC that modifies the kinetic energy, namely the PBC discussed in Section **3**.2.4.B.1. As pointed out in Remark 3.9 this modification requires an additional dynamic extension, besides the one used to inject damping. It will become clear below that to address tracking problems the PBC procedure gives first an *implicit* definition of this dynamic extension, which has to be later explicitly realized. As we will see in Section 1.2 the realization is immediate for fully actuated systems, however, for underactuated systems like the flexible joint robot discussed in Section 3.3 this implies an additional step, which involves some kind of "system inversion". The importance of this step will be further underscored in our applications to electromechanical systems in Part III.

A final difficulty for applying PBC in tracking is that, with respect to position, the EL dynamics is relative degree two. Hence, we cannot apply passivity principles (that require a unitary relative degree) directly to this signal. This obstacle can be overcomed via a suitable definition of the error signal s, for instance with the ingenious inclusion of a linear filter, as used in the robot control literature [251].

In Section 1 we address the problem of state feedback tracking control of *fully-actuated* mechanical systems. We consider the case of known parameters as well as adaptive schemes. PBC is particularly suited for the latter because, as we pointed out in Section **3**.4.2, the standard gradient estimation algorithm defines also a passive map, thus passivity (and consequently $\mathcal{L}_2$ stability) of the overall scheme are ensured.

Section 3 is devoted to the extension of PBC to *underactuated* EL systems, always assuming state–feedback. We take as case study the model of robots with flexible joints. This problem has a strong practical motivation for high performance robots where the elasticity phenomenon is no further negligible and has to be explicitly taken into account in the design, typically adding a linear torsional spring as in regulation problems. It is also a very challenging theoretical problem since the number of degrees of freedom of the system is twice the number of control actions and the matching property between nonlinearities and inputs is lost.

We finish the chapter with some results of PBC to the difficult *output feedback* case.

1   State feedback control of fully-actuated systems

The solution to the *state* feedback tracking control problem of fully actuated EL systems (in particular, rigid-robot manipulators) has been known from many years now, for a literature review see e.g. [212, 283].

In this section we study the passivity based tracking control technique for fully-actuated systems. The approach is based on the passivity property of EL systems established in Proposition 2.5. More precisely consider the following problem:

Definition 4.1 (State feedback tracking control of fully-actuated systems)
Consider the system (2.15) which we repeat here for convenience:

$$D(q)\ddot{q} + C(q,\dot{q})\dot{q} + g(q) = u \tag{4.1}$$

where $u \in \mathbb{R}^n$. Design a state feedback control law $u = u(t,q,\dot{q})$ that ensure

$$\lim_{t\to\infty} \|\tilde{q}(t)\| = \lim_{t\to\infty} \|q(t) - q_*(t)\| = 0 \tag{4.2}$$

for any desired time-varying trajectory $q_* \in \mathcal{C}^2$.

Below we revise the well known controllers of Slotine and Li [251] and the PD+ controller of Paden and Panja [216], from a passivity point of view.

1.1   The PD+ controller

The PD+ controller originally introduced in [216] was one of the first results guaranteeing global tracking for rigid-joint robots, that is global uniform asymptotic stability (GUAS) of the origin $(\tilde{q}, \dot{\tilde{q}}) = (0, 0)$. It is the natural extension to tracking control, of the energy-shaping plus damping injection PD plus gravity cancelation used for regulation (see Section **3.1**).

Motivated by the passive error dynamics Lemma 2.7 it is reasonable to aim at an *error system* of the form

$$D(q)\ddot{\tilde{q}} + C(q, \dot{q})\dot{\tilde{q}} + K_p\tilde{q} + K_d\dot{\tilde{q}} = \Psi \tag{4.3}$$

where K_d, K_p are positive definite matrices. This system defines an OSP map $\Psi \mapsto \dot{\tilde{q}}$ with storage function

$$\mathcal{H}(\dot{\tilde{q}}, \tilde{q}) = \frac{1}{2}\dot{\tilde{q}}^\top D(q)\dot{\tilde{q}} + \frac{1}{2}\tilde{q}^\top K_p\tilde{q}. \tag{4.4}$$

The proof follows straightforward, differentiating the storage function (4.4) with respect to time, along the trajectories of (4.3) to obtain

$$\dot{\mathcal{H}}(\dot{\tilde{q}}, \tilde{q}) \leq \dot{\tilde{q}}^\top K_d\dot{\tilde{q}} + \Psi^\top\dot{\tilde{q}}.$$

Hence if we can set $\Psi \equiv 0$ then $\dot{\tilde{q}} \in \mathcal{L}_2^n$, which is our first step towards the proof of convergence of the errors to zero. A control law that achieves this objective is precisely the PD+ control law of [216]

$$u = D(q)\ddot{q}_* + C(q, \dot{q})\dot{q}_* + g(q) - K_p\tilde{q} - K_d\dot{\tilde{q}} \tag{4.5}$$

which borrows its name from the last three terms that correspond to the PD controller used for set-point control purposes.

This controller was proved in [216] to globally asymptotically stabilize the closed loop system (4.3) with $\Psi \equiv 0$, by relying on Lyapunov techniques. In particular the authors of that reference used Matrosov's theorem [226] which involves tedious calculations due to the computation of $\ddot{\mathcal{H}}$ and $\mathcal{H}^{(3)}$.

It is important to stress, however, that the proof of this statement via OSP and zero-state detectability is yet to be established. The underlying obstacle to carry out this proof is that we have to reduce the relative degree (with respect to $\tilde{q}$) of the error dynamics, which in this case is two. The latter is achieved by the Slotine and Li controller presented below.

Remark 4.2 In Section 3.5 we present a controller which borrows inspiration from the PD+ algorithm, for tracking control of flexible-joint manipulators.

1.2   The Slotine and Li controller

To reduce the relative degree of the error system let us redefine the error signal as

$$s \overset{\triangle}{=} \dot{\tilde{q}} + \Lambda \tilde{q}, \tag{4.6}$$

with $\Lambda > 0$. Notice that $\tilde{q} = (pI + \Lambda)^{-1}s$, hence it is the output of a strictly proper asymptotically stable LTI system with input s. Invoking Lemma A.12 of Appendix A we have that if $s \in \mathcal{L}_2$ then $\tilde{q} \to 0$. This motivates us to aim at an error dynamics of the form

$$D(q)\dot{s} + [C(q,\dot{q}) + K_d]s = \Psi \tag{4.7}$$

which defines an OSP operator $\Psi \mapsto s$ with storage function

$$\mathcal{H}_d(q,s) = \frac{1}{2}s^\top D(q)s. \tag{4.8}$$

It is easy to see that the system (4.7) is equivalent to the system (4.1) with

$$\Psi = u - (D(q)\ddot{q}_r + C(q,\dot{q})\dot{q}_r + g(q)) + K_d s \tag{4.9}$$

where we defined $\dot{q}_r = \dot{q}_* - \Lambda \tilde{q}$. The controller that sets $\Psi \equiv 0$ is obtained then in an obvious manner. This is the celebrated Slotine and Li controller [251].

Remark 4.3 In [233] we study the disturbance attenuation properties of PBC for robots manipulators. We show that adding a proportional gain around the position error, as proposed in [229], we can achieve *arbitrarily good* disturbance attenuation (*without* compromising the convergence rate). More precisely, we prove the following proposition.

Proposition 4.4 (Disturbance attenuation) *Consider the EL system*

$$D(q)\ddot{q} + C(q,\dot{q})\dot{q} + g(q) = u + \zeta$$

where $\zeta \in \mathcal{L}_{2e}$ represents an external disturbance, in closed–loop with the PBC controller

$$u = D(q)\ddot{q}_r + C(q,\dot{q})\dot{q}_r + g(q) - K_d s - K_p \tilde{q}$$

with q_r, s as defined above, and λ, k_d, k_p positive scalars verifying

$$K_d > \frac{1}{2}\left(\frac{1}{\gamma^2} + 1 + \lambda\right)$$

$$K_p > \frac{1}{2\lambda}\left(\lambda^2 + 1 + \lambda\right)$$

for a fixed $\gamma > 0$. Under these conditions, the following inequality holds

$$\|[\dot{\tilde{q}}^\top, \tilde{q}^\top]^\top\|_{2T} \leq \gamma \|\zeta\|_{2T}$$

for all $T \geq 0$. Consequently, arbitrarily good disturbance attenuation (in the $\mathcal{L}_2$–sense) is achievable by increasing the gain k_d.   □

Remark 4.5 In the original contribution of [251], and several further applications of this algorithm, only *convergence* of the signals $\dot{\tilde{q}}$ and $\tilde{q}$ is actually proved. However it should be noted that the far stronger property of global exponential stability of the origin $(\dot{\tilde{q}}, \tilde{q}) = (0, 0)$ follows considering the Lyapunov function proposed in [254], and restricting (without loss of generality) K_d, K_p and Λ to be diagonal. This Lyapunov function will be instrumental for the analysis presented in Section 3.3 and Chapters **5** and **12**.

Remark 4.6 It is shown in [212] that via a suitable definition of the error signal s above we can recover various global tracking controllers reported in the literature. That is, in general we can set $s = H^{-1}(p)\tilde{q}$, with $H(p)$ strictly proper and asymptotically stable, and the proof will go through. In the case of [251] we choose $H^{-1}(p) = p + \lambda$, while Sadegh and Horowitz considered in [229] a PID. Another interesting controller, studied within this framework in [212], is the one proposed in [124].

Remark 4.7 A procedure, similar to the one presented above, can be used for some underactuated EL systems, e.g., robots with flexible joints and diagonal inertia matrix. However, the extension is not straightforward and one must be careful with choosing a suitable desired potential energy. See Section 3.3.

2 Adaptive trajectory tracking

It has been shown in the regulation problem of Section **3**.4.2 that one of the advantages of ensuring OSP of the closed–loop is that, in view of the the passivity of the gradient estimator, adaptive extensions are straightforward. In this section we present an adaptive implementation of the PBC of Slotine and Li. One important drawback of this scheme is that, as shown in [22], the parameter estimates may drift in the presence of measurement noise. To overcome this problem we present the robustified version reported in [22].

2.1 Adaptive controller of Slotine and Li

To derive an adaptive version of the PBC above we invoke first the Property **P2.1** of linearity with respect to the parameters of the EL dynamics of Chapter **2** to write

$$D(q)\ddot{q}_r + C(q, \dot{q})\dot{q}_r + \hat{g}(q) = \Phi(q, \dot{q}, \dot{q}_r, \ddot{q}_r)\theta$$

where $\theta \in \mathbb{R}^q$ contains the unknown parameters and the regressor matrix $\Phi(q, \dot{q}, \dot{q}_r, \ddot{q}_r)$ consists of known functions.

Then, we propose a certainty equivalent adaptive controller of the form

$$u = \Phi(q, \dot{q}, \dot{q}_r, \ddot{q}_r)\hat{\theta} + K_d s$$

with $\hat{\theta}$ the estimated parameters. This yields the error dynamics

$$D(q)\dot{s} + [C(q,\dot{q}) + K_d]s = \Phi(q,\dot{q},\dot{q}_r,\ddot{q}_r)\tilde{\theta} \overset{\triangle}{=} \Psi \qquad (4.10)$$

where, with some abuse of notation, we have used again Ψ. Finally, we propose a parameter adaptation law

$$\dot{\hat{\theta}} = -\Gamma^{-1}\Phi^{\top}(q,\dot{q},\dot{q}_r,\ddot{q}_r)s \qquad (4.11)$$

where $\Gamma = \Gamma^{\top} > 0$. Analogously to the regulation case we have an OSP operator $\Psi \mapsto s$ defined by (4.10) in negative feedback with a passive operator $s \mapsto -\Psi$ defined by the estimator (4.11). Hence, the closed–loop system is OSP, and $s \in \mathcal{L}_2$. The proof of global convergence of $\tilde{q}$ is completed, as before, invoking Lemma A.12.

2.2   A robust adaptive controller

It is well known [13] that the equilibrium set of adaptive systems is unbounded. Therefore, in underexcited conditions and in the presence of noise in the adaptation law, the instability mechanism of parameter drift appears. This instability mechanism appears, in particular, in the adaptive controller given above. To exemplify this phenomenon, consider a single link pendulum moving in the horizontal plane, that is $d\ddot{q} = u$ where the inertia $d > 0$ is unknown. Excitation is lost in regulation, therefore consider the desired position $q_* = const$. Assume further that velocity is corrupted by noise with zero mean and variance σ^2, that is, $\zeta \approx N(0, \sigma^2)$. In these circumstances the adaptation law (4.11) looks like

$$\dot{\hat{d}} = -\Gamma[-\lambda(\dot{q} + \zeta)](\dot{q} + \zeta + \lambda\tilde{q})$$

whose expectation is

$$E\{\dot{\hat{d}}\} = \Gamma\lambda\dot{q}(\dot{q} + \lambda\tilde{q}) + \Gamma\lambda\sigma^2$$

Consequently, the integral of the right hand term introduces a parameter drift, thus the controller is not robust to measurement noise. As it will become clear below, in the adaptation law we propose here, there are no quadratic terms of $\dot{q}$ and the drift phenomenon disappears.

In this section we present a controller, originally introduced in [22], which has the following features: 1) enhanced robustness with respect to velocity measurement noise, in particular, the aforementioned drift is avoided; 2) does not require high gain loops; 3) we provide a relationship between convergence rates and compensator gains that is independent of the desired trajectory. To achieve these objective we introduce two key modifications to the Slotine and Li controller: the inclusion of an additional "Coriolis–forces–dependent" term and the use of a normalization factor similar to the one presented in Section **3**.4.2.

We present below the main result and refer the reader to [22] for further details and some illustrative simulations. As standard in PBC, we start by remarking the passivity property of a desired closed loop system.

Proposition 4.8 (OSP of the error equation.) *The error equation*

$$D(q)\ddot{\tilde{q}} + C(q,\dot{q})\dot{\tilde{q}} + \lambda C(q,\tilde{q})\dot{q}_* + K_d\dot{\tilde{q}} + K_p\tilde{q} = \Psi \tag{4.12}$$

with K_p, K_d positive definite and satisfying

$$\lambda_0 < \min\left\{ \frac{k_{d_m}}{3d_M + 2k_c}, \frac{4k_{p_m}}{k_{d_M} + k_{d_m}} \right\}. \tag{4.13}$$

defines a passive operator $\Psi \mapsto s$ with storage function

$$\mathcal{H}_d(s, \tilde{q}) = \frac{1}{2} s^\top D(q) s + \frac{1}{2} \tilde{q}^\top K_p \tilde{q} \tag{4.14}$$

where $s = \dot{\tilde{q}} + \lambda\tilde{q}$, and[1]

$$\lambda \triangleq \frac{\lambda_0}{1 + \|\tilde{q}\|} \tag{4.15}$$

where $\lambda_0 > 0$. □

Proof. *(sketch)* Evaluating the time derivative of $\mathcal{H}_d$ along the trajectories of (4.12), we obtain after some bounding and using the Properties **P2.1** - **P2.2**:

$$\dot{\mathcal{H}}_d(s', \tilde{q}) \leq -\kappa_1 \|s'\|^2 - \kappa_2 \left\|\frac{\lambda\tilde{q}}{2}\right\|^2 + s^\top \Psi \tag{4.16}$$

where we have defined $s' \triangleq \dot{\tilde{q}} + \frac{\lambda}{2}\tilde{q}$ and the constants

$$\begin{aligned}
\kappa_1 &\triangleq k_{d_m} - 3\lambda_0 d_M - 2\lambda_0 k_c \\
\kappa_2 &\triangleq \frac{4k_{p_m}}{\lambda_0} - k_{d_M} - 2\lambda_0 d_M - 2\lambda_0 k_c.
\end{aligned} \tag{4.17}$$

which are positive in view of (4.17). The proof is completed integrating on both sides of (4.16). ∎

Motivated by this passivity result we construct the stabilizing control law for the known parameter case, as in previous sections, by setting $\Psi \equiv 0$ and comparing equations (4.12) and (4.1). This yields the controller

$$\begin{aligned}
u &= D(q)\ddot{q}_* + C(q, \dot{q} - \lambda\tilde{q})\dot{q}_* + g(q) - K_d\dot{\tilde{q}} - K_p\tilde{q} \\
&\triangleq \Phi(\ddot{q}_*, \dot{q}_*, q_*, q, \dot{q})^\top \theta
\end{aligned}$$

---

[1]The normalization of the term λ is due to [284].

where the second equation is introduced to stress the fact that this control law is still linear in the parameters. Therefore, in the case of parameter uncertainty, an adaptive version is simply obtained as $u = \Phi(\ddot{q}_*, \dot{q}_*, q_*, q, \dot{q})^\top \hat{\theta}$ with estimation law (4.11). The closed loop has exactly the form (4.12) with $\Psi \triangleq \Phi(\ddot{q}_*, \dot{q}_*, q_*, q, \dot{q})^\top \tilde{\theta}$, and we can apply *verbatim* the analysis of the previous section to prove global convergence.

Remark 4.9 In [22] it is shown that, in the known parameter case, the closed loop trajectories $[\dot{\tilde{q}}(t)^\top, \tilde{q}(t)^\top]^\top$ tend to the origin exponentially fast if

$$\lambda_0 < \sqrt{\frac{4d_M k_{p_M}}{d_M^2}}.$$

It is important to remark that the bounds for the design parameter λ_0 do not depend on any bounds on the desired trajectory $q_*(t)$ and its derivatives as is the case of other results, for instance in [163, 284]. Therefore tuning of the controller parameters is task independent.

3 State feedback of underactuated systems

In this section we study the problem of global tracking of underactuated EL systems via state feedback. We concentrate our attention on the problem of robots with flexible joints with block diagonal inertia matrix. First, following exactly the technique used in the previous section to derive the Slotine and Li controller for rigid robots, but with a suitably defined storage function, we design a global tracking static state–feedback PBC. Then, to provide some insight into this scheme, we compare it with the controllers obtained from the application of backstepping and decoupling principles. Some of the material in this section has been reported in [40].

3.1 Model and problem formulation

The *simplified model* of an n link robot (which assumes the angular part of the kinetic energy of each rotor is due only to its own rotation) was derived in Section 2.4.3. We recall that the EL parameters are given by

$$\mathcal{T}(q_1, \dot{q}) \triangleq \frac{1}{2}\dot{q}^\top \mathcal{D}(q_1)\dot{q}, \ \mathcal{V}(q) \triangleq \frac{1}{2}q^\top \mathcal{K}q + \mathcal{V}_g(q_1)$$

where

$$\mathcal{K} \triangleq \left[\begin{array}{cc} K & -K \\ -K & K \end{array} \right], \ \mathcal{D}(q_1) \triangleq \left[\begin{array}{cc} D(q_1) & 0 \\ 0 & J \end{array} \right]$$

$q_1 \in R^n$ and $q_2 \in R^n$ represent the link angles and motor angles, respectively, $D(q_1)$ is the $n \times n$ inertia matrix for the rigid links, J is a diagonal matrix of actuator inertias reflected to the link side of the gears, and $K > 0$ is a diagonal matrix containing the joint stiffness coefficients. For ease of reference, we repeat here the dynamical model (2.35) in its compact form

$$\mathcal{D}(q_1)\ddot{q} + \mathcal{C}(q_1, \dot{q}_1)\dot{q} + \mathcal{G}(q_1) + \mathcal{K}q = \mathcal{M}u \qquad (4.18)$$

and in the separate form

$$\begin{cases} D(q_1)\ddot{q}_1 + C(q_1, \dot{q}_1)\dot{q}_1 + g(q_1) = & K(q_2 - q_1) \\ J\ddot{q}_2 + K(q_2 - q_1) = u \end{cases} \qquad (4.19)$$

As usual $C(q_1, \dot{q}_1)\dot{q}_1$ represents the Coriolis and centrifugal forces and $g(q_1)$ represents the gravitational terms. As suggested throughout the book we define $C(q_1, \dot{q}_1)$ via the Christoffel symbols. We will refer in the future to the first and second equations above as link dynamics and motor dynamics, respectively.

We are interested here in the following problem.

Definition 4.10 (Global tracking problem) *For the system (4.19) define an internally stable control law that, for all $q_{1*}(t) \in \mathcal{C}^4 \cap \mathcal{L}_\infty^n$ and arbitrary initial conditions, ensures*

$$\lim_{t \to \infty} \|\tilde{q}_1(t)\| = \lim_{t \to \infty} \|q_1(t) - q_{1*}(t)\| = 0.$$

3.2   Literature review

The flexible-joint robot model (4.19) is globally feedback linearizable (by static state feedback) and therefore globally stable controllers can be derived with "classical" geometric techniques, see e.g. [255]. Besides the intrinsic lack of robustness of schemes based on nonlinearity cancelation, the proposed solutions suffer from the additional drawback that the control implementation relies on the availability of link acceleration and jerk. Even though these signals can be derived *without differentiation* from the systems model, this is not a desirable procedure since the accuracy in their calculation will be highly sensitive to uncertainty in the robot parameters. One way to overcome this difficulty is to use parameter adaptation techniques, unfortunately it is not clear at this point how to make these schemes adaptive preserving the global stability property.

In [195] a Lyapunov-based backstepping technique is applied to derive the first global tracking controller. In [166] a scheme that is not based on feedback linearization is presented. Availability of link acceleration and jerk is still required, but the sensitivity problems mentioned above are claimed to be overcomed by the adaptive implementation. The controller is a complicated dynamic state feedback that requires the realization of several filtering stages. Furthermore, some incorrect technical issues pointed out by Prof. Liu Hsu had to be addressed in [39], leading to the modification of the controller. In [40] a globally stabilizing PBC that does not suffer from the shortcomings of [166] was derived. The new controller, which is a simple *static state feedback*, has a clear physical interpretation, is exponentially stable, and also admits an adaptive implementation. In [40] the PBC is compared with the controllers obtained from the application of backstepping and decoupling principles using the following performance indicators: continuity properties *vis a vis* the joint stiffness, availability of adaptive implementations when the robot parameters are unknown, and robustness to "energy-preserving" (i.e., passive) unmodeled effects. Complete stability proofs of all the resulting controllers are also given.

It should be remarked that, to the best of our knowledge, the global tracking problem for the *complete* model of [195] is as yet *open*. Some results concerning dynamic feedback linearization for certain particular robot structures are given in [66]. Other efforts aimed at solving this problem may be found in [145].

3.3   A passivity–based controller

Following the PBC approach we propose to assign to the closed–loop a *desired storage function* of the form

$$\mathcal{H}_d = \frac{1}{2}s^\top \mathcal{D}(q_1)s + \frac{1}{2}\tilde{q}^\top \mathcal{K}\tilde{q} \tag{4.20}$$

where, as defined in the previous section, $s = \dot{\tilde{q}} + \Lambda\tilde{q}$, with

$$\tilde{q} \stackrel{\triangle}{=} q - \begin{bmatrix} q_{1*} \\ q_{2d} \end{bmatrix} \tag{4.21}$$

and (with an obvious partitioning) $\Lambda = \text{blockdiag}\{\Lambda_1, \Lambda_1\} > 0$, and Λ_1 *diagonal*. As we will see below this particular choice of Λ is needed for the proof.

At this point we recall the reader an important *notational convention* that we use throughout the remaining of the book. Notice that the definition of $\tilde{q}$ (4.21) includes the actual link reference q_{1*} and a signal q_{2d} to be defined below.[2] This is in contrast with the definition of $\tilde{q}$ in our previous developments (e.g., for regulation in Section **3**.2.3, and tracking with fully-actuated systems in Section 1) where it represents the error between q and a *given* reference value q_*. (Remember that in our notation $(\cdot)_*$ is used exclusively for external reference signals, while $(\cdot)_d$ denotes signals generated by the controller.) Another difference with the fully–actuated case is that, to define $\mathcal{H}_d$, we must take into account the presence of the potential energy term $q^\top \mathcal{K}q$ which cannot be removed. Similarly to the regulation problems of Chapter **3**, we propose to shape it so as to have a global and unique minimum at $\tilde{q} = 0$. It has been shown in that chapter that with different choices of the desired "potential energy" term we can obtain different PBCs, two examples for tracking purpose are given below.

---

[2]This signal will be chosen so as to insure the energy shaping, i.e. such that the closed loop is passive with storage function $\mathcal{H}_d$.

First, let the perturbed desired error dynamics be

$$\mathcal{D}(q_1)\dot{s} + [\mathcal{C}(q_1, \dot{q}_1) + K_d]s + \mathcal{K}\tilde{q} = \Psi \tag{4.22}$$

where $K_d \triangleq diag\{K_{d1}, K_{d2}\} > 0$ is a diagonal matrix that injects the damping, and the perturbation term is defined as

$$\Psi \triangleq \mathcal{M}u - \left[\mathcal{D}(q_1)\ddot{q}_r + \mathcal{C}(q_1, \dot{q}_1)\dot{q}_r + \mathcal{K} \left[\begin{array}{c} q_{1*} \\ q_{2d} \end{array} \right] + \mathcal{G}(q_1) \right] + K_d s \tag{4.23}$$

where $\dot{q}_r \triangleq [\dot{q}_{1*}^\top, \dot{q}_{2d}^\top]^\top - \Lambda\tilde{q}$. Comparing with the corresponding equations for the fully–actuated case (4.7), (4.9) we remark that the error dynamics Lemma 2.7 is not directly applicable because of the presence of the term $\mathcal{K}\tilde{q}$. However, we notice that taking the time derivative of $\mathcal{H}_d$ along the trajectories of (4.22) we get

$$\begin{aligned} \dot{\mathcal{H}}_d &= -s^\top K_d s - s^\top \mathcal{K}\tilde{q} + \dot{\tilde{q}}^\top \mathcal{K}\tilde{q} + s^\top \Psi \\ &= -s^\top K_d s - \tilde{q}^\top \Lambda \mathcal{K}\tilde{q} + s^\top \Psi \\ &\leq -s^\top K_d s + s^\top \Psi \end{aligned}$$

where the last inequality uses the fact that, for the particular choice of Λ, we have that $\Lambda \mathcal{K} \geq 0$. Hence, we have the desired OSP property.

The next step of the design procedure is to calculate, using (4.23), the control signals u and the functional relations for q_{2d}, required to assign the desired storage function, that is, to ensure that $\Psi \equiv 0$.

The perturbation term is set equal to zero with the control law

$$\begin{aligned} q_{2d} &= q_{1*} + K^{-1}u_r \\ u &= -K_{d2}s_2 + J(\ddot{q}_{2d} - \Lambda_1\dot{\tilde{q}}_2) - K(q_{1*} - q_{2d}) \end{aligned} \tag{4.24}$$

where u_r is the control signal for the rigid-joint robot case derived in Section 1.2, that is

$$u_r = D(q_1)\ddot{q}_{1r} + C(q_1, \dot{q}_1)\dot{q}_{1r} + g(q_1) - K_{d1}s_1 \tag{4.25}$$

A very important remark at this point is that $\ddot{q}_{2d}$, required for the implementation of (4.24), is computable *without differentiation*. This fundamental property of (4.19) is lost in the complete model of [195].

We are in position now to present the main result of the section.

Proposition 4.11 *The nonlinear static state feedback control (4.24) solves the global tracking problem. Furthermore, it ensures that the closed–loop system is globally exponentially stable.* □

Proof. The proof of global convergence follows exactly the same lines as in the fully-actuated case using the storage function $\mathcal{H}_d$. To prove global exponential stability we consider the Lyapunov function candidate proposed in [254]

$$V_{PB} = \frac{1}{2}s^T \mathcal{D}(q_1)s + \tilde{q}_1 \Lambda_1^T K_{d1} \tilde{q}_1 + \tilde{q}_2^T \Lambda_1^T K_{d2} \tilde{q}_2 + \frac{1}{2}\tilde{q}^T \mathcal{K}\tilde{q}$$

For which it can be shown that $\dot{V}_{PB} \leq -\alpha V_{PB}$ for some $\alpha > 0$.   ∎

3.4   Comparison with backstepping and cascaded designs

In [40], besides the PBC given above, three different global tracking controllers derived from considerations of cascaded systems or the backstepping technique [142] are presented. The former uses a cascade decomposition property of the robot model, and is motivated by the result on stability of cascaded connections of stable systems with bounded orbits of [236]. The resulting closed loop is a cascade connection. Two backstepping-based schemes, which use also the cascade decomposition property of the model but combined with the integrator augmentation stabilization of [137], are presented. Typical to backstepping designs, the closed–loops are not triangular anymore but satisfy some "anti–symmetric properties". The first scheme results from a direct application of the technique and closely resembles the one reported in [195]. We also present a robustified version, which incorporates some elements of PBC in the construction of the Lyapunov function.

In this section we compare these controllers in relation to the following practical questions: What happens in the "almost rigid" case, that is, when the joint stiffness K takes infinitely large values?; Do they yield high gain designs? The latter question is particularly important because of noise sensitivity considerations.

The four control laws derived in [40] are summarized in the equations below. In all cases u_r corresponds to the rigid robot control signal (4.25) and s and $\tilde{q}$ are defined as above.

Decoupling-based Control

$$\begin{aligned} u &= J\ddot{q}_{2d} - K_{d1}\tilde{q}_2 - K_{d2}\dot{\tilde{q}}_2 + K(q_2 - q_1) \\ q_{2d} &= K^{-1}u_r + q_1 \end{aligned} \tag{4.26}$$

Backstepping-based Control

$$\begin{aligned} u &= J[\ddot{q}_{2d} - 2\dot{\tilde{q}}_2 - 2\tilde{q}_2 - K(\dot{s}_1 + s_1)] + K(q_2 - q_1) \\ q_{2d} &= K^{-1}u_r + q_1 \end{aligned} \tag{4.27}$$

Robustified Backstepping-based Control

$$\begin{aligned} u &= J[\ddot{q}_{2d} - 2\dot{\tilde{q}}_2 - 2\tilde{q}_2 - (\dot{s}_1 + s_1)] + K(q_2 - q_1) \\ q_{2d} &= K^{-1}u_r + q_1 \end{aligned} \tag{4.28}$$

Passivity-based Control

$$\begin{aligned}
u &= J\ddot{q}_{2d} + K(q_{2d} - q_{1*}) - K_{2d}s_2 \\
q_{2d} &= K^{-1}u_r + q_{1*}
\end{aligned} \qquad (4.29)$$

The following remarks are in order:

• The backstepping-based controller (4.27) becomes *high gain* design for increasing values of the joint stiffness, due to the term $K(\dot{s}_1 + s_1)$. Notice that this effect does not appear as a consequence of a term $K(q_2 - q_1)$ because of the convergence of $q_2 - q_1$ to zero as $K \to \infty$. This drawback is removed in (4.26) and (4.28), and is conspicuous by its absence in the passivity-based designs. Notice in particular that the control signal u is *independent* of the gain K in (4.29): the dependence on K comes only from K^{-1} and $K(q_2 - q_1)$ which vanishes as $K \to +\infty$.

• As K grows unbounded the control (4.29) converges to the controller of [251] for the *complete rigid* robot (2.39). On the other hand, in the control (4.26) a term $J\ddot{q}_1$ is added to u_r.

• For large K, the decoupling and backstepping based controllers feed directly into the loop the signal $\ddot{q}_1$ that is calculated using (4.19) through $\ddot{q}_{2d}$, while (4.29) uses instead the noise-free reference $\ddot{q}_{1*}$. Therefore, it is reasonable to expect *better noise sensitivity* properties for the latter.

It is very difficult to derive a definite conclusion about the performance of the different classes of controllers out of the observations made above. This is particularly true since, as shown in [40], modifications introduced at various stages of the backstepping and decoupling designs yield significant improvements. In this respect the passivity-based technique yields robust "tuning knob free" designs in "one-shot", provided of course we can come out with the right desired storage function. One final remark is that it is not clear how to remove the noise sensitivity problem of backstepping and decoupling controllers.

3.5   A controller without jerk measurements

Motivated by the passivation property of the PD+ controller, and the passivity of a flexible-joint manipulator one would like to extend this result to the underactuated case. Intuitively, a direct extension of control law (4.5) could be

$$\mathcal{M}u = \mathcal{D}(q)\ddot{q}_d + \mathcal{C}(q,\dot{q}_d)\dot{q}_d + \mathcal{G}(q) - \mathcal{K}_p\tilde{q} - \mathcal{K}_d\dot{\tilde{q}} \qquad (4.30)$$

where $\mathcal{K}_d \triangleq \text{blockdiag}\{K_{d_1}, K_{d_2}\}$, $\mathcal{K}_p \triangleq \text{blockdiag}\{K_{p_1}, K_{p_2}\}$, are positive definite. However, let us recall that in the underactuated case, the control $u = [0;\, u_f^\top]^\top$, where $u \in \mathbb{R}^{2n}$ and $u_f \in \mathbb{R}^n$. Thus in order for (4.30) to be realized we should define

$$\begin{aligned}
u_f &= K(q_{2d} - q_{1*}) + J\ddot{q}_{2d} - K_{p2}\tilde{q}_2 - K_{d_1}\dot{\tilde{q}}_2 \qquad (4.31) \\
q_{2d} &= K^{-1}[D(q_1)\ddot{q}_{1*} + C(q_1,\dot{q}_1)\dot{q}_{1*} + g(q_1) - K_{p_1}\tilde{q}_1 - K_{d_1}\dot{\tilde{q}}_1] \qquad (4.32)
\end{aligned}$$

However, clearly the control input u_f involves link jerk measurements due to the explicit dependence of q_{2d} on $\dot{q}_1$.

The controller we present in this section, borrows inspiration from the PD+ controller and includes a dynamic extension to avoid the explicit presence of $\dot{q}_1$ in q_{2d} and thereby, jerk measurements.

We start with establishing a passivity property for the system

$$\mathcal{D}\ddot{\tilde{q}} + (\mathcal{C} + \mathcal{C}_d)\dot{\tilde{q}} + \mathcal{K}_p\tilde{q} + \mathcal{K}_d\vartheta + \mathcal{K}\tilde{q} = \Psi \tag{4.33}$$

$$\dot{q}_c = -\mathcal{A}(q_c + \mathcal{B}\tilde{q}) \tag{4.34}$$

$$\vartheta = q_c + \mathcal{B}\tilde{q} \tag{4.35}$$

where for simplicity we have omitted the arguments and $\mathcal{A} \triangleq \text{blockdiag}\{A_1, A_2\}$, $\mathcal{B} \triangleq \text{blockdiag}\{B_1, B_2\}$, are positive definite and $\mathcal{C}_d \triangleq \text{blockdiag}\{C(q_1, \dot{q}_{1*}), 0\}$.

Notice that the equations (4.34) and (4.35) correspond to the dynamics of the approximate differentiation filter which was shown to belong to the EL class of controllers. We know from the previous chapter that this system defines an OSP operator $\vartheta \mapsto \tilde{q}$. On the other hand, the error dynamics (4.33) with $\vartheta \equiv \dot{\tilde{q}}$ and $\mathcal{C}_d \equiv 0$ has the form of the closed loop system (4.3). From these facts, an immediate interesting passivity property which can be established for the system (4.33)–(4.35) is the following. Using a storage function similar to (4.4) with the additional term $\frac{1}{2}\tilde{q}^\top \mathcal{K}\tilde{q}$ and differentiating along the trajectories of (4.33)–(4.35) we obtain that

$$\dot{\mathcal{H}}(t, \dot{\tilde{q}}, \tilde{q}, \vartheta) \leq -\dot{\tilde{q}}^\top \mathcal{C}_d \dot{\tilde{q}} - \vartheta^\top \mathcal{K}_d \mathcal{B}^{-1}\mathcal{A}\vartheta + \dot{\tilde{q}}^\top \Psi. \tag{4.36}$$

Assuming that the desired trajectory $\dot{q}_{1*}(t)$ is uniformly bounded and using the Property **P2.2** we obtain that $-\dot{\tilde{q}}^\top \mathcal{C}_d \dot{\tilde{q}} \leq k_c \left\|\dot{\tilde{q}}\right\|^2$. Integrating on both sides of the inequality above from 0 to T we obtain that

$$\langle \Psi \mid \dot{\tilde{q}} \rangle_T \geq \beta - k_c \left\|\dot{\tilde{q}}\right\|^2$$

where we have defined $\beta \leq -\mathcal{H}(0) + \vartheta^\top \mathcal{K}_d \mathcal{B}^{-1}\mathcal{A}\vartheta$. Roughly speaking the inequality above states that the system (4.33)–(4.35) has a *lack* of passivity.

In the terminology of [237] we say that this system with storage (4.3) defines a map $\Psi \mapsto \dot{\tilde{q}}$ which is Output *Feedback* Passive with negative index k_c, in short OFP(k_c).

The OFP property with a negative index is important in the sense that it establishes that the map $\Psi \mapsto \dot{\tilde{q}}$ can be rendered passive if we are able to compensate for that lack of passivity with an OSP map (that is OFP with positive index). As a matter of fact, it can be proven that the OSP property of the linear approximate differentiation filter (4.34)-(4.35) does the job. The proposition below formalizes this claim.

Proposition 4.12 (Tracking without jerk measurements.) *The system (4.33)-(4.35) defines a locally output strictly passive operator $\Psi \mapsto \varepsilon[\tilde{q} - \vartheta] + \dot{\tilde{q}}$. Furthermore if $\Psi \equiv 0$ then system (4.33) is semi-globally exponentially stable with a domain of attraction including the set*

$$\{x \in \mathbb{R}^{6n} : \|x\| < c_2\} \tag{4.37}$$

where we defined $x \triangleq [\tilde{q}^{\top}, \dot{\tilde{q}}^{\top}, \vartheta^{\top}]^{\top}$.   □

Considering the passivity result established above the stabilizing passivity-based controller is constructed by setting $\Psi = 0$ and comparing equation (4.33) and (4.18). This yields

$$u_f = J\ddot{q}_{2d} + K(q_{2d} - q_{1*}) - K_{p_2}\tilde{q}_2 - K_{d_2}\vartheta_2 \tag{4.38}$$

and the desired motor-shaft trajectory is then chosen as

$$q_{2d} \triangleq K^{-1}[D(q_1)\ddot{q}_{1*} + C(q_1, \dot{q}_{1*})\dot{q}_{1*} + g(q_1) - K_{p_1}\tilde{q}_1 - K_{d_1}\vartheta_1] + q_{1*} \tag{4.39}$$

which corresponds to the controller of [163].

Notice that the calculation of u_f in (4.38) requires $\ddot{q}_{2d}$ however, in contrast with other solutions to this problem, this controller does not require the calculation of $q_1^{(3)}$. This stems from the use of $\dot{q}_{1d}$ instead of $\dot{q}_1$ in the second right hand term of q_{2d}, and the use of the filter. The second derivative of q_{2d} still needs link acceleration and velocity. Yet, only link velocity is considered to be available for measurement and acceleration can be computed using the first equation of (4.18).

Sketch of proof of Proposition 4.12. We briefly give below the main guidelines of the proof of the proposition above. For more detail we invite the reader to see [161].

First, we notice that Property **P2.2** holds true also for C_d. Now, consider the storage function

$$\mathcal{H}(t, x) = \frac{1}{2}\dot{\tilde{q}}^{\top}\mathcal{D}\dot{\tilde{q}} + \frac{1}{2}\tilde{q}^{\top}(\mathcal{K}_p + \mathcal{K})\tilde{q} + \frac{1}{2}\vartheta^{\top}\mathcal{K}_d B^{-1}\vartheta + \varepsilon\tilde{q}^{\top}\mathcal{D}\dot{\tilde{q}} - \varepsilon\vartheta^{\top}\mathcal{D}\dot{\tilde{q}} \tag{4.40}$$

which is positive definite for sufficiently small values of $\varepsilon > 0$. In a compact form and after a long but straight forward bounding it can be proven that for sufficiently large control gains (and small ε) there exist some positive constants γ_1, γ_2, γ_3 such that the time derivative of function $\mathcal{H}$ above satisfies

$$\dot{\mathcal{H}}(t, x) \leq -\varepsilon\gamma_1 \|\tilde{q}\|^2 - \varepsilon[\gamma_2 - k_c(\|\vartheta\| + \|\tilde{q}\|)] \|\dot{\tilde{q}}\|^2 - \gamma_3\vartheta^2 + \Psi^{\top}[\varepsilon(\tilde{q} - \vartheta) + \dot{\tilde{q}}]. \tag{4.41}$$

Notice to this point that $\dot{\mathcal{H}}$ above is locally negative definite, that is, if $(\|\vartheta\| + \|\tilde{q}\|) \leq \gamma_2/k_c$ which is satisfied if (4.37) holds with $c_2 = \gamma_2/2k_c$. To this point, Lyapunov stability immediately follows from standard theorems by setting $\Psi \equiv 0$ and noticing that $V(t,x) \triangleq \mathcal{H}(t,x)$ qualifies as a Lyapunov function.

Further, under the conditions above and after the completion of some squares one can find a constant $\beta > 0$ such that

$$\dot{\mathcal{H}}(t,x) \leq -\beta \left\| [\varepsilon(\tilde{q} - \vartheta) + \dot{\tilde{q}}] \right\|^2 + \Psi^\top [\varepsilon(\tilde{q} - \vartheta) + \dot{\tilde{q}}]$$

from which the OSP property follows.   ■

It is important to remark however, that in the computation of the constant β we have assumed that (4.37) holds, from which we deduce the local character of the passivity property. Interestingly enough, one can prove that by applying high control gains, one can enlarge the constant γ_2 so that the stability and passivity properties become semiglobal.

4 Output feedback of fully-actuated systems

In this final section we address the following problem:

Definition 4.13 (Output feedback tracking control) *For the system (4.1) assume that only generalized positions is available for measurement. Under this condition, define an internally stable (smooth) control law (whose gains may depend on the systems initial conditions) that ensures (4.2) for all $q_d \in \mathcal{C}^2$, $\|q_d(t)\|$, $\|\dot{q}_d(t)\|$, $\|\ddot{q}_d(t)\| < B_d$.*

The output (i.e. position) feedback tracking control problem stated above has attracted a lot of attention in the robotics literature during the last decade. From a practical point of view, the problem is clearly motivated due to the noisy velocity measurements, while from a purely theoretical perspective the problem of proving *global* uniform asymptotic stability is very challenging. This problem, as far as we know continues open.

As in the regulation control problem, a natural approach is to design an observer that makes use of position information to reconstruct the velocity signal. Then, the controller is implemented replacing the velocity measurement by its estimate. As far as we know, one of the first results in this direction was reported in [194] where it was proved that a nonlinear observer asymptotically reproduces the whole robot dynamics, in a PD plus gravity compensation scheme. The authors prove the equilibrium is *locally* asymptotically stable provided the observer gain satisfies some lower bound determined by the robot parameters and the trajectories error norms. Later, in [21] the authors presented a systematic procedure that exploits the passivity properties of robot manipulators into the design of controller-observer systems. Local asymptotic stability was proved for sufficiently high gains.

In [163], based on a computed torque plus PD+ controller first appeared in [283], we added the n-th order "approximative differentiation filter" studied in the previous chapter, to eliminate the necessity of velocity measurements. In that paper we proved semiglobal asymptotic stability of the closed loop system hence showing that the domain of attraction can be arbitrarily enlarged by increasing the filter gain. Some more recent and stronger results addressing the same problem are for instance: [155], and [196]. Lim *et al* proposed the first adaptive controller for flexible joint robots by using only position measurements. In [196] the authors proposed a globally asymptotically stable observer-based controller needing only link position feedback.

In this section we briefly present from a passivity point of view the position feedback PD+ controller of [163] which is an extension to the output feedback case, of the controller of Proposition 4.12 and to tracking control of the EL controller by approximate differentiation of Section **3.2.4.B.2**.

4.1   Semiglobal tracking control of robot manipulators

We start by recalling the passivity property of system (4.33)-(4.35). Notice that the equivalent of this system for the fully-actuated case (i.e. u, $q \in \mathbb{R}^n$) is

$$D\ddot{\tilde{q}} + (C + C_d)\dot{\tilde{q}} + K_p\tilde{q} + K_d\vartheta + K\tilde{q} = \Psi \tag{4.42}$$

$$\dot{q}_c = -A(q_c + B\tilde{q}) \tag{4.43}$$

$$\vartheta = q_c + B\tilde{q} \tag{4.44}$$

where $C_d = C(q, \dot{q}_*)\dot{q}_*$ and as before, all matrices A, B, K_p, K_d are diagonal and positive definite. Moreover it is assumed that $BD(q) + D(q)B > 0$ for all $q \in \mathbb{R}^n$.

Similar to the case when joint flexibilities cannot be neglected, a similar passivity property to that established in Proposition 4.12 holds for this system. We repeat this below for convenience.

Proposition 4.14 (Tracking without velocity measurements.) *The system (4.42)-(4.44) defines a locally output strictly passive operator* $\Psi \mapsto \varepsilon[\dot{\tilde{q}} - \vartheta] + \dot{\tilde{q}}$, *where* Ψ, $q \in \mathbb{R}^n$. *Furthermore if* $\Psi \equiv 0$ *then system (4.33) is semi-globally exponentially stable with a domain of attraction including the set*

$$\{x \in \mathbb{R}^{3n} : \|x\| < c_2\}$$

where we redefined $x \stackrel{\triangle}{=} [\tilde{q}^\top, \dot{\tilde{q}}^\top, \vartheta^\top]^\top$. □

Considering the passivity result established above the stabilizing passivity-based controller is constructed by setting $\Psi = 0$ and comparing equation (4.33) and (4.18).

This yields

$$u = D(q)\ddot{q}_{1*} + C(q, \dot{q}_*)\dot{q}_* + g(q) - K_p\tilde{q} - K_d\vartheta \tag{4.45}$$

which corresponds to the controller of [163]. Notice also that u above corresponds to the terms in brackets in the definition of q_{2d} in (4.39).

The proof of the proposition above follows along the same lines as the proof of Proposition 4.12 using the storage function

$$\mathcal{H}(t, x) = \frac{1}{2}\dot{\tilde{q}}^\top D\dot{\tilde{q}} + \frac{1}{2}\tilde{q}^\top K_p\tilde{q} + \frac{1}{2}\vartheta^\top K_d B^{-1}\vartheta + \varepsilon\tilde{q}^\top D\dot{\tilde{q}} - \varepsilon\vartheta^\top D\dot{\tilde{q}} \tag{4.46}$$

which is positive definite for a sufficiently small $\varepsilon > 0$. Then observing that its time derivative satisfies a bound like (4.41).

It is important to remark in the latter inequality the cubic terms $(\|\vartheta\|^2 + \|\tilde{q}\|) \|\dot{\tilde{q}}\|^2$ which are similar to those encountered in inequality (3.52). Clearly, as in the regulation problem of Section **3**.4.5 case the presence of these cubic terms in the time derivative of $\mathcal{H}$ is a major technical obstacle to claim OSP in the usual (global) sense, and therefore *global* asymptotic stability.

This is a common drawback of numerous articles in the literature of robot control. Even though it is beyond the scope to do an exhaustive review of the literature (see for instance [21]) we briefly discuss below some of the latest attempts in solving the global output feedback tracking control problem.

4.2   Discussion on global tracking

As an attempt to bound the cubic terms in the time derivative of the storage function we presented in [160] as far as we know, the first smooth controller which renders the *one-degree-of-freedom* (dof) EL system *globally uniformly asymptotically stable*. Our approach relies on a computed torque plus PD structure and a nonlinear dynamic extension based on the *linear* approximate differentiation filter. The main innovation in our controller, which allows us to give explicit lower bounds for the controller gains, in order to ensure GUAS, is the use of hyperbolic trigonometric functions in a Lyapunov function with cross terms.

Global uniform asymptotic stability is ensured provided the controller and filter gains satisfy some lower bound depending on the system parameters and the reference trajectory norm. Unfortunately, the performance of our approach can be ensured only for one dof systems and nothing can be claimed for the general multivariable case.

Independently, in [43] Burkov showed by using singular perturbation techniques, that a computed torque like controller plus a linear observer is capable of making a rigid joint robot track a trajectory starting from *any* initial conditions. The main drawback of this result is that no explicit bounds for the gains can be given. Thus, the author proves in an elegant way, the *existence* of an output feedback tracking controller that ensures GUAS.

Later, A.A.J. Lefeber proposed in [151] an approach which consists on applying a global output feedback *set point* control law (for instance an EL controller) from the initial time t_0 till some "switching time" t_s, at which it is supposed that the trajectories are contained in some pre-specified bounded set. At time t_s one switches to a local output feedback tracking control law (such as any among those mentioned above). The obvious drawback of this idea is that the controller is no longer smooth, furthermore, the switching time may depend on bounds on the *unmeasured* variables. The results contained in [151] concern the *existence* of the time instant t_s such that the closed loop system is GUAS.

Most recently, the authors of [294] proposed an apparent extension of the controller of [160] to the multivariable case. The Lyapunov stability proof is carried out relying on a nonlinear change of coordinates (See Eqs. 35 and 39 of that reference). Unfortunately this diffeomorphism is *not* invertible and therefore the controller the authors propose in [294] is *not* implementable without velocity measurements.

Last but not least in this brief review, we mention [23] where the author gives an elegant alternative result for one-degree-of-freedom systems. The controller proposed in [23] is based upon a *global* nonlinear change of coordinates which makes the system *affine* in the unmeasured velocities. This is crucial to define a very simple controller which has at most *linear* growth in the state variables, as a matter of fact the proposed controller is of a PD+ type. This must be contrasted with the *exponential* growth of the control law proposed in [160], due to the use of hyperbolic trigonometric functions. Hence, from a practical point of view, the controller of [23] supersedes by far that of [160]. Unfortunately, the nonlinear change of coordinates proposed in [23] does not hold true for n-degrees-of-freedom systems.

As far as we know the position tracking control problem stated at the beginning of this section for *any* initial conditions and for n-degrees-of-freedom systems remains open.

5   Simulation results

Using SIMULINK™ of MATLAB™ we tested the control algorithms of Propositions 4.12 and 4.14 in the two link robot arm model of [21], see also p. 59. In Fig. 4.1 we show the first link trajectory as well as the reference. In this case we fixed the controller gains to $K_p =$ diag([5000 6000]) and $K_d =$ diag([7000 8800]) and the filter parameters to $A =$ diag([1000 1000]) and $B =$ diag([1000 1000]). The followed reference in both cases is $q_{1*} = [\frac{1}{10\pi}Sin(10\pi t) \; ; \; \frac{1}{10\pi}Sin(10\pi t)]$. Fig. 4.2 shows the response for the case when flexibility in the joints cannot be neglected, for which we have set to K =diag([10000 10000]). In this case we have set K_{p_j} =diag([10000 10000]) K_{d_j}=diag([7000 8800]) and we have considered actuator inertias to be J =diag([0.10 0.10]).

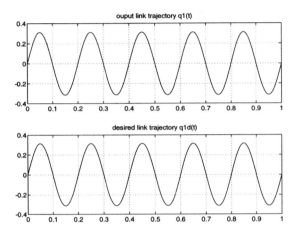

Figure 4.1: Output feedback of rigid-joint robots.

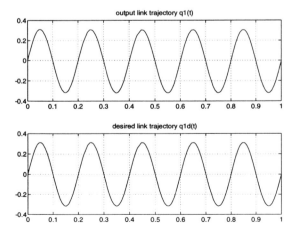

Figure 4.2: State feedback of flexible-joint robots.

6 Concluding remarks

We have shown in this chapter how to modify the passivity-based method, originally developed for regulation, to solve trajectory tracking problems. While in the former case one simply reshapes the potential energy of the EL plant, for tracking the desired storage function must also contain a suitably shaped "kinetic energy term". The key technical result to achieve this end is the passive error dynamics Lemma 2.7, which essentially allows us to treat the EL system as a double integrator. A second fundamental idea is the ingenious inclusion of the s variable[3] of [251], which allows us to conclude convergence of the tracking errors from square integrability of s. Of course, all this is possible because of the linearity with respect to s of the error dynamics in EL systems.

We have revisited from a passivity point of view some well known results in state feedback tracking control of *fully-actuated* mechanical systems such as the PD+ scheme [216] and the Slotine and Li algorithm [251] of rigid robots. We have seen, in particular, how to derive this popular scheme from the error dynamics Lemma 2.7. The idea behind this procedure is to render a closed loop map $\Psi \mapsto s$ output strictly passive. Setting $\Psi \equiv 0$ defines the known–parameter controller, while $\Psi \equiv \Phi(q, \dot{q})\tilde{\theta}$, with $\Phi(q, \dot{q})$ a regressor and $\tilde{\theta}$ the parameter error, results in the adaptive case. Strict passivity of the map $\Psi \mapsto s$ ensures that a standard gradient algorithm, which in its turn defines a passive map $-s \mapsto \Psi$, preserves the strict passivity (and consequently the $\mathcal{L}_2$ stability) of the overall scheme. The proof of convergence follows from Lemma A.12 of Appendix **A** which shows that $s \in \mathcal{L}_2 \Rightarrow \tilde{q} \to 0$.

The extension of PBC to *underactuated* EL systems is illustrated with the example of robots with flexible joints. The results are restricted to the simplified model with block diagonal inertia matrix. The main stumbling block to extend these results to the complete model is the unavailability (without differentiation) of higher order derivatives of the unactuated coordinates, which is related with the inability of rendering the system triangular and to decompose it into the feedback interconnection of passive subsystems of Proposition 2.10. This is a very interesting *open problem* that apparently requires the development of new mathematical machinery.

An output feedback tracking controller which is the natural extension of the approximate differentiation EL controller for regulation, was presented from a passivity point of view. Finally it should be remarked that in spite of recent significant advances, the *global* tracking control problem by output feedback for n degrees-of-freedom systems, remains open.

---

[3]Notice that by taking s as the output of the operator we reduce its relative degree to one.

Chapter 5

Adaptive disturbance attenuation: Friction compensation

> *"I don't know if our scientific equations correspond with reality, because I don't know what reality is. All that matters to me is that theory predicts the results obtained by the measurements."*
>
> *S. Hawking.*

[1]In this chapter we illustrate one further advantage of PBC: the possibility of attenuating the effect of bounded external disturbance via high–gain feedback. Roughly speaking, this feature stems from the fundamental property of infinite gain margin of passive maps, hence stability is preserved when placed in closed–loop with high–gain operators. The simplest application of this principle is in sliding mode control, where a passifiable (e.g., minimum phase and relative degree one) system is controlled with a relay, which defines an operator with infinite gain, albeit passive. Stability is then preserved because of the fundamental property of passivity being invariant under feedback interconnection.

To illustrate this important feature of PBC we first present a novel adaptive friction compensator based on a dynamic model recently proposed in the literature. The compensator ensures global position tracking when applied to an n degree of freedom robot manipulator perturbed by friction forces with only measurements of position and velocity, and *all* the system parameters (robot and friction model) unknown. Instrumental for the solution of the problem is the observation that friction compensation can be recasted as a disturbance rejection problem as pointed out above. The control signal is then designed in two steps, first we (strictly) passify the system with the adaptive robot controller of Slotine and Li studied in Section 4.2, and then add a relay–based outer–loop that rejects the disturbance.

---

[1]The material reported in this chapter is based on work done in collaboration with Elena Panteley and Magnus Gäfvert.

We finish the chapter with a more general theorem applicable to a broader class of passifiable state–space systems.

1   Adaptive friction compensation

It is well known that one of the major limitations to achieve good performance in mechanical systems is the presence of friction, which is a nonlinear phenomenon difficult to describe analytically. Different models have been proposed to capture this phenomenon, starting with the classical descriptions via static maps between velocity and friction force. However, it appears that classical models are unable to capture some of the behaviour experimentally observed in systems subject to strong friction effects. It is argued in [11] that *dynamic* models are necessary to describe the friction phenomena accurately. For a review of some of the existing dynamic friction models see [199], [11] and references cited therein.

Inspired by the works of [99] and [26], a new dynamical model of the friction force —based on a bristle deflection interpretation— is proposed in [52], (see also [50]). This model is briefly presented in Section 1.1 of this chapter. The friction *compensation* problem with this model is stymied by the fact that the parameters are uncertain and some of the state variables are inaccessible. Further, the structure of the equations is such that the existing techniques for adaptive output feedback global tracking —based on transformations to special forms, see e.g., [174], [142]— are (apparently) inapplicable. In this chapter we present an adaptive global tracking controller for robot manipulators perturbed by friction forces represented by this dynamical model. This scheme was first reported in [219].

The problem of adaptive friction compensation has a very long history that dates at least as far back as [289]. It is presented in [149] as an application example of model reference adaptive control. In [52] the authors treated the problem of friction compensation of a DC motor by assuming that all the friction model parameters were *known*. A further extension of this result was reported in [51] where the friction force is linearly parameterized in terms of one unknown parameter, but all the remaining parameters assumed exactly known. See also [10] where an impulsive model is considered and [8] for some preliminary results on Coulomb friction adaptive compensation. The controller we present here extends these results in several directions. First, we treat the general case of position *tracking* of an n degree of freedom (dof) robot manipulator with only position and velocity measurable. Second, we consider all the parameters of the system (friction and robot) to be unknown. Besides its theoretical significance, the proposed controller is of practical importance since the friction compensation is achieved with a very simple adaptive law.

For the sake of clarity we present first the result for the DC motor in Section 1.2, and then extend it in Section 1.3 to the general n–dof robot case. To illustrate the performance of the proposed controller we present some simulations corresponding to the experimental setup studied in [199], [51] and [90] in Section 1.4. A simulation study that compares our controller with the one reported in [4] is also given. In the concluding remarks of Section 3 we mention several immediate extensions of the present work and point out some of the open problems.

1.1 The LuGre friction model

To capture the effect of friction in mechanical systems a bristle–based dynamical model –inspired by the works of [99] and [26]–, was recently proposed in [52] as

$$F = \sigma_0 z + \sigma_1 \dot{z} + \sigma_2 \dot{q} \tag{5.1}$$

$$\dot{z} = -\sigma_0 a(\dot{q}) z + \dot{q} \tag{5.2}$$

$$a(\dot{q}) \triangleq \frac{|\dot{q}|}{\alpha_0 + \alpha_1 e^{-(\frac{\dot{q}}{\alpha_2})^2}} \tag{5.3}$$

where F is the friction force, z is the average deflection of the bristles, $\dot{q}$ is the relative velocity between the surfaces, α_i, $i = 0, 1, 2$ and σ_i, $i = 0, 1, 2$ are some positive coefficients which are typically *unknown*. We should underscore the fact that, besides the parametric uncertainty, in this model neither F nor z are measurable. As pointed out in the concluding remarks this considerably complicates the task of adaptive friction compensation.

Before proceeding with the adaptive controller design we recall an important property of the friction model that will be used in the sequel. As discussed in [52], from physical considerations it is reasonable to assume that the initial bristles deflection is bounded, that is, $|z(0)| \leq \alpha_0 + \alpha_1$. In this way we ensure that it is *uniformly bounded*, namely

$$|z(t)| \leq \Delta \triangleq \frac{1}{\sigma_0}(\alpha_0 + \alpha_1) \tag{5.4}$$

for all $t \geq 0$. This fundamental property will be used throughout the chapter.

It is difficult to assess whether the equations above (or for that matter any other mathematical model) constitute a *bona fide* friction model. Particularly because of the distressing fact that it was not known whether the model above satisfies the fundamental property of defining a passive operator $u \mapsto \dot{q}$. This property captures the dissipation nature of friction and should be reflected in any sensible model of it.[2] See [200] for a detailed discussion on this topic.

We present below the complete answer to this question which was reported in the recent paper [14]. That is, we give *necessary and sufficient* conditions for the

---

[2]It is worth pointing out that the passivity established in [52] pertains to the map $u \mapsto z$, which is not the one of physical interest.

passivity property to hold. The conditions are expressed in terms of a simple algebraic inequality involving the parameters of the model. If this inequality does not hold we construct an input signal that generates a periodic orbit along which the passivity inequality is violated. We present below the proof of sufficiency. The necessity part of the proof, being quite technical, is omitted and we refer the interested reader to [14].

Proposition 5.1 (Passivity of the friction dynamic system) *The dynamical system (5.1)–(5.3) defines a passive operator* $\Sigma : \mathcal{L}_{2e} \to \mathcal{L}_{2e} : \dot{q} \mapsto F$ *if and only if*

$$\sigma_2 - \frac{\sigma_1 \alpha_1}{\alpha_0} > 0 \tag{5.5}$$

$\square$

Proof. *(Sufficiency)*

We will prove that if (5.5) holds, then along the solutions of (5.1)–(5.3) with zero initial conditions we have

$$I(0,T) = \int_0^T \dot{q} F dt \geq 0 \tag{5.6}$$

for all $\dot{q} \in \mathcal{L}_{2e}$ and all $T \geq 0$. First, it is clear from (5.3) that

$$\frac{a(\dot{q})}{|\dot{q}|} \in \left(\frac{1}{\alpha_0 + \alpha_1}, \frac{1}{\alpha_0} \right) \tag{5.7}$$

for all $\dot{q} \neq 0$. Now, we will evaluate (5.6) splitting it into two terms $I = I_1 + I_2$ with

$$I_1(0,T) \triangleq \int_0^T \dot{q}(\sigma_1 \dot{z} + \sigma_2 \dot{q}) dt$$

and

$$I_2(0,T) \triangleq \sigma_0 \int_0^T z\dot{q} dt$$

Replacing $\dot{q}$ from (5.2) in I_2 we get

$$
\begin{aligned}
I_2(0,T) &= \sigma_0 \int_0^T z[\dot{z} + \sigma_0 a(\dot{q})z] dt \\
&= \frac{\sigma_0}{2} z^2(T) + \sigma_0^2 \int_0^T z^2 a(\dot{q}) dt \geq 0
\end{aligned}
$$

For the other term we replace $\dot{z}$ from (5.2) and use the bounds (5.4), (5.7) to get

$$
\begin{aligned}
I_1(0,T) &= \int_0^T \dot{q}[-\sigma_1\sigma_0 a(\dot{q})z + (\sigma_1 + \sigma_2)\dot{q}]dt \\
&\geq \int_0^t \dot{q}^2(-\sigma_1\sigma_0 \cdot \frac{1}{\alpha_0} \cdot \frac{\alpha_0 + \alpha_1}{\sigma_0} + \sigma_1 + \sigma_2)dt \\
&= \int_0^T \dot{q}^2(\sigma_2 - \frac{\sigma_1\alpha_1}{\alpha_0})dt \geq 0
\end{aligned}
$$

where we have used (5.5) to establish the last inequality. This completes the proof of sufficiency. ∎

Remark 5.2 In [199] the author presents some results pertaining to the passivity property of the friction model discussed above. Motivated by the inability to conclusively establish passivity, it is proposed to make the damping coefficient σ_1 decrease with increasing velocity. Physically this is motivated by the change of damping characteristics as velocity increases, due to more lubricant being forced into the interface. With this modification it is possible to prove passivity of the operator.

1.2 DC motor with friction

In this paragraph we consider a DC motor

$$J\ddot{q} = u - F \tag{5.8}$$

perturbed by a friction torque modeled by (5.1)–(5.3). We assume that position and velocity are available for measurement and consider the case when the rotor inertia J and all the parameters of the friction model are *unknown*. Our objective is to design an adaptive friction compensator that ensure global asymptotic position tracking of a given reference signal q_*.

To derive the solution to the problem, we will first rearrange the friction model (5.1)–(5.3) in such a way that the friction torque can be treated as a disturbance. The control signal is then designed in two steps, first we use in an inner–loop a classical adaptive controller for the DC motor that (strictly) passifies the system, and then we add a relay–based outer–loop that rejects this disturbance. The amplitude of the relay is adjusted on–line via a suitably designed parameter estimator.

Towards this end, let us substitute (5.2) in (5.1)

$$
\begin{aligned}
F &= (\sigma_1 + \sigma_2)\dot{q} + \sigma_0 z - \sigma_0\sigma_1 a(\dot{q})z \\
&\triangleq (\sigma_1 + \sigma_2)\dot{q} + F_z
\end{aligned}
$$

where, for convenience, we have separated the viscous friction forces from the remaining terms which we denote F_z. We now design a Slotine–Li adaptive controller (like the one presented in Section 4.2) for the motor model with viscous friction as

$$u = -K_p s - \hat{\theta}_{il}^\top \phi_{il} + u_{ol} \tag{5.9}$$

where $K_p > 0$, u_{ol} is an outer–loop signal that will take care of the remaining friction terms, and

$$
\begin{aligned}
s &= \dot{\tilde{q}} + \lambda \tilde{q} \\
\tilde{q} &= q - q_*
\end{aligned}
\tag{5.10}
$$

with $\lambda > 0$. The regressor is, as usual, defined as

$$\phi_{il} \triangleq \left[\begin{array}{c} -\ddot{q}_* + \lambda \dot{\tilde{q}} \\ -\dot{q} \end{array} \right]$$

and the parameters $\hat{\theta}_{il} \in \mathbb{R}^2$ are updated as follows

$$\dot{\hat{\theta}}_{il} = \Gamma_{il} \phi_{il} s$$

with $\Gamma_{il} = \Gamma_{il}^\top > 0$. If we now define the parameter error vector $\tilde{\theta}_{il} \triangleq \hat{\theta}_{il} - [J, \sigma_1 + \sigma_2]^\top$ and replace (5.9) in (5.8) we get

$$J\dot{s} + K_p s = -\tilde{\theta}_{il}^\top \phi_{il} + u_{ol} - F_z \tag{5.11}$$

Let us first look at the input–output properties of the inner loop. It is clear that the transfer function $\frac{1}{Jp+K_p}$ is SPR, hence is OSP in view of the Kalman–Yakubovich–Popov lemma. On the other hand, as we have seen before the operator $s \mapsto \tilde{\theta}_{il}^\top \phi_{il}$, defined by the parameter update law, is passive. Consequently the operator $(u_{ol} - F_z) \mapsto s$ is OSP.

We will see now how we can exploit this property to design the outer–loop part of the scheme. To this end we notice that F_z can be bounded as

$$|F_z(\dot{q}, z)| \triangleq |[\sigma_0 - \sigma_0 \sigma_1 a(\dot{q})]z| \leq \Delta + \frac{\Delta}{\alpha_0} \sigma_1 |\dot{q}| \tag{5.12}$$

where we have used the property (5.4) and the inequality $a(\dot{q}) \leq \frac{1}{\alpha_0} |\dot{q}|$, which follows from (5.3). Notice that the bound (5.12) is *independent* of z.

The bound above suggests to close the outer–loop with a relay with adjustable amplitude, which we parameterize as

$$u_{ol} = -\hat{\theta}_{ol}^\top \phi_{ol}$$

where $\phi_{ol} \triangleq \text{sgn}(s)[1, \, |\dot{q}|]^\top$. Defining the parameter error vector $\tilde{\theta}_{ol} \triangleq \hat{\theta}_{ol} - \Delta[1, \frac{\sigma_1}{\alpha_0}]^\top$ we obtain the error equation

$$J\dot{s} + K_p s = -\tilde{\theta}_{il}^\top \phi_{il} - \tilde{\theta}_{ol}^\top \phi_{ol} - F_z - \Delta \text{sgn}(s) - \frac{\sigma_1 \Delta}{\alpha_0} |\dot{q}| \text{sgn}(s)$$

A suitable choice for the estimator is then

$$\dot{\theta}_{ol} = \Gamma_{ol}\phi_{ol}s$$

with $\Gamma_{ol} = \Gamma_{ol}^{\mathsf{T}} > 0$. To see this consider the Lyapunov function candidate reported in [254]

$$V = \frac{1}{2}(Js^2 + 2\lambda K_p\tilde{q}^2 + \tilde{\theta}^{\mathsf{T}}\Gamma^{-1}\tilde{\theta})$$

where $\tilde{\theta} \triangleq [\tilde{\theta}_{il}^{\mathsf{T}}, \tilde{\theta}_{ol}^{\mathsf{T}}]^{\mathsf{T}}$, $\phi = [\phi_{il}^{\mathsf{T}}, \phi_{ol}^{\mathsf{T}}]^{\mathsf{T}}$, $\Gamma \triangleq \text{blockdiag}\{\Gamma_{il}, \Gamma_{ol}\}$. Taking the derivative along the trajectories of the closed loop system we get

$$\begin{aligned}
\dot{V} &= -K_p s^2 + 2\lambda K_p \tilde{q}\dot{\tilde{q}} - F_z s - \Delta|s| - \frac{\sigma_1\Delta}{\alpha_0}|\dot{q}||s| \\
&\leq -K_p \dot{\tilde{q}}^2 - \lambda^2 K_p \tilde{q}^2
\end{aligned}$$

where we have used the bound (5.12) to get the last inequality. From which we conclude that $\tilde{q}$, $\dot{\tilde{q}}$ are square integrable, which implies, that $\tilde{q} \to 0$ as $t \to \infty$.

We have thus established the following result:

Proposition 5.3 *Consider the model of a DC motor (5.8) with friction modeled by (5.1)–(5.3). Let the adaptive control law be given by*

$$u = -K_p s - \hat{\theta}^{\mathsf{T}}\phi$$

with (5.10), and the regressor defined as

$$\phi \triangleq \begin{bmatrix} -\ddot{q}_* + \lambda\dot{\tilde{q}} \\ -\dot{q} \\ \text{sgn}(s) \\ |\dot{q}|\,\text{sgn}(s) \end{bmatrix}$$

where λ, K_p are positive numbers, and the parameters $\hat{\theta} \in \mathbb{R}^4$ are updated as follows

$$\dot{\hat{\theta}} = \Gamma\phi s, \quad \Gamma = \Gamma^{\mathsf{T}} > 0$$

Then, the closed loop system ensures global asymptotic position tracking, that is,

$$\lim_{t\to\infty} |\tilde{q}(t)| = 0$$

$\square$

1.3   Robot manipulator

In this section we extend the previous result to the case of an n–dof rigid robot manipulator described by

$$D(q)\ddot{q} + C(q, \dot{q})\dot{q} + g(q) = u - F \tag{5.13}$$

where $q \in \mathbb{R}^n$ is the vector of joint angles, $D(q)$ is the inertia matrix, $C(q, \dot{q})$ is the matrix of Coriolis and centrifugal forces, $g(q)$ is the gravity, $u \in \mathbb{R}^n$ are the control torques, and $F = [F_1, \ldots, F_n]^\top$ are friction forces acting independently in each joint as

$$\begin{aligned} F_j &= \sigma_{0j}z_j + \sigma_{1j}\dot{z}_j + \sigma_{2j}\dot{q}_j \\ \dot{z}_j &= -\sigma_{0j}a_j(\dot{q}_j)z_j + \dot{q}_j \end{aligned} \tag{5.14}$$

$$a_j(\dot{q}_j) \triangleq \frac{|\dot{q}_j|}{\alpha_{0j} + \alpha_{1j}e^{-(\dot{q}_j/\alpha_{2j})^2}} \tag{5.15}$$

where $j \in \bar{n} \triangleq \{1, \ldots, n\}$, and all the remaining variables are as described above.

Again, proceeding from physical considerations, we assume that the initial bristles deflection are bounded, that is,

$$|z_j(0)| \leq \Delta_j \triangleq \alpha_{0j} + \alpha_{1j}. \tag{5.16}$$

In this way we can ensure the fundamental property that they are *uniformly bounded*, namely that $|z_j(t)| \leq \frac{1}{\sigma_{0j}}(\alpha_{0j} + \alpha_{1j})$, $j \in \underline{n}$ for all $t \geq 0$.

We will use the well–known properties mentioned in Section **2.2.3**. Namely that $D(q)$ is bounded from above and below, that $\dot{D}(q) - 2C(q, \dot{q})$ is skew-symmetric and that the robot model (5.13) can be parameterized as

$$D(q)\dot{w} + C(q, \dot{q})w + g(q) = \Phi_1(q, \dot{q}, w, \dot{w})\theta_1 \tag{5.17}$$

where $\theta_1 \in \mathbb{R}^l$ contains the unknown parameters of the manipulator and the regressor matrix $\Phi_1(q, \dot{q}, w, \dot{w})$ contains known functions.

The control problem can be now formulated as follows:

Definition 5.4 *Assume that q and $\dot{q}$ are measurable, the initial bristles deflections are bounded as (5.16), and the parameters of the friction model α_{ij}, σ_{ij}, $j \in \bar{n}$, $i = 0, 1, 2$, and the robot θ_1 are unknown. Under these conditions, design an internally stable output feedback global tracking controller that ensures*

$$\lim_{t \to \infty} \|q(t) - q_*(t)\| = 0 \tag{5.18}$$

for any twice differentiable bounded reference $q_*(t)$ with known bounded first and second order derivatives.

To derive the solution to this problem we will proceed as in the case of the DC motor, that is, first rearrange the friction model (5.14)–(5.15) in such a way that the friction force can be treated as a disturbance, and then adopt a cascaded configuration based on the passivity and gain margin considerations explained above. The decomposition of F mimics the one above, that is,

$$F = F_z(\dot{q}, z) + \Phi_2(\dot{q})\theta_2$$

where we have defined a parameterization

$$\text{diag}\{(\sigma_{1j} + \sigma_{2j})\}\dot{q} = \Phi_2(\dot{q})\theta_2 \tag{5.19}$$

where $\theta_2 \stackrel{\triangle}{=} [(\sigma_{11} + \sigma_{21}), \dots, (\sigma_{1n} + \sigma_{2n})]^\top$ and $\Phi_2(\dot{q}) \stackrel{\triangle}{=} \text{diag}\{\dot{q}_i\}$, and the elements of $F_z(\dot{q}, z)$ satisfy

$$|F_{z_j}(\dot{q}, z)| \stackrel{\triangle}{=} |[\sigma_{0j} - \sigma_{0j}\sigma_{1j}a(\dot{q}_j)]z_j| \le \Delta_j + \frac{\Delta_j}{\alpha_{0j}}\sigma_{1j}|\dot{q}_j|$$

Hence, F will be treated as a (linearly bounded) disturbance.

The design of the control law follows *verbatim* the DC motor case, and is summarized in the next proposition.

Proposition 5.5 *Consider the robot model (5.13)–(5.15) with (5.16). Let the adaptive control law be defined as*

$$u = -K_d s + \Phi\hat{\theta}$$

with (5.10) and Λ, K_d diagonal positive definite matrices. The regressor matrix is given by

$$\Phi \stackrel{\triangle}{=} [\Phi_1(q, \dot{q}, q_r, \dot{q}_r), \ \Phi_2(\dot{q}), \ \Phi_3(\dot{q}, s)] \in \mathbb{R}^{n \times (l+3n)}$$

with $q_r \stackrel{\triangle}{=} \dot{q}_* - \Lambda\tilde{q}$, (5.17), (5.19) and

$$\Phi_3(s, \dot{q}) \stackrel{\triangle}{=} -[\text{diag}\{\text{sgn}(s_i)\}, \text{diag}\{|\dot{q}_i|\text{sgn}(s_i)\}]$$

The parameters $\hat{\theta} \in \mathbb{R}^{l+3n}$ are updated as

$$\dot{\hat{\theta}} = \Gamma\Phi^\top s, \ \Gamma = \Gamma^\top > 0.$$

Then, the closed loop system ensures the global asymptotic position tracking objective (5.18) with internal stability. □

Proof. The proof considers the Lyapunov function [254]

$$V = \frac{1}{2}\left[s^\top D(q)s + 2\tilde{q}^\top\Lambda K_d\tilde{q} + \tilde{\theta}^\top\Gamma^{-1}\tilde{\theta}\right]$$

with $\tilde{\theta} = \hat{\theta} - \theta$ and θ defined in an obvious manner. Following the arguments above we can prove that

$$\dot{V} \leq -\dot{\tilde{q}}^{\top} K_d \dot{\tilde{q}} - \tilde{q}^{\top} \Lambda^{\top} K_d \Lambda \tilde{q}$$

From this inequality we conclude that all signals are bounded and $\tilde{q}$, $\dot{\tilde{q}}$ are square integrable. The latter implies, that $\tilde{q} \to 0$ as $t \to \infty$. ∎

Remark 5.6 It is interesting to underscore that

$$\Phi_3(\dot{q}, s)s = -\left[I_n, \operatorname{diag}\{|\dot{q}_i|\}\right]^{\top} \left[|s_1|, \ldots, |s_n|\right]^{\top}$$

which reveals –as in the DC motor case– the relay–based structure of the compensator.

1.4 Simulations

An experimental setup consisting of a DC motor connected to a gear box with significant friction is installed in the Laboratoire d'Automatique de Grenoble, see [199] for a detailed description. The identified parameters for the motor (5.8) with friction modeled by (5.1)–(5.3) are

$$[J, \ \sigma_0, \ \sigma_1, \ \sigma_2, \ \alpha_0, \ \alpha_1, \ \alpha_2]^{\top} = [0.0025, 280, 1, 0.017, 0.22, 0.17, 0.1]^{\top}.$$

In all simulations we consider the sinusoidal position reference $q_* = -\sin t$.

To exhibit the effect of friction we first closed the loop with a (fixed parameter) controller which neglects friction

$$u = J(\ddot{q}_* - \lambda \dot{\tilde{q}}) - K_p s$$

with $\lambda = K_p = 1$. The behaviour of position and speed, starting from the initial values $[q(0), \dot{q}(0)] = [1, 0]$, are shown in Fig.5.1 and Fig.5.2. Notice the presence of a large tracking error, particularly near the zero speed region. This error can, of course, be reduced with a larger gain K_p but at the cost of larger input signals.

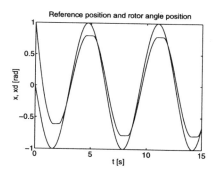

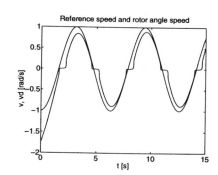

Figure 5.1: Reference position q_* and rotor angle position q *without* friction compensation.

Figure 5.2: Reference speed $\dot{q}_*$ and rotor angle speed $\dot{q}$ *without* friction compensation.

The adaptive controller of Proposition 5.3 was then implemented with $\Gamma = I$, zero initial conditions, and the ideal relay function approximated everywhere by $\tanh(\gamma s)$, where $\gamma = 1000$. The evolution of the relevant signals is shown in Fig. 5.3–Fig. 5.7. A peak of 1.9 Nm in the input torque due to the initial conditions is observed. As expected, eventhough parameter convergence is not achieved, the control signal asymptotically compensates the friction force. Actually, as depicted in Fig. 5.7, the estimates converge to values very far from the real parameters, which are given by

$$\theta = [0.0025, 109, 496.4, 1.1]^\top.$$

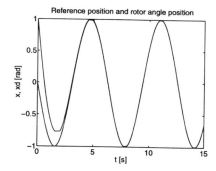

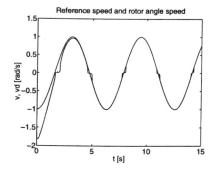

Figure 5.3: Reference position q_* and rotor angle position q *with* adaptive friction compensation.

Figure 5.4: Reference speed $\dot{q}_*$ and rotor angle speed $\dot{q}$ *with* adaptive friction compensation.

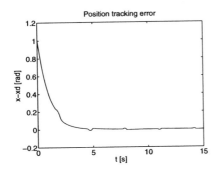

Figure 5.5: Position tracking error $q - q_*$.

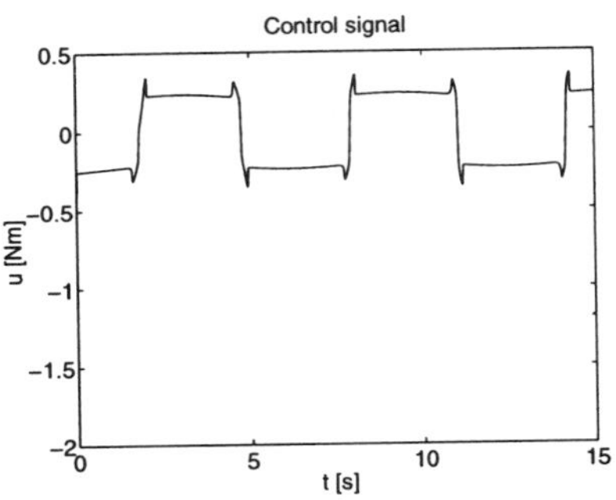

Figure 5.6: Control signal u.

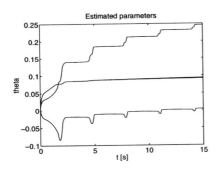

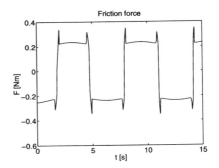

Figure 5.7: Estimated parameters $\hat{\theta}$.

Figure 5.8: Friction force F.

For the sake of comparison we have repeated the same experiment for the controller proposed in [4], with measurable velocity, namely

$$u = J(\ddot{q}_* - \lambda \dot{\tilde{q}}) - K_p s + \hat{F}$$

with $\hat{F}$ given by

$$\hat{F} = J \left(z - k|\dot{q}|^\mu \right) \operatorname{sgn}(\dot{q})$$

and

$$\dot{z} = \frac{1}{J} k\mu |\dot{q}|^{\mu-1} (u - \hat{F}) \operatorname{sgn}(\dot{q}).$$

The behaviour of the tracking error for $k = 1$ and $\mu = 1.5$ is shown in Fig. 5.9. Comparing with Fig. 5.5 we see that the performance of [4] is almost identical to the one of our controller, even though the former is designed for a simple Coulomb friction model. Note however that, while for our controller all parameters are unknown, the controller of [4] requires knowledge of J, whose choice critically affects the transient behaviour.

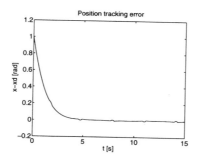

Figure 5.9: Position tracking error $q - q_*$ for the controller of [4].

We have observed that this particular reference is quite benign, in the sense that a simple relay feedback $u = -k\mathrm{sgn}(s)$, with $k > 0$ sufficiently large to overcome the friction, will provide good performance. Of course, a more clever stabilization mechanism is needed if other references are imposed, as would be the case in the robot example.

Remark 5.7 It is interesting to try to check whether the passivity condition (5.5) is satisfied or not in the present application. Unfortunately, the key passivity inequality is not satisfied. Since the difference is of an order of magnitude the violation of the inequality can hardly be attributed to (unavoidable but small) identification errors. In spite of this discrepancy the model seems to provide a good fit for the experimental data. In this respect, and keeping in mind Stephen Hawking's quote, we refer the reader to the experimental evidence reported in [90], [52] and [199].

Also, we hope the analysis presented for the proof of Proposition 5.1 provides some guidelines for the refinement of the basic LuGre model, as proposed in [200]. See also Remark 5.2.

2 State-space passifiable systems with disturbances

2.1 Background

Motivated by the results above and [164], we will present in this section a general theorem concerning the stabilization of nonlinear passifiable systems with time-dependent disturbances. These results have been established in [165].

To establish the connection with the adaptive friction compensation problem, let us recapitulate the main developments of the previous section. First, we use a nested–loop controller configuration, an ubiquitous approach throughout the book, that is substantiated by the fact that passivity is invariant under feedback interconnection. In the present problem of disturbance attenuation, the inner–loop is a PBC which strictly passifies the map $(u_{ol} - F_z) \mapsto s$ defined by the closed loop system (5.11). The outer–loop control is a relay–based operator with input s that quenches, via high-gain, the effect of the disturbances. This outer–loop operator is also passive, consequently passivity (and therefore stability) of the overall system is ensured. This is, of course, the essence of the often–quoted robustness of sliding mode control [269].

We consider nonlinear systems of the form

$$\dot{x} = f(t, x) + g(t, x)[u + d(t, x)] \tag{5.20}$$

where $x \in \mathbb{R}^n$ and we impose that the disturbance $d : \mathbb{R}_{\geq 0} \times \mathbb{R}^n \to \mathbb{R}^n$ satisfies the following assumption:

A 5.1 Each component of the vector function d is uniformly bounded in t by

$$|d_i(t, x)| \leq \theta_{1_i} \delta_i(\|x\|) + \theta_{2_i}, \quad \forall i \leq n \tag{5.21}$$

where θ_{1_i} and θ_{2_i} are *unknown* non-negative constants and $\delta_i(\|x\|) : \mathbb{R}_{\geq 0} \to \mathbb{R}_{\geq 0}$ is a *known* continuous function.

An obvious implication of (5.21) is the existence of a continuous function $\delta : \mathbb{R}_{\geq 0} \to \mathbb{R}_{\geq 0}$ and positive constants θ_1 and θ_2 such that $\|d(t, x)\| \leq \theta_1 \delta(\|x\|) + \theta_2$, or in compact form

$$\|d(t, x)\| \leq \theta^\top \Delta(\|x\|) \tag{5.22}$$

where we have defined $\theta^\top \triangleq (\theta_1, \theta_2)$ and $\Delta(\|x\|)^\top \triangleq (\delta(\|x\|), 1)$. The disturbance $d(t, x)$ may be the result of noisy measurements or parameter uncertainty affecting the plant.

Mimicking the derivations of the previous section,[3] we make the following assumption

A 5.2 A state–feedback controller $u = u^0(t, x)$ and an output function $y \in \mathcal{L}_2^m$, $y = h(t, x)$, where $h \in \mathcal{C}^1$, are known such that the system

$$\dot{x} = f(t, x) + g(t, x)[u^0(t, x) + v(t)] \tag{5.23}$$

defines an OSP operator $\Sigma : v \mapsto y$ with a positive definite and proper storage function $V^0(t, x)$.

---

[3]The same approach will be taken for interconnected systems in Chapter **12**.

We are seeking for an outer–loop state–feedback $v = v(t, x)$ guaranteeing that the trivial solution $x(t) = 0$ of (5.20) with arbitrary initial conditions t_0, $x_0 = x(t_0)$ be globally uniformly convergent, that is,

$$\lim_{t \to \infty} \|x(t)\| = 0$$

for any initial conditions (t_0, x_0).

2.2   A theorem for passifiable affine nonlinear systems

In order to formulate our main result in terms of input-output properties, we need to introduce the following concept.

Definition 5.8 (Strong zero-state detectability.) *The nonlinear system (5.20) with input $v \stackrel{\triangle}{=} u + d$ and output $y = h(t, x)$ where h is a continuous function of its arguments, is said to be strongly zero-state detectable if*

$$\{v(t) \in \mathcal{L}_\infty^n \text{ and } y(t) \equiv 0\} \Rightarrow \lim_{t \to \infty} \|x(t)\| = 0. \tag{5.24}$$

The above definition differs from the one given in Appendix A (see also [46]) in the fact that we need the property to hold for all uniformly bounded inputs.

It is well known, e.g., [287], that a detectability assumption is always needed to transfer input-output properties to state–space properties. Thus it will become clear later, that the condition (5.24) will allow us to prove our main result using *passivity* arguments.

Theorem 5.9 (Robust adaptive stabilization [165].) *Consider the system (5.20), (5.21) under the Assumptions 5.1, 5.2 and*

A 5.3 System (5.23) is strongly zero-state detectable.

Under these conditions, the controller

$$u = u^0(t, x) - \hat{\theta}^\top \Delta(\|x\|) sgn(y) \tag{5.25}$$

$$\dot{\hat{\theta}} = \Gamma \Delta(\|x\|) sgn(y)^\top y \tag{5.26}$$

where $\hat{\theta}$ is the estimate of θ and $\Gamma = \Gamma^\top > 0$, in closed loop with (5.20), (5.21) guarantees that the state $x(t)$ is globally uniformly convergent.   □

Proof.   We start by noticing that since (5.23) defines an OSP operator with storage function $V^0(t, x)$ then there exists a positive constant β such that the time derivative of V^0 along the trajectories of (5.23) satisfies

$$\dot{V}^0(t, x) \leq v^\top y - \beta \|y\|^2 . \tag{5.27}$$

Define

$$v \triangleq d(t,x) - \hat{\theta}^{\top}\Delta(\|x\|)\mathrm{sgn}(y), \qquad (5.28)$$

then the bound (5.27) on the derivative of the *storage function* V^0, coincides with the following bound on time derivative of the *Lyapunov function* V^0 along the closed loop solutions (5.20), (5.25)

$$\dot{V}^0(t,x,\tilde{\theta}) \leq -\beta\|y\|^2 - d^{\top}y - \hat{\theta}^{\top}\Delta(\|x\|)\mathrm{sgn}(y)^{\top}y. \qquad (5.29)$$

Consider on the other hand the Lyapunov function candidate

$$V(t,x,\tilde{\theta}) \triangleq V^0(t,x) + \frac{1}{2}\tilde{\theta}^{\top}\Gamma^{-1}\tilde{\theta}$$

where $\tilde{\theta} \triangleq \hat{\theta} - \theta$, which is clearly positive definite and proper. Using the inequalities (5.29) and (5.22), the time derivative of V along the closed loop solutions $(x(t), \tilde{\theta}(t))$ of system (5.20), (5.25), (5.26) is bounded by

$$\dot{V}(t,x,\tilde{\theta}) \leq -\beta\|y\|^2 - \theta^{\top}\Delta(\|x\|)(|y| - \|y\|) \qquad (5.30)$$

where $|y|$ stands for the Frobenius norm (c.f. Notations). Notice that the subtraction of the last two terms on the right hand side of (5.30) is non-positive hence

$$\dot{V}(t,x,\tilde{\theta}) \leq -\beta\|y\|^2. \qquad (5.31)$$

Inequality (5.31) implies that $V(t)$ is decreasing and hence bounded so the complete state $(x, \tilde{\theta}) \in \mathcal{L}_{\infty}^{n+2}$. Furthermore, integrating (5.31) from t_0 to ∞ one concludes that $y \in \mathcal{L}_2^m$. Next, notice from (5.28), (5.22) that since θ is constant and $\Delta(\|x\|)$ is continuous, $v(t)$ is also uniformly bounded. From the continuity of $h(t,x)$ it follows that $y \in \mathcal{L}_{\infty}^m$. Then, from the closed loop dynamics (5.23), (5.28) we obtain that $\dot{x} \in \mathcal{L}_{\infty}^n$ and since $h \in \mathcal{C}^1$ we conclude using Barbalat's lemma that $y(t) \to 0$ as $t \to \infty$. The result is finally obtained using the assumption of strong zero-state detectability. ∎

Interestingly enough the theorem above applies to non-autonomous non-linear systems. In other words, if we think of the state x as a *tracking* error Theorem 5.9 can be used to design robust tracking controllers such that the tracking error is guaranteed to converge to zero uniformly as $t \to \infty$. It may clarify the reason for introducing condition (5.24) that a proof in the spirit of the proof of Theorem 3.2 of [46] requires that $v \equiv 0$ (see also Lemma 3.2.3 of [272]) which cannot be guaranteed in our case. It may be also clear that our motivation for calling condition (5.24) "strong" zero-state detectability stems from the fact that this condition is more restrictive than the usual definition.

It is also worth remarking that in Theorem 5.9 we have implicitly assumed that all parameters of our system are known, however, notice that the perturbation $d(t,x)$ may also contain terms which come from uncertainties in the model.

More precisely, consider a nonlinear non-autonomous system defined by

$$\begin{aligned} \dot{x} &= f(t, x, \theta) + g(t, x)u \\ y &= h(t, x) \end{aligned} \qquad (5.32)$$

where $x \in \mathbb{R}^n$, and $\theta \in \mathbb{R}^l$ of constant unknown parameters. Using Theorem 5.9 a controller $u = u(t, x)$ guaranteeing that $x(t) \to 0$ as $t \to \infty$ can be designed if the dynamics (5.32) admit the following parameterization

$$\dot{x} = f'(t, x) + g(t, x)[u + d(t, x, \theta)]$$

where the nonlinearity $d(t, x, \theta)$ satisfies

$$\|d(t, x, \theta)\| \le \theta_1' \Delta(\|x\|) + \theta_2' \qquad (5.33)$$

with θ_1' and θ_2' unknown constants, and $\Delta(\|x\|)$ a continuous known function.

Notice furthermore, that the parameters vector θ does *not* need to be constant, as long as it defines continuous uniformly bounded signals, that is, if $\theta(t) \in \mathcal{L}_\infty^l$ and a bound like (5.33) can be determined.

3 Concluding remarks

We have illustrated in this chapter how, via an adaptive high-gain design, it is possible to attenuate the effect of disturbances for passifiable systems. The control task is divided into two steps: 1) A passifying control law is designed for the system without disturbance. 2) A discontinuous control loop, with adaptive gain, is added around the output of the resulting passive map of the first step. A theorem for general passifiable state–space systems is given. The principle is applied to solve the tracking control problem of robot manipulators affected by frictional forces.

The key observation that allows us to design an adaptive friction compensator is that, in view of property (5.16), the effect of the friction force can be "dominated" (with measurable functions), and hence treated as a (linearly bounded) disturbance. This, together with the fact that passivity–based control of robot manipulators ensures that the closed–loop is a passive map, allows us to implement a "high–gain" design that rejects this disturbance.

We should underscore that, even though the resulting controller is *non–smooth*, the stabilization mechanism does not rely on the generation of sliding regimes. Furthermore, it is clear that for all practical purposes the *signum* function can be replaced by a smooth function with minor (quantifiable) performance degradation.

It is interesting to note that even in the simple DC motor case, the problem of global tracking is (apparently) not solvable with the techniques reported in [142],

[174]. Leaving aside the presence of a non–smooth nonlinearity, this stems from the fact that the system, which can be written in state space form as

$$
\dot{\xi} = \begin{bmatrix} 0 & 1 & 0 \\ 0 & -\frac{1}{J}(\sigma_1 + \sigma_2) & -\frac{\sigma_0}{J}(1 - \sigma_1 a(\dot{q})) \\ 0 & 1 & -\sigma_0 a(\dot{q}) \end{bmatrix} \xi + \begin{bmatrix} 0 \\ \frac{1}{J} \\ 0 \end{bmatrix} u
$$

where $\xi \triangleq [q, \dot{q}, z]^\top$, is not transformable into any of the special forms considered in the literature for robust (or adaptive) *output* feedback tracking, see e.g., [174], [142].

Among the possible practically (and theoretically) interesting extensions to this result we have the problems of PD regulation with adaptive friction compensation, the case of robots with flexible joints and actuator dynamics, and the removal of velocity measurement. While the solution to the first three problems follow *mutatis mutandis* from the present analysis, the latter seems to require the development of a radically new approach.

Part II

Electrical systems

Chapter 6

Modeling of switched DC–to–DC power converters

We start with this chapter the second part of the book which is devoted to electrical systems, and in particular to DC–to–DC power converters. The study of these devices constitutes an active area of research and development in both power electronics and control theory. Switched DC–to–DC converters have an ubiquitous variety of industrial and laboratory applications thanks to their reduced cost, simplicity and off-the-shelf availability. This part of the book consists of two chapters. In view of the presence of the switches some new considerations with respect to those made in Chapter **2** and Appendix **B**, are needed to formalize the mathematical modeling. This is done in Chapter **6**, which is fully devoted to modeling and the exploration of the structural properties useful for PBC, which as we will see later, applies *verbatim* for this class of systems. This material is presented in Chapter **7**.

1 Introduction

Modeling of switched regulated DC–to–DC power converters was initiated by the pioneering work of Middlebrook and Ćuk in [182] and [63] in the mid seventies. The area has undergone a wealth of practical and theoretical development as evidenced by the growing list of research monographs and textbooks devoted to the subject, see e.g. [238], [117] where interesting relations between power electronics and control are discussed). The reader is also invited to see the recent text [34] where an extensive bibliography, and a rather complete historical perspective of power electronics, is presented through seminal and cornerstone articles.

In this chapter, both a Lagrangian and a Hamiltonian dynamics approach are used for deriving physically motivated models of switch regulated DC–to–DC power converters. The Lagrangian approach consists in establishing the EL parameters of the circuits associated with each of the topologies corresponding to the two possible positions of the regulating switch. This consideration immediately leads us to realize that some EL parameters remain invariant under the switching action while some others are definitely modified by either, the addition or annihilation of certain quantities. A switched model of the non-invariant parts of the EL parameters can then be proposed by their suitable inclusion through the switch position parameter. This inclusion is carried out in a *consistent* manner so that, under a particular switch position parameter value, the original EL parameters corresponding to the two intervening circuit topologies, are exactly recovered.

The switched EL parameter considerations immediately lead, through the use of the classical Lagrangian dynamics equations, to systems of differential equations with discontinuous right hand sides, describing the actual behavior of the treated converters. The obtained switch-regulated models entirely coincide with the *state models* of DC–to–DC power converters introduced in [63, 182].

The Hamiltonian modeling approach complements and generalizes the Lagrangian approach in various respects. In modeling switched power converters the use of ideal switches is sometimes insufficient due to several practical realization constraints. Also, actual DC–to–DC power converters are synthesized using isolation transformers and sometimes, magnetic couplings are usually purposely introduced in the circuit to achieve desirable smoothing effects. Ideal switches are often replaced by suitable combinations of transistors and diodes. While transistors can effectively behave as ideal switches, diodes have physical characteristics which are not suitable for Lagrangian modeling. Using some recent results contained in [74], we show in this chapter that the Hamiltonian viewpoint leads to a natural modeling methodology in which not only ideal switches have a suitable representation but also diodes, transformers, isolation transformers etc. The theoretical basis for these developments may be found in [178, 273].

Section 2 begins with some general issues about the modeling of switch-regulated EL systems. In that section, the switched models of the traditional DC–to–DC power converters, such as the Boost, the Buck-boost and the Ćuk converters, are derived from the Lagrangian viewpoint. The modeling approaches are shown to be also valid for DC–to–DC power converters with non-ideal switches including parasitics. We also present an extension to the afore-mentioned modeling approaches, which includes the multivariable version of the Boost converter constituted by the cascaded connection of two such devices. The Hamiltonian modeling viewpoint of switched power converters is presented in Section 3. In Section 4 the Lagrangian approach is applied to obtain the average state space models of pulse-width-modulation (PWM) regulated DC–to–DC power converters with ideal switching devices.

2   Lagrangian modeling

In this section, a Lagrangian approach is used for deriving mathematical models of the most commonly found switched DC–to–DC power converters. The approach is suitable to be applied to a large class of physically existing DC–to–DC power converters.

The Lagrangian modeling technique is based on a suitable parametrization, in terms of the switch position parameter, of the EL functions describing each intervening system and subsequent application of the Lagrangian formalism. The resulting system is also shown to be a EL system, hence, passivity-based regulation may be proposed as a natural controller design technique.

2.1   Modeling of switched networks

A large class of technological systems are characterized by the presence of one or several *regulating switches*, i.e., devices that can only adopt one of two possible positions, each of these giving rise to a determined dynamical behaviour of the system. Switch–regulated systems are quite common in everyday life specially eversince the commercial development of modern electronics.

We are particularly interested in dynamical systems containing a single *switch*, regarded as the only *control function* of the system. The switch position, denoted by the scalar u, is assumed to take values on a discrete set of the form $\{0, 1\}$. We assume that for each one of the switch position values, the resulting system is an EL system characterized by its corresponding EL parameters. In other words, we assume that when the switch position parameter takes the value, say, $u = 1$, the system, denoted by Σ_1, is characterized by a known set of EL parameters[1] $\{\mathcal{T}_1, \mathcal{V}_1, \mathcal{F}_1, \mathcal{Q}_1\}$. Similarly, when the switch position parameter takes the value $u = 0$, we assume that the resulting system, denoted by Σ_0 is characterized by $\{\mathcal{T}_0, \mathcal{V}_0, \mathcal{F}_0, \mathcal{Q}_0\}$.

Definition 6.1 *A function* $\phi_u(\dot{q}, q) = \phi(\dot{q}, q, u)$, *parameterized by* u, *is said to be* consistent *with the functions* $\phi_0(\dot{q}, q)$ *and* $\phi_1(\dot{q}, q)$ *whenever*

$$\phi_u|_{u=0} = \phi_0 \quad ; \quad \phi_u|_{u=1} = \phi_1.$$

We introduce the set of *switched EL parameters* $\{\mathcal{T}_u, \mathcal{V}_u, \mathcal{F}_u, \mathcal{Q}_u\}$ as a set of functions parameterized by u which are consistent, in the sense described above, with respect to the EL parameters of the systems Σ_0 and Σ_1 for each corresponding value of u.

---

[1]In the class of systems studied in this part of the book the input enters bilinearly, hence we do not use the EL parameter $\mathcal{M}$. On the other hand, there is always interaction with the environment, thus the EL parameter $\mathcal{Q}_\zeta$ will be nonzero.

A switched system Σ_u, arising from the EL systems Σ_0 and Σ_1 is a switched EL system whenever it is completely characterized by its set of switched EL parameters $\{\mathcal{T}_u, \mathcal{V}_u, \mathcal{F}_u, \mathcal{Q}_u\}$.

The basic problem in an EL approach to the modeling of switched systems, arising from individual EL systems, is the following:

Definition 6.2 (Modeling problem for switched EL systems) *Given two EL systems Σ_0 and Σ_1 characterized by EL parameters, $\{\mathcal{T}_0, \mathcal{V}_0, \mathcal{F}_0, \mathcal{Q}_0\}$ and $\{\mathcal{T}_1, \mathcal{V}_1, \mathcal{F}_1, \mathcal{Q}_1\}$, respectively, determine a consistent parameterization of the EL parameters, $\{\mathcal{T}_u, \mathcal{V}_u, \mathcal{F}_u, \mathcal{Q}_u\}$ in terms of the switch position u, such that the model obtained by direct application of the EL equations, results in a parameterized model Σ_u, which is consistent with Σ_0 and Σ_1.*

Consistent parameterizations of the EL parameters, by means of the switch position parameter u, may be, generally speaking, carried out in an infinite number of ways. However, the general rule is to carefully consider those parameterizations that not only account for the effect of the change of the switch position in a particular EL parameter but, also, those for which the obtained switched EL parameter respects the essential nature of the fundamental physical laws that intervene in the constitution of such an EL parameter. For instance, if a change in a switch position inserts, or removes, a current source of value I into a node, to which a single resistive element R is attached, along with capacitive branches that are unaffected by the switching action, then the fundamental law to be respected, in obtaining a suitable parameterization of the Rayleigh dissipation function, is Kirchoff's law of addition of currents at a node. Note that the parameterization $\mathcal{F}_u = 1/2\, R[(\sum i_j)^2 + (1-u)I^2]$, where the i_j's are the currents in capacitive branches that do not change with the switch position, is a consistent parameterization. However, the correct parameterization would be $\mathcal{F}_u = 1/2\, R(\sum i_j + (1-u)I)^2$

2.2   A variational argument

Suppose a consistent, phisically meaningful, parameterization has been carried out on the set of EL parameters in terms of the switch position u. Assume that the non-conservative switched Lagrangian function of the system has been derived as $\mathcal{L}^u(\dot{q}, q)$.

To derive the dynamical model associated with the nonconservative swtiched Lagrangian function we depart from Hamilton's principle. Recall that Hamilton's principle establishes that the trajectory of the system minimizes the *action integral*, which is defined as the integral of the Lagrangian function. The variational condition for a stationary value of the action integral is given by

$$\delta \int_{t_0}^{t_1} \mathcal{L}(\dot{q}, q, u)dt = 0 \tag{6.1}$$

for any arbitrary but fixed endpoints t_0 and t_1 of the considered time interval. The variational argument specifically establishes that for any *virtual* (i.e., arbitrary) variation of the system trajectory, $(\dot{q}(t), q(t))$, the corresponding variation of the action integral should be zero. The virtual variations in the system trajectories must be infinitely differentiable, of infinitesimal nature, and moreover, with zero values at the end points of the considered time interval.

The explicit dependance of the nonconservative Lagrangian function $\mathcal{L}^u(\dot{q}(t), q(t))$ on the switch position parameter u, plus the *causality* relation existing between u, considered as a function of time, and the system trajectories $(\dot{q}(t), q(t))$, makes one wonder if the admissible variations of the trajectories should be synthesized from variations of the control input u or, if due to their arbitrariness one should exercise them directly on the state trajectory *without* regard for the intrinsic causality relation. An examination of the issue in the class of switch regulated systems establishes that the first road must be discarded as argued in the next paragraph.

Suppose an arbitrary time-realization of the switch position u, viewed now as a function of time $u(t)$, is given. The class of *actual* admissible variations $\delta u(t)$ of the switch position function $u(t)$ is then represented by "pulses" of infinitesimal duration δt, taking values now in the set $\{-1, 0, 1\}$. These pulses satisfy the restriction that a negative pulse may only take place while $u = 1$ and a positive pulse can take place when $u = 0$. The problem with this class of discontinuous, and restricted control input variations, is that each of them produces corresponding variations $(\delta q(t), \delta \dot{q}(t))$ in the generalized coordinates $(q, \dot{q})$ which are not infinitely differentiable and furthermore, the obtained perturbed trajectory may not be itself, an infinitesimal variation of the original trajectory in the course of time.

A third obstacle is also represented by the fact that the corresponding trajectory variations may not be zero at the end points of the time interval. The first fact posses a technical problem in the application of the calculus of variation. The second fact is just inadmissible while the third would further restrict the control input variations to those that produce zero effects at the end points. Moreover, each control input variation of the described class results in an *actual* system trajectory which in itself minimizes the action integral. This argument clearly leads to the bizarre situation of having infinitely many minima of the action itegral which are infinitesimaly close to each other!

Hence the involved action integral minimization must be understood in the sense that comparisons of the values of this functional are to be performed between its evaluation on an actual system trajectory, produced say by a fixed $u(t)$, and its evaluation on infinitesimal and infinitely differentiable system trajectory variations $(q(t) + \delta q(t), \dot{q}(t) + \delta \dot{q}(t))$, that *cannot* and *should not* be synthesizable from discontinuous control input variations $\delta u(t)$ taking values in the set $\{-1, 0, 1\}$. One is then confronted with the following choice: either one carries the variational arguments, without any regard for the presence of the parameter u, in terms of trajectory variations alone, or else, one introduces arbitrary infitesimal and infinitely differentiable variations (i.e., *virtual* variations) $\delta u(t)$ on the action integral, even if they are not phisically synthesizable as a switch position function. We choose the first avenue, not only because it is conceptually simpler but also because it allows one to regard u as a constant parameter simply indicating one of two possible switch positions. This conforms to the idea of consistency in a more natural manner.

Following standard, and well known arguments, in the calculus of variations we obtain the following development. Virtual variations of the system trajectory are allowed which result in the following corresponding variation of the action integral

$$
\begin{aligned}
\delta \int_{t_0}^{t_1} \mathcal{L}^u(q(t), \dot{q}(t)) \mathrm{d}t &= \int_{t_0}^{t_1} \delta \mathcal{L}^u(q(t), \dot{q}(t)) \mathrm{d}t \\
&= \int_{t_0}^{t_1} \left[\mathcal{L}^u(q(t) + \delta q(t), \dot{q}(t) + \delta \dot{q}(t)) - \mathcal{L}^u(q(t), \dot{q}(t)) \right] \mathrm{d}t \\
&= \int_{t_0}^{t_1} \left[\frac{\partial \mathcal{L}^u}{\partial q} \delta q(t) + \frac{\partial \mathcal{L}^u}{\partial \dot{q}} \delta \dot{q} \right] \mathrm{d}t \\
&= \int_{t_0}^{t_1} \left[\frac{\partial \mathcal{L}^u}{\partial q} - \frac{d}{dt} \frac{\partial \mathcal{L}^u}{\partial \dot{q}} \right] \delta q(t) \mathrm{d}t + \left. \left(\frac{\partial \mathcal{L}^u}{\partial \dot{q}} \delta q \right) \right|_{t_0}^{t_1} = 0.
\end{aligned}
$$

From the fact that admissible trajectory variations should have no contribution from the end points, the second term in the last expression above is identically zero. The well known *Fundamental Lemma of the Calculus of Variations* (see, for instance [148]) can now be applied to obtain the following result:

Proposition 6.3 *The EL equations are valid for the nonconservative switched Lagrangian function parameterized by the switch position u, treated as a constant. The resulting dynamical model of the switched system is a dynamical model parameterized also by u which is consistent with the intervening dynamical models for each switch position parameter value, that is*

$$
\frac{d}{dt} \frac{\partial \mathcal{L}^u(q, \dot{q})}{\partial \dot{q}} - \frac{\partial \mathcal{L}^u(q, \dot{q})}{\partial q} = 0. \tag{6.2}
$$

□

2.3   General Lagrangian model: Passivity property

In this section we present a general procedure to derive the EL models of a class of switched DC–to–DC converters. We also underscore some properties of the model, in particular its passivity, which will be exploited for the design of PBCs in the next chapter.

A Structural considerations

DC-to-DC power conversion is accomplished in switched circuits through ingenious cyclic energy transfers, regulated by the position(s) of the commanding switch(es). Generally speaking, switching actions allow that energy, obtained from the external sources, be stored in "input" inductive branches by placing those inductors in the same mesh as the sources. The stored kinetic energy is then transfered to either "internal" or to "output" potential energy reservoirs (i.e., capacitors). The stored potential energy in "internal" capacitors is transfered again, in the form of kinetic energy to a different "internal" inductive branch, usually connected to output capacitors. In the case of "output" capacitors, the energy may be directly drawn by the resistive loads.

Generalized forcing parameters, or external voltage source terms, will be associated with "input" inductor current variables in the EL equations, while, generally speaking, the external voltage sources will be absent from the equations corresponding to "output" capacitor charges.

Potential energy delivery to the output loads is accomplished thanks to the fact that "output" storing capacitors will be connected in parallel to the resistive loads. This RC "output sturcture" is always fixed and the switches do not affect it directly. Thus, nodes containing output storing capacitor branches will always have the output resisitive load branch attached to them. We assume without loss of generality, that the switch-invariant capacitors are all "output" capacitors.

In summary, the switching action in DC-to-DC power converters, accomplishes one or two of the following three possibilities during the transfer phase.

1. Switchings insert (respectively remove) a constant external voltage source into (resp. from) a mesh containing an "input" inductor branch.

2. Switchings insert (respectively remove) a charged inductor branch (whether "input" or "internal" branch) into (resp. from) a node containing either an "output" RC branch or an "internal" storing capacitor branch.

3. Switchings insert (resp. remove) an "internal" storing capacitor into (resp from) a mesh containing an "internal" inductor branch.

The following developments include all of the previously treated *ideal* DC-to-DC power converters (i.e., non perturbed by parasitics). In the formulation we take into account all previous remarks. The switch position parameter, u, is regarded to be a vector so as to include the multivariable versions of the DC-to-DC power converters.

B   Energy and dissipation functions

We hypothesize a general kinetic energy EL parameter of the form

$$T_u = \frac{1}{2}\dot{q}_L^\top L \dot{q}_L \; ; \quad \dot{q}_L \in R^{n_L}$$

where $\dot{q}_L$ denotes the vector of inductor currents and L is a diagonal positive definite matrix.

The generalized potential energy EL parameter, $\mathcal{V}$, contains two terms. One describing the stored potential energy in capacitors which are *not* directly affected by the switchings and a second term including the potential energy of capacitors that are removed from an (inductive) mesh and inserted into another mesh by means of the switching action. We denote the first vector of "switch-invariant" capacitor charges as an n_{C_I}-dimensional vector, q_{C_I}. The vector of charges associated with the capacitors that migrate from one mesh to another is deemed as a vector of "non-switch-invariant" charges. These will be denoted by an n_{C_V}-dimensional vector q_{C_V}. However, these charges *are not* generalized coordinates since their values depend on the particular flowing charge associated with the inductive mesh where these capacitors happen to be for a particular switching position. In other words, $q_{C_V} = V(u)q_L$, where $V(u)$ is an $n_{C_V} \times n_L$ matrix, parameterized in terms of the switch position u. We call the matrix $V(u)$, the "Capacitor mesh insertion matrix".

$$\begin{aligned}
\mathcal{V}_u &= \frac{1}{2}q_{C_I}^\top C_I^{-1} q_{C_I} + \frac{1}{2}q_{C_V}^\top C_V^{-1} q_{C_V} \\
&= \frac{1}{2}q_{C_I}^\top C_I^{-1} q_{C_I} + \frac{1}{2}q_L^\top V^\top(u) C_V^{-1} V(u) q_L
\end{aligned}$$

The generalized Rayleigh dissipation EL parameter is obtained by considering the currents flowing through the output load resistors. These currents are given by the difference between the currents flowing from the inductive branches which are inserted into the output nodes by the switching action and the currents flowing through the "output capacitors" (which, as assumed before, constitute the set of switch-invariant capacitors). The currents through the output capacitors are described by the vector term, $\dot{q}_{C_I}$. The vector of inductive currents inserted into, or removed from, the output nodes is, evidently, a subvector of the vector q_L. This subvector is expressed in the form $Z(u)\dot{q}_L$ where the $n_o \times n_{q_L}$ matrix, $Z(u)$, is the "inductive branch node insertion matrix". The Rayleigh dissipation EL parameter is then given by

$$\mathcal{F}_u = \frac{1}{2}\left[Z(u)\dot{q}_L - \dot{q}_{C_I}\right]^\top R \left[Z(u)\dot{q}_L - N\dot{q}_{C_I}\right]$$

The matrix R is a diagonal matrix containing all the output resistance values.

Finally, the vector of external sources, $\mathcal{Q}_u$, usually contains a single nonzero element that may, or may not, be inserted into the "input" mesh by the switching action.

However, we consider that the subvector E of nonzero entries, representing the n_E external voltage sources, is of dimension $n_E = n_i$, where $n_i \leq n_L$ is the number of "input inductors" which may form "input" meshes (i.e., inductor-charging meshes) with the external sources by means of the switching action. The vector of generalized forcing functions $\mathcal{Q}$ may then be expressed as

$$\mathcal{Q}_u = \left[(Q(u)E)^\top, 0_{1\times(n_L-n_i)} \right]^\top$$

where $Q(u)$ is a (square) "input mesh external source insertion matrix", of dimension $n_i \times n_E = n_i \times n_i$.

C   Properties of the EL model

Applying the EL equations to the above set of EL parameters we obtain a rather general model of a large class of ideal DC-to-DC power converters, including the class of multivariable DC-to-DC converters.

$$\begin{aligned} L\ddot{q}_L + V^\top(u)C_V^{-1}V(u)q_L &= -Z^\top(u)R\left[Z(U)\dot{q}_L - q_{C_I}\right] + \mathcal{Q}_u \\ C_I^{-1}q_{C_I} &= R\left[Z(u)\dot{q}_L - \dot{q}_{C_I}\right]. \end{aligned}$$

A state space representation of the generalized switched circuit may be obtained by defining a composite state vector as, $x_L = \dot{q}_L$, $x_{C_V} = C_V^{-1}q_{C_V} = C_V^{-1}V(u)q_L$ and $x_{C_I} = q_{C_I}$. After some algebraic manipulations, and differentiations where the vector u is treated as a vector of constant parameters, one obtains the following general state space model for a large class of switched DC-to-DC power converters

$$\begin{aligned} &\begin{pmatrix} L & 0 & 0 \\ 0 & C_V & 0 \\ 0 & 0 & C_I \end{pmatrix} \begin{bmatrix} \dot{x}_L \\ \dot{x}_{C_V} \\ \dot{x}_{C_I} \end{bmatrix} + \begin{pmatrix} 0 & -Z^\top(u) & -V^\top(u) \\ Z(u) & 0 & 0 \\ V(u) & 0 & 0 \end{pmatrix} \begin{bmatrix} x_L \\ x_{C_V} \\ x_{C_I} \end{bmatrix} \\ &+ \begin{pmatrix} 0 & 0 & 0 \\ 0 & 0 & 0 \\ 0 & 0 & R^{-1} \end{pmatrix} \begin{bmatrix} x_L \\ x_{C_V} \\ x_{C_I} \end{bmatrix} = \begin{bmatrix} F(u)E \\ 0 \\ 0 \end{bmatrix}. \end{aligned}$$

The set of equations above can be rewritten in the following general form,

$$\mathcal{D}\dot{x} + \mathcal{G}(u)x + \mathcal{R}x = \mathcal{E}$$

where $x \triangleq [x_L^\top, x_{C_V}^\top, x_{C_I}^\top]^\top$, $\mathcal{D}$ is diagonal and positive definite, $\mathcal{G}(u)$ is a skew-symmetric matrix for any allowable values of the switching parameter vector components and $\mathcal{R}$ is a diagonal positive semidefinite matrix.

To reveal some structural properties of this model notice first that the total energy of the circuit is given by

$$\mathcal{H}_u = \mathcal{T}_u + \mathcal{V}_u = \frac{1}{2} x^\top \mathcal{D} x.$$

Differentiating $\mathcal{H}_u$, taking into account the skew–symmetry of $\mathcal{G}(u)$, and integrating back, we obtain the energy–balance equation

$$\underbrace{\mathcal{H}_u(T) - \mathcal{H}_u(0)}_{\text{stored energy}} + \underbrace{\int_0^T x^\top \mathcal{R} x(t) dt}_{\text{dissipated energy}} = \underbrace{\int_0^T x^\top(t) \mathcal{E} dt}_{\text{supplied energy}} \qquad (6.3)$$

The following observations are in order:

- As expected, the circuit dynamics defines a passive operator from the supplied voltages E to the inductance currents x_L.

- The model, though an EL system, differs from the ones considered previously, since the control signal enters bilinearly. However, action of the control signal is "workless", in the sense that it does not affect the energy balance equation.

- The system is in general only partially damped. Thus, additional damping will need to be injected to achieve strict passivity.

Let us elaborate further this points with the example of the Boost converter. In this case

$$\mathcal{D} = \begin{bmatrix} L & 0 \\ 0 & C \end{bmatrix} \quad ; \quad \mathcal{G}(u) = -u\mathcal{J} = u \begin{bmatrix} 0 & 1 \\ -1 & 0 \end{bmatrix} \quad ; \quad \mathcal{R} = \begin{bmatrix} 0 & 0 \\ 0 & 1/R \end{bmatrix} \quad ; \quad \mathcal{E} = \begin{bmatrix} E \\ 0 \end{bmatrix}$$

where $x = [x_1, x_2]^\top$, and the energy balance equation reduces to

$$\mathcal{H}_u(T) - \mathcal{H}_u(0) = \int_0^T E x_1(t) dt - \frac{1}{R} \int_0^T x_2^2(t) dt.$$

Eventhough, as pointed out above, the inductance current x_1, being a passive output, is "easy" to control, in these devices the output to be controlled is the capacitor voltage x_2. The problem is further complicated by the fact that, as we will show later, the system is *non–minimum phase* with respect to x_2. We will, therefore, control x_2 indirectly via the regulation of x_1.

In this circuit the workless role of the control is quite clear since the switch simply transfers the magnetic energy stored in the inductance to the RC circuit, with part of it being stored as electric energy in the capacitor and the rest dissipated by the resistance.

2.4 Examples

A The Boost converter

Consider the switch–regulated Boost converter circuit of Fig. 6.1.

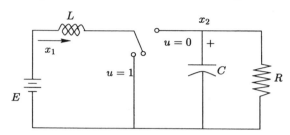

Figure 6.1: The Boost converter circuit.

The differential equations describing the circuit were derived in [182] using classic Kirchoff laws. Such set of equations are given by

$$\dot{x}_1 = -(1-u)\frac{1}{L}x_2 + \frac{E}{L} \tag{6.4}$$

$$\dot{x}_2 = (1-u)\frac{1}{C}x_1 - \frac{1}{RC}x_2 \tag{6.5}$$

where x_1 and x_2 represent, respectively, the input inductor current and the output capacitor voltage variables. The positive quantity E represents the constant voltage value of the external voltage source. The parameter u denotes the switch position. The switch position parameter takes values in the discrete set $\{0,1\}$.

We consider separately the Lagrange dynamics formulation of the two circuits associated with each of the two possible positions of the regulating switch. Our aim in carrying out this formulation is to gain some insight on the physical effects of the switching action in terms of the EL parameters of the two circuit topologies. In order to use standard notation we refer to the input current x_1 in terms of the derivative of the circulating charge q_L, as $\dot{q}_L$. Also the capacitor voltage x_2 will be written as q_C/C where q_C is the electrical charge stored in the output capacitor.

Consider then $u = 1$. In this case two separate, or decoupled circuits are clearly obtained and the corresponding Lagrange dynamics formulation can be carried out as follows. Define $\mathcal{T}_1(\dot{q}_L)$ and $\mathcal{V}_1(q_C)$ as the kinetic and potential energies of the circuit respectively. We denote by $\mathcal{F}_1(\dot{q}_C)$ the Rayleigh dissipation function of the circuit. These quantities are readily found to be

$$\begin{cases} \mathcal{T}_1(\dot{q}_L) = \frac{1}{2}L\left(\dot{q}_L\right)^2 & ; \quad \mathcal{V}_1(q_C) = \frac{1}{2C}q_C^2 \\ \mathcal{F}_1(\dot{q}_C) = \frac{1}{2}R\left(\dot{q}_C\right)^2 & ; \quad \mathcal{Q}_{q_L}^1 = E \quad ; \quad \mathcal{Q}_{q_C}^1 = 0 \end{cases} \tag{6.6}$$

where $\mathcal{F}^1_{q_L}$ and $\mathcal{F}^1_{q_C}$ are the *generalized forcing* functions associated with the coordinates q_L and q_C, respectively.

Evidently, the EL equations used on these EL parameters immediately rederive equations (6.4), (6.5), with $u = 1$, as it can be esasily verified.

Consider now the case $u = 0$. The corresponding Lagrange dynamics formulation is carried out in the next paragraphs.

Define $\mathcal{T}_0(\dot{q}_L)$ and $\mathcal{V}_0(q_C)$ as the kinetic and potential energies of the circuit, respectively. We denote by $\mathcal{F}_0(\dot{q}_L, \dot{q}_C)$ the Rayleigh dissipation function of the circuit. These quantities are readily found to be,

$$\left\{ \begin{array}{ll} \mathcal{T}_0(\dot{q}_L) = \frac{1}{2}L\left(\dot{q}_L\right)^2 & ; \quad \mathcal{V}_0(q_C) = \frac{1}{2C}q_C^2 \\ \mathcal{F}_0(\dot{q}_L, \dot{q}_C) = \frac{1}{2}R\left(\dot{q}_L - \dot{q}_C\right)^2 & ; \quad \mathcal{Q}^0_{q_L} = E \quad ; \quad \mathcal{Q}^0_{q_C} = 0 \end{array} \right. \tag{6.7}$$

where, $\mathcal{Q}^0_{q_L}$ and $\mathcal{Q}^0_{q_C}$ are the *generalized forcing* functions associated with the coordinates q_L and q_C, respectively.

The EL parameters of the two situations generated by the different switch position values result in identical kinetic and potential energies. The switching action merely changes the Rayleigh dissipation function between the values $\mathcal{F}_0(\dot{q}_C)$ and $\mathcal{F}_1(\dot{q}_L, \dot{q}_C)$. Therefore, the *dissipation structure* of the system is the only one affected by the switch position. One may then regard the switching action as a "damping injection", performed through the inductor current.

The following set of switched EL parameters are proposed for the description of the switched system,

$$\left\{ \begin{array}{ll} \mathcal{T}_u(\dot{q}_L) = \frac{1}{2}L\left(\dot{q}_L\right)^2 & ; \quad \mathcal{V}_u(q_C) = \frac{1}{2C}q_C^2 \\ \mathcal{F}_u(\dot{q}_L, \dot{q}_C) = \frac{1}{2}R\left[(1 - u)\dot{q}_L - \dot{q}_C\right]^2 & ; \quad \mathcal{Q}^u_{q_L} = E \quad ; \quad \mathcal{Q}^u_{q_C} = 0. \end{array} \right. \tag{6.8}$$

Note that in the cases where u takes the values $u = 1$ and $u = 0$, one recovers, respectively, the dissipation functions $\mathcal{F}_1(\dot{q}_C)$ in (6.6) and $\mathcal{F}_0(\dot{q}_L, \dot{q}_C)$ in (6.7) from the proposed dissipation function, $\mathcal{F}_u(\dot{q}_L, \dot{q}_C)$, of equations (6.8). The proposed EL parameters are therefore *consistent*.

The switched Lagrangian function associated with the above defined EL parameters is given by,

$$\mathcal{L}_u = \mathcal{T}_u(\dot{q}_L) - \mathcal{V}_u(q_C) = \frac{1}{2}L\left(\dot{q}_L\right)^2 - \frac{1}{2C}q_C^2$$

One then proceeds, using the EL equations (6.2) to formally obtain the switch-position parameterized differential equations defining the switch regulated system which corresponds to the proposed switched EL parameters (6.8). This results in the following set of differential equations

$$L\ddot{q}_L = -(1-u)R[(1-u)\dot{q}_L - \dot{q}_C] + E$$
$$q_C/C = R[(1-u)\dot{q}_L - \dot{q}_C]$$

which can be rewritten, after substitution of the second equation into the first, as

$$\ddot{q}_L = -(1-u)\frac{q_C}{LC} + \frac{E}{L}$$
$$\dot{q}_C = -\frac{1}{RC}q_C + (1-u)\dot{q}_L$$

or using the state-space coordinates $x_1 = \dot{q}_L$ and $x_2 = q_C/C$ one obtains

$$\dot{x}_1 = -(1-u)\frac{1}{L}x_2 + \frac{E}{L} \tag{6.9}$$
$$\dot{x}_2 = (1-u)\frac{1}{C}x_1 - \frac{1}{RC}x_2. \tag{6.10}$$

The proposed switched dynamics (6.9), (6.10) coincides with the classical state space model developed in [182] and [63].

B   The Buck–boost converter

The circuit of the Buck-boost converter is shown in Fig. 6.2. We summarize all the relevant formulae and equations, leading to the switched model of the Buck–boost converter circuit, through an EL formulation, in the Table 6.1.

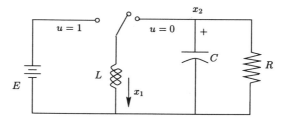

Figure 6.2: The Buck–boost converter circuit.

Remark 6.4 The Lagrangian approach to modeling of the Buck-boost converter reveals that only the "dissipation structure" and the "external forcing functions" are non-invariant with respect to the switching action.

BUCK – BOOST CONVERTER

EL parameters for possible switch positions

u	$u = 0$	$u = 1$
Kinetic energy	$\mathcal{T}_0(\dot{q}_L) = \frac{1}{2}L\left(\dot{q}_L\right)^2$	$\mathcal{T}_1(\dot{q}_L) = \frac{1}{2}L\left(\dot{q}_L\right)^2$
Potential energy	$\mathcal{V}_0(q_C) = \frac{1}{2C}q_C^2$	$\mathcal{V}_1(q_C) = \frac{1}{2C}q_C^2$
Rayleigh dissipation	$\mathcal{F}_0(\dot{q}_L, \dot{q}_C) = \frac{1}{2}R\left(\dot{q}_L + \dot{q}_C\right)^2$	$\mathcal{F}_1(\dot{q}_C) = \frac{1}{2}R\left(\dot{q}_C\right)^2$
External forces	$\mathcal{Q}_{q_L}^0 = 0$; $\mathcal{Q}_{q_C}^0 = 0$	$\mathcal{Q}_{q_L}^1 = E$; $\mathcal{Q}_{q_C}^1 = 0$

Switched EL Parameters

Kinetic energy	$\mathcal{T}_u(\dot{q}_L) = \frac{1}{2}L\left(\dot{q}_L\right)^2$
Potential energy	$\mathcal{V}_u(q_C) = \frac{1}{2C}q_C^2$
Dissipation function	$\mathcal{F}_u(\dot{q}_L, \dot{q}_C) = \frac{1}{2}R\left[(1-u)\dot{q}_L + \dot{q}_C\right]^2$
External Forces	$\mathcal{Q}_{q_L}^u = u\,E$; $\mathcal{Q}_{q_C}^u = 0$

Lagrangian for the Buck – boost converter model

$$\mathcal{L}_u = \mathcal{T}_u(\dot{q}_L) - \mathcal{V}_u(q_C) = \frac{1}{2}L\left(\dot{q}_L\right)^2 - \frac{1}{2C}q_C^2$$

Switched model in generalized coordinates

$$L\ddot{q}_L = -(1-u)R\left[(1-u)\dot{q}_L + \dot{q}_C\right] + uE$$
$$\frac{q_C}{C} = -R\left[(1-u)\dot{q}_L + \dot{q}_C\right]$$

Definition of state variables

$$x_1 = \dot{q}_L \;\; ; \;\; x_2 = q_C/C$$

Switched model for the Buck – boost converter

$$\dot{x}_1 = (1-u)\frac{1}{L}x_2 + u\frac{E}{L}$$
$$\dot{x}_2 = -(1-u)\frac{1}{C}x_1 - \frac{1}{RC}x_2$$

Table 6.1: An EL approach for the modeling of the Buck-boost converter.

C   The Ćuk converter

The Ćuk converter model is shown in Fig. 6.3. In Tables 6.2 and 6.3 we summarize all the relevant formulae and equations leading to the switched model of the Ćuk converter circuit through our proposed EL formulation.

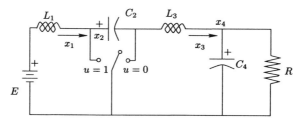

Figure 6.3: The Ćuk converter circuit.

Remark 6.5 The Lagrangian approach to modeling of the Ćuk converter reveals that only the "potential energy" structure of the system is non-invariant with respect to the switching action.

D   Multivariable DC–to–DC power converters

A series of cascaded DC-to-DC power converters constitutes an interesting and new class of multivariable switch-regulated converters with potential practical value and which nonetheless, seem to have been overlooked in the literature.

Consider the boost–boost converter circuit of Fig. 6.4.

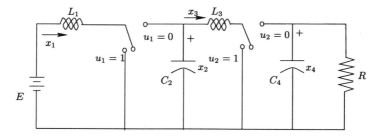

Figure 6.4: The Boost–boost converter circuit.

In Table 6.4 we present the EL parameters for the four possible combinations of the switch position parameters u_1 and u_2. The Table 6.5 contains the switched EL parameters, the system equations in generalized coordinates as well as the switch regulated state equations.

ĆUK CONVERTER.

EL Parameters for possible switch positions

	$u = 0$	$u = 1$
Kinetic energy	$\mathcal{T}_0(\dot{q}_{L_1}, \dot{q}_{L_3}) = \frac{1}{2}L_1(\dot{q}_{L_1})^2 + \frac{1}{2}L(\dot{q}_{L_3})^2$	$\mathcal{T}_1(\dot{q}_{L_1}, \dot{q}_{L_3}) = \frac{1}{2}L_1(\dot{q}_{L_1})^2 + \frac{1}{2}L(\dot{q}_{L_3})^2$
Potential energy	$\mathcal{V}_0(q_{L_1}, q_{C_4}) = \frac{1}{2C_2}q_{L_1}^2 + \frac{1}{2C_4}q_{C_4}^2$	$\mathcal{V}_1(q_{L_3}, q_{C_4}) = \frac{1}{2C_2}q_{L_3}^2 + \frac{1}{2C_4}q_{C_4}^2$
Rayleigh dissipation	$\mathcal{F}_0(\dot{q}_{L_3}, \dot{q}_{C_4}) = \frac{1}{2}R(\dot{q}_{L_3} - \dot{q}_{C_4})^2$	$\mathcal{F}_1(\dot{q}_{L_3}, \dot{q}_{C_4}) = \frac{1}{2}R(\dot{q}_{L_3} - \dot{q}_{C_4})^2$
External forces	$\mathcal{Q}^0_{q_{L_1}} = E$; $\mathcal{Q}^0_{q_{L_3}} = 0$; $\mathcal{Q}^0_{q_{C_4}} = 0$	$\mathcal{Q}^1_{q_{L_1}} = E$; $\mathcal{Q}^1_{q_{L_3}} = 0$; $\mathcal{F}^1_{q_{C_4}} = 0$

Switched EL parameters

u	
Kinetic energy	$\mathcal{T}_u(\dot{q}_{L_1}, \dot{q}_{L_3}) = \frac{1}{2}L_1(\dot{q}_{L_1})^2 + \frac{1}{2}L(\dot{q}_{L_3})^2$
Potential energy	$\mathcal{V}_u(q_{L_1}, q_{L_3}, q_{C_4}) = \frac{1}{2C_2}[(1-u)q_{L_1} + uq_{L_3}]^2 + \frac{1}{2C_4}q_{C_4}^2$
Dissipation function	$\mathcal{F}_u(\dot{q}_{L_3}, \dot{q}_{C_4}) = \frac{1}{2}R(\dot{q}_{L_3} - \dot{q}_{C_4})^2$
External forces	$\mathcal{Q}^u_{q_{L_1}} = E$; $\mathcal{F}^u_{q_{L_3}} = 0$; $\mathcal{F}^u_{q_{C_4}} = 0$

Table 6.2: An EL approach for the modeling of the Ćuk converter. EL parameters

ĆUK CONVERTER
Lagrangian for the "Ćuk" converter model
$\mathcal{L}_u(\dot{q}_{L_1}, \dot{q}_{L_3}, q_{L_1}, q_{L_3}, q_{C_4}) = \mathcal{T}_u(\dot{q}_{L_1}, \dot{q}_{L_3}) - \mathcal{V}_u(q_{L_1}, q_{L_3}, q_{C_4})$ $= \frac{1}{2}L_1(\dot{q}_{L_1})^2 + \frac{1}{2}L(\dot{q}_{L_3})^2 - \frac{1}{2C_2}[(1-u)q_{L_1} + uq_{L_3}]^2 - \frac{1}{2C_4}q_{C_4}^2$
Switched model in generalized coordinates
$L_1\ddot{q}_{L_1} = -(1-u)\frac{1}{C_2}[(1-u)q_{L_1} + uq_{L_3}] + E$ $L_3\ddot{q}_{L_3} = -u\frac{1}{C_2}[(1-u)q_{L_1} + uq_{L_3}] - R(\dot{q}_{L_3} - \dot{q}_{C_4})$ $\frac{q_{C_4}}{C_4} = R(\dot{q}_{L_3} - \dot{q}_{C_4})$
Definition of state variables
$x_1 = \dot{q}_{L_1}$; $x_2 = \frac{1}{C_2}[(1-u)q_{L_1} + uq_{L_3}]$; $x_3 = \dot{q}_{L_3}$; $x_4 = \frac{q_4}{C_4}$
Switched model for the Ćuk converter
$\dot{x}_1 = -(1-u)\frac{1}{L_1}x_2 + \frac{E}{L_1}$ $\dot{x}_2 = (1-u)\frac{1}{C_2}x_1 + u\frac{1}{C_2}x_3$ $\dot{x}_3 = -u\frac{1}{L_3}x_2 - \frac{1}{L_3}x_4$ $\dot{x}_4 = \frac{1}{C_4}x_3 - \frac{1}{RC_4}x_4$

Table 6.3: An EL approach for the modeling of the Ćuk converter. Dynamic equations.

BOOST – BOOST CONVERTER

(u_1, u_2)	EL Parameters for possible switch positions			
	(0,0)	(1,0)	(0,1)	(1,1)
Kinetic energy	$T_{00} = \frac{1}{2}L_1\dot{q}_{L_1}^2 + \frac{1}{2}L_3\dot{q}_{L_3}^2$	$T_{10} = \frac{1}{2}L_1\dot{q}_{L_1}^2 + \frac{1}{2}L_3\dot{q}_{L_3}^2$	$T_{01} = T_{00}$	$T_{11} = T_{10}$
Potential energy	$V_{00} = \frac{1}{2C_2}(q_{L_1} - q_{L_3})^2 + \frac{1}{2C_4}q_{C_4}^2$	$V_{10} = \frac{1}{2C_2}q_{L_3}^2 + \frac{1}{2C_4}q_{C_4}^2$	$V_{01} = V_{00}$	$V_{11} = V_{10}$
Dissipation function	$\mathcal{F}_{00} = \frac{1}{2}R(q_{L_3} - \dot{q}_{C_4})^2$	$\mathcal{F}_{10} = \mathcal{F}_{00}$	$\mathcal{F}_{01} = \frac{1}{2}R\dot{q}_{C_4}^2$	$\mathcal{F}_{11} = \mathcal{F}_{01}$
Generalized force	$Q^{00}_{q_{L_1}, q_{L_3}, q_{C_4}} = \begin{bmatrix} E \\ 0 \\ 0 \end{bmatrix}$	$Q^{10}_{q_{L_1}, q_{L_3}, q_{C_4}} = \begin{bmatrix} E \\ 0 \\ 0 \end{bmatrix}$	$Q^{01}_{q_{L_1}, q_{L_3}, q_{C_4}} = \begin{bmatrix} E \\ 0 \\ 0 \end{bmatrix}$	$Q^{11}_{q_{L_1}, q_{L_3}, q_{C_4}} = \begin{bmatrix} E \\ 0 \\ 0 \end{bmatrix}$

Table 6.4: A Lagrangian approach to modeling of the multivariable Boost-boost converter. EL parameters.

BOOST – BOOST CONVERTER

Switched EL parameters	
Kinetic energy	$\mathcal{T}_{u_1 u_2} = \frac{1}{2} L_1 \dot{q}_{L_1}^2 + \frac{1}{2} L_3 \dot{q}_{L_3}^2$
Potential energy	$\mathcal{V}_{u_1 u_2} = \frac{1}{2C_2}\left[(1-u_1)q_{L_1} - q_{L_3}\right]^2 + \frac{1}{2C_4}q_{C_4}^2$
Dissipation function	$\mathcal{F}_{u_1 u_2} = \frac{1}{2} R\left[(1-u_2)\dot{q}_{L_3} - \dot{q}_{C_4}\right]^2$
External forces	$\mathcal{Q}_{q_{L_1}, q_{L_3}, q_{C_4}}^{u_1 u_2} = [E\ 0\ 0]^T$

System equations in generalized coordinates

$$L_1 \ddot{q}_{L_1} + \frac{1}{C_2}(1-u_1)\left[(1-u_1)q_{L_1} - q_{L_3}\right] = E$$
$$L_3 \ddot{q}_{L_3} - \frac{1}{C_2}\left[(1-u_1)q_{L_1} - q_{L_3}\right] = -(1-u_2)R\left[(1-u_2)\dot{q}_{L_3} - \dot{q}_{C_4}\right]$$
$$\frac{q_{C_4}}{C_4} = R\left[(1-u_2)\dot{q}_{L_3} - \dot{q}_{C_4}\right]$$

State variables definition

$$x_1 = \dot{q}_{L_1} \;\; ; \;\; x_2 = \frac{1}{C_2}\left[(1-u_1)q_{L_1} - q_{L_3}\right] \;\; ; \;\; x_3 = \dot{q}_{L_3} \;\; ; \;\; x_4 = \frac{q_{C_4}}{C_4}$$

Switched system state equations

$$\dot{x}_1 = -(1-u_1)\frac{1}{L_1}x_2 + \frac{E}{L_1}$$
$$\dot{x}_2 = (1-u_1)\frac{1}{C_2}x_1 - \frac{1}{C_2}x_3$$
$$\dot{x}_3 = \frac{1}{L_3}x_2 - (1-u_2)\frac{1}{L_3}x_4$$
$$\dot{x}_4 = (1-u_2)\frac{1}{C_4}x_3 - \frac{1}{RC_4}x_4$$

Table 6.5: A Lagrangian approach to modeling of the multivariable Boost-boost converter. Dynamic equations.

E   Modeling of non-ideal switches

In this section we extend the modeling approach presented in the previous section to include more realistic models of the switching device and of the circuit elements. Ideal switches do not exist in practise and these must be replaced by suitable arrangements of a transistor and a diode (see [238], [117]). Such physical components exhibit non idealities generally addressed as "parasitics". For instance, these are represented by a small "ON resistance" in the transistor and a small "forward resistance" of the diode. An important parasitic element is constituted by an "offset" voltage source for the diode. Also, real life inductors and capacitors exhibit small resistances. An additional parasitic effect which, in the interest of simplicity, will not be included in this study, is the so called "storage time modulation" effect. This topic is treated at length in [181] and also in [281].

Consider the more realisitic Boost converter model shown in Fig. 6.5.

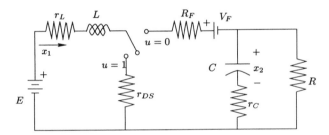

Figure 6.5: Boost converter model with parasitic components.

This model, which includes an ideal switch, has been proposed in [118]. The parasitic resistances r_{DS} and R_F represent, respectively, the ON resistance of the transistor and the forward resistance of the diode. V_F represents the offset voltage in the diode model. The resistances r_L and r_C are the resistances associated with the inductor and the capacitor of the circuit. The model, however, neglects the output capacitances of the transistor and of the diode.

Consider first the switch position parameter value $u = 1$. As in the ideal case, two separate, or decoupled, circuits are clearly obtained and the corresponding Lagrange dynamics formulation can be carried out as follows. The EL parameters of the circuit are readily obtained as

$$\left\{ \begin{array}{l} \mathcal{T}_1(\dot{q}_L) = \frac{1}{2}L\left(\dot{q}_L\right)^2 \quad ; \quad \mathcal{V}_1(q_C) = \frac{1}{2C}q_C^2 \\ \mathcal{F}_1(\dot{q}_C) = \frac{1}{2}(r_C + R)\left(\dot{q}_C\right)^2 + \frac{1}{2}(r_L + r_{DS})\dot{q}_L^2 \\ \mathcal{Q}_{q_L}^1 = E \quad ; \quad \mathcal{Q}_{q_C}^1 = 0 \end{array} \right. \tag{6.11}$$

Then, using EL equations and letting, $x_1 = \dot{q}_L$, $x_2 = q_C/C$, one obtains after some elementary algebraic manipulations, the following set of differential equations,

$$\dot{x}_1 = -\frac{1}{L}\left(r_L + r_{DS}\right)x_1 + \frac{E}{L}$$

$$\dot{x}_2 = -\frac{1}{(r_C + R)C}x_2$$

and the output load voltage, denoted now by x_o, is readily obtained as

$$x_o = \frac{R}{r_C + R}\,x_2. \tag{6.12}$$

Consider now the when case $u = 0$. The corresponding EL parameters of the resulting circuit are given by

$$\begin{cases} \mathcal{T}_0(\dot{q}_L) = \frac{1}{2}L\left(\dot{q}_L\right)^2 \quad ; \quad \mathcal{V}_0(q_C) = \frac{1}{2C}q_C^2 \\ \mathcal{F}_0(\dot{q}_L, \dot{q}_C) = \frac{1}{2}(r_L + R_F)\dot{q}_L^2 + \frac{1}{2}r_C(\dot{q}_C)^2 + \frac{1}{2}R\left(\dot{q}_L - \dot{q}_C\right)^2 \\ \mathcal{Q}_{q_L}^0 = E - V_F \quad ; \quad \mathcal{Q}_{q_C}^0 = 0. \end{cases} \tag{6.13}$$

Note that the appearance of the diode offset voltage source V_F modifies the generalized force $\mathcal{Q}_{q_L}$. Alternatively, this voltage source could have also been taken into account as part of the potential energy term, as $\mathcal{V}_0(q_C) = \frac{1}{2C}q_C^2 - V_F q_L$ (see [179]). The resulting dynamical model equations are identical in any case.

Using the EL equations and the previous definitions for x_1 and x_2, one obtains, after some rearrangement, the following set of differential equations

$$\dot{x}_1 = -\frac{1}{L}\left[r_L + R_F + (r_C\|R)\right]x_1 - \frac{R}{L(r_C + R)}x_2 + \frac{E}{L} - \frac{V_F}{L}$$

$$\dot{x}_2 = \frac{R}{(r_C + R)C}x_1 - \frac{1}{(r_C + R)C}x_2$$

where the symbol $(r_C\|R)$ denotes the resistance of the parallel arrangement of r_C and R and the output load voltage, x_o, is now obtained as,

$$x_o = \frac{R}{r_C + R}\,x_2 + \frac{r_C R}{r_C + R}x_1. \tag{6.14}$$

Following the procedure proposed in the previous sections one obtains the following set of switched EL parameters

$$\begin{cases} \mathcal{T}_u(\dot{q}_L) = \frac{1}{2}L\left(\dot{q}_L\right)^2 \quad ; \quad \mathcal{V}_u(q_C) = \frac{1}{2C}q_C^2 \\ \mathcal{F}_u(\dot{q}_L, \dot{q}_C) = \frac{1}{2}r_L(\dot{q}_L)^2 + \frac{1}{2}\left[(1-u)R_F + ur_{DS}\right](\dot{q}_L)^2 + \\ \qquad\qquad \frac{1}{2}r_C(\dot{q}_C)^2 + \frac{1}{2}R\left[(1-u)\dot{q}_L - \dot{q}_C\right]^2 \\ \mathcal{Q}_{q_L}^u = E - (1-u)V_F \quad ; \quad \mathcal{Q}_{q_C}^u = 0 \end{cases} \tag{6.15}$$

Note that in the cases where u takes the values $u = 1$ and $u = 0$, one recovers, respectively, the Rayleigh dissipation functions $\mathcal{F}_1(\dot{q}_C)$ in (6.11) and $\mathcal{F}_0(\dot{q}_L, \dot{q}_C)$ in (6.13) from the proposed dissipation function, $\mathcal{F}_u(\dot{q}_L, \dot{q}_C)$, in equations (6.15).

We define the Lagrangian function associated with the above defined switched EL parameters as,

$$\mathcal{L}_u = \mathcal{T}_u(\dot{q}_L) - \mathcal{V}_u(q_C) = \frac{1}{2} L \left(\dot{q}_L \right)^2 - \frac{1}{2C} q_C^2 \tag{6.16}$$

Using the EL equations, on (6.15), (6.16), one obtains the differential equations describing the switched system which corresponds to the proposed EL parameters (6.15)

$$L\ddot{q}_L = -R(1 - u)\left[(1 - u)\dot{q}_L - \dot{q}_C\right] + E - (1 - u)V_F - \left[r_L + (1 - u)R_F + ur_{DS}\right]\dot{q}_L$$
$$q_C/C = R\left[(1 - u)\dot{q}_L - \dot{q}_C\right] - r_C\dot{q}_C$$

which can be rewritten, after substitution of the second equation into the first, as

$$\ddot{q}_L = -\frac{1}{L}r\dot{q}_L - (1 - u)\frac{R}{r_C + R}\frac{q_C}{LC} + \frac{E}{L} - (1 - u)\frac{V_F}{L}$$
$$\dot{q}_C = (1 - u)\frac{R}{r_C + R}\,\dot{q}_L - \frac{1}{r_C + R}\,q_C$$

where r denotes an "equivalent switched resistance" defined as

$$r = \left[r_L + ur_{DS} + +(1 - u)R_F + (1 - u)^2(r_C \| R)\right]$$

Using $x_1 = \dot{q}_L$ and $x_2 = q_C/C$, as the state variables representing the inductor current and the capacitor voltage, one obtains

$$\dot{x}_1 = -\frac{1}{L}r\,x_1 - (1 - u)\frac{R}{L(r_C + R)}\,x_2 + \frac{E}{L} - (1 - u)\frac{V_F}{L} \tag{6.17}$$
$$\dot{x}_2 = (1 - u)\,\frac{R}{(r_C + R)C}\,x_1 - \frac{1}{(r_C + R)C}\,x_2 \tag{6.18}$$

It may be easily verified that if the values of the parasitic elements, r_C, r_L, r_{DS}, R_F and V_F are all set to zero in (6.17), (6.18) one obtains the ideal switched model for the Boost converter presented in (6.9), (6.10).

On the basis of this switched circuit model, the output voltage accross the load resistance, denoted by x_o, is readily obtained as

$$x_o = \frac{R}{r_C + R}\,x_2 + (1 - u)\frac{r_C R}{r_C + R}\,x_1$$

which is in complete accordance with expressions (6.12) and (6.14), describing the load voltages found for each one of the circuit topologies.

3   Hamiltonian modeling

Broadly speaking, a generalized Hamiltonian system is a lossless nonlinear system, provided with external control inputs, which is defined in a traditional state space framework. However, a generalized Hamiltonian system is identified by an explicit inclusion of the contribution of the gradient of the total energy of the system in the differential equations describing the controlled evolution of the state vector. In general, electric circuits are not considered to be Hamiltonian due to the presence of resistive elements and some non dissipative elements which, nevertheless, disrupt the fundamental "canonical structure" of a lossless system. Thus, circuits with transformers, magnetic couplings between inductors, diodes, etc. are not *per se* considered to be truly generalized Hamiltonian systems.

Yet the Hamiltonian formalism and much of its rich geometric structure and appealing physical properties, can still be suitably preserved in the case of electric circuits by allowing a basic lossless structure, constituted by the underlying *LC* circuit, while considering the rest of the elements to be modeled as external ports or inputs along with constitutive relations describing their particular input-output properties. In this manner, resistive circuits, switched circuits and other interesting circuits can be systematically modeled with an extended Hamiltonian viewpoint in a relatively straightforward manner (see [178, 273]).

The Hamiltonian approach allows for the systematic modeling of networks including resistors, transformers, diodes and switches, something that is not easy, and in some cases not even possible, to achieve within the Lagrangian formalism. These non-energetic terms are first extracted from the circuit leaving an energy- conserving LC circuit with ports corresponding to the various extracted elements. The LC circuit with ports can be represented as a generalized linear Hamiltonian system with external input variables. By terminating the ports of the Hamiltonian system with the constitutive relations characterizing each of the extracted terms, one obtains a mathematical description of the original switched circuit in forced generalized Hamiltonian form with external inputs. One of the advantages of the proposed approach is that for all operating modes of the switched circuit, the description of the system uses the same state variables, the same Hamiltonian and the same dissipation functions. The obtained mathematical model is readily suitable for the application of the PBC feedback regulation design methodology.

Generalized Hamiltonian systems have been extensively treated in the literature. Fundamental references, with plenty of physical examples, are constituted in the books [62, 197] and the recent monograph [272]. The background results for the systematic treatment of this topic, within the realm of general circuit theory, is found in [178, 273]. The results in this section follow the work in [74].

3.1   Constitutive elements

Within the generalized Hamiltonian modeling approach, all circuit elements are represented as electrical *ports*, in which currents are identified as *flow* variables, denoted by f, and voltages are identified as *effort* variables, denoted by e.

We consider first the *continuous* constitutive elements, such as resistors, capacitors, inductors and transformers. Next, we discuss the *discontinuous* elements, such as switches and diodes. Denote by q_C and ϕ_L the circulating charge in the capacitor and the linkage flux in the inductor, respectively.

A   Continuous constitutive elements

The constitutive laws for (linear constant) resistive, capacitive and inductive elements are then given by,

- Resistance:

$$e_R = R f_R$$

where f_R is the current circulating through the resistance.

- Capacitor:

$$f_c = \dot{q}_C \;\; ; \;\; e_c = \frac{q_C}{C}$$

- Inductor:

$$f_L = \frac{\phi_L}{L} \;\; ; \;\; e_L = \dot{\phi}_L$$

In terms of the *natural* coordinates q_C and ϕ_L the electric energy in the capacitor (playing the role of a potential energy V) and the magnetic energy in the inductance (playing the role of a kinetic energy T) are given, respectively, by,

$$\mathcal{V} = \frac{1}{2C} q_C^2 \;\; ; \;\; \mathcal{T} = \frac{1}{2L} \phi_L^2$$

The Hamiltonian, which represents the total energy in an LC circuit, is simply given by

$$\mathcal{H} = \mathcal{T} + \mathcal{V} = \frac{1}{2C} q_C^2 + \frac{1}{2L} \phi_L^2$$

- The ideal transformer:

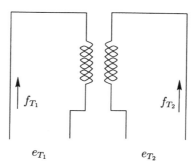

Figure 6.6: An ideal transformer circuit.

The case of an *ideal transformer*, shown in Fig. 6.6 requires the consideration of the primary and secondary terminals as two *separate* ports, each with its corresponding definition of flow and effort variables. The ports, however, interact with each other, in an external fashion, according to one of the following constitutive relations

$$\text{(a)} \quad \begin{bmatrix} e_{T1} \\ f_{T2} \end{bmatrix} = \mathcal{J}(n) \begin{bmatrix} f_{T1} \\ e_{T2} \end{bmatrix} \qquad \text{(b)} \quad \begin{bmatrix} f_{T2} \\ e_{T1} \end{bmatrix} = \mathcal{J}_T(n) \begin{bmatrix} e_{T2} \\ f_{T1} \end{bmatrix} \qquad (6.19)$$

where

$$\mathcal{J}_T(n) = \begin{bmatrix} 0 & n \\ -n & 0 \end{bmatrix}$$

is a skew symmetric matrix parameterized by n, thus reflecting the power-conservation property of the transformer circuit.

B Discontinuous constitutive elements

The only discontinuous devices that we consider here are the *ideal switch*, and the diode, although the approach is suitable for the modeling of some other discontinuous devices (such as transistors, for instance).

- The ideal switch:

An *ideal switch* can be considered as a *lossless* element, due to the fact that it can conduct current at zero voltage (while being closed) and it is capable of holding a voltage with zero current (when it is open). The variable describing the switch is u which takes values from the discrete set $\{0, 1\}$. The following relations among the flow and effort variables describe the behaviour of a switching element,

$$u = \begin{cases} 1 & \Rightarrow \quad ; \quad f_{SW} \in \mathbb{R} \; , \quad e_{SW} = 0 \\ 0 & \Rightarrow \quad ; \quad f_{SW} = 0 \; ; \quad e_{SW} \in \mathbb{R}. \end{cases}$$

Thus the ideal switch can, thus, conduct current in both directions.

- The ideal diode:

An *ideal diode* is a particular case of a unidirectional controlled switch. Its input-output behaviour is represented by the following relations

$$\begin{cases} \text{Mode 1} & e_D \leq 0, \quad f_D = 0 \\ \text{Mode 2} & f_D \geq 0, \quad e_D = 0, \quad f_D e_D = 0. \end{cases}$$

3.2   LC circuits

As pointed out above in the Hamiltonian modeling approach the non-energetic elements such as resistors, transformers, diodes and switches are first extracted from the circuit, thereby leaving an energy-conserving LC circuit with ports corresponding to the various extracted elements. This LC circuit with ports can be represented in an intrinsic way as a Hamiltonian system with port variables. The representation of the original circuit is then obtained by terminating the ports of this Hamiltonian system by the extracted non-energetic elements. We briefly review in this subsection the modeling of LC circuits.

It has been shown in [178] that an n-element LC circuit with m external ports, and total energy given by

$$\mathcal{H}(x) = \frac{1}{2} x^\top Q x$$

where $x \in \mathbb{R}^n$ is the state vector of the circuit, consisting of inductance fluxes ϕ_{Li} and capacitor charges q_{Ci} and Q being a diagonal symmetric matrix containing the circuit parameters $1/C_i$, $1/L_i$ can always be written in the form,

$$\dot{x} \;=\; JQx + Gu \qquad\qquad (6.20)$$

where $u \in R^m$ is the vector of external inputs to the system. The matrix G is called the *input matrix* and $\mathcal{J}$ is an $n \times n$ skew-symmetric matrix, which is called the *structure matrix*. The matrices G and $\mathcal{J}$ are determined from Kirchoff's laws.

It can actually be shown that the *natural* outputs of the generalized Hamiltonian system (6.20) can always be written in the form

$$y \;=\; G^\top Q x + D u \;\; ; \;\; y \in R^m$$

where D is a skew-symmetric matrix, called the *throughput matrix*, that appears whenever there are static relations between port variables. The skew-symmetry of these matrices stems from the fact that the interconnections are all energy conserving.

Furthermore, it immediately follows that along the trajectories of the system, we have $\dot{\mathcal{H}} = u^\top y$, which expresses *energy conservation*, then integrating from 0 to T we verify the passivity property of the circuit.

Remark 6.6 It is important to mention that the linear Hamiltonian systems we are dealing with are a special class of a more general class of systems called the *generalized Hamiltonian systems* which include systems which are described in local coordinates by the following equations,

$$\dot{x} = J(x)\frac{\partial H}{\partial x} + G(x)u$$
$$y = G^{\top}(x)\frac{\partial H}{\partial x} + D(x)u$$

where the matrices, $J(x)$, $G(x)$ and $D(x)$ are not constant, but functions of the state. In such a general case the Hamiltonian need not be a *quadratic* function of x.

3.3 Examples

A Boost converter

Let us consider the ideal Boost converter circuit shown in Fig. 6.1. The circuit coordinates are $x = [\phi_L, q_C]^{\top}$, and the electric and magnetic energies are given by

$$\mathcal{V} = \frac{1}{2C}q_C^2$$
$$\mathcal{T} = \frac{1}{2L}q_L^2,$$

So the Hamiltonian is then given by

$$\mathcal{H} = \frac{1}{2C}q_C^2 + \frac{1}{2L}q_L^2 = \frac{1}{2}[\phi_L, q_C]^{\top}\begin{bmatrix} \frac{1}{L} & 0 \\ 0 & \frac{1}{C} \end{bmatrix}\begin{bmatrix} \phi_L \\ q_C \end{bmatrix} = \frac{1}{2}x^{\top}Qx$$

and the gradient vector of the Hamiltonian is simply

$$\frac{\partial \mathcal{H}}{\partial x} = Qx = \begin{bmatrix} \phi_L/L \\ q_C/C \end{bmatrix}.$$

We replace the single switching element in the Boost circuit by two ideal *dependent*, or *conjugated*, switches as shown in Fig. 6.7

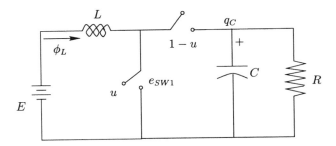

Figure 6.7: Boost converter model with conjugate switches.

Using the concept of extracting the switches and the resistors from the circuit we have,

$$\begin{bmatrix} \dot{\phi}_L \\ \dot{q}_C \end{bmatrix} = \begin{bmatrix} 1 \\ 0 \end{bmatrix} (-e_{SW1}) + \begin{bmatrix} 0 \\ 1 \end{bmatrix} (f_{SW2}) + \begin{bmatrix} 0 \\ 1 \end{bmatrix} (-f_R) + \begin{bmatrix} 1 \\ 0 \end{bmatrix} E$$

and the corresponding outputs for the source, resistance and switches are given by

$$\begin{bmatrix} f_s \\ f_{SW1} \\ e_{SW2} \\ e_R \end{bmatrix} = \begin{bmatrix} 1 & 0 \\ 1 & 0 \\ 0 & 1 \\ 0 & 1 \end{bmatrix} Qx + \begin{bmatrix} 0 & 0 & 0 & 0 \\ 0 & 0 & -1 & 0 \\ 0 & 1 & 0 & 0 \\ 0 & 0 & 0 & 0 \end{bmatrix} \begin{bmatrix} E \\ -e_{SW1} \\ f_{SW2} \\ -f_R \end{bmatrix}. \tag{6.21}$$

From the constitutive relations for resistances and equation (6.21) we have

$$f_R = \frac{1}{R} e_R = \frac{1}{RC} q_C$$

thus for each switch position we have

$$u = \begin{cases} 0 \;\Rightarrow\; \begin{cases} e_{SW1} = q_C/C & e_{SW2} = 0 \\ f_{SW1} = 0 & f_{SW2} = \phi_L/L \end{cases} \\ 1 \;\Rightarrow\; \begin{cases} e_{SW1} = 0 & e_{SW2} = q_C/C \\ f_{SW1} = \phi_L/L & f_{SW2} = 0. \end{cases} \end{cases}$$

Notice that even though the number of switches is two we have only two modes of operation instead of four due to the fact that the ideal switches are dependent (when one is closed the other is open and vice-versa). We can then write

$$e_{SW1} = (1 - u)q_C/C \;\;;\;\; f_{SW2} = (1 - u)\phi_L/L$$

Summarizing the derivations presented above, we obtain the dynamical equations for the Boost converter model

$$\begin{bmatrix} \dot{\phi}_L \\ \dot{q}_C \end{bmatrix} = \begin{bmatrix} 0 & -(1 - u) \\ (1 - u) & 0 \end{bmatrix} \begin{bmatrix} \phi_L/L \\ q_C/C \end{bmatrix} + \begin{bmatrix} 0 & 0 \\ 0 & -\frac{1}{R} \end{bmatrix} \begin{bmatrix} \phi_L/L \\ q_C/C \end{bmatrix} + \begin{bmatrix} 1 \\ 0 \end{bmatrix} E.$$

Letting $x_1 = \phi_L/L$ and $x_2 = q_C/C$ we recover the traditional switched model (6.9), (6.10) for the Boost converter derived using the Lagrangian approach.

B   Ćuk converter

Consider the Ćuk converter, shown in Fig. 6.3 provided with an ideal switch. The system coordinates are given by $x = [\phi_{L1}, q_{C2}, \phi_{L3}, q_{C4}]^\top$ and the electric and magnetic energies are given by

$$\mathcal{V} = \frac{1}{2C_2}q_{C2}^2 + \frac{1}{2C_4}q_{C4}^2$$

$$\mathcal{T} = \frac{1}{2L_1}\phi_{L1}^2 + \frac{1}{2L_3}\phi_{L3}^2.$$

As before, we introduce two single ideal conjugate switches that replace the single switch in order to obtain a graphical model from which the switch extraction procedure is more clearly accomplished (see Fig. 6.8).

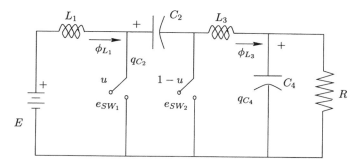

Figure 6.8: Ćuk converter model with conjugate switches.

Then we have

$$
\begin{bmatrix} \dot{\phi}_{L1} \\ \dot{q}_{C2} \\ \dot{\phi}_{L3} \\ \dot{q}_{C4} \end{bmatrix} = \begin{bmatrix} 0 & -1 & 0 & 0 \\ 1 & 0 & 0 & 0 \\ 0 & 0 & 0 & -1 \\ 0 & 0 & 1 & 0 \end{bmatrix} \begin{bmatrix} \phi_{L1}/L_1 \\ q_{C2}/C_2 \\ \phi_{L3}/L_3 \\ q_{C4}/C_4 \end{bmatrix} + \begin{bmatrix} 1 \\ 0 \\ 0 \\ 0 \end{bmatrix} E
$$

$$
+ \begin{bmatrix} 0 \\ 1 \\ 0 \\ 0 \end{bmatrix} (-f_{SW1}) + \begin{bmatrix} 1 \\ 0 \\ -1 \\ 0 \end{bmatrix} (-e_{SW2}) + \begin{bmatrix} 0 \\ 0 \\ 0 \\ 1 \end{bmatrix} (-f_R)
$$

with the outputs for the source, switches and resistance given by

$$
\begin{bmatrix} f_s \\ e_{SW1} \\ f_{SW2} \\ e_R \end{bmatrix} = \begin{bmatrix} 1 & 0 & 0 & 0 \\ 0 & 1 & 0 & 0 \\ 1 & 0 & -1 & 0 \\ 0 & 0 & 0 & 1 \end{bmatrix} Qx + \begin{bmatrix} 0 & 0 & 0 & 0 \\ 0 & 0 & -1 & 0 \\ 0 & 1 & 0 & 0 \\ 0 & 0 & 0 & 0 \end{bmatrix} \begin{bmatrix} E \\ -f_{SW1} \\ -e_{SW2} \\ -f_R \end{bmatrix}
$$

where $f_R = (1/R)e_R = (1/RC_4)q_{C4}$. The effort and flow variables for the two possible switch positions are given by

$$
u = \begin{cases} 1 & \Rightarrow & \begin{cases} f_{SW2} = 0 \\ e_{SW1} = 0 \end{cases} & \begin{cases} e_{SW2} = -q_{C2}/C_2 \\ f_{SW1} = \phi_{L1}/L_1 - \phi_{L3}/L_3 \end{cases} \\ 0 & \Rightarrow & \begin{cases} e_{SW2} = 0 \\ f_{SW1} = 0 \end{cases} & \begin{cases} f_{SW2} = \phi_{L1}/L_1 - \phi_{L3}/L_3 \\ e_{SW1} = q_{C2}/C_2 \end{cases} \end{cases}
$$

then we can write

$$
\begin{aligned}
e_{SW2} &= -u q_{C2}/C_2 \\
f_{SW1} &= u\left(\phi_{L1}/L_1 - \phi_{L3}/L_3\right).
\end{aligned}
$$

In summary we obtain the following dynamical equations

$$
\begin{bmatrix} \dot{\phi}_{L1} \\ \dot{q}_{C2} \\ \dot{\phi}_{L3} \\ \dot{q}_{C4} \end{bmatrix} = \begin{bmatrix} 0 & -(1-u) & 0 & 0 \\ (1-u) & 0 & u & 0 \\ 0 & -u & 0 & -1 \\ 0 & 0 & 1 & 0 \end{bmatrix} \begin{bmatrix} \phi_{L1}/L_1 \\ q_{C2}/C_2 \\ \phi_{L3}/L_3 \\ q_{C4}/C_4 \end{bmatrix} + \begin{bmatrix} E \\ 0 \\ 0 \\ 0 \end{bmatrix} + \begin{bmatrix} 0 \\ 0 \\ 0 \\ -\frac{q_{C4}}{RC_4} \end{bmatrix}
$$
$$
f_s = \phi_{L1}/L_1.
$$

C   A Boost converter model with switches and diodes

The ideal switches considered above allows the flow of current in both directions. In actual converters, this is not the case due to the fact that ideal switches are usually realized by means of transistors and diodes.

We illustrate the main differences with the above developments by considering the inclusion of a clamping diode in the Boost converter circuit as illustrated in Fig. 6.9

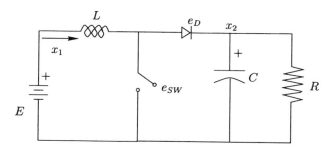

Figure 6.9: Boost converter model with a diode.

As before, one obtains the model by extracting both, diodes and switches, from the circuit and treating them as external ports. This procedure leads to

$$\left[\begin{array}{c}\dot{\phi}_L \\ \dot{q}_C\end{array}\right] = \left[\begin{array}{cc}0 & -1 \\ 1 & 0\end{array}\right]\left[\begin{array}{c}\phi_L/L \\ q_C/C\end{array}\right] + \left[\begin{array}{c}1 \\ 0\end{array}\right]E + \left[\begin{array}{c}0 \\ 1\end{array}\right](-f_R) + \left[\begin{array}{cc}1 & 0 \\ 0 & 1\end{array}\right]\left[\begin{array}{c}-e_D \\ -f_{SW}\end{array}\right]$$

$$e_R = [0\ 1]Qx = q_C/C$$

with outputs for the port variables given by,

$$\left[\begin{array}{c}f_D \\ e_{SW}\end{array}\right] = \left[\begin{array}{cc}1 & 0 \\ 0 & 1\end{array}\right]\left[\begin{array}{c}q_L/L \\ q_C/C\end{array}\right] + \left[\begin{array}{cc}0 & -1 \\ 1 & 0\end{array}\right]\left[\begin{array}{c}e_D \\ f_{SW}\end{array}\right],$$

and the output current, flowing through the resistor given by

$$f_R = \frac{1}{R}e_R = \frac{1}{RC}q_C$$

Analyzing now each one of the two positions of the switch $u \in \{0,1\}$, we have

$$u = \begin{cases} 0 & \Rightarrow & f_{SW} = 0 & \Rightarrow & \begin{cases} e_D = e_{SW} - q_C/C \\ f_D = \phi_L/L \end{cases} \\ 1 & \Rightarrow & e_{SW} = 0 & \Rightarrow & \begin{cases} e_D = -q_C/C \\ f_{SW} = \phi_L/L - f_D. \end{cases} \end{cases}$$

The values of f_{SW} and e_D can then be parameterized in terms of u as

$$\left[\begin{array}{c}-e_D \\ -f_{SW}\end{array}\right] = \left[\begin{array}{c}uq_C/C - (1-u)e_D \\ u(-\phi_L/L + f_D)\end{array}\right]$$

which yields the following switched model placed in terms of the effort and flow variables of the diode:

$$\left[\begin{array}{c}\dot{\phi}_L \\ \dot{q}_C\end{array}\right] = \left[\begin{array}{cc}0 & -(1-u) \\ 1-u & 0\end{array}\right]\left[\begin{array}{c}\phi_L/L \\ q_C/C\end{array}\right] + \left[\begin{array}{c}E \\ 0\end{array}\right] + \left[\begin{array}{c}0 \\ -\frac{q_C}{RC}\end{array}\right] + \left[\begin{array}{c}-(1-u)e_D \\ uf_D\end{array}\right].$$

For each switch position, there are two modes of operation depending on whether or not the diode is in conduction mode. Thus we can distinguish four modes of operation.

Mode 1 : $u = 0, e_D = 0$ $\Rightarrow f_D = \phi_L/L$ $\Rightarrow \begin{cases} \dot{\phi}_L = -q_C/C + E \\ \dot{q}_C = \phi_L/L - \frac{1}{RC}q_C \end{cases}$

Mode 2 : $u = 1, f_D = 0$ $\Rightarrow e_D = -q_C/C$ $\Rightarrow \begin{cases} \dot{\phi}_L = E \\ \dot{q}_C = -q_C/C \end{cases}$

Mode 3 : $u = 0, f_D = \phi_L/L = 0 \Rightarrow e_D = E - q_C/C$ $\Rightarrow \begin{cases} \dot{\phi}_L = 0 \\ \dot{q}_C = -(1/RC)q_C \end{cases}$

Mode 4 : $u = 1, e_D = q_C/C = 0 \Rightarrow f_D = 0$ $\Rightarrow \begin{cases} \dot{\phi}_L = E \\ \dot{q}_C = 0. \end{cases}$

In order to know the mode which the circuit is operating in, one must firstly observe the switch position value and then look at the states of the system. For $u = 0$ there are two modes, if the signal $\phi_L/L > 0$ then the system is in mode 1, and if $\phi_L/L = 0$ then it is in mode 3. For $u = 1$, if $q_C/C > 0$ then the system is in mode 2, but if $q_C/C = 0$ then it is operating in mode 4.

D   Converter circuits with switches, diodes and transformers

To conclude this section, we consider the circuit depicted in Fig. 6.10 commonly known as the *Flyback*.

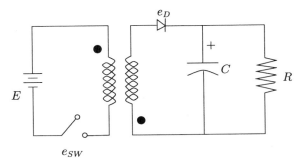

Figure 6.10: The Flyback converter circuit.

It is in fact a Buck-boost converter with an ideal transformer isolating the output circuit from the input circuit.

Contrary to previous cases we proceed to establish the generalized Hamiltonian model of the switched circuit for each switch position. This demonstrates the flexibility of the method.

Case $u = 1$. The equations for this case are given by

$$\begin{bmatrix} \dot{\phi}_L \\ \dot{q}_C \end{bmatrix} = \begin{bmatrix} 1 \\ 0 \end{bmatrix} e_s + \begin{bmatrix} 0 \\ 1 \end{bmatrix} (-f_R) + \begin{bmatrix} 0 \\ f_D \end{bmatrix}$$

with the corresponding outputs

$$\begin{aligned} e_R &= q_C/C \\ e_D &= -q_C/C - \frac{1}{n}e_s = -q_C/C - \frac{1}{n}E \\ f_s &= \phi_L/L + \frac{1}{n}f_D \end{aligned}$$

where f_s is the current flowing through the source and

$$f_R = e_R/R = \frac{1}{RC}q_C.$$

If the converter is in the normal mode of operation, i.e., $q_C/C \geq 0$, then $e_D < 0$. This in turn, implies that $f_D = 0$, and we have only one mode of operation. For this case then, the model results in

$$\text{Mode 1 :} \quad \begin{bmatrix} \dot{\phi}_L \\ \dot{q}_C \end{bmatrix} = \begin{bmatrix} 1 \\ 0 \end{bmatrix} e_s + \begin{bmatrix} 0 \\ 1 \end{bmatrix} (-\frac{1}{RC}q_C).$$

Case $u = 0$. The equations are

$$\begin{bmatrix} \dot{\phi}_L \\ \dot{q}_C \end{bmatrix} = \begin{bmatrix} 0 \\ 1 \end{bmatrix} (-f_R) + \begin{bmatrix} 0 \\ f_D \end{bmatrix} + \begin{bmatrix} e_{T1} \\ 0 \end{bmatrix}$$

with corresponding outputs

$$\begin{aligned} e_R &= q_C/C \\ e_D &= -q_C/C + e_{T2} \\ f_{T1} &= \phi_L/L. \end{aligned}$$

and from the constitutive relations of the ideal transformer, we have

$$e_{T1} = -ne_{T2} \quad ; \quad f_{T2} = nf_{T1}$$

From this and the fact that $f_D = f_{T2}$ we obtain that

$$e_{T1} = -ne_D - nq_C/C \quad ; \quad f_D = n\phi_L/L$$

whereas for the resistance,

$$f_R = \frac{1}{RC}q_C.$$

Substituting these expressions in the model, we obtain

$$\begin{bmatrix} \dot{\phi}_L \\ \dot{q}_C \end{bmatrix} = \begin{bmatrix} 0 & -n \\ n & 0 \end{bmatrix} \begin{bmatrix} \phi_L/L \\ q_C/C \end{bmatrix} + \begin{bmatrix} 0 \\ 1 \end{bmatrix} (-\frac{1}{RC}q_C) + \begin{bmatrix} -ne_D \\ 0 \end{bmatrix}.$$

As we can see for the case $u = 0$ there are two modes depending on whether the diode is conducting or not.

$$\text{Mode 2}: \quad e_D = 0 \quad \Rightarrow \quad f_D = n\phi_L/L$$

hence

$$\begin{bmatrix} \dot{\phi}_L \\ \dot{q}_C \end{bmatrix} = \begin{bmatrix} 0 & -n \\ n & 0 \end{bmatrix} \begin{bmatrix} q_L/L \\ q_C/C \end{bmatrix} + \begin{bmatrix} 0 \\ 1 \end{bmatrix} (-\frac{1}{RC}q_C)$$

$$\text{Mode 3}: \quad f_D = 0 \quad \Rightarrow \quad e_D = -q_C/C \quad \Rightarrow \quad \begin{bmatrix} \dot{\phi}_L \\ \dot{q}_C \end{bmatrix} = \begin{bmatrix} 0 \\ 1 \end{bmatrix} (-\frac{1}{RC}q_C)$$

The mode in which the system is operating is found out by first observing in which position is located the switch. If $u = 0$ then the circuit is in mode 1 of operation. If $u = 1$ then we should observe the state of the circuit. If the quantity $\phi_L/L > 0$ then we are in mode 2, if, on the other hand, it is exactly zero then we are in mode 3.

4   Average models of PWM regulated converters

In this section a Lagrangian dynamics approach is used for deriving a physically oriented models of the *average* behavior of pulse width modulation (PWM) regulated DC-to-DC power converters. As in the switched case, the approach consists in first establishing the EL parameters of the circuits associated with each one of the topologies corresponding to the two possible positions of the regulating switch. Secondly, an average PWM model of the non-invariant EL parameters can then be proposed by their suitable modulation through the duty ratio function. This modulation is done in a *consistent* fashion so that, under extreme duty ratio saturation conditions, the original EL parameters, corresponding to the two intervening circuit topologies, are exactly recovered. Secondly, an *intermediary* condition which is consistent with the physically plausible interpretation of the average value of the EL parameter guides the choice of the weighting of each intervening EL parameter by means of a simple linear function of the duty ratio function.

The average EL parameter considerations lead, through use of the classical Lagrangian dynamics equations, to systems of continuous differential equations, describing the average PWM converter behavior. These equations are interpretable in terms of ideal equivalent circuit realizations obtained by replacing the switching device by a suitable ideal transformer. This particular result, first reported in [241], is in accordance with well-known circuit equivalents of PWM switches already derived in

[150] and [281]. The obtained average PWM models entirely coincide with the *state average models* of DC-to-DC power converters introduced in [182] and also with the *infinite switching frequency* model reported in [240].

We begin this section by considering some general remarks about PWM controlled nonlinear systems. We then present some considerations about the average modeling of discontinuously controlled EL systems. The general modeling procedure is applied to obtaining the average models, from the Lagrangian formalism viewpoint, of DC-to-DC power converters of the "Boost" , the "Buck-boost" and the "Buck" type, equipped with ideal switching devices.

4.1 General issues about pulse-width-modulation

PWM is a widely used regulation technique for switch commanded systems. Typically, a switch regulated system exhibits as a control input called the *switch position function*, whose values can always be constrained to be found in the discrete set $\{0, 1\}$. In such a context, a "pulse" refers to a scalar time signal corresponding to the switch position function which is zero everywhere except on a finite interval of time where it takes the value of one. The PWM regulation technique is used in conjunction with a sampling process carried out at a fixed frequency. At each sampling instant, the state of the controlled system is determined and, on its basis, the width of a pulse is specified for the current sampling interval. The width T of the sampling interval is usually referred to as the *duty cycle* and the fraction of this time interval occupied by the pulse is usually referred to as the *duty ratio* and denoted by $\mu(\cdot)$. Although many particular forms of PWM schemes do exist, we assume here that the pulse begins precisely at the sampling instant t_k and that it ends before the sampling interval is over, i.e, before $t_k + T$. If the pulse occupies all of the sampling interval, or no part of it, the duty ratio is said to be operating under *saturation conditions*. Hence, the duty ratio function is a feedback function of the state, taking values in the open interval $(0, 1)$.

Basic work related to PWM control schemes, in linear and nonlinear dynamical systems, can be found in [220, 250]. Seminal work is due to the efforts of Tsypkin [268]. For a more recent account with a rather complete survey of early work see [144]. Also, some generalizations and applications of PWM control in the area of DC-to-DC power conversion can be found in [239, 240].

Consider a nonlinear single-input single output system of the form,

$$\dot{x} = f(x) + g(x)u \qquad (6.22)$$

where u is a switch position function taking values in the discrete set $\{0, 1\}$. Generally speaking, a PWM regulation policy is specified as follows,

$$u = \begin{cases} 1 & \text{for } t_k < t \le t_k + \mu(x(t_k))T \\ 0 & \text{for } t_k + \mu(x(t_k))T \le t < t_k + T = t_{k+1} \end{cases} \qquad (6.23)$$

The exact determination of the state $x(t)$ of the controlled system (6.22)- (6.23), at the sampling instants t_k ; $k = 0, 1, 2, \ldots$, for a given duty ratio function is an extremely difficult problem, except for the simplest drift vector fields $f(x)$ and control input fields $g(x)$. For the linear systems case, the problem admits only approximate solutions even for low dimensional systems.

Continuous time average models of PWM controlled systems have been used as simple and useful substitutes of exactly discretized switch regulated systems. Here we explore one such average model usually known as *average state model*. It consists in the obtained continuous time description of the PWM regulated system when the sampling interval is reduced to zero by a limiting argument.

Consider an arbitrary sampling time interval $[t_k, t_{k+1}] = [t_k, t_k + \mu(x(t_k))T] \cup [t_k + \mu(x(t_k))T, t_{k+1}]$, of length T. We shall determine, on the basis of (6.22) and (6.23), the value of the state, at the end of the sampling interval, by rewriting the differential equation as an integral equation.

$$x(t_k + \mu(x(t_k))T) = x(t_k) + \int_{t_k}^{t_k+\mu(x(t_k))T} [f(x(\sigma)) + g(x(\sigma))]\, d\sigma. \qquad (6.24)$$

On the other hand, at time $t_k + T = t_{k+1}$ the state is given by

$$x(t_k + T) = x(t_k + \mu(x(t_k))T) + \int_{t_k+\mu(x(t_k))T}^{t_k+T} f(x(\sigma))d\sigma \qquad (6.25)$$

Substituting (6.24) into (6.25) one obtains,

$$
\begin{aligned}
x(t_k + T) &= x(t_k) + \int_{t_k}^{t_k+\mu(x(t_k))T} f(x(\sigma)) + g(x(\sigma))d\sigma + \int_{t_k+\mu(x(t_k))T}^{t_k+T} f(x(\sigma))d\sigma \\
&= x(t_k) + \int_{t_k}^{t_k+T} f(x(\sigma))d\sigma + \int_{t_k}^{t_k+\mu(x(t_k))T} g(x(\sigma))d\sigma
\end{aligned}
$$

The difference of the states at the beginning and at the end of the sampling interval $[t_k, t_k + T]$, divided by the sampling period T is given by

$$\frac{1}{T}[x(t_k + T) - x(t_k)] = \frac{1}{T}\int_{t_k}^{t_k+T} f(x(\sigma))d\sigma + \frac{1}{T}\int_{t_k}^{t_k+\mu(x(t_k))T} g(x(\sigma))d\sigma$$

Taking limits as $T \to 0$ and letting the time instant t_k take the generic value t one obtains,

$$
\begin{aligned}
\lim_{T\to 0}\frac{1}{T}[x(t + T) - x(t)] &= \lim_{T\to 0}\left[\frac{1}{T}\int_{t}^{t+T} f(x(\sigma))d\sigma + \frac{1}{T}\int_{t}^{t+\mu(x(t))T} g(x(\sigma))d\sigma\right] \\
&= \dot{x}(t) = f(x(t)) + g(x(t))\mu(x(t)). \qquad (6.26)
\end{aligned}
$$

The continuous average model of a PWM regulated system is thus given by the same nonlinear model except that the discrete control input u is substituted by the continuous, but limited, duty ratio function μ. In order to establish a difference between the actual nonlinear model and its continuous time average (6.26), we denote the average state vector by $z(t)$ rather than by $x(t)$. In summary we have the following.

Proposition 6.7 *The continuous state average model of a PWM regulated system,*

$$\dot{x} = f(x) + g(x)u \tag{6.27}$$

with,

$$u = \begin{cases} 1 & \text{for } t_k < t \le t_k + \mu(x(t_k))T \\ 0 & \text{for } t_k + \mu(x(t_k))T \le t < t_k + T = t_{k+1} \end{cases} \tag{6.28}$$

is given by

$$\dot{z} = f(z) + g(z)\mu \tag{6.29}$$

□

The continuous time approximation (6.29) of PWM regulated systems (6.27)-(6.28) has been extensively used in the literature devoted to the control of nonlinear PWM regulated systems to transform an essentially discrete-time nonlinear control synthesis problem into an equivalent continuous time nonlinear control problem. Once the continuous time control-constrained regulation problem has been solved for the average approximation, in the sense that a specification of the duty ratio has been found in the form of a sate feedback function, then the obtained average closed loop duty ratio synthesis law is used in the actual PWM regulated model using the actual sampled state variables instead of their averaged values. The nature of the approximation depends on how close is the actual fixed sampling interval to the ideal value of zero. Generally speaking, the higher the sampling frequency the more accurate is the average continuous time approximation.

4.2 Examples

In this subsection we derive EL models to describe the average behaviour of PWM–controlled DC–to–DC converters. We will see that the average models exactly coincide with the exact switched models derived in the two previous sections, simply replacing the switch position u by the duty ratio μ. For the sake of brevity we present only the example of the boost and the Buck–boost converters.

A Boost converter

Consider the switch–regulated Boost converter circuit of Fig. 6.1. The differential equations describing the circuit are given by (6.4), (6.5), and are repeated here for convenience as

$$\dot{x}_1 = -(1 - u) \frac{1}{L} x_2 + \frac{E}{L} \tag{6.30}$$

$$\dot{x}_2 = (1 - u) \frac{1}{C} x_1 - \frac{1}{RC} x_2. \tag{6.31}$$

A PWM policy regulating the switch position function u, may be specified as follows,

$$u(t) = \begin{cases} 1 \text{ for } \quad t_k \leq t < t_k + \mu(t_k)T \\ 0 \text{ for } \quad t_k + \mu(t_k)T \leq t < t_k + T \end{cases}$$
$$t_{k+1} = t_k + T \quad ; \quad k = 0, 1, \dots \tag{6.32}$$

where t_k represents a sampling instant; the parameter T is the fixed sampling period, also called the *duty cycle*; the sampled values of the state vector $x(t)$ of the converter are denoted by $x(t_k)$. The function, $\mu(\cdot)$, is the *duty ratio function*, truly acting as an external control input to the average PWM model of the converter (see [240]). The value of the duty ratio function, $\mu(t_k)$, determines, at every sampling instant, t_k, the width of the upcoming "ON" pulse as $\mu(t_k)T$ (during this period the switch is fixed at the position represented by $u = 1$). The actual duty ratio function, $\mu(\cdot)$, is evidently a function limited to take values on the closed interval $[0, 1]$ of the real line.

We recall, from previous sections, the EL parameters associated with each one of the two electric circuits obtained as a result of fixing the switch at each one of the two possible regulating positions.

For $u = 1$ we have,

$$\begin{cases} \mathcal{T}_1(\dot{q}_L) = \frac{1}{2} L \, \dot{q}_L^2 \; ; \; \mathcal{V}_1(q_C) = \frac{1}{2C} q_C^2 \\ \mathcal{F}_1(\dot{q}_C) = \frac{1}{2} R \dot{q}_C^2 \; ; \; \mathcal{Q}_{q_L} = E \; ; \; \mathcal{Q}_{q_C} = 0 \end{cases} \tag{6.33}$$

For $u = 0$, we have,

$$\begin{cases} \mathcal{T}_0(\dot{q}_L) = \frac{1}{2} L \, \dot{q}_L^2 \; ; \; \mathcal{V}_0(q_C) = \frac{1}{2C} q_C^2 \\ \mathcal{F}_0(\dot{q}_L, \dot{q}_C) = \frac{1}{2} R \left(\dot{q}_C - \dot{q}_L \right)^2 \; ; \; \mathcal{Q}_{q_L} = E \; ; \; \mathcal{Q}_{q_C} = 0 \end{cases} \tag{6.34}$$

Note that, according to the PWM switching policy (6.32), on every sampling interval of period T, the Rayleigh dissipation function $\mathcal{F}_1(\dot{q}_C)$ is valid only a fraction of the sampling period given by $\mu(t_k)$ while the Rayleigh dissipation function $\mathcal{F}_0(\dot{q}_L, \dot{q}_C)$ is valid a fraction of the sampling period equal to $(1 - \mu(t_k))$.

There are, of course, a variety of ways in which one could reasonably propose an average value of the Rayleigh dissipation function for a circuit of the form (6.30), (6.31) undergoing a switching policy of the form (6.32). One possible and perhaps natural way is to propose the following set of average EL parameters,

$$
\begin{cases}
\mathcal{T}_\mu(\dot{q}_L) = \frac{1}{2} L \dot{q}_L^2 \;\; ; \;\; \mathcal{V}_\mu(q_C) = \frac{1}{2C} q_C^2 \\
\mathcal{F}_\mu(\dot{q}_L, \dot{q}_C) = \frac{1}{2} R \left[\dot{q}_C - (1 - \mu)\dot{q}_L \right]^2 \;\; ; \;\; \mathcal{Q}_{q_L}^\mu = E \;\; ; \;\; \mathcal{Q}_{q_C}^\mu = 0
\end{cases}
\tag{6.35}
$$

Note that in the cases where μ takes the extreme saturation values $\mu = 1$, or $\mu = 0$, one recovers, respectively, the dissipation functions $\mathcal{F}_1(\dot{q}_C)$ in (6.33) and $\mathcal{F}_0(\dot{q}_L, \dot{q}_C)$ in (6.34) from the proposed average dissipation function, $\mathcal{F}_\mu(\dot{q}_L, \dot{q}_C)$, of equation (6.35). Indeed, such a "consistency" condition is verified by noting that,

$$
\mathcal{F}_\mu(\dot{q}_L, \dot{q}_C)\big|_{\mu=0} = \mathcal{F}_0(\dot{q}_L, \dot{q}_C) \;\; ; \;\; \mathcal{F}_\mu(\dot{q}_L, \dot{q}_C)\big|_{\mu=1} = \mathcal{F}_1(\dot{q}_C).
$$

Also, it is easy to see that the proposed average Rayleigh dissipation function satisfies an important "intermediary" condition of the form,

$$
\min \{ \mathcal{F}_0(\dot{q}_L, \dot{q}_C), \mathcal{F}_1(\dot{q}_C) \} \; < \; \mathcal{F}_\mu(\dot{q}_L, \dot{q}_C) \; < \; \max \{ \mathcal{F}_0(\dot{q}_L, \dot{q}_C), \mathcal{F}_1(\dot{q}_C) \}
$$

for any μ lying in the open interval $(0, 1)$.

We note that the Lagrangian function associated with the above defined average EL parameters is actually *invariant* with respect to the switch position function. Nevertheless, to keep the notation consistent, we denote it by,

$$
\mathcal{L}_\mu = \mathcal{T}_\mu(\dot{q}_L) - \mathcal{V}_\mu(q_C) = \frac{1}{2} L \dot{q}_L^2 - \frac{1}{2C} q_C^2.
$$

One then proceeds, using the EL equations to obtain the differential equations defining the average PWM model as

$$
L\ddot{q}_L \;=\; (1 - \mu) R \left[\dot{q}_C - (1 - \mu)\dot{q}_L \right] + E \tag{6.36}
$$
$$
\frac{q_C}{C} \;=\; -R \left[\dot{q}_C - (1 - \mu)\dot{q}_L \right] \tag{6.37}
$$

which can be rewritten, after substitution of (6.37) into (6.36), as

$$
\ddot{q}_L \;=\; -(1 - \mu) \frac{q_C}{LC} + \frac{E}{L}
$$
$$
\dot{q}_C \;=\; -\frac{1}{RC} q_C + (1 - \mu)\dot{q}_L.
$$

Using $z_1 = \dot{q}_L$ and $z_2 = q_C/C$ one finally obtains

$$
\dot{z}_1 \;=\; -(1 - \mu) \frac{1}{L} z_2 + \frac{E}{L} \tag{6.38}
$$
$$
\dot{z}_2 \;=\; (1 - \mu) \frac{1}{C} z_1 - \frac{1}{RC} z_2 \tag{6.39}
$$

where we denote by z_1 and z_2 the *average input current* and the *average output capacitor voltage*, respectively, of the PWM regulated Boost converter. We establish this distinction with the non–averaged variables x_1 and x_2 so that the state variables associated with the average PWM model are not mistakingly confused with the *actual* PWM regulated circuit variables.

Note that the proposed average dynamics (6.38), (6.39) coincides with the state average model developed in [182], and with the infinite switching frequency model, or Filippov average model, found in [5, 240]. To obtain the average model (6.38), (6.39), one simply replaces the switch position function, u, in (6.30), (6.31) by the duty ratio function μ and the actual state variables x_1, x_2 by their averaged values, z_1, z_2. The developments above are formalized in the following proposition.

Proposition 6.8 *The state average model of the Boost converter (see [182]), given by (6.38), (6.39) is an EL system corresponding to the set of average EL parameters given by (6.35). These parameters are, in turn, obtained by suitable modulation, through the duty ratio function μ, of the EL parameters, given by (6.33) and (6.34), which are associated to each one of the intervening circuit topologies arising from a particular value of the switch position function.* □

Similarly to the developments of Section 2, for ease of reference we will be using the following, more compact, matrix representation of (6.38), (6.39)

$$\mathcal{D}_B \dot{z} - (1 - \mu)\mathcal{J}z + \mathcal{R}_B z = \mathcal{E}_B \qquad (6.40)$$

where

$$\mathcal{D}_B = \begin{bmatrix} L & 0 \\ 0 & C \end{bmatrix} \;\; ; \;\; \mathcal{J} = \begin{bmatrix} 0 & -1 \\ 1 & 0 \end{bmatrix} \;\; ; \;\; \mathcal{R}_B = \begin{bmatrix} 0 & 0 \\ 0 & 1/R \end{bmatrix} \;\; ; \;\; \mathcal{E}_B = \begin{bmatrix} E \\ 0 \end{bmatrix} (6.41)$$

Remark 6.9 It is easy to realize that the average model (6.38), (6.39) has a circuit–theoretic interpretation by letting the quantity $(1 - \mu)z_2$, in the first equation, represent a *controlled voltage source* while also letting the quantity $(1 - \mu)z_1$, in the second equation, represent a *controlled input current source*. Fig. 6.11 depicts the ideal equivalent circuit describing the average PWM model.

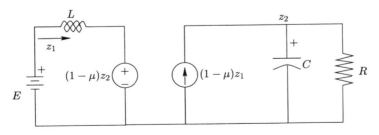

Figure 6.11: Equivalent circuit of the average PWM model of the Boost converter circuit.

In such a circuit, a quadrapole (shown in Fig. 6.12) is connecting the "input" and "output" circuits which effectively replaces, in an average sense, the actual switching device.

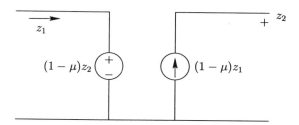

Figure 6.12: Ideal transformer representing the average PWM switch position function.

With some elementary power considerations we can establish that the quadrapole is a lossless, ideal (average) power transferring device, where the average input voltage to the quadrapole, $(1 - \mu)z_2$, is amplified to the value z_2 at the output, while the input current to the quadrapole, z_1, is attenuated to the value $(1-\mu)z_1$ at the output. The switching element has been thus effectively replaced by an *ideal transformer* with *turn ratio* parameter given by $(1 - \mu)$.

B   Buck–boost converter

Following exactly the same considerations as for the Boost converter we have the proposition below for the switch–regulated Buck–boost converter.

Proposition 6.10 *The state average model of the Buck-boost converter (see [182]) given by*

$$\dot{z}_1 = (1 - \mu)\frac{1}{L}z_2 + \mu\frac{E}{L} \tag{6.42}$$

$$\dot{z}_2 = -(1 - \mu)\frac{1}{C}z_1 - \frac{1}{RC}z_2 \tag{6.43}$$

is an EL system corresponding to the following set of average EL parameters

$$\mathcal{T}_\mu(\dot{q}_L) = \frac{1}{2}L\,\dot{q}_L^2 \;\; ; \;\; \mathcal{V}_\mu(q_C) = \frac{1}{2C}q_C^2$$

$$\mathcal{F}_\mu(\dot{q}_L, \dot{q}_C) = \frac{1}{2}R\,[\dot{q}_C + (1 - \mu)\dot{q}_L]^2 \;\; ; \;\; \mathcal{Q}_{q_L}^\mu = \mu\,E \;\; ; \;\; \mathcal{Q}_{q_C}^\mu = 0$$

obtained by suitable modulation, through the duty ratio function μ. $\square$

Fig. 6.13 depicts the equivalent circuit of the average PWM regulated dynamics for the Buck–boost converter circuit.

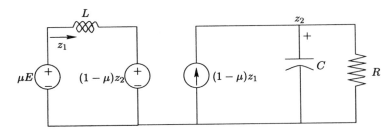

Figure 6.13: Equivalent circuit of the average PWM model of the Buck–boost converter circuit.

We will be using the following matrix representation of (6.42), (6.43)

$$\mathcal{D}_{BB}\dot{z} + (1 - \mu)\mathcal{J}z + \mathcal{R}_{BB}z = \mu\mathcal{E}_{BB}$$

where

$$\mathcal{D}_{BB} = \begin{bmatrix} L & 0 \\ 0 & C \end{bmatrix} ; \quad \mathcal{J} = \begin{bmatrix} 0 & -1 \\ 1 & 0 \end{bmatrix} ; \quad \mathcal{R}_{BB} = \begin{bmatrix} 0 & 0 \\ 0 & 1/R \end{bmatrix} ; \quad \mathcal{E}_{BB} = \begin{bmatrix} E \\ 0 \end{bmatrix} \tag{6.44}$$

4.3   Some structural properties

In Section 2 we have shown that the switched model of the DC–to–DC converters satisfies the energy balance equation (6.3). Obviously this equation also holds true for the averaged model, thus establishing the *passivity* of the operator mapping external voltage to the current. Besides this fundamental property, DC–to–DC converters have some other structural properties that will be considered for the design of the PBC. We present in this section those pertaining to the Boost and the Buck–boost converter.

A   Boost converter

We will first prove below that the system (6.38), (6.39) is *non–minimum phase* with respect to the output z_2. This poses a serious difficulty for the PBC designs, since as we have pointed out before, the implementation of the controller relies on some kind of the system inversion. Fortunately, as we will see later, the equilibria of z_2 and z_1 are in a one–to–one correspondence, and furthermore the zero dynamics with respect to z_1 are stable. This suggests to control z_2 indirectly via the regulation of z_1.

A.1 Zero dynamics

A straightforward elimination of z_1 from the set of differential equations (6.38), (6.39) leads to the following nonlinear input–output differential representation,

$$\ddot{z}_2 + \left(\frac{1}{RC} + \frac{\dot{\mu}}{1-\mu}\right)\dot{z}_2 + \frac{1}{LC}\left[(1-\mu)^2 + \frac{L}{R}\frac{\dot{\mu}}{1-\mu}\right]z_2 = (1-\mu)\frac{E}{LC}$$

The "zero dynamics" at a desired equilibrium point $z_2 = z_{2*}$, associated with this input–output representation, is obtained by letting $\dot{z}_2 = 0$ and $\ddot{z}_2 = 0$ (see [81]). The resulting differential equation describing the "remaining dynamics" of the duty ratio function μ is simply obtained as

$$\dot{\mu} = \frac{R(1-\mu)^2}{Lz_{2*}}\left[E - (1-\mu)z_{2*}\right]. \tag{6.45}$$

The equilibrium points of (6.45) are given by

$$\mu = 1 \quad ; \quad \mu = 1 - \frac{E}{z_{2*}}$$

among which the equilibrium value $\mu = \mu_* = 1 - E/z_{2*}$ has physical significance, provided $z_{2*} > E$. This fact confirms the "amplifying" features of the Boost converter. However, the phase-plane diagram of equation (6.45), shown in Fig. 6.14 , readily reveals that this equilibrium point is unstable. We conclude that the average PWM model of the Boost converter with ouput, the average capacitor voltage z_2, is actually a *non–minimum phase* system.

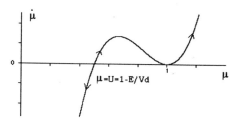

Figure 6.14: Zero-dynamics of Boost converter corresponding to constant average output voltage.

Consider now the output of the circuit to be represented by the average input current, z_1. One obtains the following differential input–output representation for the average system

$$\ddot{z}_1 + \left(\frac{1}{RC} + \frac{\dot{\mu}}{1-\mu}\right)\dot{z}_1 + \left[(1-\mu)^2\frac{1}{LC}\right]z_1 = \frac{E}{L}\left(\frac{1}{RC} + \frac{\dot{\mu}}{1-\mu}\right). \tag{6.46}$$

The "zero dynamics" at an equilibrium point $z_1 = z_{1*}$, associated with the input–output representation (6.46), is obtained as,

$$\dot{\mu} = \frac{1-\mu}{RCE}\left[(1-\mu)^2 R z_{1*} - E\right].\qquad(6.47)$$

and the equilibrium points of (6.45) are given by

$$\mu = 1 \quad ; \quad \mu = 1 - \sqrt{\frac{E}{R z_{1*}}}; \quad \mu = 1 + \sqrt{\frac{E}{R z_{1*}}}$$

The equilibrium value, $\mu = \mu_* = 1 - \sqrt{E/R z_{1*}}$, has physical significance provided $R z_{1*}$, the average steady state voltage across the load resistor, satisfies $R z_{1*} > E$. This fact confirms, once more, the "amplifying" character of the Boost converter. The phase-plane diagram of equation (6.47), shown in Fig. 6.15, reveals that this equilibrium point is now locally stable. We conclude that the average PWM model of the Boost converter, with output represented by the average input inductor current $y = z_1$, is a *minimum phase* system.

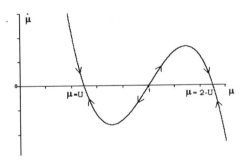

Figure 6.15: Zero-dynamics of Boost converter corresponding to constant average input current.

A.2 Equilibria

We proceed now to establish the relationship between the equilibria of the average output voltage and the average input current. To this end assume a constant duty ratio function $\mu = \mu_*$. It easily follows from the average PWM model equations (6.38), (6.39) that the corresponding stable equilibrium values for the average input current, denoted by z_{1*}, and the average output voltage, denoted by z_{2*}, are given by

$$z_{1*} = \frac{E}{(1-\mu_*)^2 R} \quad ; \quad z_{2*} = \frac{E}{1-\mu_*}.$$

Henceforth, given a desired equilibrium value z_{2*} for the output voltage, which corresponds to a constant value of the duty ratio function $\mu = \mu_* = 1 - E/z_{2*}$, the *unique* corresponding equilibrium value for the average input current is given by

$$z_{1*} = \frac{z_{2*}}{R(1 - \mu_*)} = \frac{1}{RE} z_{2*}^2. \tag{6.48}$$

This means that if we desire to regulate z_2 towards an equilibrium value z_{2*} which is known to correspond to a steady state value μ_* of the duty ratio function μ, then, such a regulation can be *indirectly* accomplished by stabilizing the average input current z_1 towards the corresponding equilibrium value z_{1*} computed from (6.48).

B   Buck–boost converter

Similarly to the Boost converter case, one can easily establish the non–minimum phase character of the average model of the PWM regulated Buck–boost converter system when the output of the system is taken as the average capacitor voltage z_2. When the output of the system is taken to be the average input inductor current, z_1, the resulting input–output system is seen to be locally minimum phase (see [244]).

Given a constant duty ratio function $\mu = \mu_*$, it easily follows from the average PWM model equations (6.42), (6.43) that the corresponding stable equilibrium values for the average input current, denoted by z_{1*}, and the average output voltage, denoted by z_{2*}, are given by

$$z_{1*} = \left[\frac{\mu_*}{(1 - \mu_*)^2} \right] \frac{E}{R} \quad ; \quad z_{2*} = -\left(\frac{\mu_*}{1 - \mu_*} \right) E \tag{6.49}$$

This means that, depending on the particular value of the steady state duty ratio function, μ_*, the Buck–boost converter can accomplish, in steady state, either source voltage "amplification" or "attenuation", modulo a polarity inversion, at the load.

It follows from (6.44) that, given a desired equilibrium value, z_{2*}, for the output voltage, which corresponds to a constant value μ_* of the duty ratio function μ, then the *unique* corresponding equilibrium value for the average input current, z_{1*}, is given by

$$z_{1*} = -\frac{z_{2*}}{R(1 - \mu_*)} = \left(\frac{z_{2*}}{RE} - \frac{1}{E} \right) z_{2*} \tag{6.50}$$

Hence, if we desire to regulate z_2 towards an equilibrium value z_{2*} which corresponds to a steady state value $\mu_* = z_{2*}/(z_{2*} - E)$ of the duty ratio function μ, then, such a regulation can be *indirectly* accomplished by stabilizing the average input current z_1 towards the corresponding equilibrium value z_{1*} computed from (6.49).

5   Conclusions

We have shown that well-known models of DC–to–DC power converters constitute a special class of EL systems with switch-dependent EL parameters. Ideal switching devices were first considered and the corresponding switched models of the traditional converters structures were derived by appropriately combining the EL parameters associated with the intervening circuit topologies. The Lagrangian formalism was also extended to handle multivariable versions of switched-regulated power converters and realistic models of traditional switch–regulated power supplies including parasitic resistances and parasitic voltage sources.

It has also been shown that using the generalized Hamiltonian formalism of [178] and [273] there exist possibilities for the systematic derivation of mathematical models that describe a wider range of DC-to-DC power converters. In particular, the Hamiltonian viewpoint naturally allows for the introduction of some continuous, as well as discontinuous, devices which are not easy to model from the Lagrangian viewpoint. Specifically, external sources, resistors, diodes, isolating transformers and parasitic sources may be systematically included in lossless LC circuits as port elements along with their characterizing constitutive relations. This procedure results in a more realistic description of switched power converters in general.

Classical state average models, or infinite switching frequency models, of DC-to-DC power converters were shown to be EL systems for a suitable set of average EL parameters. The derived average PWM models were also shown to be interpretable in terms of ideal circuit realizations including internal controlled sources and modulated external inputs.

As it has been discussed throughout the book the Lagrangian formulation is consistent with our PBC approach, which has emerged as an advantageous physically motivated controller design technique which exploits the energy structure of EL systems. Thus, this chapter shall be regarded as a first step towards the formalization and development of PBC for a variety of switched regulated models of DC–to–DC power converters, which is addressed in the next Chapter 7.

Chapter 7

Passivity-based control of DC–to–DC power converters

1 Introduction

The feedback regulation of DC–to–DC power supplies is, broadly speaking, accomplished through either PWM feedback strategies, or by inducing appropriate stabilizing sliding regimes. PWM control of these devices is treated in several books, among which we cite [117, 238]. The topic has been also extensively treated, among many others, by the third author an collaborators in [244, 249], where emphasis has been placed in using advanced nonlinear feedback control design techniques for the regulation of average PWM models of the various converters.

Sliding mode (SM) control of switched power supplies was first treated in [278] from a linearized model viewpoint. The subject benefited from the advances in the geometric understanding of sliding modes in nonlinear systems. Some references adopting this last viewpoint are constituted in [242, 243] and [247]. More recently, in the context of motion control systems, the problem has been successfully cast as part of a larger regulation problem in [228].

The feedback controller design approaches mentioned above entirely overlook the, energy related, physical properties of either the original converter circuit or of its closed loop structure. The controller design philosophy, primordially insists on a mathematically motivated, average closed–loop linearization, geared to solve the stabilization or tracking task.

The main advantage of underscoring the often overlooked physical properties of DC–to–DC power converters in an EL or Hamiltonian modeling context is that, in this way, we can advantageously exploit these properties at the feedback controller design stage. In particular, we explore the relevance and implications of a "passivity-based" approach in the feedback duty ratio synthesis problem. In this chapter we describe two approaches for the passivity based regulation of DC to DC power converters within the context of a stabilization task. First, we prescribe feedback duty ratio designs based on the average PWM model of the circuit. In the second part of the section a combination of passivity based regulation and sliding mode control is explored.

In order to effectively deal with a sometimes realistic issue of constant but unknown resistive loads for the converters under study, an *adaptive* feedback regulation scheme has also been explored within the various design methodologies available for nonlinear systems and in particular within the PBC methodology. Adaptive feedback regulation of power supplies has been our subject of study in [249], and [248]. More recently the passivity based regulation schemes for switched power converters have also been extended to the adaptive case in [245].

In Section 2 we study the passivity based stabilization of average models of PWM regulated DC–to–DC power converters. An extension of the PBC design which suitably combines the energy shaping plus damping injection methodology with sliding mode control is introduced in Section 3. Section 4 is devoted to extend the PBC design for the adaptive regulation of DC–to–DC power converters. The last section of this chapter, Section 5, reports an experimental comparison of several linear and nonlinear regulators for the stabilization of DC–to–DC power converters, including linear control, PBC and feedback linearization. The chapter is closed with some conclusions and suggestions for further study in this area.

2 PBC of stabilizing duty ratio

In this section we develop PBCs for DC–to–DC power converters described by the average models derived in Section **6**.4. Therefore, for the sake of validation of the model, we implicitly assume that the sampling frequency is sufficiently high. Even though we work out the details only for the Boost and the Buck–boost converter the application of the technique to other converters follows *mutatis–mutandis*.

Due to the non-minimum phase nature of the average output voltage variable, a direct application of the passivity based design method, aimed primarily at output voltage regulation, leads to an unstable dynamical feedback controller. This is due to an underlying partial inversion of the average system model, carried out at the controller design stage. For this reason, an *indirect* approach, consisting of output voltage regulation through inductor current stabilization is undertaken. Indirect controller design for non-minimum phase systems has been justified in the context of DC–to–DC power converters, in [244]. The indirect control technique also naturally arises, from module-theoretic results, in sliding mode control of linear multivariable non-minimum phase systems, as inferred from [84].

The performance of the derived, indirect, dynamical state feedback controllers is successfully tested, via computer simulations, for the Boost converter example.

The model used for the switched Boost converter included an unmodeled stochastic perturbation input, directly affecting the external voltage source, as well as unmodeled parasitic resistances attached to each one of the circuit elements.

2.1 The Boost converter

We consider the average model of the Boost converter 6.40-6.41, which we repeat here for ease of reference

$$\mathcal{D}_B \dot{z} - (1 - \mu)\mathcal{J}z + \mathcal{R}_B z = \mathcal{E}_B \tag{7.1}$$

$$\mathcal{D}_B = \begin{bmatrix} L & 0 \\ 0 & C \end{bmatrix} \; ; \quad \mathcal{J} = \begin{bmatrix} 0 & -1 \\ 1 & 0 \end{bmatrix} \; ; \quad \mathcal{R}_B = \begin{bmatrix} 0 & 0 \\ 0 & 1/R \end{bmatrix} \; ; \quad \mathcal{E}_B = \begin{bmatrix} E \\ 0 \end{bmatrix} \tag{7.2}$$

The control objective is to regulate the output capacitor voltage z_2 to a constant value, $z_{2*} > E$.

A Strict passivation

Following the PBC methodology we will achieve the control objective by making the closed–loop passive with respect to a desired storage function. Motivated by the form of the total energy function of the average system model, which, as it was shown before, is given by $\mathcal{H} = \frac{1}{2}z^\top \mathcal{D}_B z$, we propose as desired storage function

$$\mathcal{H}_d = \frac{1}{2}\tilde{z}^\top \mathcal{D}_B \tilde{z} \tag{7.3}$$

where $\tilde{z} \stackrel{\triangle}{=} z - z_d$, and z_d is a "desired" value for z, yet to be defined. It is important to observe at this point that the system is underactuated, therefore as for the flexible joint robot problem of Section 4.3.3, we cannot select arbitrary functions for the "desired" signals, they will result instead from the definition of the error dynamics.

The desired error dynamics associated to the storage function (7.3) is then

$$\mathcal{D}_B \dot{\tilde{z}} + (1 - \mu)\mathcal{J}_B \tilde{z} + \mathcal{R}_{Bd} \tilde{z} = \Psi \tag{7.4}$$

where we have added the required damping by choosing a desired Rayleigh error dissipation term,

$$\mathcal{F}_d = \frac{1}{2}\tilde{z}^\top \mathcal{R}_{Bd} \tilde{z} = \frac{1}{2}\tilde{z}^\top (\mathcal{R}_B + \mathcal{R}_{1B}) \tilde{z}$$

where

$$\mathcal{R}_{1B} = \begin{bmatrix} R_1 & 0 \\ 0 & 0 \end{bmatrix} \; ; \quad R_1 > 0.$$

It is clear that the error dynamics (7.4) defines an OSP map $\Psi \mapsto \tilde{z}$ with storage function (7.3). Notice that in this particular example the OSP is obtained with respect to the full state vector, therefore we do not need to invoke detectability arguments to prove asymptotic stability as in previous examples. Actually, in this simple example it is easy to show that the unperturbed error dynamics

$$\mathcal{D}_B \dot{\tilde{z}} + (1 - \mu)\mathcal{J}_B \tilde{z} + \mathcal{R}_{Bd}\tilde{z} = 0 \tag{7.5}$$

is *exponentially* convergent, by taking the time derivative of $\mathcal{H}_d$ along the solutions of (7.5) to get

$$\dot{\mathcal{H}}_d = -\tilde{z}^\top \mathcal{R}_{Bd}\tilde{z} \leq -\frac{\alpha}{\beta}\mathcal{H}_d < 0 \quad \forall\, \tilde{z} \neq 0$$

where α may be taken to be $\alpha = \min\{R_1, 1/R\}$ and $\beta = \max\{L, C\}$.

The next step is then to derive the controller dynamics by setting $\Psi \equiv 0$. It is easy to see from (7.1) and (7.4) that the perturbation term

$$\Psi \triangleq \mathcal{E}_B - (\mathcal{D}_B \dot{z}_d + (1 - \mu)\mathcal{J}_B z_d + \mathcal{R}_B z_d - \mathcal{R}_{1B}\tilde{z})$$

hence, setting $\Psi \equiv 0$ we get

$$\mathcal{D}_B \dot{z}_d + (1 - \mu)\mathcal{J}_B z_d + \mathcal{R}_B z_d - \mathcal{R}_{1B}\tilde{z} = \mathcal{E}_B$$

Using (7.2) the equations above are explicitly written as

$$L\dot{z}_{1d} + (1 - \mu)z_{2d} - (z_1 - z_{1d})R_1 = E \tag{7.6}$$

$$C\dot{z}_{2d} - (1 - \mu)z_{1d} + \frac{1}{R}z_{2d} = 0 \tag{7.7}$$

The equations above give an *implicit* definition of our controller. To obtain an *explicit* expression we use the degree of freedom that we have because there are three free variables (μ, z_{1d}, z_{2d}) and only two equations to be satisfied.

B   Direct output voltage regulation

At this stage, one is tempted to fix $z_{2d} = z_{2*}$, in this case $\tilde{z} \to 0$ would automatically ensure the control objective. Thus, the control problem is the following: given a desired constant output voltage value $z_{2d} = z_{2*}$, find a bounded function $z_{1d}(t)$ and a suitable duty ratio function μ, such that (7.6) and (7.7) be satisfied. We proceed to eliminate the variable $z_1(t)$ from these equations as follows: From (7.7) we obtain

$$z_{1d}(t) = \frac{z_{2*}}{R(1 - \mu(t))}.$$

Substituting this expression into (7.6) we obtain after some algebraic manipulations, an expression for the dynamical feedback duty ratio synthesizer of the form,

$$\dot{\mu} = \frac{R(1-\mu)^2}{Lz_{2*}} \left[E - (1-\mu)z_{2*} + R_1 \left(z_1 - \frac{z_{2*}}{R(1-\mu)} \right) \right]. \tag{7.8}$$

This controller stabilizes z_1 and z_2 towards their desired values z_{1d} and $z_{2d} = z_{2*}$, respectively. Unfortunately, the controller (7.8) is not feasible due to its lack of stability. Indeed the "remaining", or zero-dynamics associated with the above controller results in

$$\dot{\mu} = \frac{R(1-\mu)^2}{Lz_{2*}} \left[E - (1-\mu)z_{2*} \right],$$

which coincides with the zero-dynamics already found in (6.45) and shown to be unstable around its only physically meaningful equilibrium point.

C Indirect output voltage regulation

In the previous paragraph we have shown that a *direct* output voltage control scheme is not feasible (with internal stability). In this paragraph we provide a feasible regulation alternative based on an *indirect* output capacitor voltage control, achievable through the regulation of the input current.[1]

Suppose it is desired to regulate z_1 towards a constant value $z_{1d} = z_{1*}$. In order to find a suitable feedback controller for this task, one proceeds now to eliminate the variable z_{2d} from the set of equations (7.6), (7.7). Using (7.6), $z_{2d}(t)$ is given by

$$z_{2d}(t) = \frac{E + (z_1 - z_{1*})R_1}{(1 - \mu(t))} \tag{7.9}$$

Substituting (7.9) into (7.7), we obtain after some algebraic manipulations,

$$\dot{\mu} = \frac{(1-\mu)}{C[E + (z_1 - z_{1*})R_1]} \left\{ (1-\mu)^2 z_{1*} - \frac{E + (z_1 - z_{1*})R_1}{R} - \frac{R_1 C}{L} [E - (1-\mu)z_2] \right\} \tag{7.10}$$

The zero-dynamics associated with the controller (7.10) is obtained by letting z_1 and z_2 coincide with their respective desired values thet is,

$$\dot{\mu} = \frac{1-\mu}{RCE} \left[(1-\mu)^2 R z_{1*} - E \right]. \tag{7.11}$$

---

[1]We remark that some other possible alternatives include proposing a different error energy function for the system. In this instance, we have just chosen to explore the implications of using the *most natural* energy function for the system.

The zero-dynamics (7.11) coincides with the zero-dynamics derived in equation (6.47), which was shown to be locally stable around the only physically meaningful equilibrium point. Therefore, the indirect controller (7.10) is feasible.

We will now complete the proof that the equilibrium point $(z_1, z_2, \mu) = (z_{1*}, z_{2*}, \mu_*)$ of the overall system (7.1), (7.10) is locally asymptotically stable. To this end, we introduce the following auxiliary variable

$$\xi = \frac{1}{2}\left(\frac{E + (z_1 - z_{1*})R_1}{1 - \mu}\right)^2 - \frac{z_{2*}^2}{2} \tag{7.12}$$

which is well defined for μ in a neighborhood of the equilibrium point, $\mu = \mu_* \neq 1$. It is easy to show that ξ satisfies the following *linear* differential equation,

$$\dot{\xi} = -\frac{2}{RC}\,\xi + \frac{z_{2*}^2 R_1}{RCE}(z_1 - z_{1*}). \tag{7.13}$$

Recalling that $\tilde{z}_1 = z_1 - z_{1*} \to 0$ exponentially fast, we conclude that $\xi \to 0$ as well. It follows that $z_{2d} \to z_{2*}$ locally which in turn implies that $\mu \to \mu_*$.

The derivations above are summarized in the following claim.

Proposition 7.1 *Given a desired constant value, $z_{2*} > E$, for the output capacitor voltage of a Boost converter. The dynamically generated duty ratio function (7.10), with z_{1*} given by (6.48), locally asymptotically stabilizes the state trajectories of the average PWM model (7.1) towards the desired equilibrium point (z_{1*}, z_{2*}, μ_*) with μ converging to a constant value given by $\mu = \mu_* = 1 - E/z_{2*}$.* □

D   Discussion

The passivity–based dynamical duty ratio synthesizer design is carried out under the assumption that the average PWM model (6.38), (6.39) of the converter captures the essential behavior of the actual switch-regulated circuit, described by (6.30), (6.31). This assumption has shown to be only approximately valid due to the fact that in practice, infinite sampling frequency and corresponding infinitely fast switchings are impossible to achieve. Yet, for *sufficiently high* sampling frequencies, feedback controllers designed on the basis of average models can indeed be used to regulate the actual switched converter, with rather satisfactory results (see [117]). The scheme, shown in Fig. 7.1 is based on this philosophy. The underlying approach has been extensively used for similar nonlinear dynamical feedback controllers and its validity has been justified both, from a theoretical viewpoint as well as through extensive computer simulation results (see for instance [244] and the references therein).

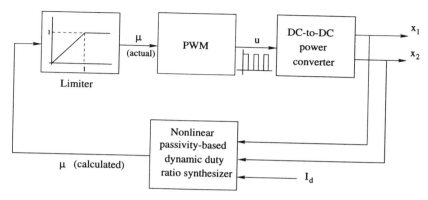

Figure 7.1: PWM feedback control scheme for indirect PBC output voltage regulation of DC-to-DC power converters.

Two additional remarks are in order, regarding the use of a feedback PWM scheme such as that of Fig. 7.1

- The average–based duty ratio synthesizer produces a *computed* duty ratio function. As such, it is entirely possible that these computed values exceed the physical bounds of the required *actual* duty ratio function, which is necessarily limited to the closed interval $[0, 1]$. For this reason, a *hard limiter* must be used in conjunction with the derived dynamical feedback regulator, as shown in Fig. 7.1. As a consequence of this limitation, only *local asymptotic stability* of the closed loop system may actually be guaranteed. Large initial state deviations may induce destabilizing saturation effects which are not accounted for in the previous developments.

- The duty ratio synthesizer (7.10) requires the on–line values of the average PWM circuit states z_1 and z_2. These average states can be approximately obtained by *low pass filtering* the actual circuit states, x_1 and x_2. However, note that in the presented scheme depicted in Fig. 7.1, the *actual* circuit states x_1, x_2, are being used for feedback, rather than their averaged or filtered versions z_1 and z_2. In this respect, it is worth pointing out again, that for large sampling frequencies the differences between using one or the other set of states is entirely negligible, due to the underlying low pass filtering effects of the system itself.

2.2 The Buck–boost converter

Following exactly the same procedure as in the previous case one concludes that, for the Buck–boost converter, a direct regulation policy of the output voltage is unfeasible due to non–minimum phase phenomena. We, thus, summarize in a proposition the dynamical feedback regulation scheme achieving indirect output capacitor voltage regulation, towards a given desired equilibrium value z_{2*}, through input current stabilization towards a desired constant value z_{1*}, computable in terms of z_{2*}, as given by equation (6.50).

Proposition 7.2 *Given a desired constant value z_{2*} for the output capacitor voltage of a Buck-boost converter, the dynamically generated duty ratio function given by*

$$\dot{\mu} = \frac{1-\mu}{C[E+(z_1-z_{1*})R_1]}\Big\{(1-\mu)^2 z_{1*} - \frac{\mu E+(z_1-z_{1*})R_1}{R} - \frac{R_1 C}{L}[\mu E+z_2(1-\mu)]\Big\} \tag{7.14}$$

locally asymptotically stabilizes the state trajectories of the average PWM model (6.42), (6.43) towards the desired equilibrium point (z_{1*}, z_{2*}) with μ converging to a constant value given by, $\mu = \mu_* = 1 - \sqrt{E/Rz_{1*}}$, with z_{1*} obtained from z_{2*} from equation (6.50). □

Note that the zero-dynamics associated with the controller (7.14) is given by

$$\dot{\mu} = \frac{(1-\mu)}{RCE}\big\{(1-\mu)^2 Rz_{1*} - \mu E\big\}$$

which has three equilibrium points given by,

$$\mu = 1 \; ; \; \mu = 1 + \frac{E}{2Rz_{1*}} \pm \sqrt{\Big(\frac{E}{2Rz_{1*}}\Big)^2 + \frac{E}{Rz_{1*}}}.$$

Two of the equilibrium points ($\mu = 1$, and the one corresponding to the plus sign of the square root) are unstable while the remaining one, which is the only physically significant, is locally asymptotically stable.

2.3   Simulation results

Simulations were performed for the closed loop behaviour of a Boost circuit regulated by means of the passivity-based indirect PWM controller (7.10). In order to test the effectiveness and robustness of the proposed feedback controller with respect to unmodeled parasitic resistances and unmodeled realistic switching devices, the following stochastically perturbed version of a Boost converter circuit, taken from [64], was used for the simulations.

$$\dot{x}_1 = -\frac{1}{L}r(u) - (1-u)\frac{R}{L(R+r_C)}x_2 + \frac{E+\eta}{L} - (1-u)\frac{V_F}{L}$$

$$\dot{x}_2 = (1-u)\frac{R}{(R+r_C)C}x_1 - \frac{1}{(R+r_C)C}x_2$$

where $r(u) = r_L + ur_{DS} + (1-u)(R_F + r_C\|R)$ with r_L being the resistance associated with the inductor, r_{DS} is the resistance associated with the "ON" state of the transistor used in the realization of the switching element constituted by a transistor-diode arrangement. R_F is the forward resistance of the diode, r_C is the resistance associated with the output capacitor while $r_C\|R$ denotes the resistance of a parallel arrangement of r_C and R. The voltage V_F represents a small constant voltage drop associated with the conducting phase of the diode. The signal η, added to the external source voltage, represents an external stochastic perturbation input affecting the system behaviour.

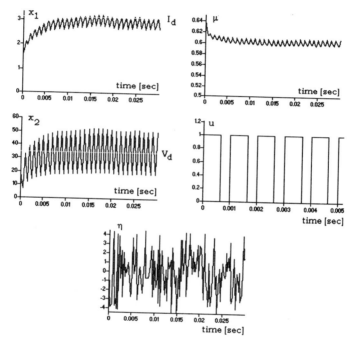

Figure 7.2: Closed loop performance of PBC in a stochastically perturbed Boost converter model including parasitics.

Note that the perturbation input η is of the "unmatched" type. i.e., it enters int the system equations through an input channel vector field, given by $[1/L \quad 0]^T$ which is *not* in the range space of the control input channel, given by the vector field

$$\begin{bmatrix} -\dfrac{r_{DS}+R_F+r_C\|R}{L}x_1 + \dfrac{R}{L(R+r_C)}x_2 - \dfrac{V_F}{L} \\ \dfrac{R}{(R+r_C)C}x_1 \end{bmatrix}.$$

The peak-to-peak magnitude of the noise was chosen to be, approximately, 20% of the value of E.

The circuit parameter values were taken to be the following "typical" values,

$$C = 20 \ \mu\text{F} \ ; \ R = 30 \ \Omega \ ; \ L = 20 \ \text{mH} \ ; \ E = 15 \ \text{V} \ ; \ V_F = 0.7 \ \text{V}$$

$$r_L = 0.05 \ \Omega \ ; \ r_C = 0.2 \ \Omega \ ; \ r_{DS} = 0.1 \ \Omega \ ; \ R_F = 0.05 \ \Omega$$

The sampling frequency for the PWM policy was set to 1 KHz. The duty ratio function is obtained from a sampling process carried out on the output $\mu(t)$, of the smooth dynamical duty ratio synthesizer (7.10). To avoid the use of low pass filters, instead of using the averaged state variables, z_1, z_2, for feedback on the duty ratio synthesizer, we used as it is customarily done, the actual PWM controlled states x_1, x_2, on the controller expressions. The desired ideal average input inductor current was set to be $z_{1*} = 3.125$ Amp., with a steady state duty ratio of $\mu_* = 0.6$. This corresponds to an ideal average output voltage, $z_2 = z_{2*} = 37.5$ V. Fig. 7.2 shows the closed loop state trajectories as well as the duty ratio function and a realization of the computer generated stochastic perturbation signal η.

As it can be seen from Fig. 7.2, the proposed dynamical feedback controller (7.10) achieves the desired indirect stabilization of the output voltage for the non-ideal stochastically perturbed model around the desired equilibrium value. The average steady state errors, with respect to the desired equilibrium values, approximately range from 2.5 % in the average inductor current variable to a 2.6 % in the average capacitor voltage variable. The ideal duty ratio is achieved within less that 0.5 % error. The controller performance also exhibits a high degree of robustness with respect to the external stochastic perturbation inputs.

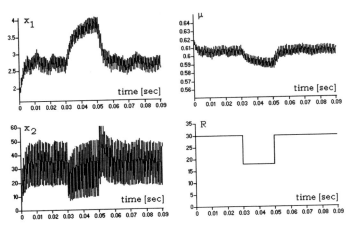

Figure 7.3: Robustness test of controller performance to sudden load variations.

Unknown load resistance variations generally affect the behaviour of the closed loop performance of the controlled converter. Simulations, shown in Fig. 7.3, were performed to depict the sensitivity of the regulated input current, the output capacitor voltage and the duty ratio with respect to abrupt, but temporary unmodeled changes in the load resistance R. An unmodeled sudden change of the load resistance, was set to an 80% of its nominal value. As seen from the figures, the controller manages to rapidly restore the desired steady state conditions, right after the load perturbation disappears. As expected, the state variable most affected by such a perturbation is the output voltage. The duty ratio function, on the contrary, is seen to be little unaffected by such sudden load changes.

In this section we have derived some physically-motivated dynamical feedback duty ratio synthesizers for the indirect average output voltage stabilization of DC–to–DC power converters of the Boost and Buck–boost types. In accordance with the passivity-based methodology undertaken in this book, the dynamical feedback controllers are based on the modification of the total energy function of the average converter circuit model. In the following section we explore the possibility of combining classical techniques of sliding mode control together with the passivity-based design developed here.

3 Passivity based sliding mode stabilization

3.1 Introduction

The main objective of this section is to extend the PBC methodology to include switch-regulated models of DC–to–DC power converters, without appealing to approximate ("average" PWM) models. For this kind of models, where the control variable is either 0 or 1, sliding mode control policies are naturally defined. Therefore we propose a new mixed passivity-based sliding mode controller in short, SM+PBC. We only treat in detail the regulator design for the Boost case, however, our results can be extended to other DC–to–DC power converters such as the Buck, Buck-boost and the Ćuk converter.

The advantages of the passivity-based sliding mode controller over the traditional sliding feedback regulator are established on the basis of an evaluation of the total error energy, which is a weighted version of the integral square state stabilization error of the system. Depending on the choice of the controller's initial state, the passivity-based sliding mode controller is shown to possess a smoother and, in fact, finite weighted integral square state stabilization error. The corresponding measure of performance of the traditional sliding controller is shown to be infinite. Further, it is shown that (in any given finite interval) the total energy absorbed by the circuit from the external source is strictly smaller for the proposed mixed controller than for the classical sliding mode scheme.

In this section we revisit the traditional "current mode" sliding mode controller for the ideal version of the Boost converter circuit. Under this mode of operation, the controller adopts as a sliding surface a desired *constant* inductor current corresponding to a desired equilibrium value of the capacitor voltage. It should be remarked, as already found in [243], that a sliding surface based on a desired constant equilibrium capacitor voltage leads to an *unstable* closed loop dynamics for both converters. This phenomenon is due to the underlying non-minimum phase zero-dynamics associated with the capacitor voltage output discussed in the previous section. The error energy behavior is analytically derived and shown to have an unbounded time integral. The time derivative of this error energy is shown to also exhibit an infinitely large discontinuity. Such a discontinuity occurs, precisely, at the moment of reaching the sliding mode phase from a typical zero state "start-up" condition.

We present the derivation of the passivity based sliding mode controller for the Boost converter. As in the traditional sliding mode controller, our proposed regulator also results in an output feedback scheme but, this time, of dynamic rather than static nature. However, as a counterbalance to the controller complexity, it is shown that for suitable set of controller initial conditions (a possibility denied in the traditional static sliding mode controllers), the performance index is not only smooth but it also exhibits a *finite* value as long as the controller initial state does not coincide with the plant initial state. Some simulations are presented depicting the satisfactory closed loop behaviour of the derived PBC with respect to unmodeled parasitic resistances and parasitic voltage sources. Such parasitic elements are typical of the switching devices usually constituted by a suitable transistor-diode arrangement.

3.2   Sliding mode control of the Boost converter

In the previous chapter we have derived the model of the *switch–regulated* Boost converter circuit as (6.4), (6.4), which we repeat here for ease of reference given by

$$\dot{x}_1 = -(1-u)\frac{1}{L}x_2 + \frac{E}{L} \qquad (7.15)$$

$$\dot{x}_2 = (1-u)\frac{1}{C}x_1 - \frac{1}{RC}x_2 \qquad (7.16)$$

Notice that, in contrast to the *average* model, (7.1), here x_1 and x_2 represent, respectively, the actual input inductor current and the output capacitor voltage variables, and u denotes the switch position function, which takes values in the discrete set $\{0,1\}$.

A   Direct sliding mode control

In this section we show the unfeasibility, due to instability, of a "voltage mode" sliding mode control scheme (see also [243]). This is the switching–mode analog of the unfeasability analysis of Section 2, which applied to the PWM–controlled system.

In the sequel, we denote by $\bar{x}_1$ and $\bar{x}_2$ the state variables of the system under *ideal sliding mode* conditions. In other words, $\bar{x}_1$ and $\bar{x}_2$ represent the "average" values of the state variables under sliding mode operation. The "equivalent control", denoted by u_{eq}, represents an ideal, i.e., a *virtual* feedback control action that smoothly keeps the controlled state trajectories of the system evolving on the sliding surface, provided motions are started, precisely, at the sliding surface itself (see [270] for definitions). The equivalent control, for the case of switch-regulated systems, is not synthetizable in practise due to the discrete-character of the control input. However, its use on the system equations provides a convenient way of analysis for the "average" closed-loop sliding motions behavior.

Proposition 7.3 *Consider the switching line $s = x_2 - V_d$, where $V_d > 0$ is a desired constant capacitor voltage value. The switching policy, given by*

$$u = 0.5 \left[1 + \mathrm{sgn}(s) \right] = 0.5 \left[1 + \mathrm{sgn}(x_2 - V_d) \right] \tag{7.17}$$

locally creates an unstable sliding regime on the line $s = 0$ with ideal sliding dynamics characterized by

$$\bar{x}_2 = V_d \; ; \; \dot{\bar{x}}_1 = \frac{E}{L} \left[1 - \frac{V_d^2}{RE\,\bar{x}_1} \right] \; ; \; u_{eq} = 1 - \frac{\bar{x}_2}{\bar{x}_1}$$

$\square$

Proof. Evidently, the closed loop system (7.15)-(7.17) satisfies the well known sliding mode condition $s\dot{s} < 0$ (see [270]) in the vicinity of the sliding surface $s = 0$, provided $x_1 > V_d/R$, hence the local character of the sliding mode. The ideal sliding dynamics is easily seen to have a unique but *unstable* equilibrium point at $\bar{x}_1 = V_d^2/(RE)$. $\blacksquare$

B Asymptotic stability of indirect sliding mode control

As discussed in Section 2, in order to avoid the non-minimum phase behavior we must proceed to *indirectly* regulate the capacitor voltage x_2.

Proposition 7.4 *Consider the switching line $s = x_1 - V_d^2/RE$, where $V_d > 0$ is a desired constant capacitor voltage value. The switching policy, given by*

$$u = 0.5 \left[1 - \mathrm{sgn}(s) \right] = 0.5 \left[1 - \mathrm{sgn}(x_1 - V_d^2/RE) \right] \tag{7.18}$$

locally creates a stable sliding regime on the line $s = 0$ with ideal sliding dynamics characterized by

$$\bar{x}_1 = \frac{V_d^2}{RE} \; ; \; \dot{\bar{x}}_2 = -\frac{1}{RC} \left[\bar{x}_2 - \frac{V_d^2}{\bar{x}_2} \right] \; ; \; u_{eq} = 1 - \frac{E}{\bar{x}_2}. \tag{7.19}$$

Moreover, the ideal sliding dynamics behaviour of the capacitor voltage variable, described by (7.19), can be explicitly computed as

$$\bar{x}_2(t) = \left[V_d^2 + \left[\bar{x}_2^2(t_h) - V_d^2\right]e^{-\frac{2}{RC}(t-t_h)}\right]^{1/2} \tag{7.20}$$

where t_h stands for the reaching instant of the sliding line $s = 0$ and $\bar{x}_2(t_h)$ is the capacitor voltage at time t_h.   □

Proof.   Note that the switching policy (7.18) creates a sliding regime on $s = 0$ provided $x_2 > E$, which is a well known "amplifying" property of the Boost converter. The ideal sliding dynamics has stable equilibrium points at $\bar{x}_2 = V_d$ for all initial values of $\bar{x}_2$ which are strictly positive, and at $\bar{x}_2 = -V_d$ for all strictly negative initial conditions of $\bar{x}_2$. Note, however, that this second equilibrium point is not achievable when the constant value of the voltage source E is strictly positive. To show that the expression (7.20) is a solution of the differential equation in (7.19) one simply differentiates (7.20).   ■

C   Simulation results

We show below in simulations, the "current-mode" sliding feedback controlled state trajectories of a typical Boost converter. The circuit parameter values were taken as in the simulation of Subsection 2.3 without the parasitics. Fig. 7.4 depicts the state variables evolution of a typical zero initial conditions "start up" response of a current–mode controlled Boost converter, using the controller of Proposition 7.4.

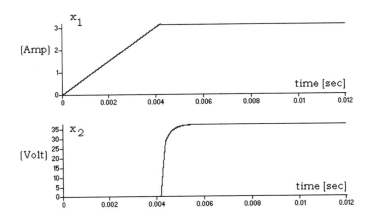

Figure 7.4: Typical sliding "current-mode" controlled state responses for the Boost converter.

Note that the capacitor voltage x_2 remains at its initial value of zero, while the switch position u is maintained at the value $u = 1$. The inductor current grows, during this "reaching" phase as a linear function of time given by

$$x_1(t) = \frac{E}{L} t.$$

When the sliding line $s = x_1 - V_d^2/RE = 0$ is reached, at time $t_h = (L/R)(V_d/E)^2$, by the controlled inductor current, the voltage capacitor x_2 rapidly grows from zero towards its equilibrium value V_d, ideally governed by

$$\bar{x}_2(t) = V_d\sqrt{1 - e^{-\frac{2}{RC}(t-t_h)}}.$$

Note that $\bar{x}_2$ exhibits an infinite positive time derivative at the sliding mode reaching time $t = t_h$.

D Performance

The stored error energy of the controlled system is defined as

$$\mathcal{H}(t) = \frac{1}{2}\left[L\left(x_1(t) - \frac{V_d^2}{RE}\right)^2 + C\left(x_2(t) - V_d\right)^2\right]. \tag{7.21}$$

Proposition 7.5 *Suppose the Boost converter circuit is initially at the zero state. Then if the switching policy (7.18) is applied to the converter, then during the reaching phase of the sliding mode ($0 \le t < t_h$) the stored energy $\mathcal{H}(t)$ decreases quadratically with respect to time. After the sliding mode is achieved ($t \ge t_h$), then the stored energy of the circuit further decreases to zero, as the ideal sliding motions of the capacitor voltage asymptotically approach the equilibrium value.* $\square$

Proof. The control dependent time derivative of the stored stabilization error energy $\mathcal{H}(t)$ is given by

$$\frac{d\mathcal{H}}{dt} = -(1 - u)V_d\left[x_1 - \frac{V_d x_2}{ER}\right] + \left(x_1 - \frac{V_d^2}{RE}\right)E - (x_2 - V_d)\frac{x_2}{R} \tag{7.22}$$

During the sliding mode reaching phase, with $x_2(t) = 0 \ \forall \ t < t_h$ and $u = 1$, the above expression reduces to

$$\frac{d\mathcal{H}}{dt} = \left(x_1 - \frac{V_d^2}{RE}\right)E \le 0$$

and the stored error energy $\mathcal{H}(t)$ decreases from the value

$$\mathcal{H}(0) = 0.5[LV_d^4/(R^2E^2) + CV_d^2]$$

towards the value

$$\mathcal{H}(t_h) = 0.5CV_d^2$$

At the reaching instant, t_h, the time derivative of $\mathcal{H}(t)$ is also zero.

During the reaching stage, the stored error energy, is a *quadratically* decreasing function of time, given by

$$\mathcal{H}(t) = 0.5\{L[(E/L)t - V_d^2/RE]^2 + CV_d^2\}$$

Its time derivative, at time $t = 0$, is given by $\dot{\mathcal{H}}(0) = -V_d^2/R$.

Right after the sliding mode is reached the time derivative of the stored stabilization error energy is obtained in an average sense, by substituting in (7.22) the switch position function u by the equivalent control expression found in (7.19). After some algebraic manipulations, we obtain that the time derivative of $\mathcal{H}(t)$ for all $t \geq t_h$, is given by

$$\frac{d\mathcal{H}}{dt} = -\frac{1}{R}\left(1 + \frac{V_d}{\bar{x}_2}\right)(\bar{x}_2 - V_d)^2 < 0 \tag{7.23}$$

Since at time $t = t_h$, the capacitor voltage, $x_2(t_h) = \bar{x}_2(t_h)$, is zero, according to (7.23), the stored error energy instantaneously exhibits an infinitely large negative time derivative. The total stored error energy thus asymptotically decreases to zero as the average voltage approaches its desired equilibrium value V_d. ∎

A measure of the performance of the sliding mode controlled system, described above, is obtained by using the integral of the stored stabilization error energy. This quantity is given by

$$\mathcal{I}_B = \int_0^\infty \mathcal{H}(\tau)d\tau = \int_0^\infty \frac{1}{2}\left[L\left(x_1(\tau) - \frac{V_d^2}{RE}\right)^2 + C\left(x_2(\tau) - V_d\right)^2\right]d\tau. \tag{7.24}$$

Such a performance criterion can also be regarded as a weighted integral square state stabilization error for the state vector. We simply address such an index as the "WISSSE" index.

Proposition 7.6 *The WISSSE index, computed along the sliding mode controlled trajectories of the Boost converter, is unbounded for all initial conditions of the converter.* □

Proof. We compute the WISSSE index by first evaluating the integral during the sliding mode reaching phase in the time interval $0 \leq t < t_h$. For the evaluation of the index during the sliding mode phase, we consider the ideal sliding dynamics for the variable $\bar{x}_2$. For the sake of simplicity, to perform the calculation we assume that the initial states of the converter are set to zero. Nevertheless, the result is valid for arbitrary initial conditions, including starting the converter at the equilibrium values.

$$
\begin{aligned}
\mathcal{I}_B(t) &= \int_0^{t_h} \frac{1}{2}\left[\left(\frac{E}{L}\tau - \frac{V_d^2}{RE}\right)^2 + CV_d^2\right] d\tau + \int_{t_h}^{t} \frac{1}{2}CV_d^2\left(\sqrt{1 - e^{-\frac{2}{RC}(\tau - t_h)}} - 1\right)^2 d\tau \\
&= \frac{LV_d^2}{RE}\left[C + \frac{6LV_d^2}{R^2E^3}\right] - \frac{1}{4}\left[-4\frac{t - t_h}{RC} - e^{-2\frac{t-t_h}{RC}} - 4\sqrt{1 - e^{-2\frac{t-t_h}{RC}}}\right. \\
&\quad \left. - \ln\left(\frac{-2 + e^{-\frac{2(t-t_h)}{RC}} + 2\sqrt{1 - e^{-\frac{2(t-t_h)}{RC}}}}{e^{-\frac{2(t-t_h)}{RC}}}\right)^2\right] RC^2V_d^2
\end{aligned}
\tag{7.25}
$$

It is easy to verify that

$$
\lim_{t\to\infty} \mathcal{I}_B(t) = \infty.
$$

This limit operation requires using L'Hôpital's rule in the last term in (7.25). ∎

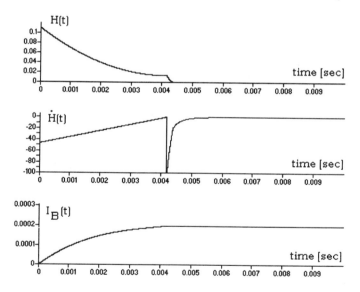

Figure 7.5: Weigthed integral square stabilization error behavior for the Boost converter.

It is easy to see that the above result is independent of the initial conditions. In fact, the transient towards the sliding regime contributes with a finite quantity to the

WISSSE index behaviour. The sliding regime phase, reached always in finite time, contributes with the infinite portion of the WISSSE index value.

Fig. 7.5 depicts the behaviour of the total stored stabilization error energy, its time derivative, and its time integral for the sliding mode controlled state vector of the converter, started at zero initial conditions.

The unbounded growth of the WISSSE index is hardly remarkable for such a small time interval. As evidenced from (7.25), this is due to the fact that the dependence on t, of the above integral, is scaled by a constant factor which is of the order of CV_d^2. For the parameters used in our simulated Boost converter, previously defined, this value is of the order of 10^{-3}.

We compute, also for the purpose of comparison, the integral square error associated with the difference between the applied control input $u(t)$ and its constant steady state value, denoted by $u_{eq}(\infty) = 1 - \frac{E}{V_d}$. Since during the reaching phase the input u is at a fixed value, either $u = 1$ or $u = 0$, the integral square error of the control input, during this transient, is finite. For this reason we concentrate on the computation of the performance index during the sliding phase. We again assume, just for simplicity, that the converter initial states are set at zero.

$$
\begin{aligned}
W(t) &= \frac{1}{2} \int_0^\infty [u(\tau) - u_{eq}(\infty)]^2 \, d\tau \\
&= \int_0^{t_h} \left[0.5\mathrm{sgn}(x_1(\tau) - \frac{V_d^2}{E}) - 1 + \frac{E}{V_d} \right]^2 d\tau + \int_{t_h}^t \left[u_{eq}(\tau) - 1 + \frac{E}{V_d} \right]^2 d\tau \\
&\geq \frac{E}{V_d} \int_{t_h}^t \left[1 - \frac{V_d}{\bar{x}_2(\tau)} \right]^2 d\tau = \frac{E}{V_d} \int_{t_h}^t \left[1 - \frac{1}{\sqrt{1 - e^{-\frac{2}{RC}(\tau - t_h)}}} \right]^2 d\tau \\
&= \frac{E^2 RC}{V_d^2} \left\{ \frac{2}{RC}(t - t_h) + \frac{1}{2} \ln \left[1 - e^{-\frac{2}{RC}(t - t_h)} \right] \right. \\
&\qquad \left. + \ln \left[\frac{-2 + e^{-\frac{2}{RC}(t - t_h)} + 2\left(1 - e^{-\frac{2}{RC}(t - t_h)}\right)}{e^{-\frac{2}{RC}(t - t_h)}} \right] \right\}
\end{aligned}
$$

It is also easy to verify that the limit of this last term is unbounded as $t \to \infty$ therefore, we conclude that

$$
\lim_{t \to \infty} W(t) = \infty
$$

3.3   Passivity-based sliding controller

In the following developments we introduce an auxiliary state vector, denoted by x_d. The basic idea is to take x_d as a "desired" vector trajectory for the converter state vector x. This auxiliary vector variable will be determined on the basis of the energy shaping plus damping injection considerations of the passivity–based approach. The feedback regulation of the auxiliary state x_d, towards the desired constant equilibrium value of the state x, will in fact result in the specification of a *dynamical output feedback controller* for the original converter state. We will be using a sliding mode control viewpoint for the regulation of x_d towards the desired equilibrium value of x.

A Asymptotic stability

We rewrite the Boost converter equations (7.15), (7.16) in matrix-vector form as

$$\mathcal{D}_B \dot{x} - (1 - u)\mathcal{J}x + \mathcal{R}_B x = \mathcal{E}_B$$

where $\mathcal{D}_B$, $\mathcal{J}$, $\mathcal{R}_B$, $\mathcal{E}_B$ were defined in (7.2).

Proceeding analogously to the case of the averaged model of Section 2 we define the desired closed–loop storage function

$$\mathcal{H}_d = \frac{1}{2}(x - x_d)^\top \mathcal{D}_B(x - x_d) \triangleq \tilde{x}^\top \mathcal{D}_B \tilde{x}$$

where x_d satisfies the following controlled differential equation

$$\mathcal{D}_B \dot{x}_d - (1 - u)\mathcal{J}x_d + \mathcal{R}x_d - \mathcal{R}_1(x - x_d) = \mathcal{E}_B \tag{7.26}$$

Notice that, as done in Section 2, we have added damping via $R_1 > 0$.

We now concentrate our efforts in regulating from u, by means of a sliding mode control policy, the auxiliary system (7.26) towards the desired equilibrium state of the converter. This is summarized in the following proposition.

Proposition 7.7 *Consider the switching line* $s = x_{1d} - V_d^2/RE$, *where* $V_d > 0$ *is a desired constant capacitor voltage value for the auxiliary variable* x_{2d} *and for the converters capacitor voltage* x_2. *The switching policy, given by*

$$u = 0.5[1 - \text{sgn}(s)] = 0.5[1 - \text{sgn}(x_{1d} - V_d^2/RE)] \tag{7.27}$$

locally creates a sliding regime on the line $s = 0$. *Moreover, if the sliding-mode switching policy (7.27) is applied to* both *the converter and the auxiliary system, the converter state trajectory* $x(t)$ *converges towards the auxiliary state trajectory* $x_d(t)$ *and, in turn,* $x_d(t)$ *converges towards the desired equilibrium state. i.e.,*

$$(x_1, x_2) \;\rightarrow\; (x_{1d}, x_{2d}) \;\rightarrow\; \left(\frac{V_d^2}{RE}, V_d\right).$$

The ideal sliding dynamics is then characterized by

$$\overline{x}_{1d} \;=\; \frac{V_d^2}{RE} \tag{7.28}$$

$$\dot{\overline{x}}_{2d} \;=\; -\frac{1}{RC}\left[\overline{x}_{2d} - \left(\frac{V_d^2}{E}\right)\frac{E + R_1(\overline{x}_1 - V_d^2/RE)}{\overline{x}_{2d}}\right] \tag{7.29}$$

$$u_{eq} \;=\; 1 - \frac{E + R_1(\overline{x}_1 - V_d^2/RE)}{\overline{x}_{2d}} \tag{7.30}$$

where $\overline{x}_1$ *is the converters inductor current under sliding mode conditions, primarily occurring in the controller's state space and induced, through the control input* u *on the controlled system state space.*

Furthermore, the ideal sliding dynamics for $\overline{x}_2(t)$ *for* $t \geq t_h$, *can be explicitly computed in terms of the inductor current error signal* $(x_1(t) - Vd^2/RE)$ *as,*

$$\overline{x}_{2d}(t) = \left[e^{-\frac{2}{RC}(t-t_h)} \overline{x}_{2d}^2(t_h) + \frac{2V_d^2}{RC} \int_{t_h}^{t} e^{-\frac{2}{RC}(t-\sigma)} \left[1 + \frac{R_1}{E}(\overline{x}_1(\sigma) - x_{1d}(\sigma)) \right] d\sigma \right]^{1/2}$$

(7.31)

□

Proof. That the state trajectory $x(t)$ converges towards $x_d(t)$ follows from the previous proposition. The sliding mode locally exists on $s = 0$ provided $x_{2d} > E + R_1(x_1 - V_d^2/RE) > 0$. Note that since the same switching policy u is being applied to both the converter and the auxiliary system, the state error $x_1 - V_d^2/RE$ is actually decreasing, in absolute value, during the sliding mode reaching phase. The positive value E will eventually overcome the term $R_1(x_1 - V_d^2/RE)$ and the hitting of the sliding line $s = 0$ is thus guaranteed in finite time. The fact that x_{1d} converges in finite time V_d^2/RE implies that x_1 will also converge towards V_d^2/RE. Hence, the feasible steady state equilibrium for $\overline{x}_{2d}$ from (7.29), is given by $x_{2d} = V_d$. By virtue of the previous proposition, x_2 also converges towards V_d. After all transients have elapsed, the auxiliary system is just a copy of the converter dynamics. Since, under such steady state conditions it is always true for the Boost converter, that the "amplifier" relation $x_2 (= x_{2d}) > E$ is valid, the sliding mode existence on $s = 0$ is indefinitely guaranteed. ∎

The validity of (7.31) as a solution of (7.29) readily follows upon time differentiation of the proposed solution.

Fig. 7.6 represents the passivity-based sliding "current-mode" control scheme for the Boost converter. The auxiliary system is regarded as a dynamical feedback controller synthesizing a suitable sliding line for the converter dynamics.

Fig. 7.7 depicts some simulations of the closed loop state behaviour of the regulated converter system for several initial conditions of the dynamical controller. When the initial states of the converter are chosen to be zero, in coincidence with the initial states of the controlled converter, the closed loop response of the PBC coincides with that of the traditional sliding mode controller. As the initial conditions for the controller are chosen "closer" to the desired equilibrium state of the controlled plant, the state responses of the plant become smoother with slightly larger settling times but with a much better behaved transient shape.

Remark 7.8 We remark that the sliding mode created on the controller's state space, on the basis of output feedback, is also induced at the same instant on the controlled system state space by the actively switching control input u, also shared by the plant. An analytic expression for the three-dimensional *integral manifold* of the redundant fourth order closed loop ideal sliding dynamics, using as control the nonlinear equivalent control input defined in (7.30), is too difficult to obtain. The form of the intersection of this integral manifold with the two-dimensional controlled converter state space would help in explaining the obtained smoothness of the controlled system state responses.

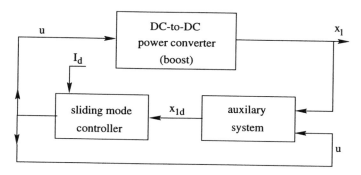

Figure 7.6: SM+PBC scheme for regulation of the Boost converter.

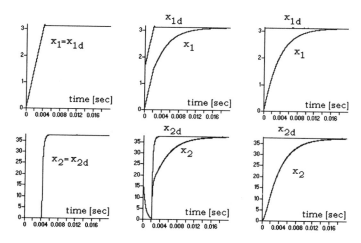

Figure 7.7: Controller and plant state responses for different controller initial conditions; (—) $(x_{1d}(0), x_{2d}(0)) = (0, 0)$; $(\cdots)$ $(x_{1d}(0), x_{2d}(0)) = (1.6, 19)$ (-·-·) $(x_{1d}(0), x_{2d}(0)) = (3.125, 37.5)$.

If initial conditions for the dynamical controller at $t = 0$, are chosen precisely at the desired equilibrium values of the controlled plant state, $(x_{1d}(0), x_{2d}(0)) = (V_d^2/RE, V_d)$, a sliding regime is immediately created on $s = x_{1d} - V_d^2/RE = 0$ from time $t = 0$ on. The ideal sliding dynamics for $\bar{x}_2(t)$ is then simply given by

$$\bar{x}_{2d}(t) = V_d \left[e^{-\frac{2}{RC}(t-t_h)} + \frac{2}{RC} \int_0^t e^{-\frac{2}{RC}(t-\sigma)} \left[1 + \frac{R_1}{E} \left(\bar{x}_1(\sigma) - \frac{V_d^2}{RE} \right) \right] d\sigma \right]^{1/2} .(7.32)$$

Note however that under these circumstances, at time $t = 0$ the time derivative of $\bar{x}_{2d}(t)$ is actually negative, as it can be inferred from (7.29) or from (7.32) itself. As a consequence, if the controller initial states are set at the equilibrium value $(V_d^2/RE, V_d)$, the value of $\bar{x}_{1d}(t)$ remain constant at the value of V_d^2/RE, but the values of $\bar{x}_{2d}(t)$ actually decrease from V_d to later on recover and converge towards V_d again. Moreover, from the nature of the single-mode exponential nature of $\bar{x}_{2d}(t)$, no overshoot of V_d ever occurs. This fact allows us to conclude that during the transient of $\bar{x}_{2d}(t)$, $V_d \leq \bar{x}_{2d}(t)$ for all t.

B   Performance

Consider again the stored stabilization error energy $\mathcal{H}(t)$ defined in (7.21). The control-dependent time derivative of the stored stabilization error energy $\mathcal{H}(t)$ is identical to the expression already given in (7.22), except for the fact that the control input u is now synthesized on the basis of the switching line $s = x_{1d} - V_d^2/RE$, defined on the controller's state space, rather than on the basis of the switching line $s = x_1 - V_d^2/RE$, defined on the systems state space.

$$\frac{d\mathcal{H}}{dt} = -(1 - u)V_d \left(x_1 - \frac{x_2 V_d}{ER} \right) + \left(x_1 - \frac{V_d^2}{RE} \right) E - (x_2 - V_d) \frac{x_2}{R}. \qquad (7.33)$$

A second fundamental difference between time derivative (7.33) and that of (7.22) is that now the instant t_h, beyond which it is valid to perform the substitution of u by the "equivalent control", u_{eq}, depends quite heavily on the initial conditions of the dynamical feedback controller $(x_{1d}(0), x_{2d}(0))$.

Suppose one sets the initial conditions of the dynamical PBC to be exactly the same as those of the converter. Then, it can be easily seen that the term $R_1(x_1 - x_{1d})$ in the controller dynamics (7.26) is identically zero during the reaching phase since the controller dynamics becomes the same as the plant dynamics and the same input u is being applied to *both* the "plant" and the controller from identical initial states. As a consequence, as in the traditional sliding mode controller of Subsection 3.2, a large discontinuity is to be expected in the time derivative of $\mathcal{H}$ at the hitting of the sliding line, occurring at time t_h.

Suppose now that the initial conditions of the controller are set, precisely, at the desired equilibrium value of the converter state, i.e., at $(x_{1d}, x_{2d}) = (V_d^2/RE, V_d)$, while the converter initial states are placed, as customarily, at zero. The passivity-based switching policy immediately creates a sliding regime on the controller's "inductor current" variable $x_{1d}(t)$. As a consequence, the "equivalent control" (7.30) can be substituted in the time derivative expression of $\mathcal{H}$, right from the very initial instant $t = t_h = 0$. The corresponding average value of the stored stabilization error energy satisfies then $\frac{d\mathcal{H}}{dt} \leq -\alpha\mathcal{H}$, and we can prove the following result.

Proposition 7.9 *The passivity-based sliding current-mode controller described in Proposition 7.7 yields a finite WISSSE index (7.24) provided the initial states of the converter satisfy $x_{1d}(0) \neq x_1(0)$ and $x_{2d}(0) \leq V_d$.* □

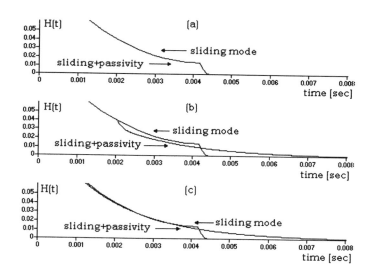

Figure 7.8: Comparison of total stored stabilization error energy for traditional and SM+PBC Boost converter.

Fig. 7.8 depicts the behaviour of the total stored error energy $\mathcal{H}(t)$ for different initial states of the passivity-based dynamical sliding mode feedback controller. As it is customary, zero initial conditions were taken for the "start-up" phase of the converter in all four computational runs. Note that as the controller initial conditions were chosen closer to the desired constant equilibrium value of the plant, the performance index becomes smoother and similar to an exponential behaviour. The controller initial states in Fig. 7.8a through 7.8c are $(x_{1d}(0), x_{2d}(0)) = (0,0)$; $(x_{1d}(0), x_{2d}(0)) = (1.6, 19)$; $(x_{1d}(0), x_{2d}(0)) = (3.125, 37.5)$ respectively.

Another performance indicator of the closed loop system is the total energy absorbed from the voltage source by the system at time $\infty > t \geq 0$, that is, $\int_0^t Ex_1(s)ds$.

Now, the inductor current in the standard sliding mode scheme converges to its final value in finite time, while in the new controller the convergence is only exponential without overshoot. Consequently, it is clear that the latter outperforms the sliding mode scheme also from the energy consumption viewpoint.

C   Simulation results: Robustness to unmodeled parasitics

We have performed some simulations in order to test the robustness and performance of the proposed passivity-based sliding current-mode control scheme for a realistic model of a Boost converter. The derived controller, obtained from a non-perturbed version of the converter model, was directly applied to a power converter "plant" containing typical modeling errors and stochastic perturbations. In Fig. 6.5, we show the rather realistic Boost converter model including parasitic resistances and parasitic voltage sources commonly considered in modeling the transistor-diode switching arrangement. This converter, originally proposed in [64], was also additionally perturbed by including a (computer generated) stochastic noise source, denoted by ξ, of significant voltage amplitude (approximately 20%) in relation to the constant voltage source value E.

Our controller was applied then to the following converter model

$$\dot{x}_1 = -\frac{1}{L}\left[r_L + ur_{DS} + (1-u)\left(R_F + r_C||R_L\right)\right]x_1 - (1-u)\frac{1}{L}\left(\frac{R_L}{R_L + r_C}\right)x_2$$
$$+\frac{E+\eta}{L} - (1-u)\frac{V_F}{L} \tag{7.34}$$

$$\dot{x}_2 = (1-u)\frac{1}{C}\left(\frac{R_L}{R_L + r_C}\right)x_1 - \frac{1}{(r_C + R_L)\,C}\,x_2 \tag{7.35}$$

where as before x_1 and x_2 represent the Boost converter state variables. The parasitic resistance of the inductor is denoted by r_L. The parameter r_{DS} represents the "ON" resistance for the transistor integrating the transistor-diode switching arrangement. The parasitic resistance R_F represents the "forward" resistance, or "ON" resistance, of the diode in the switch realization. V_F is a parasitic voltage source associated with the diode operation, and r_C is the capacitor's parasitic resistance. The signal η represents the stochastic perturbation input associated with the constant voltage source E.

The parameters of the perturbed version of the Boost converter were set to

$$r_L = 0.1 \ \Omega \ ; \ r_{DS} = 0.03 \ \Omega \ ; \ R_F = 0.1 \ \Omega \ ; \ r_C = 0.01 \ \Omega \ ; \ R = 30 \ \Omega \ ; \ V_F = 0.7 \text{V}.$$

In Fig. 7.9 we depict the controlled inductor current and the controlled capacitor voltage responses of the perturbed Boost converter (7.34), (7.35) regulated by means of the passivity-based sliding "current mode" controller, derived on the basis of the non-perturbed model.

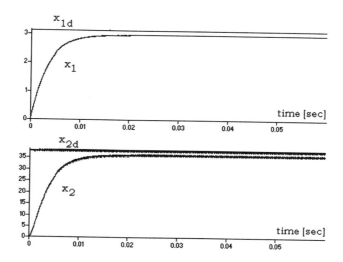

Figure 7.9: Closed loop state response of the passivity-based controlled perturbed "Boost" converter.

The obtained equilibrium value for the output capacitor voltage was $x_2(\infty) = 36.11$ V which represents a steady state error, due to unmodeled perturbations and parasitics, of 3.7 % The proposed controller may be regarded to be reasonably robust with respect to unmodeled parasitics elements, parasitic voltage sources and stochastic inputs.

Remark 7.10 It is worth pointing out that even though traditional static sliding mode controllers are obviously simpler in nature than the proposed dynamical passivity-based sliding "current-mode" regulators, our results indicate that there is a definite advantage in the use of the more complex controller. This advantage is related to their superior performance, as measured by a weighted integral square stabilization error performance index and the energy consumption. While the traditional sliding mode controller has an unbounded weighted integral square stabilization error, that of the PBC can be made to be finite for a wide range of the controller initial conditions, as long as they do not exactly coincide with those of the regulated plant. Furthermore, this is achieved consuming less energy from the external voltage source.

Remark 7.11 This study has only explored the implications of the passivity based approach in the sliding mode control of a particular class of DC–to–DC power converters, namely the Boost converter. Similar developments can, in principle, be worked out for more complex types of converters, such as the Buck–boost converter and the popular Ćuk converter.

4   Adaptive stabilization

A frequent assumption in the design of feedback regulators for DC–to–DC power supplies is that the converter loads and the parameters associated with the various circuit components are perfectly known. In practice, however, lack of precise knowledge about these parameters arises from inescapable measurement errors, unavoidable aging effects and imperfectly modeled loads. These facts motivate the adoption of an adaptive feedback approach for the design of regulation loops in DC–to–DC power supplies.

Adaptive control of DC–to–DC power supplies has been treated from an approximate linearization viewpoint in [231]. Their approach relies on Lyapunov stability and passivity considerations for the linear feedback controller design. A full adaptive feedback input–output linearization viewpoint for DC–to–DC power supplies was proposed in [249]. An adaptive feedback design technique that suitably combines input–output linearization, through generalized observability canonical forms as developed by Fliess in [81], and the backstepping design procedure, was recently presented in [248].

In this section we extend our previous developments by exploring the viability of applying passivity–based control for the *adaptive* stabilization of a class of average models of PWM regulated DC–to–DC power converters. We treat the two basic types of switched power supplies: the Boost and Buck–boost converters.

4.1   Controller design

The procedure to construct the adaptive versions mimics the one used for mechanical systems in Sections **3**.4.2 and **4**.2. Namely, we use a certainty equivalent version of the control law, and taking into account the fact that the uncertain load enters *linearly* in this signal, we generate the parameter error which appears as an additive disturbance to the error dynamics. Since by definition, the error dynamics defines an OSP operator, the construction is completed with a gradient estimator, which in its turn is passive.

A   The Boost converter

Proposition 7.12 *Consider the averaged dynamics of the Boost converter, where $C > 0$, $L > 0$, $E > 0$ are known constants representing the capacitance, inductance and external voltage respectively, and $R > 0$ is the unknown load charge resistance. Let $\tilde{z}_2 \stackrel{\triangle}{=} z_2 - z_{2d}$ and define an adaptive nonlinear dynamic state feedback controller* as,

$$\dot{z}_{2d} = -\frac{\hat{\theta}}{C}\left\{ z_{2d} - \frac{z_{2*}^2}{E z_{2d}}\left[E + R_1\left(z_1 - \hat{\theta}\frac{z_{2*}^2}{E}\right) + L\frac{z_{2*}^2}{E}z_{2d}\left(\tilde{z}_2\right)\right]\right\} \qquad (7.36)$$

$$\mu = 1 - \frac{1}{z_{2d}}\left[E + R_1\left(z_1 - \hat{\theta}\frac{z_{2*}^2}{E}\right) + L\frac{z_{2*}^2}{E}z_{2d}\tilde{z}_2\right] \qquad (7.37)$$

$$\dot{\hat{\theta}} = -z_{2d}\tilde{z}_2 \qquad (7.38)$$

where the dynamical controller initial condition is chosen so that, $z_{2d}(0) > 0$ and $\hat{\theta}(0) > 0$. The constant reference value for z_2, denoted by z_{2*}, is a strictly positive quantity. The quantity $\hat{\theta}$ denotes the estimate of $\frac{1}{R}$. The parameter R_1 is a designer chosen constant with the only restriction of being strictly positive. Under these conditions, it is always possible to chose the controller's initial state $z_{2d}(0)$ and $\hat{\theta}(0)$, such that the closed loop system (6.38), (6.39), (7.36)–(7.38) has an equilibrium point given by,

$$(z_1, z_2, z_{2d}, \hat{\theta}) = \left(\frac{1}{R}\frac{z_{2*}^2}{E}, z_{2*}, z_{2*}, \frac{1}{R}\right) \qquad (7.39)$$

which is asymptotically stable.   □

Proof. It can be verified, by direct substitution, that (7.39) represents an equilibrium point for the closed loop system.

Let us define the certainty equivalent version

$$z_{1d} = \hat{\theta}\frac{z_{2*}^2}{E} \qquad (7.40)$$

which is the same than that considered for the known parameter z_{1d} in Section 2. Hence, z_{1d} and z_1 coincide at the equilibrium point. Let, again, $z - z_d$ stand for the error vector $\tilde{z}$. In terms of the error signals, (6.38), (6.39) are rewritten as,

$$\mathcal{D}_B\dot{\tilde{z}} + (1 - \mu)\mathcal{J}_B\tilde{z} + \mathcal{R}_{Bd}\tilde{z} = \psi$$

where

$$\psi = \mathcal{E}_B - \left[\mathcal{D}_B\dot{z}_d + (1 - \mu)\mathcal{J}_B z_d + \mathcal{R}_B z_d - \begin{bmatrix} R_1\tilde{z}_1 \\ 0 \end{bmatrix}\right] \qquad (7.41)$$

and $\mathcal{R}_{Bd}$ is a positive definite matrix given by,

$$\mathcal{R}_{Bd} = \begin{bmatrix} R_1 & 0 \\ 0 & 1/R \end{bmatrix} \quad ; \quad R_1 > 0$$

Expression (7.41) is explicitly written as

$$\psi_1 = -L\dot{z}_{1d} - (1-\mu)z_{2d} + R_1\tilde{z}_1 + E$$
$$\psi_2 = -C\dot{z}_{2d} + (1-\mu)z_{1d} - \frac{1}{R}z_{2d}.$$

Using (7.40) and (7.36)–(7.38) we have $\psi_1 = 0$ and $\psi_2 = \tilde{\theta}z_{2d}$, where $\tilde{\theta} = \hat{\theta} - \frac{1}{R}$.

The resulting stabilization error system is then given by the following perturbed dynamics,

$$\mathcal{D}_B\dot{\tilde{z}} + (1-\mu)\mathcal{J}_B\tilde{z} + \mathcal{R}_{Bd}\tilde{z} = \begin{bmatrix} 0 \\ \tilde{\theta}z_{2d} \end{bmatrix}. \tag{7.42}$$

Using as a Lyapunov function the storage function of the error system plus the standard term to handle the parameter estimation error,

$$\mathcal{H}_d(t) = \frac{1}{2}\left[\tilde{z}^\top \mathcal{D}_B\tilde{z} + \tilde{\theta}^2\right]$$

we verify that the relation

$$\dot{\mathcal{H}}_d(t) = -\tilde{z}^\top \mathcal{R}_{Bd}\tilde{z} + \tilde{\theta}\left[\dot{\tilde{\theta}} + z_{2d}(z_2 - z_{2d})\right]$$

is satisfied along the trajectories of (7.42). Using (7.38) and the fact that $\dot{\tilde{\theta}} = \dot{\hat{\theta}}$ we obtain

$$\dot{\mathcal{H}}_d(t) = -\tilde{z}^\top \mathcal{R}_{Bd}\tilde{z} \leq -\alpha||\tilde{z}^2||$$

where α may be taken to be min $\left\{R_1, \frac{1}{R}\right\}$. We conclude that $\tilde{z}$ and $\tilde{\theta}$ are bounded and that $\tilde{z}$ is square integrable. To actually show that $\tilde{z} \to 0$ asymptotically, it must be verified that $\tilde{z}$ is uniformly continuous. For this, it suffices to show that $\dot{\tilde{z}}$ is bounded. From the perturbed error dynamics (7.42), and the established boundedness of $\tilde{\theta}$ and $\tilde{z}$, it follows that $\dot{\tilde{z}}$ is bounded if, and only if, z_{2d} is bounded. In order to prove that z_{2d} is bounded, note first that its associated zero-dynamics, given by

$$\dot{z}_{2d} = -\frac{\hat{\theta}}{C}\left(z_{2d} - \frac{z_{2*}^2}{z_{2d}}\right), \tag{7.43}$$

is asymptotically stable towards the equilibrium point located at $z_{2d} = z_{2*}$, for all initial conditions satisfying $z_{2d}(0) > 0$, provided $\hat{\theta} > 0 \; \forall \; t$. The dynamics (7.43) is also asymptotically stable towards a second equilibrium point, located at $z_{2d} = -z_{2*}$, for all initial conditions satisfying $z_{2d}(0) < 0$, provided $\hat{\theta} > 0 \; \forall \; t$.

Take as a Lyapunov function candidate for the controller dynamics, $V_2 = \frac{C}{2}(z_{2d} - z_{2*})^2$. The time derivative of V_2 along the trajectories of (7.36)–(7.38) results in the following expression,

$$\dot{V}_2 = -\hat{\theta}(z_{2d} - z_{2*})\left\{z_{2d} - \frac{z_{2*}^2}{z_{2d}}\left[\left(1 + \frac{R_1}{E}\tilde{z}_1\right) + L\frac{z_{2*}^2}{E}z_{2d}\tilde{z}_2\right]\right\}. \tag{7.44}$$

Then, by virtue of the boundedness of $\tilde{z}_1$, $\tilde{z}_2$, and $\hat{\theta}$, and the fact that initial conditions for such variables can be entirely chosen at will and, also, provided that $\hat{\theta} > 0 \ \forall \ t$, it follows that given positive constants β_1 and β_2, with,

$$0 \ < \ \beta_1 \ < \ \frac{E}{R_1} \ \ ; \ 0 \ < \ \beta_2 \ < \ \frac{2E^2}{Lz_{2*}^3}\sqrt{\frac{R_1}{E}\beta_1} \tag{7.45}$$

such that initial conditions for the error vector components satisfy, $|\tilde{z}_1| < \beta_1$; $|\tilde{z}_2| < \beta_2$ then, the time derivative of V_2 given by (7.44), is strictly negative *outside* the closed interval $[Z_m, Z_M]$ of the real line, containing in its interior the equilibrium point, z_{2*}, for z_{2d}, where

$$Z_m \ = \ z_{2*}\left\{ \left[1 - \frac{R_1}{E}\beta_1 + \frac{L^2 z_{2*}^6}{4E^4}\beta_2^2\right]^{1/2} - \frac{Lz_{2*}^3}{2E^2}\beta_2 \right\}$$

$$Z_M \ = \ z_{2*}\left\{ \left[1 + \frac{R_1}{E}\beta_1 + \frac{L^2 z_{2*}^6}{4E^4}\beta_2^2\right]^{1/2} + \frac{Lz_{2*}^3}{2E^2}\beta_2 \right\}.$$

We conclude that $\tilde{z}$ is absolutely continuous and hence $\lim_{t\to\infty} \tilde{z}(t) = 0$. Moreover, given that z_2 asymptotically converges to the same equilibrium point of z_{2d}, given by z_{2*}, it follows that z_1 converges to its corresponding equilibrium value, z_{2*}^2/RE. Since z_1 and z_{1d} asymptotically converge to the same equilibrium point it follows that necessarily, $\hat{\theta} \to 1/R$.

■

B   The Buck–boost converter

The following proposition summarizes the properties of a passivity–based nonlinear adaptive dynamical controller for the Buck–boost converter. The proof follows using similar arguments to those used in the proof of the previous proposition.

Proposition 7.13 *Consider the averaged dynamics of the Buck–boost converter circuit, where $C > 0$, $L > 0$, $E > 0$ are known constants representing the capacitance, inductance and external voltage respectively, and $R > 0$ is the unknown load charge resistance.*

Define an adaptive nonlinear dynamic state feedback controller as

$$\dot{z}_{2d} \ = \ -\frac{\hat{\theta}}{C}\left\{ z_{2d} + z_{2*}\left(\frac{z_{2*}}{E} + 1\right) \times \right.$$

$$\left. \times \left[\frac{E + Lz_{2*}\left(\frac{z_{2*}}{E} + 1\right)z_{2d}\tilde{z}_2 + R_1\left(z_1 - z_{2*}(\frac{z_{2*}}{E} + 1)\hat{\theta}\right)}{E - z_{2d}(t)}\right] \right\} \tag{7.46}$$

$$\mu(t) \ = \ \frac{z_{2d}(t) + Lz_{2*}\left(\frac{z_{2*}}{E} + 1\right)z_{2d}\tilde{z}_2 - z_d + R_1\left(z_1 - z_{2*}(\frac{z_{2*}}{E} + 1)\hat{\theta}\right)}{z_{2d}(t) - E} \tag{7.47}$$

where $\tilde{z}_2 \overset{\triangle}{=} z_2 - z_{2d}$ and the dynamical controller initial condition is chosen so that, $z_{2d}(0) < E$ and $\hat{\theta}(0) > 0$. The constant reference value for z_2, denoted by $-z_{2*}$, is a strictly negative quantity. The quantity $\hat{\theta}$ denotes the estimate of $\frac{1}{R}$. The parameter R_1 is a designer chosen constant with the only restriction of being strictly positive. Under these conditions,, it is always possible to chose the controller's initial state $z_{2d}(0)$ and $\hat{\theta}(0)$, such that the closed loop system (6.42), (6.43), (7.46), (7.47) has an equilibrium point given by,

$$(z_1, z_2, z_{2d}, \hat{\theta}) = \left(\frac{z_{2*}}{R} \left(\frac{z_{2*}}{E} + 1 \right), -z_{2*}, -z_{2*}, \frac{1}{R} \right)$$

which is asymptotically stable. $\qquad\square$

C The multivariable Boost converter

The following proposition summarizes the properties of a passivity–based nonlinear adaptive dynamical controller for the cascaded Boost converter that, indirectly, regulates the average output capacitor voltage towards a desired value z_{2*} by regulating the average inductor currents towards a given set of desired steady state values uniquely determined from the arbitrarily chosen sequence of growing steady state capacitor voltages.

Proposition 7.14 *Consider the averaged dynamics of the multivariable cascaded Boost converter circuit, where $C_i, L_i > 0$; $i = 1, 2, \ldots, n$, and $E > 0$ are known constants representing the capacitances, inductances and voltage of the external source, respectively. The constant parameter, $R_L > 0$, is the unknown load charge resistance.*

Consider the adaptive nonlinear dynamic state feedback controller

$$\mu_1 = 1 - \frac{1}{z_{2d}} \left[E + R_1 \left(z_1 - \hat{\theta} \frac{z_{2*}^2}{E} \right) + L_1 \frac{z_{2*}^2}{E} z_{2nd} \tilde{z}_{2n} \right] \qquad (7.48)$$

$$\mu_2 = 1 - \frac{1}{z_{4d}} \left[z_{2d} + R_3 \left(z_3 - \hat{\theta} \frac{z_{2*}^2}{\overline{z}_2} \right) + L_2 \frac{z_{2*}^2}{\overline{z}_2} z_{2nd} \tilde{z}_{2n} - z_{2nd} \right]$$

$$\vdots$$

$$\mu_n = 1 - \frac{1}{z_{2nd}} \left[z_{(2n-2)d} + R_{2n-1} \left(z_{2n-1} - \hat{\theta} \frac{V_d^2}{\overline{z}_{2n-2}} \right) + L_n \frac{V_d^2}{\overline{z}_{2n-2}} z_{2nd} \tilde{z}_{2n} \right]$$

$$\dot{z}_{2d} = -\frac{\hat{\theta}}{C_1} \left\{ 1 - \frac{\overline{z}_2}{E z_{2d}} \left[E + R_1 \left(z_1 - \hat{\theta} \frac{z_{2*}^2}{E} \right) + L_1 \frac{z_{2*}^2}{E} z_{2nd} \tilde{z}_{2n} \right] \right\} \frac{z_{2*}^2}{\overline{z}_2} + \frac{1}{R_2 C_1} \tilde{z}_2$$

$$\dot{z}_{4d} = -\frac{\hat{\theta}}{C_2} \left\{ 1 - \frac{\overline{z}_4}{\overline{z}_2 z_{4d}} \left[z_{2d} + R_3 \left(z_3 - \hat{\theta} \frac{z_{2*}^2}{\overline{z}_2} \right) + L_2 \frac{z_{2*}^2}{\overline{z}_2} z_{2nd} \tilde{z}_{2n} \right] \right\} \frac{z_{2*}^2}{\overline{z}_4} + \frac{\tilde{z}_4}{R_4 C_2}$$

$$\vdots \qquad (7.49)$$

$$\dot{z}_{2nd} = -\frac{\hat{\theta}}{C_n}\left\{z_{2nd} - \frac{1}{z_{2nd}}\left[z_{(2n-2)d} + R_{2n-1}\left(z_{2n-1} - \hat{\theta}\frac{V_d^2}{\bar{z}_{2n-2}}\right)\right.\right.$$

$$\left.\left.+L_n\frac{V_d^2}{\bar{z}_{2n-2}}z_{2nd}\tilde{z}_{2n}\right]\frac{V_d^2}{\bar{z}_{2n-2}}\right\}$$

$$\dot{\hat{\theta}} = -z_{2nd}\tilde{z}_{2n} \tag{7.50}$$

where the dynamical controller initial conditions are chosen so that, $z_{(2i)d}(0) > 0$; $i = 1, 2, \ldots, n$, and $\hat{\theta}(0) > 0$. V_d, the constant reference value for the output voltage z_{2nd}, is chosen to be a strictly positive quantity. The set of constants, $\bar{z}_2, \bar{z}_4, \ldots, \bar{z}_{2n-2}, \bar{z}_{2n}$, satisfy the restriction

$$E < \bar{z}_2 < \bar{z}_4 < \ldots < \bar{z}_{2n-2} < \bar{z}_{2n} = z_{2*}$$

but they are, otherwise, completely arbitrary. The scalar variable $\hat{\theta}$ denotes the estimate of $\frac{1}{R_L}$. The parameters $R_1, R_2, \ldots, R_{2n-1}$ are designer chosen constants with the only restriction of being strictly positive. Under these conditions, it is always possible to choose the initial state of the controller $z_{(2j)d}$ $j = 1, 2, \ldots, n$ and $\hat{\theta}(0)$, such that the closed loop system (7.48)–(7.50) has an equilibrium point given by,

$$(z_1, z_2, z_3, z_4, \ldots, z_{2n-1}, z_{2n}, z_{2d}, z_{4d}, \ldots, z_{(2n-2)d}, z_{2nd}, \hat{\theta}) =$$
$$\left(\frac{V_d^2}{R_L E}, \bar{z}_2, \frac{V_d^2}{R_L\bar{z}_2}, \bar{z}_4, \ldots, \frac{z_{2*}^2}{R_L\bar{z}_{2n-2}}, z_{2*}, \bar{z}_2, \bar{z}_4, \ldots, \bar{z}_{2n-2}z_{2*}, \frac{1}{R_L}\right)$$

which is asymptotically stable. $\qquad\square$

Remark 7.15 A passivity–based adaptive controller corresponding to the single stage Boost converter case can be immediately obtained as a particular case of Proposition (7.14) For this, one simply sets $n = 1$ and $\bar{z}_0 = E$ in (7.50).

4.2 Simulation results

Simulations of the closed loop behaviour of the average Boost converter and the passivity based indirect adaptive feedback controller were performed on the following perturbed version of the Boost converter circuit,

$$\dot{z}_1 = -(1 - \mu)\frac{1}{L}z_2 + \frac{E + \eta}{L}$$
$$\dot{z}_2 = (1 - \mu)\frac{1}{C}z_1 - \frac{1}{RC}z_2$$

where η represents an external stochastic perturbation input affecting the system behaviour directly through the external voltage source value. Note that this perturbation input is of the "unmatched" type. i.e., it enters the system equations through an input channel vector field, given by $[1/L \quad 0]^\top$ which is *not* in the range space of the control input channel, given by the vector field $[z_2/L \quad z_1/C]^\top$. The magnitude of the noise was chosen to be, approximately, 5 % of the value of E. The circuit parameter values were taken as in the simulation of Subsection 2.3, and we set $z_{1d} = 3.125$ Amp., with a steady state duty ratio of $\mu = 0.6$. This corresponds to a nominal average output voltage, $z_2 = z_{2*} = 37.5$ V. Fig. 7.10 shows the closed loop state trajectories corresponding to the feasible adaptive duty ratio synthesizer derived for the Boost converter. This figure also presents the trajectory of the duty ratio function, the trajectory of the parameter estimation values and a realization of the computer-generated stochastic perturbation signal η, addressed to as the "source perturbation noise".

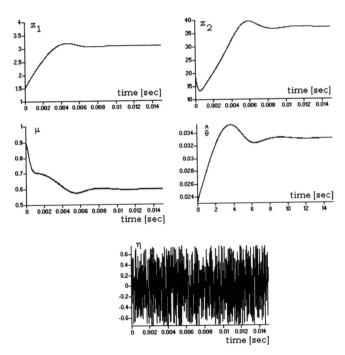

Figure 7.10: Performance evaluation of the indirect adaptive controller in a perturbed average "Boost" converter.

The simulations show that the proposed controller achieves the desired indirect stabilization of the output voltage around the desired equilibrium, while exhibiting high robustness with respect to the external stochastic perturbation input.

5   Experimental comparison of several nonlinear controllers

DC-to-DC power converters are regulated in applications by means of simple linear lead–lag compensators designed using the averaged linear approximation of the model. This induces an obvious hard constraint on the achievable performance for this class of controllers. It is then natural to ask if performance can be improved by means of nonlinear control. For most of the nonlinear control algorithms, so far proposed in the literature, we can establish stability properties by means of appropriate theoretical analysis. However, it is difficult to assess the merits and drawbacks of a particular control scheme, not to mention the potential performance improvement with respect to the linear designs, based solely on such theoretical analysis.

In this final section of the chapter we present some of the *experimental* results on a Boost converter reported in [75]. We compare five controllers, including the controller based on a linear approximation of the average PWM circuit, in terms of their ease of implementation and their closed–loop performance. For all these algorithms, local asymptotic stability of the desired equilibrium is guaranteed. The motivation of the present study is not just to illustrate the validity of the corresponding theoretical derivations, but to test the performance of the various schemes when confronted with a real physical application where unpredicted situations inevitably arise.

The performances of the various schemes are compared in the light of the following basic criteria: transient and steady state response to steps and sinusoidal references, attenuation of disturbances affecting the power supply and sensitivity to unknown loads. Particular emphasis is placed throughout on the flexibility provided by the tuning parameters in the shaping of the circuit responses. Even though this issue is not always appreciated in theoretical studies, we have found it to be of primordial importance in actual experimentation.

5.1   Feedback control laws

We now present five control laws whose performance was experimentally compared in a laboratory facility. The first controller is constituted by a simple state–feedback pole–placement scheme based on a first order linear approximation of the average PWM model. This control scheme may be regarded as the industry standard (see [117]). The other control schemes are nonlinear and have been studied in the previous sections. They rely, respectively, on linearization [244], passivation [246], sliding mode control [278] and a combination of sliding modes and passivity [211]. In the absence of external disturbances and parameter uncertainty, they all achieve (local) asymptotic stabilization, that is, they guarantee that for suitable initial conditions, the capacitor voltage converges towards a desired prespecified constant value with internal stability. We refer the reader to the above references for further details on the theoretical background of the schemes.

We consider the well–known "Boost" circuit model in its *switched–mode* representation (7.15) and (7.16), or the averaged PWM–controlled model (7.1). Two perturbations are considered: an unknown (time–varying) disturbance η, (which satisfies $|\eta| < E$) is added to the voltage source, and uncertainty in the output resistance ΔR.

The control laws that we consider are classified into two groups, depending on whether they directly generate the switching signal u, or they generate duty ratio function, $\mu(\cdot)$ that is fed to an auxiliary PWM circuit. This means also that for the control design, they use the continuous averaged model or the switched, exact, model, respectively. For lack of a better terminology we use these qualifiers to classify the schemes below.

A   Continuous control laws

A.1 Linear Averaged Controller (LAC)

This controller is based on the linearization of the averaged model around an equilibrium point

$$\dot{\tilde{z}} = A\tilde{z} + B\tilde{\mu}$$

where A and B are given by

$$A = \begin{bmatrix} 0 & -\frac{1-\bar{\mu}}{L} \\ \frac{1-\bar{\mu}}{C} & -\frac{1}{RC} \end{bmatrix} = \begin{bmatrix} 0 & -\frac{E}{V_d L} \\ \frac{E}{V_d C} & -\frac{1}{RC} \end{bmatrix}$$

$$B = \begin{bmatrix} \frac{\bar{z}_2}{L} \\ -\frac{\bar{z}_1}{C} \end{bmatrix} = \begin{bmatrix} \frac{V_d}{L} \\ -\frac{V_d^2}{REC} \end{bmatrix}$$

and $(\tilde{\cdot})$ denotes the deviation with respect to the equilibrium value.

Some simple calculations show that the pair $(A, \; B)$ is controllable. Hence, the poles of $(A - BK)$ can be located arbitrarily with a suitable choice of the state feedback gains $K = [k_1 \; k_2]$.

Taking the averaged error voltage as the circuit output we obtain the following transfer function

$$H(s) = \frac{N_1(s)}{D(s)} = \frac{\mathcal{K}(s - \mathcal{Z}_1)}{(s - \mathcal{P}_1)(s - \mathcal{P}_2)} \tag{7.51}$$

As expected from the discussion of Section **6**.3 and the commutativity of the operations of linearization and zero–dynamics extraction [110], the linearized system transfer function has a zero in the right hand side of the complex plane[2] given by $\mathcal{Z}_1 = \frac{RE^2}{LV_d^2}$. The two stable poles are located in

$$\mathcal{P}_{1,2} = -\frac{1}{2RC}\left(1 \pm \left[1 - 4\frac{E^2R^2C}{V_d^2L}\right]^{1/2}\right)$$

On the other hand, if we take as an output the averaged error current $\tilde{z}_1$, then we obtain a transfer function as in (7.51) but with a left half plane zero located in $\mathcal{Z} = -\frac{2}{RC}$.

A.2 Feedback Linearizing Controller (FLC)

In [242] and [230] we proposed the following nonlinear static state feedback controller that linearizes the input-output behaviour of the system, with linearizing output taken to be the circuit total energy:

$$\mu = \frac{z_2}{\frac{E}{L} + \frac{2z_1}{RC}}\left\{\left(\frac{2}{R^2C} - \frac{a_1}{R} + \frac{a_2C}{2}\right)z_2^2 + \left(a_1E + \frac{a_2L}{2}z_1\right)z_1 + \frac{E^2}{L} - a_2\mathcal{H}_d\right\} \quad (7.52)$$

where $a_1, a_2 > 0$ are the design parameters, and

$$\mathcal{H}_d \triangleq \frac{V_d^2}{2}\left(C + \frac{L}{R^2E^2}V_d^2\right).$$

More precisely, it is shown in [242], that the converters total energy satisfies the linear equation

$$\ddot{\mathcal{H}} + a_1\dot{\mathcal{H}} + a_2\mathcal{H} = a_2\mathcal{H}_d$$

Notice that $\mathcal{H}_d$ is the energy level required to ensure that as $\mathcal{H} \to \mathcal{H}_d$ we have $z_2 \to V_d$, as desired. Since the dynamics is now linear, the convergence rate can be fixed arbitrarily with a suitable choice of the controller parameters a_1, a_2.

The advantage of having a linear closed–loop dynamics, expressed in some physically meaningful variables, can hardly be overestimated. It allows us to easily predict the effect of the tuning parameters and simplify the controller commissioning. However, as it will be shown by our experiments, the existence of unmodeled nonlinearities, and in particular input saturation, limits the validity of these predictions.

A.3 Passivity-Based Controller (PBC)

---

[2]The system will display an initial undershoot because it has an odd number of real zeros in the open right hand plane.

In Section 4 the following nonlinear dynamic controller that preserves passivity of the closed loop circuit was proposed in slightly different but equivalent terms,

$$\mu = -\frac{1}{z_{2d}}\left[E + R_1\left(z_1 - \frac{V_d^2}{RE}\right)\right]$$

where the controller dynamics is given by

$$\dot{z}_{2d} = -\frac{1}{RC}\left\{z_{2d} - \frac{V_d^2}{Ez_{2d}}\left[E + R_1\left(z_1 - \frac{V_d^2}{RE}\right)\right]\right\}, \quad z_{2d}(0) > 0 \qquad (7.53)$$

where $R_1 > 0$ is a design parameter. As shown in Section 4 the desired storage function $\mathcal{H}_d \triangleq \frac{1}{2}\tilde{z}^\top \mathcal{D}\tilde{z}$, where $\tilde{z} = z - [\frac{V_d^2}{RE}, z_{2d}]^\top$, satisfies

$$\dot{\mathcal{H}}_d = -\tilde{z}\begin{bmatrix} R_1 & 0 \\ 0 & 1/R \end{bmatrix}\tilde{z} \leq -\alpha\mathcal{H}_d, \quad \alpha \triangleq \frac{\min(R_1, 1/R)}{\max(L, C)} > 0$$

we see that R_1 injects the damping required for asymptotic stability, and that the convergence rate of $\tilde{z}$ to zero is improved by increasing R_1. From these observations one might be tempted to try a high-gain design, but a more careful analysis and experimentation reveals this not to be convenient. To see this, notice that $\tilde{z}_2 \to 0$ does not imply that $z_2 \to V_d$ as desired, unless $z_{2d} \to V_d$ as well. To study the behaviour of the latter consider the signal $\rho \triangleq 1/2(z_{2d}^2 - V_d^2)$, which satisfies

$$RC\dot{\rho} = -2\rho + R_1 V_d^2 / E\tilde{z}_1$$

This equation clearly shows two important limitations of the scheme: first, that the speed of convergence is essentially determined by the natural time constant of the converter. We remark that even if $\tilde{z}_1$ converges to zero very fast, z_{2d} (and consequently z_2) evolves according to this time constant. Second, increasing the damping will induce a "peaking" in ρ, and consequently a slower convergence of $z_{2d} \to V_d$.

Furthermore, we have that the convergence rate of $\tilde{z}$ is bounded from below by the undamped dynamics, i.e. $\alpha \geq \frac{1}{RC}$. The sluggishness of this scheme has been observed in our experiments. To overcome this drawback we have tried to add some damping in the subsystem associated with z_{2d}, that is, we modify (7.53) to

$$\dot{z}_{2d} = -\frac{1}{RC}\left\{z_{2d} - \frac{V_d^2}{Ez_{2d}}\left[E + R_1\left(z_1 - \frac{V_d^2}{RE}\right) + G_2(z_2 - z_{2d})\right]\right\}$$

with $z_{2d}(0) > 0$ and $R_2 > 0$ the new damping coefficient. This gives a closed–loop of the form

$$\mathcal{D}\dot{\tilde{z}} - (1 - \mu)\mathcal{J}\tilde{z} + \mathcal{R}_d\tilde{z} = 0$$

but with a new damping matrix

$$\begin{bmatrix} R_1 & 0 \\ 0 & \frac{1}{R} + G_2 \end{bmatrix}$$

Hence the convergence of $\tilde{z}$ to zero can be made arbitrarily fast. Unfortunately, this does not change significantly the dynamics of z_{2d}, since ρ now satisfies

$$RC\dot{\rho} = -2\rho + \frac{V_d^2}{E}(R_1\tilde{z}_1 + G_2\tilde{z}_2)$$

Another possibility is to use cross terms in the damping matrix R_d. All these modifications were tried experimentally but no significant improvement was obtained.

B Switched control laws

B.1 Sliding Mode Controller (SMC)

The indirect sliding mode controller of Proposition 7.4 was also tested. From the analysis given there we see that when the sliding line $s = x_1 - V_d^2/RE = 0$ is reached at time $t_h = (L/R)(V_d/E)^2$, the voltage capacitor x_2 grows from zero towards its equilibrium value V_d, ideally governed by

$$\overline{x}_2(t) = V_d\left[1 - e^{-\frac{2}{RC}(t-t_h)}\right]^{1/2}$$

Notice that, similarly to PBC, it is the open–loop time constant that regulates this dynamics. Furthermore, we will show in our experiments that this remarkably simple approach is, unfortunately, extremely sensitive to parameter uncertainty and noise. Finally, as usual with sliding mode strategies, the energy consumption is very high.

B.2 Sliding mode plus PBC (SM+PBC)

We tried also the sliding mode plus PBC controller of Proposition 7.7 was also tested. Consistent with the analysis carried out in that section, our experiments will show that as the initial conditions for the controller are chosen "closer" to the desired equilibrium state of the controlled plant, the state responses of the plant become smoother with slightly larger settling times but with a much better behaved transient shape. Unfortunately, it suffers from the same drawback as PBC of providing no freedom to shape the response of the output voltage.

C Adaptive schemes

All the control strategies presented above are of the indirect type, where we control the capacitor voltage via regulation of the inductor current and invoke the one-to-one correspondence between their equilibria to achieve the output regulation objective. This strategy is clearly very sensitive to parameter uncertainty, in particular load resistance changes. To overcome this drawback we have tested adaptive modifications of the control laws. As done already in Section 4 some stability analysis can be performed.

C.1 Adaptive PBC

First, we tried the adaptive PBC of Proposition 7.12 in Section 4.

C.2 Adaptive SMC

To add adaptation to the basic SMC we propose to modify the switching line as

$$s = x_1 - \hat{\theta} \frac{V_d^2}{E}$$

with the parameter $\hat{\theta}$ estimated by

$$\dot{\hat{\theta}} = -\gamma V_d \left(x_2 - V_d \right), \qquad \gamma < \frac{E^2}{V_d^4 L} \tag{7.54}$$

To understand the main idea behind this scheme, note that in the switching line, the term $1/R$ has been replaced by its estimation $\hat{\theta}$. Moreover, the adaptation law (7.54) was motivated by the form of the corresponding adaptive version for the case PBC where z_{2d} has been substituted by V_d, because there is no auxiliary dynamics, as before.

C.3 Adaptive SM+PBC

An adaptive version for the SM+PBC can be similarly obtained considering the same switching line as above, but using the estimator

$$\dot{\hat{\theta}} = -\gamma x_{2d} \left(x_2 - x_{2d} \right)$$

The controller auxiliary dynamics is modified accordingly to

$$\dot{x}_{1d} = -\frac{1}{L}(1-u)x_{2d} + \frac{R_1}{L}(x_1 - x_{1d}) + \frac{E}{L}$$

$$\dot{x}_{2d} = \frac{1}{C}(1-u)x_{1d} - \frac{\hat{\theta}}{C}x_{2d}$$

Again, the idea behind the above proposed scheme is to substitute $1/R$ by its estimated value $\hat{\theta}$ whenever it occurs, and take the form of the adaptive law to estimate this value as in the PBC case, but now using x_2, x_{2d} instead of z_2, z_{2d}, respectively.

5.2   Experimental configuration

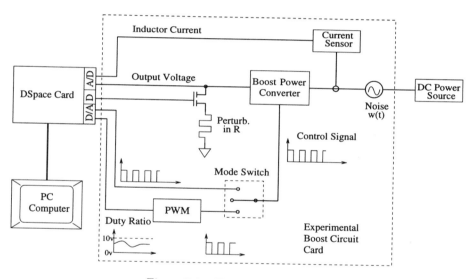

Figure 7.11: Experimental setup.

The experimental card was assembled using low cost commercial electronic elements placed on a card designed in the laboratory. In Fig. 7.11 we show the experimental set-up consisting of the Boost circuit card that receives control signals from a D/A converter of a dSpace card placed in a PC. The dSpace card acquires, using an A/D converter, the output voltage and inductor current signals previously conditioned from the Boost card. Two DC power supplies are necessary to operate the whole system, one to provide energy to the Boost system (we will refer to it as the power supply in the rest of the section), and the other one to feed the electronic part of the card.

Element	Value	Units
Capacitance	1000	μF
Inductance	170	mH
Resistance	100	Ω
Power supply	10	V

Table 7.1: Parameters for the power converters experimental setup.

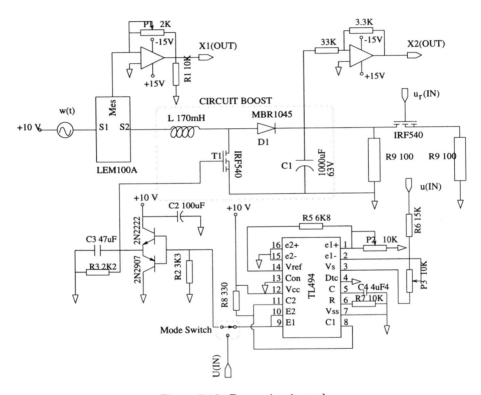

Figure 7.12: Boost circuit card

In Fig. 7.12 we show the main circuit which is constituted by a Boost circuit, a PWM circuit, and some signal conditioners. This design is very close to that of [5].

The Boost circuit is basically composed by an inductor, a capacitor, a resistive charge and a switch, the last one is implemented by interconnecting a FET transistor and a rapid diode in a suitable manner, all this elements fed by a DC power supply.

The values of its elements are shown below.

A current sensor together with a current-to-voltage converter are introduced to obtain a measurement of the inductor current x_1. In this way, a voltage signal is generated which we can feed it into the dSpace card to be used in the control law. In the case of the output voltage x_2 we put a voltage divisor so that we can reduce the level of this signal in such a way that its final value is always in the 0-10 V range. To compensate the nonlinear characteristic of the current sensor and some offset introduced by the linear amplifier circuits, we computed their characteristic functions and implemented their inverse.

In this card we have the choice between controlling the Boost circuit by means of a PWM generated signal or by directly introducing a switching signal coming from the dSpace card. This selection is done depending on the position of the *mode switch*. If a PWM control is selected, then we should feed a continuous signal that represents the duty ratio in the range 0-10 V, corresponding to 0-100 %. The PWM circuit was designed using a commercial integrated circuit and it was programmed to have a sampling rate of 50 KHz. On the other hand, if *switch* control-mode is selected, then we should provide a switching pulse signal with amplitudes of 0 and 5 V.

Another interesting feature of the card is the possibility to connect or disconnect a second resistive charge placed in parallel to the output load. This is done by means of a digital output signal, obtained from the dSpace card, which commands the gate of a MOSFET acting as a switch.

To study the effect of disturbances η in the power source, a driver circuit has been interconnected between the power supply and the Boost circuit. In this way we can introduce disturbance signals from a signal generator or even from the PC.

Programs were written for a PC using C language. The programs containing the description of the controllers are translated and down–loaded into the dSpace memory as assembler programs by means of a suitable software. Time derivatives in some of the control laws are discretized using a rectangular approximation with a sampling period of 6 x 10^{-5} sec.

5.3 Experimental results

The five control laws described in the previous subsection have been implemented in the above Boost circuit card. Their behaviour is compared with the following basic criteria:

i) Transient and steady state response to steps and sinusoidal output voltage references,

ii) Attenuation of step and sinusoidal disturbances in the power supply,

iii) Response to pulse changes in the output resistance.

To gain some further insight into the behaviour of the converter, and motivate the need for closed–loop control, we also present the responses of the open–loop system.

Unless indicated otherwise, in all the experiments we consider, as desired output voltage, the value $V_d = 20$ V. This corresponds to a desired inductor current $V_d^2/RE = 0.4$ Amp. and to an equivalent duty ratio of $\overline{\mu} = 1 - E/V_d = 0.5$.

A   Response to output voltage references

A.1   Step references

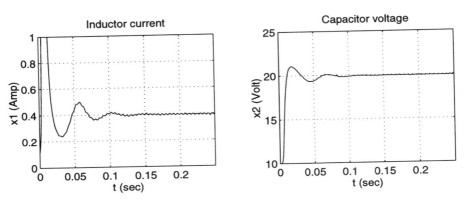

Figure 7.13: Open loop step response

In Fig. 7.13 we show the typical behaviour of the *open loop* system introducing a step in the duty ratio μ of 0.5. As we can see, the behaviour of the output voltage x_2 is quite good, it is fast and not too oscillatory. On the other hand, the current x_1 through the inductor has a large overshoot that exceeds the limit of available current in the power source for a considerable time. This behaviour is not desirable because it could trigger the safety elements that disable the power source. We also observe that there exists a small undershoot due to the nonminimum phase character of the average system.

In Fig 7.14 we show a family of step responses of the system under LAC for different pole placements, namely $(\mathcal{P}_1, \mathcal{P}_2) = (-22.08, -477.9), (-20.49, -989.5), (-21.58, -321.75)$, etc. As expected, for faster poles we obtain faster responses. However, due to the presence of nonlinearities, specially the saturation of the inductor current, the time responses do not correspond to the proposed pole placement. In particular,

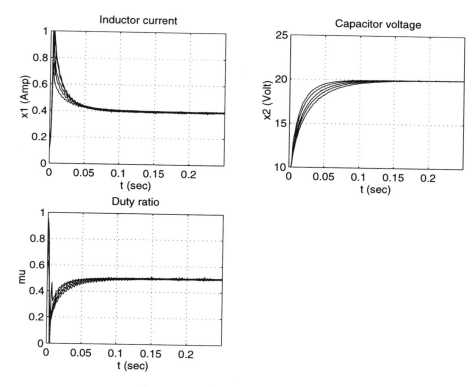

Figure 7.14: Step responses for LAC

the response of x_2 is quite slow, and we can not obtain oscillatory responses that are predicted by the linear approximation theory. The former can be explained using root locus analysis which reveals that for large k_1 one pole approaches -20 while the other approaches $-\infty$, so that the time response is dominated by the slowest pole. We also observed that for relatively small gains a significant steady state error appears, while the undershoot amplitude increases for faster responses.

In Fig. 7.15 we present typical responses of the system under FLC for different values of a_1 and a_2, namely $(a_1, a_2) = (30, 900), (75, 3000), (100, 2500)$, etc. These corresponds to different pole locations of the closed–loop system described in the coordinates $[\mathcal{H}, \ \tilde{\mathcal{H}}]$. Again, faster responses in x_2 are obtained with faster poles, which yields, in turn, higher peaks in x_1. This limits the speed of convergence of x_2 due to controller saturation. Notice, however, that for comparable convergence rates there is a significant reduction on the peak size with respect to LAC.

In Fig. 7.16 we show the responses of the PBC for different values of R_1, namely

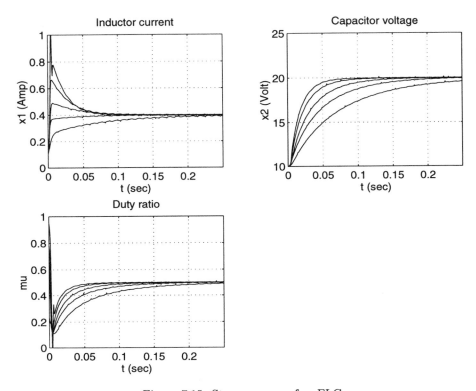

Figure 7.15: Step responses for FLC

$R_1 = 1, 5, 10, 25, 35$, etc. From the plots we see that the design parameter R_1 affects only the behaviour of x_1, while x_2 remains almost invariant. Actually, for small dampings, the current x_1 varies slowly, with a large overshoot, but, as we increase the damping x_1 converges faster -with no overshoot-. Finally, for large values of R_1 the response exhibits a fast peaking. As explained above, damping determines the speed of convergence of x_2 towards z_{2d}, the oscillatory responses in Fig. 7.16 correspond to small damping. However, z_{2d} remains essentially invariant with respect to R_1. We should note that the peaking in z_{2d}, predicted by the theory, was not observed. A reason for this is that it is actually filtered by the dominant pole.

In Fig. 7.17 we present the response of the system controlled with a SMC. As we can see, the sliding regime is reached almost instantaneously, thus x_1 reaches its desired value very fast. From the equations describing the sliding dynamics (7.19) we know that the response of x_2 is governed by the natural time constant $\frac{2}{RC}$, which makes the response slow. Notice that there are no tuning parameters in this control

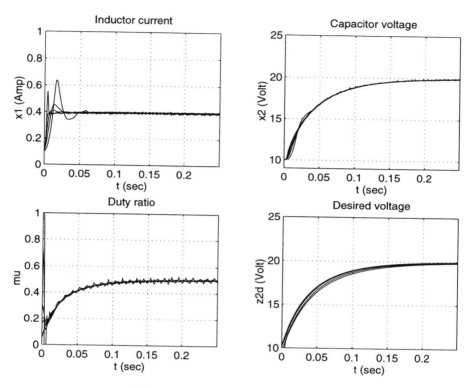

Figure 7.16: Step responses for PBC

law.

In Fig. 7.18 we present the responses of SM+PBC for different values of the design parameter R_1. Again, as in the PBC, the voltages x_2 and x_{2d} remain almost invariant with respect to R_1. Only the currents x_1 and x_{1d} are affected by this parameter and both exhibit a peaking phenomenon during the transient part of the response. This peaking becomes higher as R_1 is taken to be larger.

As a conclusion for this part it is found that only LAC and FLC provide some flexibility to shape the step response. Two important advantages of FLC over LAC is that it achieves the same convergence rates with smaller inductor currents. Hence, less energy is consumed. Furthermore, as reported in [293], the steady–state error becomes systematically smaller. As expected, the predictions of the theory are accurate only up to the point that the saturation levels are reached.

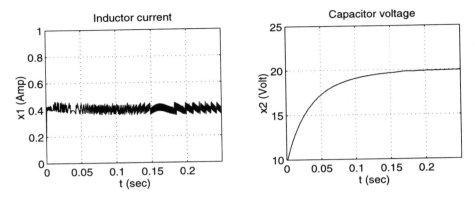

Figure 7.17: Step response for SMC

Strategy	Cut-off frequency range (Hz)
LAC	* * **
FLC	[2.0, 10.5]
PBC	3.1
SMC	3.0
SM+PBC	[2.1, 3]

Table 7.2: Comparison of cut-off frequency ranges.

A.2 Sinusoidal references

Even though the controllers are designed for a specific regulation objective, a very important characteristic to study in the controlled circuit is its ability to follow a time varying desired output signal $V_d(t)$. Obviously, this characteristic is closely related with the circuits frequency response. Specifically, it is highly dependent upon the circuit bandwidth. In Fig. 7.19 we show the closed–loop frequency responses for the five control laws. Due to reasons related to the actual physical construction of the circuit we can only follow reference signals of the form $V_d = V_{d0} + A_{Vd}sin(2\pi ft) > E$. The frequency responses were obtained assuming that we were placed at the equilibrium point corresponding to the DC-component $V_{d0} = 20$ V. As an approximation, we proceeded to take only the alternating part of the response. To generate the gain plot one divides, for each frequency value, the amplitude of the alternating output signal by the amplitude $A_{Vd} = 5$ V. In LAC, we observe that there appears a problem of steady state gain, that is, for higher values of k_1, for instance $k_1 = 2.5, k_2 = 0.01$, the steady state gain almost reaches 1 which is the expected value, but which corre-

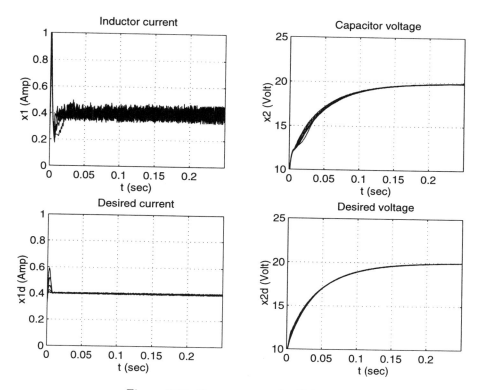

Figure 7.18: Step responses for SM+PBC

sponds to a bandwith of 6 Hz. On the contrary, for lower values of k_1, for instance $k_1 = 0.5, k_2 = 0$, the steady state gain decreases to 0.4615 with a bandwith of 20 Hz.

This problem is due to the fact that the system has been linearized around an equilibrium point, and now we are carrying the system far from this point putting in evidence its nonlinearities that are not considered in the control law.

In the case of FLC there is no problem of variations in the steady state gain and in this case the bandwidth can be enlarged choosing a_1 and a_2 such that the corresponding damping coefficient is small and the natural frequency high. For example, choosing $a_1 = 60, a_2 = 3600$ we have a bandwidth of 10.5 Hz and for $a_1 = 90, a_2 = 900$ we get a bandwidth of 2 Hz.

In PBC this frequency response is not affected when R_1 is changed, so the bandwith is fixed to 3.1 Hz. This is also the case for SMC. For SM+PBC the bandwith varies slightly depending on the value of R_1, for an $R_1 = 50$ we can enlarge the band-

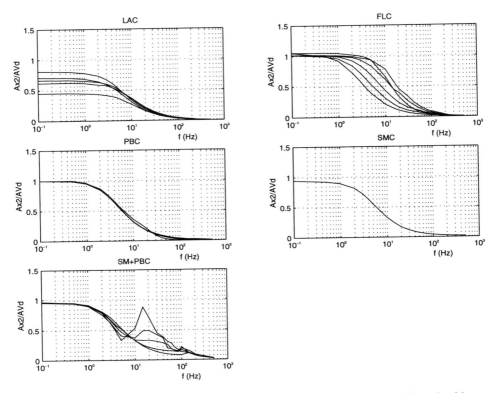

Figure 7.19: Frequency responses to periodic reference voltage, $\delta V_d(t) \mapsto \delta x_2(t)$

with to 3 Hz, and for $R_1 = 5$ we reduce it to 2.1 Hz. We have observed also some resonance phenomena in the circuit for low values of R_1, this is manifested in the form of peaks in the frequency response which disappear for higher values of R_1. We do not have at this point a physical or theoretical explanation of this phenomenon.

The results of these experiments are summarized in Table 7.2.

Besides the bandwidth we are, of course, also interested in the phase shift introduced in the loop. To assess this characteristic we show in Fig. 7.20 some typical time responses of the output voltage for the various control strategies and a reference signal of 2Hz. We have chosen the tuning that gives the largest bandwidth. We can see that the smallest phase shift is achieved with FLC, which also provides the best achievable bandwidth. We should underscore the poor performance of LAC in this respect. The remaining responses behave quite similar.

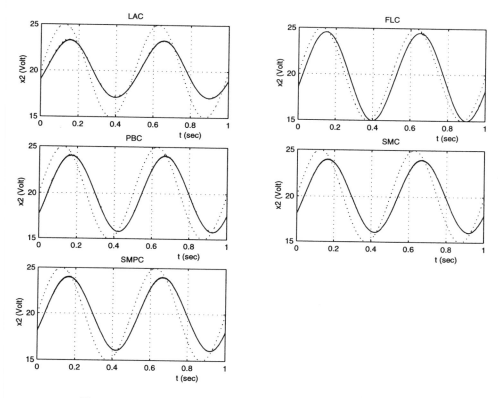

Figure 7.20: Time responses to a periodic reference signal $V_d(t)$

B   Disturbance attenuation

In this section we study the behaviour of the control laws in the face of a disturbance in the power supply. We consider two classes of disturbances steps and sinusoids.

B.1   Step disturbance

In this experiment we propose to add a pulse disturbance w to the power supply (obtained from a signal generator) of amplitude 3 V and duration 0.1 sec. In Fig. 7.21 we show the behaviour of the output voltage x_2 for each control law and different tunings.

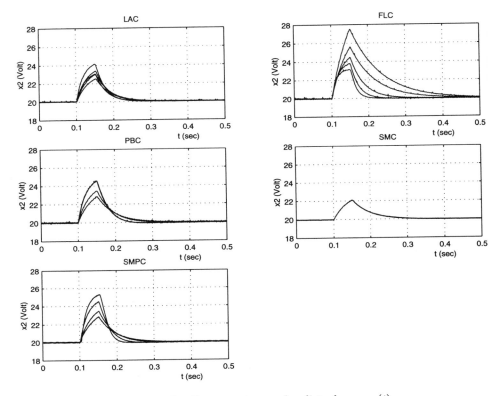

Figure 7.21: Response to a pulse disturbance $w(t)$

We can see that FLC is quite sensitive to step disturbances, while SMC is almost insensitive to it. To obtain some quantitative measure we evaluated the energy amplification of the circuit, that is we calculated the ratio

$$\gamma = \frac{||\tilde{x}_2||_2}{||w||_2} = \frac{||\tilde{x}_2||_2}{3\sqrt{0.1}}$$

where $|| \cdot ||_2^2 \triangleq \int_0^\infty (\cdot)^2 dt$.

This number provides a lower bound to the $\mathcal{L}_2$-gain of the operator $T_{w\tilde{x}_2} : w \mapsto \tilde{x}_2$. See [233] for some theoretical evaluation of bounds on this norm for FLC and PBC.

In LAC, γ can be reduced using higher values of k_1 which implies that the dominant pole is slow, so for a $\gamma = 0.9059$ we have chosen $k_1 = 2.5$ and $k_2 = 0.01$ which results in a dominant pole near 20.

Strategy	Ranges of γ
LAC	[0.5715, 0.9059]
FLC	[0.6855, 1.9787]
PBC	[0.6506, 0.9910]
SMC	0.4738
SM+PBC	[0.6409, 1.602]

Table 7.3: Comparison of amplification ratios.

Strategy	Steady-state gain range	Cut-off frequency range (Hz)
Open loop	2	14
LAC	[1.5625, 1.125]	[10.0, 10.1]
FLC	[3.10, 1.16]	[2.4, 14.0]
PBC	[1.375, 1.875]	[7.0, 10.1]
SMC	1.06	10
SM+PBC	[1.3125, 1.75]	[7, 10.2]

Table 7.4: Comparison of gain and cut-off frequency ranges.

In FLC, γ can be reduced proposing high values of a_2 and small values of a_1, this corresponds to poles with high real and imaginary parts and damping coefficients less than 1.

In PBC, big values of R_1 reduce γ, for instance, a value of $R_1 = 50$ corresponds to a $\gamma = 0.4738$. This is consistent with the theoretical results reported in [233]. The same behaviour was observed for SM+PBC. For example, taking $R_1 = 50$ we obtain a $\gamma = 0.6409$. The range of the gains that we obtained in our experiments is summarized in Table 7.3.

B.2 Sinusoidal disturbances

In these experiments we obtain the frequency responses of the output voltage under periodic perturbations introduced in the power supply. To this end, we add to the voltage source E a perturbation $w = A_w sin(2\pi f t)$ where $A_w = 3$ V and we scan different values of f.

The magnitudes of the Bode plots ($w \mapsto \tilde{x}_2$) for the open–loop system and those of the controllers (for different tuning parameters) are given in Fig. 7.22. We see that in all cases the closed–loop behaves like a low pass filter and the question is what is the effect of the tuning gains on the steady-state gains and on the bandwidth and

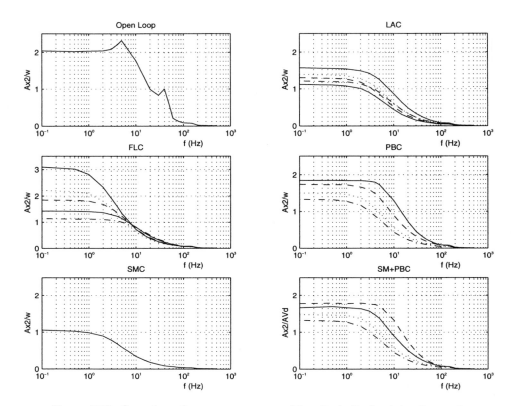

Figure 7.22: Open loop frequency response. Magnitude Bode plots of $w \mapsto \tilde{x}_2$

roll–off of the frequency responses.

Note that in these curves for each controller, variations in the parameters imply variations in both, steady state gain and cut-off frequency. Of course the best curve is that with the minimum steady state gain and cut-off frequency.

In open loop the cut–off frequency is of 14Hz, hence extremely bad disturbance attenuation. For the LAC controller, the steady state gain can be reduced if the gain k_1 is decreased, this corresponds also to a very small increasing in the cut-off frequency. For high values of k_1 and k_2 ($k_1 = 2.5, k_2 = 0.01$) the steady state gain could arrive up to 1.5625 with 10 Hz of cut-off frequency and for smaller values ($k_1 = 0.5, k_2 = 0$) this gain is 1.125 with 10.1 Hz of cut-off frequency.

In the case of FLC controller, depending on the pole locations, we can have steady state gains that go from 1.16 for fast poles ($a_1 = 60, a_2 = 3600$) until 3.1 for a dominant slow pole ($a_1 = 120, a_2 = 1600$). These pole locations correspond to cut-off frequencies of 14 Hz and 2.4 Hz respectively. Thus appearing a compromise between steady-state gain and cut-off frequency depending on the pole locations.

For PBC we have smaller steady-state gain and cut-off frequency for bigger values of R_1. For instance, a steady-state gain of 1.3775 with a cut-off frequency of 7 Hz, correspond to an $R_1 = 50$. For an $R_1 = 5$, the corresponding values are 1.875, 10.1 Hz.

In SMC, since there is no design parameter, the steady-state gain and the cut-off frequency are unique and they take respectively the values 1.06 and 10 Hz . Which means that disturbance rejection in this controller is quite good. For SM+PBC, as in the case of PBC, we have larger steady-state and cut-off frequencies for bigger values of R_1. For instance, with an $R_1 = 5$ we have an steady-state gain of 1.3125 and a cut-off frequency of 7 Hz, and for $R_1 = 50$ we have 1.75 and 10.2 Hz, respectively.

In Table 7.4 we show the steady-state gain and cut-off frequency ranges we obtained experimentally for each control strategy.

It's important to remark that this cut–off frequencies are relatively small compared with the possible perturbations caused by the natural line frequency noise (50/60 Hz), so the rejection of this kind of natural perturbations is assured.

C   Robustness to load uncertainty

In this experiment we introduce a load change that reduces the effective resistance from its nominal value of $R = 100\Omega$ to $R + \Delta R = 50\Omega$ during the interval [0.5, 1] sec. To implement this effect a digital signal generated in a dSpace card is sent to the gate of a MOSFET transistor to turn it on or off. This transistor is actuating as a switch that connects or disconnects a resistance of 100 Ω placed in parallel with the nominal load, which is also of 100 Ω.

The open–loop response is shown in Fig. 7.23. As we can see, there appears a steady state error in the voltage output x_2, that even if it's small, there is no way to reduce it.

As discussed at the beginning of this section the fact that all control strategies are indirect makes them extremely sensitive to this kind of disturbance, introducing in particular a large steady–state error. To remove this error we tried an heuristic approach of adding an integral loop around the output voltage error (for continuous control laws) as well as the adaptive versions. Notice that we do not dispose of an adaptive scheme for FLC, however for simple step changes in the load the integral action corrected the steady–state error. It is an interesting open question how to provide FLC with adaptation capabilities to track time–varying parameters, as done for PBC.

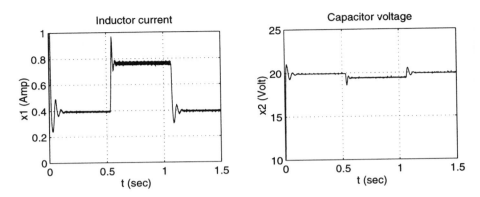

Figure 7.23: Open-loop response to a step change in the output resistance

C.1 Non adaptive versions

We show in Fig. 7.24 the behaviour of the output voltage signal x_2 of the system been controlled for each strategy when there appears a pulse change in the output resistance, which is clearly inadmissible.

C.2 Adaptive versions

The adaptive version of PBC with $\gamma = 1$ and $R_1 = 25$ is shown in Fig. 7.25. Observe that the parameter estimate converges very close to the true values, i.e. 0.1 and 0.2 Ω^{-1} (remember that the algorithm estimates $1/R$), but that this small discrepancy induces an steady state error both in the inductor current (which should be $\bar{x}_1 = 0.8$ for the new load resistance) and the output voltage. The error however vanishes when we come back to the nominal resistance, where now the estimate exactly converges to the true value. Given the proof of asymptotic stability of the desired equilibrium, the existence of this steady-state error for higher currents is particularly distressing. The only explanation we have is that since more current is passing through the electronic elements their parasitic losses become more significant. It may also be that the additional computations demanded by the adaptation law induce numerical errors. This critical issue of numerical sensitivity will be discussed within the context of induction motor control in Chapter 11.

As usual in adaptive control, even though we started the experiment with the right value of the load resistance, the estimate moves initially away from it. This, in some way, speeds-up the step response of the output voltage.

In SM+PBC we took again $\gamma = 1$ and observed a behaviour very similar to PBC,

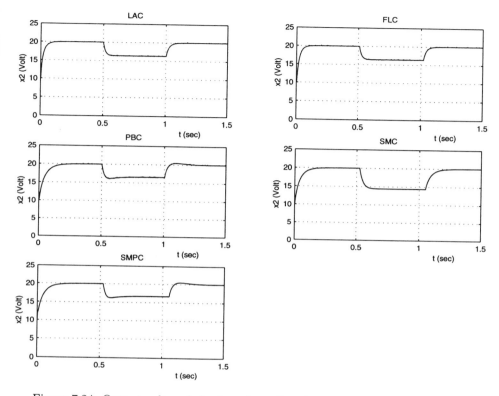

Figure 7.24: Output voltage behaviour for a disturbance in the output resistance

with the addition of the high frequency oscillations in the current mentioned above.

For adaptive SMC we take again $\gamma = 1$ and observe the same phenomenon of lack of parameter convergence when the load is reduced. In this case, however, there is no steady state error in the output voltage, this because as we see from the equations, adaptation introduces an integral term in the output voltage error. The inductor current exhibits very high frequency components due to the low frequency used to generate the switching signal u.

C.3 Adding an integral term

In Figs. 7.26 – 7.27 we present for the three laws employing a PWM, LAC, FLC and PBC, the responses to a pulse disturbance in the output resistance. In all cases the steady state error vanishes in a relatively short time with small overshoot. There is

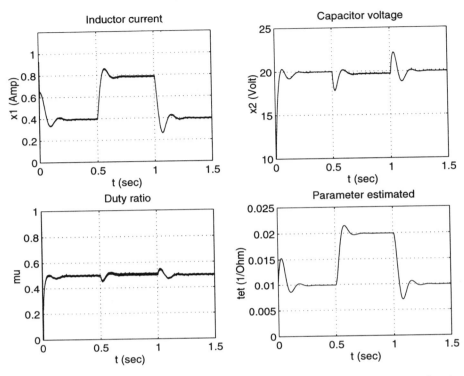

Figure 7.25: Response to a pulse disturbance in the output resistance for adaptive PBC

however some degradation in the quality of the first step response, a large overshoot, that could not be reduced via tuning without seriously degrading the transient and steady–state performances.

It is worth noting the oscillatory behaviour of LAC and PBC for both load values. To dampen the oscillations the integral term had to be considerably reduced with the ensuing increase in the settling time. While for PBC this destabilizing effect of the integral action is not a serious problem, because we dispose of an adaptive version, for LAC it casts some doubts for its practical application.

5.4   Conclusions

The following conclusions of our experimental study are in order:

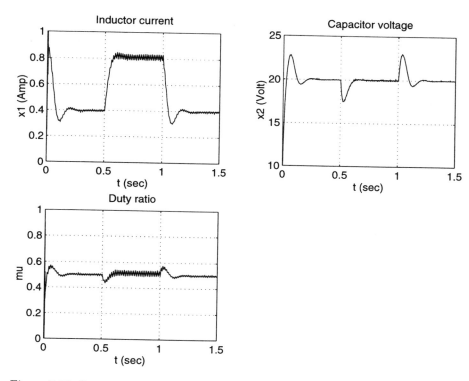

Figure 7.26: Response to a disturbance in the output resistance for LAC + integral term

- Nonlinear designs provide a promising alternative to classical lead–lag controllers. In particular, the linear scheme LAC performed very badly in tracking time–varying references, and exhibited an undesirable oscillation when an integral term was added to compensate for load uncertainty.

- FLC performed very well in output regulation and tracking but exhibited a higher sensitivity to voltage disturbances than the other schemes. Incorporating an integral action effectively compensated for a step change in load resistance, even though no theory is available to substantiate this. To handle other type of load changes (e.g., slowly time–varying) the integral action will not be sufficient and some kind of adaptation should be incorporated into the controller.

- The main drawback of PBC, which is shared also by SMC and SM+PBC, is the inability to shape the output response, which evolves according to the open–loop dynamics. This, of course, stems from the fact that we cannot inject

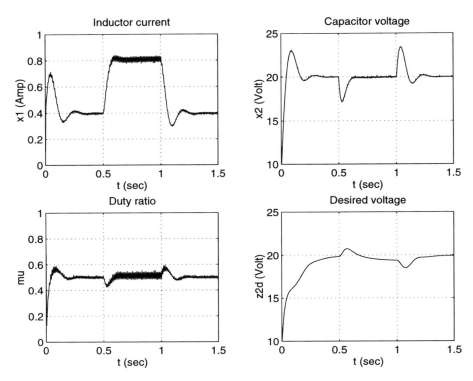

Figure 7.27: Response to a disturbance in the output resistance for PBC + integral term

damping to the voltage subsystem without nonlinearity cancelation. On the other hand, PBC achieved a better disturbance attenuation, hence it may be a viable candidate for applications where rise time is not of prime concern. We should stress that, as shown in motor control, this is not a limitation intrinsic to passivity–based designs, rather it pertains to our ability to inject (pervasive) damping to the controlled variable.

• SMC and SM+PBC proved very robust to source disturbances but highly sensitive to parameter uncertainty. The latter could be alleviated incorporating a novel adaptation mechanism. The lack of flexibility of SMC is somehow alleviated in SM+PBC, at least to shape the disturbance attenuation characteristic. Unfortunately, both schemes suffer from the main drawback mentioned above.

• In contrast to the study carried out in Section **11**.4 —see also[132]– we were not interested here in issues pertaining to numerical sensitivity and computational complexity, this might prove important in a low–cost implementation.

- Average models of DC–to–DC power converters have been known to be *differentially flat* (see [82]) i.e., all system variables are differential functions of the total energy of the system, which then qualifies as a *linearizing output*. As such, an interesting line of research can be proposed which exploits the differential flatness property of the system in connection with the possibilities of energy shaping and damping injection, i.e., passivity, controller design techniques.

Part III

Electromechanical systems

Chapter 8

Nested–loop passivity–based control: An illustrative example

We start with this chapter the third part of the book which is dedicated to PBC of electromechanical systems. Particular emphasis will be given to AC electrical machines, to which Chapters **9–11** are devoted. Chapter **12** treats robots with AC drives, hence connecting the material of the next three chapters with our previous developments on mechanical systems of Chapter **4**.

We will show throughout this third part that for electromechanical systems the PBC approach can be applied in at least *two different ways*, leading to different controllers. In the first, more direct form, a PBC is designed for the whole electromechanical system using as storage function the total energy of the full system. This is the way we designed PBCs for mechanical and electrical system in the previous chapters, and we will refer to it as *PBC with total energy shaping*.

Another route stems from the application of the passive subsystems decomposition of Section **2.2.4** to the electromechanical system. Namely, we show that (under some reasonable assumptions) we can decompose the system into its electrical and mechanical dynamics, where the latter can be treated as a "passive disturbance". We design then a PBC *for the electrical subsystem* using as storage function only the electrical part of the systems total energy. An outer–loop controller (which as shown in Chapter **12** can also be a PBC, but here is a simple pole–placement) is then added to regulate the mechanical dynamics. The so–designed controller will be called *nested–loop PBC*. There are at least three motivations for this approach: firstly, using this feedback–decomposition leads to simpler controllers, which in general do not require observers. Secondly, typically there is a time–scale separation between the electrical and the mechanical dynamics. Finally, since the nested–loop configuration is the prevailing structure in practical applications, we can in some important cases establish a clear connection between our PBC and current practice.

1 Introduction

To facilitate the presentation of the new concept of nested–loop PBC we consider in this chapter the simple problem of stabilization of a magnetically levitated ball. However, for the sake of continuity and to establish the connection with the previous material, we present first the PBC with total energy shaping, and afterwards we design the new nested–loop PBC. Similarly to the case of electrical and mechanical systems for electromechanical systems the EL model derived in Section 2.4.2 is instrumental to guide us in the design of PBC. We use this model to derive the total energy shaping PBC. On the other hand, in this simple example it is possible to exhibit a passivity property of the system without invoking the Lagrangian formalism. Therefore, we use a standard state–space description to design the nested–loop PBC. Our motivation to adopt this notation is twofold, first by simplifying the notation we believe the reader can concentrate better on the structural aspects of the design. Second, expressing the PBC in these coordinates permits to establish some interesting connections and similarities with classical feedback linearization and the widely publicized backstepping designs, with which the reader is probably more familiar. In fact, an additional objective that we pursue in this chapter is to compare in a simple example the three controller design techniques.

Application of these, seemingly unrelated, methodologies will typically lead to the definition of different control schemes. For some specific examples these differences blur and some interesting connections and similarities between the controller design techniques emerge. The careful study of such cases will improve our understanding of their common ground fostering cross–fertilization. In this chapter we investigate these questions, both analytically and via simulations, for the simple problem of stabilization of a magnetically levitated ball. Our motivation in choosing this particular example stems, not only from the fact that due to its simplicity the connections between the controllers are best revealed, but also that such an equipment is available in many engineering schools, hence experimental work can be easily carried out to complement our studies.

Before closing this introduction a word on style is in order. Since the main objectives in this chapter are the introduction of the basic idea of nested–loop PBC, and the clarification of its connections with backstepping and feedback linearization, we have adopted an informal format of presentation without theorems and proofs. For this reason we have also favored simpler, as opposed to more powerful or novel, solutions to the problem.

1.1 Model and control problem

In Section **2**.4.2 we have derived the model of the magnetically levitated ball of Fig. 2.4. We repeat here for ease of reference the EL model

$$\mathcal{D}(q_m)\ddot{q} + \mathcal{C}(\dot{q}_e, q_m, \dot{q}_m)\dot{q} + \mathcal{R}\dot{q} + \mathcal{G} = \mathcal{M}u \tag{8.1}$$

where the generalized coordinates are $q = [q_e, q_m]^\top$, and we have defined

$$\mathcal{D}(q_m) \triangleq \begin{bmatrix} \frac{c}{1-q_m} & 0 \\ 0 & m \end{bmatrix}, \; \mathcal{C}(\dot{q}_e, q_m, \dot{q}_m) \triangleq \frac{c/2}{(1-q_m)^2}\begin{bmatrix} \dot{q}_m & \dot{q}_e \\ -\dot{q}_e & 0 \end{bmatrix}$$

$$\mathcal{R} \triangleq \begin{bmatrix} R_e & 0 \\ 0 & 0 \end{bmatrix}, \; \mathcal{G} \triangleq \begin{bmatrix} 0 \\ -mg \end{bmatrix}, \; \mathcal{M} \triangleq \begin{bmatrix} 1 \\ 0 \end{bmatrix}$$

Notice that we have added a constant $c > 0$ in the definition of the inductance $L(q_m) = \frac{c}{1-q_m}$. As pointed out in Section **2**.4.2 this model can also be written in state space form as

$$\Sigma: \left\{ \begin{array}{rcl} \dot{\lambda} &=& -\frac{R_e}{c}(1 - q_m)\lambda + u \\ F &=& \frac{1}{2}\lambda^{\frac{2}{}} \\ m\ddot{q}_m &=& F - mg \end{array} \right. \tag{8.2}$$

where

$$\lambda = L(q_m)\dot{q}_e = \frac{c}{1 - q_m}\dot{q}_e$$

is the flux in the inductance $L(q_m)$, and we have defined the force of electrical origin F, which will be extensively used later.

The *control objective* is to track a bounded reference signal $q_{m*}(t)$ (with known and bounded first, second and third order derivatives) with all signals bounded. We will assume in the first instance that the *full state* is available for measurement. Then we will show that for the nested–loop PBC we only require the measurement of q_m and $\dot{q}_m$.

Remark 8.1 There is a huge body of literature on control of magnetic bearing systems with one or several electromagnets. We refer the readers to the special issue of September 1996 in the *IEEE Trans. Control Systems Technology* for an exhaustive list of references. It has, in particular, been studied from the perspective of flatness and feedback linearization in [154] and with a backstepping approach in [223]. An adaptive backstepping controller for the levitated ball studied here, albeit with a simpler model with linear electrical dynamics, has been reported in [142]. The scheme does not require the measurement of $\dot{q}_m$, it requires however q_m and $\dot{q}_e$. In [267] several PBCs are developed for this system. In particular, we present a PBC that measures instead q_m, $\dot{q}_m$, but obviates the need for $\dot{q}_e$. Here, we adapt for our simplified example the schemes of [154] for feedback linearization, of [223] for backstepping, and of [267] for PBC. We refer the reader to [110], [197] for background material on feedback linearization, and to [142] for an exhaustive coverage of backstepping.

Remark 8.2 Writing the model (8.1) in terms of λ yields the differential–algebraic representation

$$\dot{\lambda} = \frac{R_e}{c}\dot{q}_e + u$$
$$\lambda = L(q_m)\dot{q}_e$$

$$F = \frac{1}{2}\frac{\partial L}{\partial q_m}(q_m)\dot{q}_e^2$$
$$m\ddot{q}_m = F - mg$$

This model clearly reveals the passivity properties of the feedback decomposition of Proposition 2.10. That is, the operators $\Sigma_e : [u, -\dot{q}_m] \mapsto [\dot{q}_e, F]$ and $\Sigma_m : F \mapsto \dot{q}_m$ are (locally) passive with storage functions $\frac{1}{2}L(q_m)\dot{q}_e^2$ and $\frac{1}{2}m\dot{q}_m^2 + mg(1-q_m)$, respectively.

2   Passivity–based control with total energy-shaping

In this section we derive a PBC mimicking the approach taken for mechanical and electrical systems. In particular we follow closely the developments carried out to solve the tracking problem of robots with flexible joints in Section 4.3. That is, motivated by the passivity of the operator $u \mapsto \mathcal{M}^\top \dot{q}$ with storage function the total energy

$$\mathcal{H} = \frac{1}{2}\dot{q}^\top \mathcal{D}(q_m)\dot{q} + mg(1 - q_m)$$

we propose to assign to the closed–loop a storage function $\mathcal{H}_d = \frac{1}{2}s^\top \mathcal{D}(q_m)s$, where, as in Section 4.3, we define

$$s = \dot{\tilde{q}} + \Lambda\tilde{q}, \quad \dot{\tilde{q}} \stackrel{\triangle}{=} \dot{q} - \begin{bmatrix} \dot{q}_{ed} \\ \dot{q}_{m*} \end{bmatrix}, \quad \Lambda \stackrel{\triangle}{=} \begin{bmatrix} 0 & 0 \\ 0 & \Lambda_m \end{bmatrix}$$

Notice that, since we are not interested in controlling the charges q_e, but only the currents $\dot{q}_e$ and the positions q_m, we have chosen Λ with just one non–zero term. Also, remark that the first error signal $\dot{\tilde{q}}_e$ is defined in terms of a signal $\dot{q}_{ed}$ to be defined below. This signal plays the same role as q_{2d} for flexible joint robots in Section 4.3. We now write the perturbed desired error dynamics as

$$\mathcal{D}(q_m)\dot{s} + [\mathcal{C}(\dot{q}_e, q_m, \dot{q}_m) + \mathcal{R}_d]s = \Psi$$

where $\mathcal{R}_d \triangleq \mathcal{R} + K_d$, $K_d = \text{diag}\{k_{de}, k_{dm}\} > 0$ is the damping injection matrix, and the perturbation term is

$$\Psi = \mathcal{M}u + K_d s - [\mathcal{D}(q_m)\ddot{q}_r + (\mathcal{C}(\dot{q}_e, q_m, \dot{q}_m) + \mathcal{R})\,\dot{q}_r - \mathcal{G}] \tag{8.3}$$

where we have defined

$$\dot{q}_r = \begin{bmatrix} \dot{q}_{ed} \\ \dot{q}_{m*} \end{bmatrix} - \Lambda\tilde{q}$$

and $\Lambda > 0$. Setting $\Psi \equiv 0$ we obtain the controller equations as

$$u = \frac{c}{(1 - q_m)}\ddot{q}_{ed} + \frac{1}{2}\frac{c}{(1 - q_m)^2}\dot{q}_m\dot{q}_{ed} + \frac{1}{2}\frac{c}{(1 - q_m)^2}\dot{q}_e(\dot{q}_{m*} - \Lambda_m\tilde{q}_m) + R_e\dot{q}_{ed} - k_{de}\dot{\tilde{q}}_e$$

$$m(\ddot{q}_{m*} - \Lambda_m\dot{\tilde{q}}_m) - \frac{1}{2}\frac{c}{(1 - q_m)^2}\dot{q}_e\dot{q}_{ed} - k_{dm}(\dot{\tilde{q}}_m + \Lambda_m\tilde{q}_m) + mg = 0.$$

To obtain an *explicit* realization of the PBC we solve the second equation as[1] $\ddot{q}_{ed} = f_0(\dot{q}_e, \dot{q}_m, q_m, t)$. Then, we calculate u from the first equation, which requires $\ddot{q}_{ed}$ and consequently the measurement of the full state. Observe that, as in the case of the simplified model of the flexible joint robots, the diagonal structure of $\mathcal{D}(q_m)$ allows us to calculate $\ddot{q}_e$ and $\ddot{q}_m$ from knowledge of the state q and $\dot{q}$. Another important point is that we have to ensure that there are no algebraic loops in the calculation of u (like in the controller of [166]). This is (locally) the case here because one can show that u satisfies an equation of the form

$$\left(1 + \frac{\dot{q}_{ed}}{\dot{q}_e}\right)u = f_1(\dot{q}_e, \dot{q}_m, q_m, t)$$

hence, it is defined in a neighborhood of the operating point.

The stability of this PBC can be established along the, by now standard, lines showing that $\dot{\mathcal{H}}_d = -s^\top \mathcal{R}_d s$ and chasing the signals (see e.g. Chapter 4). In any case, it is clear that this controller is extremely complicated for the purposes of this application and furthermore it requires measurement of the full state, an assumption that we want to avoid in PBC.

3 Nested–loop passivity–based control

In this section we present a controller structure which is motivated from a physical partition of the system and allows us to overcome the drawbacks of the PBC above.

---

[1]Notice that there is a division by $\dot{q}_e$ in this expression that introduces a singularity.

3.1   Control structure

Notice that the system Σ can be decomposed as a feedback interconnection of an electrical subsystem $\Sigma_1 : u \mapsto \lambda$, and a mechanical subsystem $\Sigma_2 : (F - mg) \mapsto q_m$ as depicted in Fig. 8.1. We will prove below that Σ_1 is (locally) passive, hence λ (and consequently F, see (8.2)) are "easy" to control. On the other hand, since Σ_2 is just a double integrator it is reasonable to concentrate our attention on the problem of controlling F, in the understanding that if it is is suitably regulated then position q_m will be easy to control with classical linear techniques, e.g., PI control. An additional motivation for this approach stems from the fact that typically there is a natural time–scale decomposition between the electrical and mechanical dynamics. For these reasons the so–called nested–loop control is the prevailing structure in practical applications of electromechanical systems which will be adopted throughout this chapter.

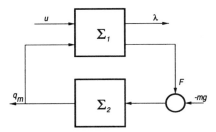

Figure 8.1: Feedback decomposition.

The nested–loop control configuration is shown in Fig. 8.2, where $\mathcal{C}_{il}$, $\mathcal{C}_{ol}$ are the inner and outer–loop controllers, respectively. The inner–loop controller is designed to achieve asymptotic tracking of the desired force F_d. The outer–loop controller takes care of the position tracking, and essentially generates a signal F_d such that if $F \to F_d$ then position tracking, i.e., $q_m \to q_{m*}$, is ensured. In applications $\mathcal{C}_{ol}$ is typically a simple PI around position errors. (In some cases, for instance if Σ_2 is not LTI as in [217], an interlaced design of $\mathcal{C}_{il}$ and $\mathcal{C}_{ol}$ is needed.)

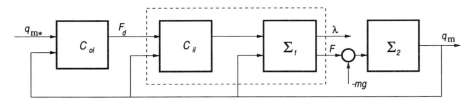

Figure 8.2: Nested–loop controller structure.

Remark 8.3 As pointed out in Remark 8.2 above it is easy to check that the Lagrangian of the levitated ball model satisfies the conditions of Proposition 2.10 for feedback decomposition into passive subsystems. This decomposition, though similar to the one introduced above, is not defined in terms of the same operators. That is, Σ_1, Σ_2 above, and Σ_e, Σ_m in Proposition 2.10 are *not the same*. In the passive subsystems decomposition, that we will use for the PBC of electrical machines, the essential property is that both operators are *passive*. This feature, which is crucial for the successful design of PBC of *underactuated* electrical subsystems, is not required in this simple example with fully actuation.

3.2 Passivity–based controller design

As discussed above we will adopt the nested loop (*i.q.* cascaded) controller structure of Fig. 8.2 to design our PBC. The inner–loop controller is designed following the principles of energy shaping and damping injection as *applied to the electrical subsystem*. The outer–loop controller C_{ol} is chosen to place the poles of the mechanical subsystem.

The procedure to design the cascaded PBC consists of the following steps: 1) Choice of u to ensure flux tracking, i.e., $\lambda \to \lambda_d$; 2) Choice of λ_d so that $\lambda \to \lambda_d \Rightarrow F \to F_d$; 3) Prove that $F \to F_d \Rightarrow q_m \to q_{m*}$ with all signals bounded. These steps are detailed below.

A   Flux tracking

As usual in PBC we first prove that the electrical subsystem satisfies some *passivity* properties. This fundamental step will identify the output which is "easy to control" and its corresponding storage function. An incremental version of this function will be the desired storage function used to enforce passivity to the closed–loop.

This step is usually carried out invoking the Lagrangian structure of the system however, by simple inspection in this example we propose as storage function candidate[2]

$$\mathcal{H}_\lambda = \frac{1}{2}\lambda^2$$

whose derivative along the trajectories of (8.2) satisfies

$$\dot{\mathcal{H}}_\lambda = -\frac{R_e}{c}(1 - q_m)\lambda^2 + \lambda u$$
$$\leq -\alpha\lambda^2 + \lambda u$$

---

[2]Note that this function is some kind of scaled magnetic energy in the inductance.

with $\alpha \stackrel{\triangle}{=} \frac{R_e}{c}\epsilon > 0$. Integrating the last inequality from 0 to t and using the fact that $\mathcal{H}_\lambda \geq 0$ yields

$$\int_0^t u(s)\lambda(s)ds \geq \alpha \int_0^t \lambda^2(s)ds + \beta$$

where we have set $\beta \stackrel{\triangle}{=} -\mathcal{H}(0)$. This proves (local) *output strict passivity* of the map $u \mapsto \lambda$.

The passivity property above suggests to try to assign to the closed–loop the *desired storage function*

$$\mathcal{H}_d = \frac{1}{2}\tilde{\lambda}^2 \tag{8.4}$$

where $\tilde{\lambda} \stackrel{\triangle}{=} \lambda - \lambda_d$, with λ_d the desired flux that induces on the mechanical subsystem the force required to regulate q_m. This signal is defined in the next step. With some abuse of notation this step is commonly referred to as *energy shaping*.

It is easy to see that if we set $u = u_{\mathrm{PB}}$ with

$$u_{\mathrm{PB}} = \dot{\lambda}_d + \frac{R_e}{c}(1 - q_m)\lambda_d + v \tag{8.5}$$

then we get the error dynamics

$$\dot{\tilde{\lambda}} = -\frac{R_e}{c}(1 - q_m)\tilde{\lambda} + v \tag{8.6}$$

Proceeding as above we can establish the bound

$$\dot{\mathcal{H}}_d \leq -\alpha\tilde{\lambda}^2 + \tilde{\lambda}v \tag{8.7}$$

which proves that the map $v \mapsto \tilde{\lambda}$ is output strictly passive as desired. Of course, if we set $v \equiv 0$ we have that $\tilde{\lambda} \to 0$, exponentially fast. Given that the rate of convergence α may be small to enhance performance it is convenient to add a so–called *damping injection* term into the electrical subsystem, setting for instance

$$v = -R_{\mathrm{DI}}\tilde{\lambda}$$

with $R_{\mathrm{DI}} > 0$. It is clear that this additional term will speed up the electrical transient. However, as will be shown in the simulations later, increasing R_{DI} does not necessarily lead to transient performance improvement. This stems from the fact that the overall closed–loop system under PBC is still nonlinear and we are confronted with unpredictable *peaking* effects which may unstabilize the system (see e.g. [237]).

B From flux tracking to force tracking

The force of magnetic origin F can be written in terms of the flux error $\tilde{\lambda}$ and the desired flux λ_d as

$$F = \frac{1}{2c}[\lambda_d^2 + \tilde{\lambda}(\tilde{\lambda} + 2\lambda_d)] \tag{8.8}$$

From the analysis above we have that $\tilde{\lambda} \to 0$, hence it is reasonable to choose λ_d as the solution of

$$F_d = \frac{1}{2c}\lambda_d^2 \tag{8.9}$$

where $F_d > 0$ is some desired force, whose derivative is assumed to be known. We can now write the control (8.5) in terms of F_d and $\dot{F}_d$ as

$$u_{\mathrm{PB}} = \sqrt{\frac{c}{2F_d}}\dot{F}_d + R(1 - q_m)\sqrt{\frac{2F_d}{c}} + v \tag{8.10}$$

If F_d is bounded we can conclude from (8.8) and (8.9) that the force tracking controller (8.10) ensures $F \to F_d$. The problem is that in position tracking problems F_d will not be *a priori* bounded, since it is generated by $\mathcal{C}_{ol}$. This leads to the third step of our design.

C From force tracking to position tracking

In this final step we have to design $\mathcal{C}_{ol}$ and prove convergence of the position error to zero. We start by defining F_d as

$$F_d = m[\ddot{q}_{m*} - k_2\dot{\tilde{q}}_m - k_1\tilde{q}_m - k_0\int_0^t \tilde{q}_m(s)ds] \tag{8.11}$$

where we have defined the tracking error $\tilde{q}_m \overset{\triangle}{=} q_m - q_{m*}$, and we choose $k_0, k_1, k_2 > 0$ to ensure

$$d(p) \overset{\triangle}{=} p^3 + k_2p^2 + +k_1p + k_0 \tag{8.12}$$

is a Hurwitz polynomial. Notice that $\dot{F}_d$ is used in (8.10). From the definition of F_d above, this implies that $\dddot{q}_m$ (or equivalently λ or F) is required for implementation of the PBC (8.10), (8.11). Remark also that we have introduced in (8.11) an integral term. This takes care of the steady-state error due to the constant disturbance mg, but it could be removed and simply replaced by g. As we will see in the simulations, and is well known in applications, the inclusion of the integral term robustifies the loop. Furthermore it simplifies the comparison of the various schemes because in this way the closed–loop mechanical dynamics will be the same order.

Let us now analyze the *stability* of the overall system (8.2), (8.10), (8.11) with $u = u_{\mathrm{PB}}$ and $v \equiv 0$ (or $v = -R_{\mathrm{DI}}\tilde{\lambda}$) . We know already from the analysis of the flux tracking step that $\tilde{\lambda} \to 0$ exponentially fast. To study the mechanical dynamics let us replace (8.8), (8.9), and (8.11) in (8.2) to get the error equation

$$G(p)\tilde{q}_m = \frac{1}{2mc}\tilde{\lambda}(\tilde{\lambda} + 2\lambda_d) - g$$

where $G(p) \triangleq \frac{1}{p}d(p)$, and $d(p)$ as in (8.12). This system admits a state–space realization of the form[3]

$$\dot{x} = Ax + B\frac{1}{2mc}\tilde{\lambda}(\tilde{\lambda} + 2\lambda_d) \tag{8.13}$$

with $x \triangleq [\tilde{q}_m, \ \dot{\tilde{q}}_m, \ z_2]^\top$, $\dot{z}_2 = \tilde{q}_m$, $B \in \mathbb{R}^3$, and the characteristic polynomial of $A \in \mathbb{R}^{3\times3}$ equal to $d(p)$. Now, $\tilde{\lambda}$ is an exponentially decaying term and from (8.9), (8.11) and boundedness of $\ddot{q}_{m*}$ we see that λ_d satisfies the bounds

$$
\begin{aligned}
|\lambda_d| &\leq \alpha_1 + \alpha_2|F_d| \\
&\leq \alpha_3 + \alpha_4|x|
\end{aligned}
$$

for some suitable constants $\alpha_i > 0$, $i = 1, \cdots, 4$. Consequently the forcing term satisfies the bound

$$|\tilde{\lambda}(\tilde{\lambda} + 2\lambda_d)| \leq (\alpha_5 + \alpha_6|x|)\tilde{\lambda} \tag{8.14}$$

for some α_5, $\alpha_6 > 0$. The proof of asymptotic stability of (8.13) follows considering the quadratic function

$$V_1 = \frac{1}{2}x^\top P x \tag{8.15}$$

with P the symmetric positive–definite solution of

$$A^\top P + PA = -Q, \ Q = Q^\top > 0$$

and invoking standard arguments for exponentially stable systems perturbed by linearly bounded terms multiplied by an exponential; see, e.g., [127].

Remark 8.4 The proof of stability above proceeds in two steps and exploits the fact that the closed–loop is the cascade connection of two exponentially stable systems. It is actually possible to obtain a proof in one shot with a single Lyapunov function

---

[3]Notice the presence of the zero at the origin that takes care of the constant term g.

that establishes the stronger *exponential* stability property.[4] To this end, we consider the Lyapunov function

$$V = V_1 + \frac{a}{2}\tilde{\lambda}^2 + \frac{b}{4}\tilde{\lambda}^4$$

with $a, b > 0$. Its derivative along the solutions of (8.13), (8.6) satisfies the bound

$$\dot{V} \le -\frac{1}{2}x^\top Q x - \epsilon(a\tilde{\lambda}^2 + b\tilde{\lambda}^4) + \frac{1}{2mc}x^\top PB\tilde{\lambda}(\tilde{\lambda} + \lambda_d)$$

The key observation here is that, using Young's inequality and a suitably weighted triangle inequality, the sign indefinite term can be bounded as

$$|x^\top PB\tilde{\lambda}(\tilde{\lambda} + \lambda_d)| \le d\|x\|^2 + \alpha_7(\tilde{\lambda}^2 + \tilde{\lambda}^4)$$

for some $\alpha_7 > 0$ and an arbitrarily small $d > 0$. Hence, with a suitable choice of a, b we get $\dot{V} \le -\alpha_8 V$ for some $\alpha_8 > 0$.

Remark 8.5 The controller (8.10) has a singularity when F_d crosses through zero, hence it is only locally defined. This problem is, of course, related with the poor controllability properties of levitation systems with only one electromagnet. Since the focus of the present chapter is on the nested–loop PBC design procedure and *structural* similarities of various controllers we will concentrate on *local* properties and assume throughout that the system operates away from singularity regions. In [267] we propose a controller that takes into account the presence of the singularity. Another interesting way to avoid the singularities is to add to the PBC (or for that matter to any other controller) a trajectory planning stage using the fact that q_m is a *flat* output. The utilization of this fundamental concept to enhance performance is a topic of current research, which is illustrated for the magnetic levitation system in [154].

4 Output–feedback passivity–based control

As we have discussed throughout the book, one of the main advantages of PBC is the possibility of avoiding full-state measurements. In this subsection we develop a PBC that requires only the measurement of q_m and $\dot{q}_m$ while preserving asymptotic stability. To this end, we recall that $\ddot{q}_m$ is needed only in the evaluation of $\dot{F}_d$, hence we propose to replace F_d in (8.11) by

$$F_d = m[\ddot{q}_{m*} - k_2 z_1 - k_1 \tilde{q}_m - k_0 \int_0^t \tilde{q}_m(s)ds] \tag{8.16}$$

---

[4]This proof, as many other nice refinements scattered throughout the book, was suggested by Laurent Praly.

where z_1 is the *approximate derivative* of $\tilde{q}_m$, that is

$$\dot{z}_1 = -az_1 + b\dot{\tilde{q}}_m \tag{8.17}$$

with a, $b > 0$. This modification does not affect the flux tracking property $\tilde{\lambda} \to 0$, nor the control signal (8.10). The key point is that now u_{PB} can be implemented *feeding–back only q_m and $\dot{q}_m$*.

The stability analysis of this new scheme mimics the developments above. That is, we obtain a state space realization of the mechanical dynamics of the form (8.13) but with the augmented state $x \triangleq [\tilde{q}_m, \dot{\tilde{q}}_m, z_1, z_2]^\top$. The new A matrix is

$$A = \begin{bmatrix} 0 & 1 & 0 & 0 \\ -k_1 & 0 & -k_2 & -k_0 \\ 0 & b & -a & 0 \\ 1 & 0 & 0 & 0 \end{bmatrix}$$

and the forcing term still satisfies the bound (8.14). To study the stability of the unforced equation let us introduce the partition

$$A \triangleq \begin{bmatrix} A_1 & \vdots & B_1 \\ \cdots & \cdot & \cdots \\ -C_1^\top & \vdots & 0 \end{bmatrix}$$

with $A_1 \in \mathbb{R}^{3\times3}$, $B_1, C_1 \in \mathbb{R}^{3\times1}$ defined in an obvious manner. We can then prove that

$$\det(sI - A) = \det(sI - A_1)[C_1^\top(sI - A_1)^{-1}B_1 + s]$$

and $\det(sI - A_1)$ is a Hurwitz polynomial for all $k_1, k_2, a, b > 0$. Hence it only remains to choose the integral gain k_0 to ensure stability of A. Stability of the forced equation follows *verbatim* from the analysis of the full state feedback case.

5   Comparison with feedback linearization and back-stepping

Among the various controller design techniques for stabilization of nonlinear systems that have emerged in the last few years we can distinguish three which are probably the most widely applicable and systematic: Feedback Linearization Control (FLC), Integrator Backstepping Control (IBC) and Passivity–based Control (PBC). The three techniques have evolved from different considerations and exploit different properties of the system. While in FLC we aim at a linear system, in PBC we are satisfied with assigning a certain passivity property to a suitably defined map, –consequently enforcing a certain storage function to the closed–loop dynamics–. In

IBC we aim at a somehow intermediate objective (more ambitious than passivation but less demanding than linearization) of assigning a *strict* Lyapunov function to the closed–loop.

It is clear then that the three techniques will typically lead to the definition of different control schemes. For some specific examples these differences blur and some interesting connections and similarities between the controller design techniques emerge. It is our belief that the detailed study of such cases will improve our understanding of their common ground fostering cross–fertilization. With this motivation in mind we have carried out in Section 4.3 a comparative study of PBC, cascaded and back-stepping control for flexible joint robots. In this section we compare PBC, FLC and IBC design techniques in the levitated ball example, which is simple enough to reveal the connections between the controllers, but yet requires the application of nonlinear control to achieve a satisfactory performance.

5.1 Feedback–linearization control

We start our study with FLC since it is clearly the best well–known and, in this particular example, the simplest to derive and understand. To this end, we find convenient to introduce a change of coordinates and we write the system dynamics in terms of force, position and velocity as

$$
\begin{aligned}
\dot{F} &= -\tfrac{2R}{c}(1 - q_m)F + \sqrt{\tfrac{2F}{c}}u \\[2mm]
m\ddot{q}_m &= F - mg
\end{aligned}
\tag{8.18}
$$

Differentiating once more $\ddot{q}_m$ we get

$$
\begin{aligned}
mq_m^{(3)} &= \dot{F} \\[2mm]
&= -R(1 - q_m)\frac{2F}{c} + \sqrt{\frac{2F}{c}}u
\end{aligned}
$$

It is clear from the equations above that, taking as output q_m and provided $F > 0$, the system has relative degree 3 and no zero dynamics. Thus, an FLC can be immediately defined as $u = u_{\mathrm{FL}}$ with

$$
u_{\mathrm{FL}} = \sqrt{\frac{c}{2F}}mv_{\mathrm{FL}}(q_m, \dot{q}_m, F) + R(1 - q_m)\sqrt{\frac{2F}{c}}
\tag{8.19}
$$

which yields

$$
q_m^{(3)} = v_{\mathrm{FL}}(q_m, \dot{q}_m, F)
$$

A suitable choice for the outer loop signal is then

$$
v_{\mathrm{FL}}(q_m, \dot{q}_m, F) = q_{m*}^{(3)} - k_2[\underbrace{(\tfrac{1}{m}F - g)}_{\ddot{q}_m} - \ddot{q}_{m*}] - k_1\dot{\tilde{q}}_m - k_0\tilde{q}_m
\tag{8.20}
$$

The FLC (8.19), (8.20) is a static state feedback law which ensures in closed–loop the third order linear dynamics

$$d(p)\tilde{q}_m = 0$$

with $d(p)$ as in (8.12).

Remark 8.6 Since in closed–loop $\dot{F} = m v_{\text{FL}}(q_m, \dot{q}_m, F)$, it is useful –for the sake of comparison with the other controllers– to think of $m v_{\text{FL}}(q_m, \dot{q}_m, F)$ as a desired value for $\dot{F}$.

5.2 Integrator backstepping control

As pointed out in Section 8.3 of [142], where a simplified version of our model is considered, the system (8.2) is *not in any of the special forms* required for application of integrator backstepping. However, as shown in [223] (see also [142]) this does not preclude a backstepping–like design. Since almost any design can be interpreted as backstepping–like we point out from the outset that we will take–off from the approach used in [223] for a magnetic bearing system as applied to our simple levitated ball. Motivated by our final objective of comparison of techniques we will also present a variation of the basic scheme of [223] that incorporates features of the PBC. A similar combination of PBC and IBC has been reported in [105] where ideas of [209] are added to an IBC to remove singularities in induction motor control.

In IBC we start from the mechanical equation of (8.2) and, *assuming F is the control*, define an F_d that stabilizes this subsystem. For the sake of comparison we choose F_d as in FLC and PBC, that is (8.11). Adding and subtracting $\dot{F}_d$ to $m q_m^{(3)} = \dot{F}$ yields the first error equation

$$\dot{x} = Ax + \frac{1}{m} B (F - F_d)$$

where x, A and B are as in (8.13). We start constructing our Lyapunov function with the quadratic function (8.15), whose derivative yields

$$
\begin{aligned}
\dot{V}_1 &= -\frac{1}{2} x^\top Q x + \frac{1}{m} x^\top P B (F - F_d) \\
&= -\frac{1}{2} x^\top Q x + \frac{1}{2mc} x^\top P B \tilde{\lambda}(\tilde{\lambda} + 2\lambda_d)
\end{aligned}
$$

where we have used the definitions of F and λ_d in (8.9) to get the last equation. Let us now look at the dynamic equation of $\tilde{\lambda}$, that is

$$\dot{\tilde{\lambda}} = -\frac{R}{c}(1 - q_m)\lambda + u - \dot{\lambda}_d.$$

At this stage we decide a control u that stabilizes $\tilde{\lambda}$ plus a term to compensate for the cross term in $\dot{V}_1$ above. We have here the choice of adopting an FLC–like approach cancel the term $-\frac{R}{c}(1 - q_m)\lambda$ and add a damping term $-R_{\mathrm{DI}}\tilde{\lambda}$, or follow the PBC route, use the existing damping, and set $u = u_{\mathrm{IB}}$ with

$$u_{\mathrm{IB}} = u_{\mathrm{PB}} \qquad (8.21)$$

and u_{PB} as in (8.5). We can complete our Lyapunov function with $\mathcal{H}_d$ (8.4) as

$$V = V_1 + \frac{1}{\beta}\mathcal{H}_d$$

where we have added a tuning parameter $\beta > 0$. Differentiating V and using (8.7) we get

$$\dot{V} \le -\frac{1}{2}x^\top Qx - \frac{\alpha}{\beta}\tilde{\lambda}^2 + \tilde{\lambda}[\frac{1}{\beta}v + \frac{1}{2mc}x^\top PB(\tilde{\lambda} + 2\lambda_d)].$$

The IBC design is completed setting

$$v = -\frac{\beta}{2mc}x^\top PB(\tilde{\lambda} + 2\lambda_d) \qquad (8.22)$$

which removes the cross terms and yields the desired strict Lyapunov function

$$\dot{V} \le -\frac{1}{2}x^\top Qx - \frac{\alpha}{\beta}\tilde{\lambda}^2.$$

Remark 8.7 Following the line of reasoning used in Remark 8.4 to generate a strict Lyapunov function for PBC, we can think of *dominating* –instead of cancelling– the cross term in the bound of $\dot{V}$. Although this idea is theoretically very attractive, our experience in applications is quite disappointing because the new control involve higher order terms of the error signals which amplify the noise and saturate the actuators. In our opinion, this undesirable phenomenon –also present in FLC– is an important drawback of IBC which is clearly illustrated in Sections **10**.6 and **11**.4 for the induction motor and in **12**.6 for robots with AC drives.

5.3 Comparison of the schemes

We will now compare FLC (8.19), (8.20), PBC (8.10), (8.11) (without damping injection, i.e., with $v = 0$), and IBC (8.10), (8.21), (8.22). To this end, we notice that, (8.11), which we repeat here for ease of reference

$$F_d = m[\ddot{q}_{m*} - k_2\dot{\tilde{q}}_m - k_1\tilde{q}_m - k_0\int_0^t \tilde{q}_m(s)ds]$$

is related to (8.20) as

$$\dot{F}_d = mv_{\mathrm{FL}}(q_m, \dot{q}_m, F).$$

Hence, the three control inputs can be written as

$$u_{\text{FL}} = \sqrt{\frac{c}{2F}} m v_{\text{FL}}(q_m, \dot{q}_m, F) + R(1 - q_m)\sqrt{\frac{2F}{c}}$$

$$u_{\text{PB}} = \sqrt{\frac{c}{2F_d}} m v_{\text{FL}}(q_m, \dot{q}_m, F) + R(1 - q_m)\sqrt{\frac{2F_d}{c}}$$

$$u_{\text{IB}} = u_{\text{PB}} - \frac{\beta}{\sqrt{2mc}}[\tilde{q}_m, \dot{\tilde{q}}_m, \int_0^t \tilde{q}_m(s)ds]PB(\sqrt{F} + \sqrt{F_d})$$

The following remarks are in order

- FLC and PBC differ only on the utilization of the forces in some terms: desired forces F_d for PBC instead of the actual forces F for FLC.

- One additional difference comes from the generation of F_d, which involves the inclusion of an additional state. Hence, PBC and IBC are *dynamic* state feedback controllers while FLC is static.

- The IBC equals the PBC plus the signal that achieves the cancelation of the cross terms in $\dot{V}$. The two schemes approach as $\beta \to 0$.

- In contrast with FLC, the closed–loop equations for PBC and IBC are nonlinear. In PBC we get a *cascade* structure where the first subsystem is exponentially stable and some growth conditions on the input to the second subsystem allows us to complete the stability proof. In IBC we get a closed–loop system of the form

$$\begin{bmatrix} \dot{\tilde{\lambda}} \\ \dot{x} \end{bmatrix} = \begin{bmatrix} -\frac{R}{c}(1 - q_m) & -\frac{\beta}{\sqrt{2c}}(\sqrt{F} + \sqrt{F_d})B^{\mathsf{T}}P \\ \frac{1}{\sqrt{2c}}(\sqrt{F} + \sqrt{F_d})B & A \end{bmatrix} \begin{bmatrix} \tilde{\lambda} \\ x \end{bmatrix}$$

$$\stackrel{\triangle}{=} A_{\text{DI}} + A_{\text{SK}}) \begin{bmatrix} \tilde{\lambda} \\ x \end{bmatrix}$$

where the block–diagonal matrix A_{DI} is "stable" and A_{SK} satisfies the skew–symmetry property

$$\begin{bmatrix} \frac{1}{\beta} & 0 \\ 0 & P \end{bmatrix} A_{\text{SK}} = -A_{\text{SK}}^{\mathsf{T}} \begin{bmatrix} \frac{1}{\beta} & 0 \\ 0 & P \end{bmatrix}.$$

This pattern is present in all IBC designs.

- Given that their closed–loop dynamics are nonlinear, predicting the effect of the tuning parameters for PBC and IBC is less obvious than for FLC.

- While there is an *output feedback* version of PBC, the presence of the extra term in IBC hampers this possibility. However, if we want to add damping to the PBC we need the full state.

- As explained in the previous section an FLC–like version of IBC may be obtained as

$$u_{\mathrm{IB}} = \sqrt{\frac{c}{2F_d}} m v_{\mathrm{FL}}\left(q_m, \dot{q}_m, F\right) + R(1 - q_m)\sqrt{\frac{2F}{c}}$$

$$-\frac{\beta}{\sqrt{2c}}\left[\tilde{q}_m, \dot{\tilde{q}}_m, \int_0^t \tilde{q}_m(s)ds\right] PB(\sqrt{F} + \sqrt{F_d}) - R_{\mathrm{DI}}(\sqrt{F} - \sqrt{F_d}).$$

Notice that it combines terms from PBC and FLC plus the damping injection and the additional "decoupling" term.

5.4 Simulation results

In this section we present some simulation results of the levitated ball model (8.2) in closed–loop with FLC, PBC and IBC. To facilitate the comparison, and unless otherwise stated, we have in all cases fixed $d(p) = (p + 10)^3$ and show the plots of position and current. We tried a series of small step references in position. As expected, when we applied larger steps we cross through controller singularities. We have to point out that for the feedback linearization controller the "admissible" references were larger than in the other cases. Away from singularities the control signals do not differ significantly, therefore we show in all cases only the position and current responses.

To motivate the utilization of nonlinear control we designed first a pole placement linear controller for the linearized approximation of (8.2) around the zero equilibrium. In Fig. 8.3 we show the responses for various 0.1 m step references. As expected the performance is degraded as we move away from the domain of validity of the modeling approximation. The response of FLC (8.19), (8.20), which is shown for the sake of reference in Fig. 8.4 is of course the same for all reference levels. It is interesting to note that we observed very little stability degradation when the actual parameters R and c were replaced by estimates within practically meaningful ranges of $\pm 50\%$. Of course, some steady state error was observed. Comparable robustness properties were observed for the other schemes.

PBC was tried in different variations. Fig. 8.5 shows the response of (8.10), (8.11) without damping injection, i.e., with $v = 0$ and without integral action $k_0 = 0$. In this case the mechanical dynamics is order two and we fixed the characteristic polynomial at $(p + 10)^2$. Even though the errors eventually converge to zero, the transient response is extremely poor even when the poles of the mechanical dynamics were pushed farther left. Fig. 8.6 shows the effect of adding an integral action, while for

Fig. 8.7 we included also the damping injection. For the latter we have observed that, for large values of R_{DI}, the response overshoots, due to the peaking phenomenon. The response of the output–feedback version of PBC (8.16), (8.17) (with the bandwidth of the approximate differentiator five times faster than the mechanical dynamics) is essentially indistinguishable from the state–feedback case, hence is not shown here for brevity. Finally in Fig. 8.8 we present the responses of the IBC (8.10), (8.21), (8.22) for a value of $\beta = 0.05$.

In summary we can say that in most of the cases, and as long as we kept away from the singularity region, the observed responses were consistent with the theoretical predictions. Also, despite the significant (analytical) differences of the three controllers that we pointed out above we could not observe big discrepancies in the simulation responses. We believe this stems from the fact that the system is rather benign (it is passive with respect to λ, and y is a flat output!) particularly for the choice of parameters given here.

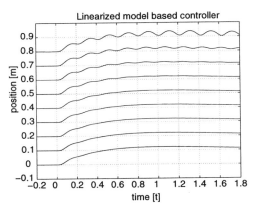

Figure 8.3: Linear controller.

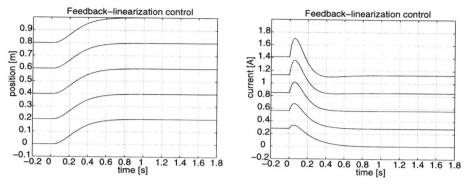

Figure 8.4: Feedback-linearization controller.

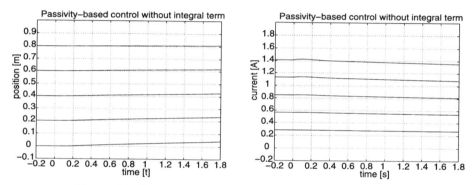

Figure 8.5: Passivity-based controller without integral term.

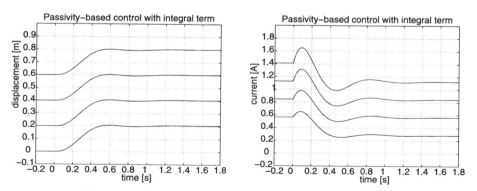

Figure 8.6: Passivity-based controller with integral term.

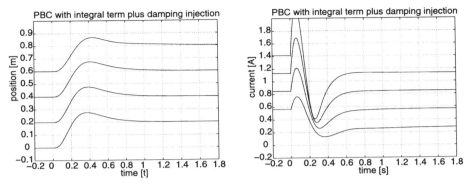

Figure 8.7: Passivity-based controller with integral term plus damping injection.

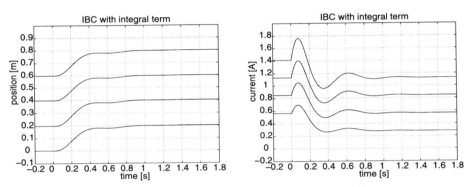

Figure 8.8: Integrator backstepping-based controller.

5.5   Conclusions and further research

The following concluding remarks concerning general aspects of FLC, PBC and IBC are in order

- While in FLC we aim at a linear system, in PBC we are satisfied with assigning a certain passivity property to a suitably defined map, –consequently enforcing a certain storage function to the closed–loop dynamics–. In IBC we aim at a somehow intermediate objective (more ambitious than passivation but less demanding than linearization) of assigning a *strict* Lyapunov function to the closed–loop.

- Both, in IBC and PBC, the main question is "Which storage–Lyapunov function you (can) want to assign?" In PBC, at least of Euler–Lagrange systems, the answer is provided by the systems energy function (either of the full system or only a part of it). The power of IBC is that, for a certain class of cascaded systems, the Lyapunov function can be *recursively* constructed. The example above shows that IBC can be profitably combined with PBC to answer this question.

- It is often argued and reasonable to expect, though hard to prove rigorously, that avoiding cancelation of nonlinearities enhances the robustness of the scheme in the face of parameter uncertainties.[5] Our simulation evidence in the present example, however, showed that all schemes are highly insensitive to these uncertainties.

The three schemes will in general exhibit different transients and posses different robustness properties, a challenging research problem is to establish some common framework to compare their robustness and performance properties. One particularly interesting question is to assess the degrees of freedom provided to the designer to enhance the systems response.

The skew–symmetry property of IBC is related, though not so clear how, with the so–called workless–forces of Euler–Lagrange systems. In PBC we essentially disregard these forces and concentrate on the damping injection to increase the convergence rate. In this class of systems these coupling terms are related with the transformation of energy from one form to the other, e.g., from electric to magnetic in electrical circuits or from potential to kinetic in mechanical systems. In the recent interesting paper [16] the possibility of "shaping" also these forces for performance improvement is explored. This is an intriguing, and certainly quite reasonable, proposition whose effect cannot be captured with our classical convergence rate analysis.

---

[5]See Section **11**.4 for an induction motor example where the superiority of PBC is unquestionably established, both theoretically and experimentally.

Chapter 9

Generalized AC motor

"Man muß immer generalisieren."

C. G. J. Jacobi

In the second part of the book we pursue our research on development of PBC for EL systems as applied to electromechanical systems. In this chapter we restrict our attention to the practically very important class of *the generalized rotating electric machines* [179, 285]. The main contribution is the definition of a class of machines for which the *output feedback torque tracking* problem can be solved with PBC. Roughly speaking, the class consists of machines whose non-actuated (rotor) dynamics is *suitably damped*, and whose electrical and mechanical dynamics can be partially *decoupled* via a coordinate transformation. Machines satisfying the latter condition are known in the electric machines literature as Blondel-Park transformable [157]. In practical terms this requires that the air-gap magneto motive force can be suitably approximated by the first harmonic in a Fourier expansion. These two conditions, stemming from the construction of the machine, have clear physical interpretations in terms of the couplings between electrical, magnetic and mechanical dynamics, and are satisfied by a large number of practical machines.

1 Introduction

1.1 AC motors

The importance of the class of physical systems chosen in this chapter can hardly be overestimated. The field within motion control called mechatronics has become a very important technology for industrial automation. This technology merges mechanics (precision mechanics, coupled electromechanical systems) and electronics (microelectronics, power electronics), using among other tools control theory, and is of high economic importance. Control of electrical drives holds a central position within this technology.

Due to their simplicity from a control point of view, DC motors have been the traditional choice where *high dynamic performance*, i.e. extremely rapid and accurate torque, speed or position control in all four quadrants and for a wide speed range (including zero speed), is required (machining tools, robots). These machines are expensive and difficult to construct for high power/speed applications, even if slot-less armature designs have increased their power range. High torque standstill operation is also difficult, due to brushes being welded to the commutator. DC motors are heavy with high rotor inertia and large dimensions, and they have failure prone brushes[1] which are exposed to corrosion and wear, hence regular maintenance is required. Because of the brushes they are not suited for hazardous environments where electric sparks are not allowed (oil and gas industry), unless they are especially encapsulated in material or by inexplosive gases under higher pressure than the surroundings. To summarize, the DC motor can be thought of as an expensive device, but with a cheap controller.

AC machinery has been the choice for high power constant speed industry applications (compressors, fans, mills, and pumps), or in assembly lines with several machines connected to the same power supply. These machines do not suffer from the typical disadvantages associated with DC motors (no brushes, less complicated rotor construction), but due to their inherent nonlinear dynamics, they have been considered difficult to control and not suited for high dynamic performance applications. Compared with DC motors, they can be thought of as cheap devices, but with expensive controllers. The recent years advances in power electronics and microprocessor technology have enabled implementation of advanced nonlinear control schemes using DSPs, and AC machines have replaced DC machines in a large variety of low and medium power applications, leading to higher reliability and lower costs.

The new advances in AC motor technology have also led to a re-examination of the control schemes used in the traditional (uncontrolled) constant speed high power industry drives, due to demands on higher product quality (tighter speed control, faster recovery from disturbances). In addition, these advances have led to an increasing number of new applications. Examples of high power, high dynamic performance applications are ships, where conventional diesel-/turbine-mechanical propulsion has reached the top of its evolution cycle, and is now being replaced with diesel-/turbine-electric propulsion [109]. It is even claimed that in future permanent magnet AC motors will take over in application areas where hydraulic and pneumatic actuators, which are bulky and failure prone, have been the traditional choice. Such applications include robotics, aircrafts, and spacecrafts. In many of these applications, and also for ships and vehicles, which carry their own fuel (e.g. batteries), economy is of high

---

[1]Tesla pointed out already in 1888 that the commutator is a highly complicated device which is the source of most of the problems experienced with DC machines.

importance, and the concept of *power efficiency* could be equally important to high dynamic performance.

It is estimated that in the United States more than 60% of the generated energy is converted to other forms by electrical drives [37], and only 8% by DC drives. With such amounts of energy being converted, power efficiency of electrical drives in general has become an area of increasing research interest. In fact, it is pointed out in a book from the American Council for an Energy-Efficient Economy, that "adjustable speed drives and other controls are the largest potential source for motor system energy savings" [28]. This does not only include large industrial drives, where a 2% improvement of speed regulation can give significant long term cost reductions despite higher initial investments. It includes all kinds of electrical drives, in a huge number of applications, including households.

To some extent high dynamic performance and power efficiency (typically more than 90% at rated conditions) of a drive can be obtained by the design of the motor itself. However, for motor drives with a highly varying range of operating conditions, motor design alone cannot ensure high performance and efficiency for all conditions, and advanced control schemes must be used together with power electronics.

From a control point of view, AC drives pose the following research problems:

- Transfer of electromagnetic into mechanical energy is essentially described by nonlinear models, making standard linear control theory inferior to nonlinear schemes for control of such systems.

- They are multivariable systems, with several input voltages or currents and one or more outputs (torque/speed, flux level) to be controlled.

- Varying parameters (e.g. temperature dependent resistances and friction parameters, inductances depending on flux level).

- Unknown load disturbances.

- Only partial state measurement (e.g. unmeasurable fluxes and rotor quantities).

- In some cases failure prediction and prevention is also needed (supervisory control system to monitor defects in windings or bearings).

Because of these factors AC drives, and especially induction motors, have become interesting *benchmark problems* for testing of new nonlinear control schemes [70]. The challenging control problems and the rapidly growing number of applications, are also highly motivating factors for working with this class of physical systems.

1.2   Review of previous work

Due to the importance of electrical drives in industry, thousands of papers and numerous textbooks presenting research in this field, have been published during the last 30 years. It would be a time consuming and difficult task to go into the details of all these approaches, and that is not the intention of this summary. In this section the most important control approaches are explained, with emphasis on recent results within the *nonlinear control theory* direction this research has taken. For additional information about this direction, the reader should consult [263] and the references therein. A recent summary of results along the more *application oriented* branch can be found in [35].

A   Classical stationary control (Scalar control)

The classical methods for control of AC machinery have been based on linearizations of the nonlinear equations at steady state operating points. This approach has the advantage that classical linear theory can be used for controller design. Typically, this has resulted in schemes where amplitude and frequency of sinusoidal stator voltages or currents are the basic control variables. Such designs give varying dynamic performance when applied to nonlinear systems, depending on to which degree the underlying small-signal assumption is fulfilled, i.e. depending on how far from a nominal operating point the system is driven. Another disadvantage of applying such methods to multivariable systems, is the problem of coupling between inputs and outputs, making independent control of outputs difficult. For instance, in a voltage fed induction motor, both torque and air gap flux are functions of voltage amplitude and frequency, giving considerable coupling and slow response when trying to control only torque [36]. Even with a well tuned scheme it is difficult to match the performance of a DC drive.

B   Vector control

Some of the deficiencies of classical linear methods when applied to control of AC machinery, were overcome by the vector control methods introduced in the period 1969 – 1972. These methods aimed at making the AC motor behave like a DC motor, with asymptotic decoupling of torque and flux control. To achieve this, the nonlinear model of the motor had to be used in the design.

The vector methods[2] consist essentially of a nonlinear coordinate transformation (a rotation), followed by a nonlinear decoupling feedback. At the time of introduction, implementation of the schemes (especially the rotations) was computationally heavy and difficult with analog electronics. These schemes were therefore considered to be

---

[2]Generically also known as field–oriented control (FOC) methods.

rather "academic" [152], and did not gain their wide popularity until the introduction of the digital microprocessor around 1980. The basic control variables were now the individual components of the rotated two-dimensional stator current *vector*, either controlled directly, or indirectly through a nonlinear feedback voltage control law. Vector control in its various implementations is now the *de facto* standard for high dynamic performance control of AC drives, and its superior dynamic performance as compared to the use of classical linear methods is widely accepted [152].

One of the most common implementations of vector control, *rotor-flux-oriented control*, is discussed in greater detail in Section **10**.5 with application to induction motors. The basic drawback of these schemes is the assumption of full state measurement (flux measurement). Sometimes it is also claimed that it would be desirable to achieve exact, instead of only asymptotic, decoupling between torque and flux control. This can be achieved with linearizing controllers, which are discussed below.

See [275] for the theoretical part of vector control, and [120, 158] for implementation aspects.

C   Modern nonlinear control

During the last 10 years, there have been significant advances in nonlinear control theory, and among many other applications for which linear theory cannot give satisfactory solutions, its application to electric machines has gained widespread interest. Of course, these schemes are also based on nonlinear models and end up with a specification of the current or voltage input vector. However, since they are derived from different and sometimes purely mathematical nonlinear control theories, aiming at formal proofs of stability, they are named *nonlinear* schemes in this book instead of *vector control* schemes.

The following review of modern nonlinear control theory applied to electric machines is based on [263], with the addition of recent results. To limit the number of references, only the recent theoretical and experimental applications of these methods to induction motor control are included here, since this machine will be particularly emphasized in Chapters **10** and **11** of this book.

According to Taylor's recent overview, the methods can be broadly classified into the three following categories:

1. Exact linearization design

2. Backstepping and manifold designs

3. Passivity–based design[3]

---

[3]In Taylor's survey paper, this approach is referred to as "Energy Shaping Design". We prefer to use the more general PBC term since, as we have seen throughout the book "energy–shaping" is just one step in the whole procedure.

Each of these approaches are explained in the following text, and some comments are given to methods not belonging to these classes.

C.1 Exact linearization design

The goal of these schemes is to transform the system into a linear input-output relation between external inputs and controlled outputs, using an inner nonlinear decoupling loop. Controllers can then be designed to ensure stability and performance of the resulting linear system, by use of conventional linear theory. Fundamental to this approach is the choice of coordinates for the representation of the system, together with the design of the inner loop decoupling control in these new coordinates. In contrast to field-oriented control, which also has a coordinate transformation and an inner decoupling loop, the decoupling between outputs is no longer only asymptotic. The coordinate transformations used are also generally more complicated than the rotation used in field-oriented control.

Disadvantages of these schemes are that measurement of the full state is needed, exact cancelation of dynamics is necessary, and controller singularities are introduced, typically for zero rotor flux norm. It also seems to be difficult to apply this method in a general way to a broader class of machines, and then derive controllers for each machine in particular from a general controller. The dynamic equations for stator and rotor quantities must be transformed to a common frame of reference (typically the stator fixed frame) before the differential geometric tools can be used. Symbolic software is often necessary to answer the question of whether the system, with a given set of inputs and outputs, can be transformed into a linear system or not. It is well known that the development of these software tools has enjoyed very limited success, hence this remains an important stumbling block for the application of this theory.

In the feedback linearization technique, all nonlinearities of the system are cancelled to obtain in closed–loop a linear input-output relation. There are two drawbacks to this approach, first the cancelation, if not exact, brings along very serious robustness problems.[4] Second, even from the mathematical point of view it is clear that not all of the nonlinearities are harmful to the closed loop dynamics.[5] These arguments have systematically been invoked in PBC designs as a critique to feedback linearization, see in particular Sections **11**.2 and the experimental results in Sec-

---

[4]It is claimed that adaptation may alleviate, to some extent, the lack of robustness of feedback linearizing schemes. However, the issue of robustness of adaptive systems in general is far from being settled.

[5]In fact, by use of Lyapunov theory, it can be shown that some nonlinearities are useful for system stabilization, and hence they should not be cancelled by the controller. For instance, consider the system $\dot{x} = -x^3 + \cos x + u$. For stabilization around zero, it is of interest to cancel the cosine term with the input u, but not the third order nonlinearity, which is helpful. See also Example 2.5 in [142].

tion **11**.4. They have recently been fervently adopted in the so–called backstepping literature [142, 237] which we review below.

For the fifth order induction motor model[6] case, recent contributions belonging to this class may be summarized in chronological order as follows:

• The exact input-state linearization under the assumption of a slowly varying speed presented in [67], which essentially was a linearizing controller design for the reduced fourth order electromagnetic model. This early paper, though not very often cited, triggered a lot of interest in the community and should be considered as an important precursor to later developments.

• The exact input-output linearization from stator voltages to torque or rotor speed and square of rotor flux norm proposed in [143], for both the rotor-flux-oriented model and the stator fixed model. This is a fundamental paper for this approach, which was later substantially clarified and extended in [172] to handle unknown constant rotor resistance and load torque by the use of adaptation, local stability results were also derived. The resulting zero dynamics with these outputs, is the dynamics of the rotor flux angle, which is weakly minimum phase, actually periodic. In this paper it was also shown that the fifth order induction motor model is not input-state exactly linearizable, and that the largest input-state feedback linearizable subsystem has dimension four.

• In [58] it was shown that the extended sixth order system obtained by augmenting the fifth order model by an integrator in one of the inputs (or the higher order system obtained by one integrator in one input and two in the other), can be exactly input-state linearized, but only for speed, not position control. The result is only locally valid, and the control structure is computationally heavy and requires switching between two transformations to avoid singularities. A nonsingular feedback linearizing transformation was shown to exist only for nonzero torque. The result has recently been extended in [59] for the case of a third order rotor-flux-oriented reduced dq-model, for which stator dynamics is neglected and stator currents are considered as inputs, while rotor flux angle, amplitude, and rotor speed are states. This model is not input-state exactly feedback linearizable. In the work by Chiasson it was shown that if one integrator is added to the input in the d-axis, a single feedback linearizing transformation and controller exist. The controller has a dynamic singularity condition, and if flux is kept constant, it must be nonzero to avoid this singularity. Otherwise, the dynamic condition sets a limit on how fast it can be decreased. This limits the flux tracking capabilities of the scheme. The controller structure is however computationally feasible. Adding instead one integrator to the q-axis input,

---

[6]In this model the two rotor currents (or fluxes), two stator currents, and rotor speed are considered to be the states of the system, while the two stator voltages are the inputs. The outputs to be controlled are torque or rotor speed and rotor flux norm. In the results reported here, the stator fixed frame of reference has been used for model representation, unless something else is explicitly stated. The model of the induction motor is also described in Section **10**.1.

results in the fifth order system with position included being feedback linearizable. Unfortunately the required controller is singular for zero torque, and another controller structure is needed when the torque is required to change sign. This limits the practical usefulness of the result.

The results in [58, 59] clearly reveal how fragile are the geometric properties upon which feedback linearization is based. Namely, they change with coordinate transformations, hence care has to be taken when model representation, inputs and outputs are to be chosen. Otherwise the linearizing approach can result in controllers having singularity conditions which are difficult to interpret and avoid.

• Experiments validating the practical importance of the feedback linearizing approach have been presented in [129, 130] and [30]. Results from extensions to nonlinear magnetics with saturation and speed observers, have also been presented in [31, 32].

A more basic property than feedback linearizability is the property of *flatness*.[7] Roughly speaking, we say that a system has a flat output if its relative degree is the same as the systems degree, hence it has no zero dynamics. A first immediate corollary is that systems with flat outputs are feedback linearizable, –but this is by far not the must important consequence of this feature. In the authors' opinion this is a fundamental concept that induces, in a natural way, a classification of nonlinear systems into flat and non-flat. It has been shown in, e.g. [83] that many physical systems are flat with, furthermore, physically meaningful flat outputs. For instance, we have seen already in Chapter 6 that the total energy is a flat output for the boost converter, while the joint velocity vector is a flat output for robots with flexible joints. A lot of research has been devoted recently to further elucidate this characterization. In the important paper [176] it was shown that the induction motor is flat, with flat outputs the slip angle of the flux and the rotor position. In the same paper an interesting extension of the flux observer of [280] is given and a predictive controller–like approach is proposed. A FOC–like controller and some experimental results along this line of research have been reported in [56].

Quite a lot of effort has been devoted to the estimation problem, and linear and nonlinear observer theory [280], extended Kalman filters [54], and more physically based adaptive observers [198] have been proposed as solutions. Recently, an observer which is adaptive with respect to rotor resistance have been proposed in [170]. In this work exponential convergence of flux and rotor resistance errors to zero, is proved using a Lyapunov type argument, under reasonable assumptions on persistency of excitation.

---

[7]We say that a system is flat if there exists certain special outputs, called the flat outputs, equal in the number to the inputs, which are functions of the state vector and of a finite number of its derivatives. Additionally, the flat outputs are such that every variable in the system can, in turn, be expressed as functions of the flat outputs and a finite number of their time derivatives. See [83] for a detailed explanation of this concept and its application to a wide variety of physical systems.

The incorporation of estimated states in the control law is however in most cases based on a "nonlinear separation principle". Even if there are rigorous theoretical proofs of stability in the case of full state measurement, generally no proofs are given for stability of the closed loop system when *ad hoc* estimates have been substituted for real states in the controller. As explained in Section **11**.2, see also [132], it is important to consider the effect of the convergence rate of the observation error on the performance of the total system, since there is only exact decoupling after the estimation error has converged to zero.

• For the reduced third order model of the induction motor, an interlaced output feedback controller and observer design has been reported in [169]. Rotor speed is the only measurement, and the controller provides singularity free (under the assumption of suitable initial conditions for the flux estimates) speed and rotor flux norm tracking. The algorithm is also adaptive with respect to a constant load torque. An implementation taking advantage of using the rotor flux reference frame for digital calculations, has been reported in [186].

• Much of the work along this direction has now evolved, via adaptive extensions of feedback linearizing controllers, into adaptive output feedback controllers, which are derived using nonlinear tools like Lyapunov theory. An important result is the globally defined adaptive output feedback controller presented in [171]. The controller is adaptive with respect to constant load torque and rotor resistance, and gives asymptotic flux and speed tracking, with convergence of estimated parameters and states to true values, under an assumption of persistency of excitation.

C.2 Backstepping and manifold designs

Backstepping is a recursive Lyapunov procedure for controller design whose origins, –as pointed out in Chapter 3 of [142]–, date back to the 10 year old work of Tsinias, Byrnes and Isidori, and Sontag and Sussmann. The first step in this approach is to choose the output to be controlled and derive its dynamic equation. A *fictitious* control signal is then chosen from this equation. Using a first Lyapunov theory approach, a desired function for this fictitious control is found, such that the the control objective of the first subsystem can be asymptotically achieved. In the original versions of backstepping the control was chosen with cancelation of dynamics in mind, however some elements of PBC have been recently incorporated at this step to develop more robust designs. If the fictitious control is the real input to the system, which can directly be specified to be the desired function, then the design ends here. This is generally not the case, and there will be a deviation between the fictitious control and its desired behavior. The dynamic equation for this error is then derived, an integrator is added, and the design above is repeated with the aim of forcing the error to zero by the use of a new fictitious control. Stability and convergence of this new subsystem can be proved by adding a term square in the error to the previous Lyapunov function. The procedure is then repeated until finally the real control can be specified to be a desired function, and the desired control properties can be proved using a final Lyapunov function, which is a sum of the previous functions. The number of steps needed, is equal to the *relative degree* between the output to be controlled, and the input of the system. For multivariable systems, the design is done separately for each of the outputs. This results in a linear combination of inputs being equal to desired functions, and inversion of a matrix is necessary to specify the real inputs, hence control singularities are likely to occur.

There are now several results from the application of backstepping to induction motor control. In all of the results listed below, the fifth order stator fixed model representation has been used, and unless something else is explicitly stated, observers have been used to avoid flux measurement.

• In [116] the first application of backstepping to speed and rotor flux norm tracking for induction motors was presented. Exact model knowledge was assumed, and a proof of exponential stability was given for the total system with an observer based on the rotor flux equations. The stability result was only *regional* in the sense that the invertibility of a matrix needed to calculate the controls, restricted the initial conditions to be in a set which could be estimated a priori. The basic idea used for the interlaced design, was to add nonlinear damping terms in the controller to compensate for interactions due to observation errors. These terms made it possible to dominate cross-terms containing observation errors in the derivative of the Lyapunov function, and it could consequently be made non-positive.

The advantage of this scheme, as compared to conventional schemes, which do not compensate for observation errors, was demonstrated by simulations, and showed that not compensating for observation errors could have significant negative impact on transient responses.

• This result was extended in [115] to compensate for unspecified modeling imperfections and external disturbances, by addition of PI-loops for speed and rotor flux norm tracking. The regional property of stability was still present, imposing restrictions on initial conditions and reference functions. In the observer structure used in that paper, additional nonlinear terms were introduced, as compared to the previous design. In the stability analysis, it was shown that with a proper choice of these terms, they could be used for eliminating instead of dominating cross terms stemming from observation errors, in the derivative of the Lyapunov function.

The introduction of nonlinear damping terms in the observer/controller, gave a systematic method for handling of estimation errors and other disturbances in stability proofs. Later designs have taken advantages of these ideas.

• The above results were changed to semi-global uniform ultimately bounded position tracking error and rotor flux norm regulation in [107]. To avoid controller singularities for zero flux estimates, a control parameter had to be made sufficiently large relative to initial conditions.

• Departing somewhat from the previous results with interlaced controller and observer designs, adaptive and robust controllers, which could compensate for parametric uncertainty in all parameters (robust case also had additive norm bounded disturbances), were presented in [108]. There were controller singularities for zero rotor flux norm, and proofs of asymptotic position/speed tracking (adaptive case) and uniform ultimate boundedness of position tracking error (robust case), were derived under the assumption of full state measurement.

• While all the above approaches required speed measurement, a design with both speed and flux observers was proposed in [106]. Local exponential rotor flux norm and position tracking was proved, with measurement of only stator currents and position. The local nature of this result was again due the the invertibility of a matrix needed for control calculations. To avoid the singularity, certain restrictions had to be imposed on rotor flux norm reference and initial conditions. The result was novel in the way that speed and flux observers were both taken into the stability analysis.

• An adaptive controller which could compensate for parameter uncertainty in both rotor resistance and the mechanical subsystem, was presented in [104]. The observer was based on the adaptive observer from [170], but to obtain an interlaced design of observer and controller, additional terms were added to the observer. The proof of asymptotic position/speed tracking and norm of estimated fluxes converging to a desired function, was only locally valid, due to a singularity in the controller for zero flux. A drawback of this scheme is that asymptotic convergence of flux estimates to real values was not proved (not even for exactly known rotor resistance), hence rotor flux norm tracking cannot be claimed. However, this was *the first asymptotically stable output feedback scheme with adaptation of rotor resistance*, which has been reported.

• We have shown already in Chapter **8** how we can combine backstepping ideas with PBC to robustify the design. Other examples are reported in [277], where global results were achieved by incorporation of some of the ideas presented in this book into a backstepping design. The clue to avoid singularities was to use, as in [209], the desired rotor currents and fluxes instead of the actual ones in the controller design.

In the first paper a result on global asymptotic position/speed tracking and rotor flux norm regulation was presented. The scheme included adaptation of mechanical parameters. The result in the second paper was a scheme with globally exponential velocity tracking and rotor flux norm regulation. Experimental results were included in both papers.

Drawbacks of these schemes are that they do not provide rotor flux norm tracking, electrical parameters are not compensated for, and they are computationally heavy. This last point limits the value of the experimental results, since responses are restricted to be very slow because of the high sampling period needed.

• Some of the above problems have been solved in [276], where an adaptive singularity-free controller for asymptotic rotor position and rotor flux tracking has been proposed. Rotor flux estimates were calculated by integration of voltage and current measurements using Stanley's equations [258]. This allows for adaptation of rotor flux resistance and mechanical parameters, but relies on zero initial conditions and use of open loop integration for flux calculation. This approach is not numerically robust, and an *ad hoc* integration method with a forgetting factor has to be used for implementation. The performance of the controller was demonstrated by experiments. Unfortunately the same points as above regarding response times were present. It is believed that this will be solved in future work, either by use of controller simplifications or special hardware (ASICs).

It must be pointed out that this scheme (as well as many other backstepping schemes) has been derived with the stator voltages as basic control variables, and it is not clear how the controller can be applied to the reduced order model, with currents as inputs.

In the above schemes the full order model has been used for controller design, and there are no approximations or implicit assumptions about time scale separations between interacting parts of the dynamics.

The rationale behind manifold designs based on *singular perturbation* or *integral manifold* theory, is to take advantage of inherent time scale separations between different parts of dynamics in a system. Such a separation exists for instance between the high frequency current dynamics and its low frequency average dynamics, because of small inductances or high-gain current control. Another example is the relatively slow mechanical dynamics for high rotor inertias, as compared to the fast electrical dynamics. See [176] for a nice presentation of these practical considerations.

These methods essentially consist in first designing a fast control, which steers the fast dynamics to the manifold of the slow dynamics (i.e. makes it attractive), and is inactive once the fast states hit the slow manifold. A second slow control is then designed to give the desired behavior of the slow dynamics, assuming that the reduced order slow model is a sufficiently good approximation of the system dynamics. These methods can be applied in combination with other approaches, to analyze and implement approximations of for instance schemes based on feedback linearization.

Another related approach is to force the systems dynamics to evolve on a manifold called a *sliding mode* with discontinuous controls, having analogy to classical bang-bang control. The behavior on the sliding mode is specified to be the desired one, giving for instance zero speed and flux tracking error. Once on the sliding mode, the system states will stay on it, due to the controls being discontinuous across it. Any small deviation from the sliding mode will activate controls to force the states back to it. Theoretically this is equivalent to infinitely high gain. The inherently high frequency discontinuous on-off switching of controls have made this approach very appealing for control of power converters and electric machines, where thyristors or other switches are used.

For the design of sliding mode motor controllers, the converter is taken into the model, and the switching pattern follows directly from the controller, not from a PWM block with outputs from the controller scheme as reference inputs. This sets requirements to speed of computation, if high switching frequencies are needed for satisfactory control. Unfortunately, analysis of differential equations with discontinuous right-hand sides is technically difficult, due to the fact that classical theorems of existence and uniqueness are not necessarily satisfied for such systems. This has motivated the development of special tools for rigorous analysis of sliding mode systems.

Examples of these schemes and their combinations for motor control can be found in [271], and [228]. Experimental results have been reported in [260]. The application of backstepping for control of electric machinery is explained in detail in [65].

C.3 Passivity–based design

As pointed out in Chapter **1** (see also [263]), this approach has evolved from considerations of physical properties like energy conservation and passivity. This should be contrasted with feedback linearization, which results from purely mathematical considerations. Briefly, in PBC design it is aimed at reshaping the energy of the system in a way leading to the desired asymptotic output tracking properties. The main goal is to drive the system to a desired dynamics, leaving the closed loop system nonlinear, without cancelling dynamics or introducing controller singularities.

To give some historical perspective to the developments of PBC as applied to electrical machines, we briefly summarize them here in a chronological order. *Cela va sans dire* that these results are the core of this chapter, and will therefore be presented in full detail here.

• In [205] the controller design method used in robot motion control to solve the output tracking problem for a class of underactuated Euler-Lagrange systems, was extended to torque regulation of the induction motor, with all internal states bounded. There were no controller singularities, but exact model knowledge and full state measurement had to be assumed. It was also indicated how to follow sinusoidally varying torque references. A model representation in a *dq*-frame of reference was used, and this model became the basis for later designs.

• The previous design was extended to a globally stable controller for torque regulation without measurements of rotor variables in [206]. This globally defined and globally stable interlaced design of controller and observer, was *the first such result reported in the control literature* [263]. Exact model knowledge was assumed, but it was indicated how to compensate for unknown rotor resistance and load torque, unfortunately under the assumption of full state measurement. The torque reference was restricted to be below a certain upper limit depending on motor and controller parameters, but again it was shown how this could be avoided in the case of full state measurement. This is a foundational paper for PBC that first illustrated the application of this technique for electromechanical systems.

• Torque regulation with a globally defined and stable controller without measurements of rotor variables, was extended to torque tracking with adaptation of unknown linearly parameterized load torque in [204].

• In [77] the passivity-based controllers were extended to include the important case of rotor flux norm regulation without rotor variable measurements. The coordinate independent properties of this approach were also rigorously explained. It follows that PBC can be derived in any frame of reference chosen for model representation, hence clarifying some erroneous claims made in [169]

• Recently, a new approach to the induction motor control problem was presented in [76], where it was shown that global torque tracking and rotor flux norm regulation could be done without flux measurement or estimation. This was accomplished by the fundamental observation that the mechanical part of the induction motor dynamics defines a passive feedback around the electrical subsystem, which is also passive. Hence, instead of shaping the energy of the total system as in previous designs, the control goal could be achieved by shaping only the energy of the electrical subsystem, with the mechanical subsystem as a passive disturbance. It was also shown in this paper how to extend the controller for speed tracking with adaptation of a constant load torque.

Drawbacks of this scheme are that it is open loop in speed, and that the convergence rate of the speed tracking error is bounded from below by the mechanical time constant, relying on a positive damping of the mechanical system. However, this paper gave a first rigorous solution to the longstanding problem of avoiding rotor flux estimates in induction motor control, still with global stability results, but unfortunately under the assumption of known parameters.

• The problems with the convergence rate and the speed controller were solved in [209]. In this paper mechanical damping was injected into the closed loop by use of linear filtering of the speed tracking error, giving a globally stable observer-less speed (or position) tracking controller with flux regulation.

• In [79] it was shown how the controller controller could be extended from regulation to tracking of rotor flux norm, an important result for power efficient operation of induction motor drives.

• The results cited above are specific to the induction motor, and it was of interest to see if these results could be extended to other types of electric machines. An answer to this question was given in the seminal paper [193], where it is shown that passivity-based controllers can be designed for a large class of electric machines, including synchronous-, stepper- and reluctance motors.

• In [215] the first globally stable *discrete–time* induction motor PBC was presented. The controller was based on the *exact* discrete–time model of a current–fed induction motor.

• Experimental results from the application of passivity-based controllers, have been presented in several publication, e.g. [48, 95, 96, 132, 192, 262].

Other results along the line of PBC for electromechanical systems will be presented in Chapters **10–12**.

C.4 Other approaches

It might be tempting to include also a fourth class, consisting of all those schemes based on other approaches in nonlinear control theory which do not fit into the above framework. Among these we find the so-called "intelligent" schemes, based on expert systems, fuzzy logic and neural networks. Common for the schemes are that they do not provide formal proofs of stability for the resulting system, not even under the assumption of full state measurement. Such proofs are important goals for the other three classes. In lack of theoretical results, the issues of performance and stability are instead demonstrated by simulations or experimental results. An overview of these techniques is given in [38], and examples of recent applications can be found in [264] and [98]. Some of the problems with the derivation of proofs for stability and performance of "intelligent" schemes, are due to their model free structure, which also seems to be their advantage when the model is "fuzzy" or missing. It is not yet clear if these nonlinear function approximation techniques have some advantages over the above approaches for high-performance control of electromechanical systems with a structured model but uncertain varying parameters.

1.3   Outline of the rest of this chapter

The outline of the rest of this chapter is as follows. In Section 2 the Lagrangian model of the generalized electric machine is presented, and the control problem is defined for AC machines. Some remarks to the dynamic model are also given in this section. An approach for PBC design is proposed in Section 3, and in Section 4 a torque tracking controller is derived. In Section 5 we revisit the PBC of electric machines from a geometric perspective. Section 6 presents examples of controller design for some typical AC machines. Finally, concluding remarks to the PBC design is given in Section 7.

2   Lagrangian model and control problem

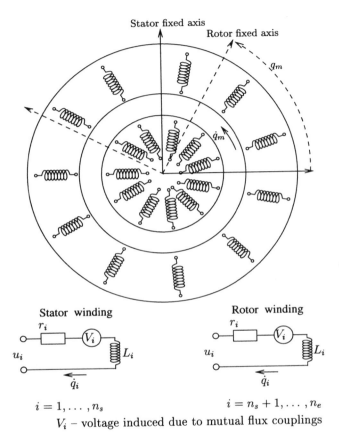

Stator fixed axis

Rotor fixed axis

q_m

$\dot{q}_m$

$\dot{q}_m$

Stator winding

r_i

u_i V_i L_i

$\dot{q}_i$

$i = 1, \ldots, n_s$

Rotor winding

r_i

u_i V_i L_i

$\dot{q}_i$

$i = n_s + 1, \ldots, n_e$

V_i – voltage induced due to mutual flux couplings

Figure 9.1: Cross-section and schematic circuits of generalized electric machine.

In this section we apply the variational technique to derive the model of the generalized rotating machine considered in [157], (see also [288, 292]) and present the torque tracking problem. The advantages of using variational principles, particularly for the purposes of PBC design, have already been thoroughly discussed in previous chapters. As pointed out in these chapters the model is, of course, the same one would obtain by applying first principles, see also Remark 9.2.

It is widely recognized that the fundamental control problem in electrical machines is to regulate the torque [153], [36]. Therefore we concentrate in this chapter on PBC of torque. A good behavior of the mechanical subsystem can be expected by use of a simple (e.g., PI) outer speed loop due to the fact that PBC ensures passivity in closed loop. These observations are rigorously formalized in the remaining of the chapter, and in Chapter **10**, where speed and/or position are actually controlled for the induction motor. Further, in Chapter **12** we show that this approach is still applicable for the far more challenging problem of controlling robot manipulators actuated by AC drives.

2.1 The Euler–Lagrange equations for AC machines

The machine consists of in all $n_e = n_s + n_r$ windings on stator and rotor, as shown in Fig. 9.1. Even if it is not shown in the figure, there could also be permanent magnets or a salient rotor in the machine.

Ideal symmetrical phases and sinusoidally distributed phase windings are assumed. The permeability of the fully laminated cores is assumed to be infinite, and saturation, iron losses, end winding- and slot effects are neglected. Only linear magnetic materials are considered, and it is further assumed that all parameters are constant and known.

Under the assumptions above, application of Gauss' law and Ampere's law leads to the following affine relationship between the *flux linkage* vector λ and the *current* vector $\dot{q}_e$

$$\lambda = D_e(q_m)\dot{q}_e + \mu(q_m) \tag{9.1}$$

with $\lambda = [\lambda_1, \ldots, \lambda_{n_e}]^\top$, $\dot{q}_e = [\dot{q}_1, \ldots, \dot{q}_{n_e}]^\top$, $q_m \in \mathbb{R}$ the *mechanical angular position* of the rotor, and $D_e(q_m) = D_e^\top(q_m) > 0$ the $n_e \times n_e$ multiport inductance matrix of the windings. The vector $\mu(q_m)$ represents the flux linkages due to the possible existence of *permanent magnets*. Both being bounded and periodic in q_m with period $2\pi/N$, $N \in \mathbb{N}$.

If the *generalized coordinates* of the system are defined as the total amounts of moving electric charge that has passed any point on the different phase windings, q_i, $i = 1, \ldots, n_e$, and the angular position of the rotor q_m, the magnetic-field coenergy (with ' denoting the variable of integration) can be computed as [179]

$$\mathcal{T}_e^*(\dot{q}_e, q_m) = \sum_{i=1}^{n_e} \int_0^{\dot{q}_i} \lambda_i(\dot{q}_i')d\dot{q}_i' = \frac{1}{2}\dot{q}_e^\top D_e(q_m)\dot{q}_e + \mu^\top(q_m)\dot{q}_e$$

and the mechanical kinetic coenergy as $\mathcal{T}_m^*(\dot{q}_m) = \frac{1}{2}D_m\dot{q}_m^2$, where $D_m > 0$ is the rotational inertia of the rotor.

Neglecting the capacitive effects in the windings of the motor, and considering a rigid shaft, the potential energy $\mathcal{V}$ of the system is only due to the interactions between the magnetic materials in stator and rotor, i.e. $\mathcal{V} = \mathcal{V}(q_m)$. This energy contribution is zero if there are magnetic materials in only one part (stator or rotor), and the reluctance properties of the other part is uniform. The *Lagrangian* consists then of the sum of the electrical

$$\mathcal{L}_e(\dot{q}_e, q_m) \triangleq \frac{1}{2}\dot{q}_e^\top D_e(q_m)\dot{q}_e + \mu^\top(q_m)\dot{q}_e - \mathcal{V}(q_m) \tag{9.2}$$

and the mechanical Lagrangian

$$\mathcal{L}_m(\dot{q}_m) \triangleq \frac{1}{2}D_m\dot{q}_m^2 \tag{9.3}$$

leading to

$$\mathcal{L}(\dot{q}_e, \dot{q}_m, q_m) = \frac{1}{2}\dot{q}^\top D\dot{q} + \mu^\top(q_m)\dot{q}_e - \mathcal{V}(q_m) \tag{9.4}$$

where we have defined $q \triangleq [q_e^\top, q_m^\top]^\top$ and $D \triangleq \mathrm{diag}\{D_e(q_m), D_m\}$.

To model the *external forces*, it will be assumed that the dissipative effects are linear, time-invariant and only due to the resistances in the windings $r_i \geq 0$, $i = 1, \cdots, n_e$, and the mechanical viscous friction coefficient $R_m \geq 0$. Therefore the corresponding Rayleigh dissipation function takes the form

$$\mathcal{F}(\dot{q}) \triangleq \frac{1}{2}\dot{q}^\top R\dot{q}$$

where $R \triangleq \mathrm{diag}\{r_1, \cdots, r_{n_e}, R_m\} \geq 0$.

The control forces are the voltages applied to the windings $u \in \mathbb{R}^{n_s}, n_s \leq n_e$. In this work fully actuated as well as underactuated machines, that is, machines where the voltages can be applied only to stator windings (e.g. induction motor), or to both stator and rotor windings (e.g. synchronous motor with field windings) will be considered. Hence, it is convenient to partition the vector of generalized electrical coordinates as $q_e = [q_s^\top, q_r^\top]^\top \in \mathbb{R}^{n_e}$, $q_s \in \mathbb{R}^{n_s}$, $q_r \in \mathbb{R}^{n_r}$, $n_r = n_e - n_s$, where the subscripts s, r are used to denote variables related to windings with and without actuation respectively.[8] In the case of underactuated machines the partition coincides with stator and rotor variables as well. Notice however, that there are also machines, like the PM synchronous, PM stepper and variable reluctance motors, where q_e *consists only of stator variables* which are directly actuated by the stator voltages, see the examples in Section 2.4. Finally, it is assumed that the load torque τ_L in the mechanical subsystem is of the form[9]

$$\tau_L(q_m, \dot{q}_m) = [c_1 + c_2\dot{q}_m^2]\tanh(\frac{\dot{q}_m}{\varsigma}) \tag{9.5}$$

---

[8]See Definition 2.2 in Chapter 2.

[9]The presence of a load torque τ_L of this form ensures that to every bounded τ there exists a bounded $\dot{q}_m$. Except from this, as will be shown below, the load torque τ_L plays no role in the torque tracking problem. Also, as shown in [76] it can be treated as an external disturbance for the speed tracking problem.

with a scaling parameter $\varsigma > 0$ and $c_1, c_2 \in \mathbb{R}_{\geq 0}$.

With the considerations above, and by applying the EL equations (2.1) of Chapter 2 to (9.4), the *equations of motion* of the generalized machine are derived as

$$D_e(q_m)\ddot{q}_e + W_1(q_m)\dot{q}_m\dot{q}_e + W_2(q_m)\dot{q}_m + R_e\dot{q}_e = M_e u \qquad (9.6)$$

$$D_m\ddot{q}_m - \tau(\dot{q}_e, q_m) + R_m\dot{q}_m = -\tau_L \qquad (9.7)$$

where, without loss of generality, we have assumed $r_s \triangleq r_1 = \cdots = r_{n_s}$ and $r_r \triangleq r_{n_s+1} = \cdots = r_{n_r}$, and we have defined

$$W_1(q_m) \triangleq \frac{dD_e(q_m)}{dq_m}, \ W_2(q_m) \triangleq \frac{d\mu(q_m)}{dq_m}, \ R_e \triangleq \mathrm{diag}\{r_s I_{n_s}, r_r I_{n_r}\}, \ M_e \triangleq \begin{bmatrix} I_{n_s} \\ 0 \end{bmatrix}$$

Here τ is the *generated torque*, which as discussed in Chapter **2** couples the electrical and mechanical subsystems according to

$$\tau = \frac{\partial \mathcal{L}}{\partial q_m}(q, \dot{q})$$

which in our case reduces to

$$\tau = \frac{1}{2}\dot{q}_e^\top W_1(q_m)\dot{q}_e + W_2^\top(q_m)\dot{q}_e + \eta(q_m) \qquad (9.8)$$

where

$$\eta(q_m) \triangleq -\frac{d\mathcal{V}}{dq_m}(q_m)$$

which is also bounded and periodic in q_m.

Notice that, as pointed out in Chapter **2**, the machine is *fully characterized* by what we called the EL parameters, that is the quadruple $\{\mathcal{T}(q, \dot{q}), \mathcal{V}(q), \mathcal{F}(\dot{q}), \mathcal{M}\}$, where $\mathcal{M} \triangleq \begin{bmatrix} M_e \\ 0 \end{bmatrix}$.

2.2 Control problem formulation

It will be *assumed* here that the currents of the actuated windings $\dot{q}_s$, rotor position q_m, and velocity $\dot{q}_m$ are *available for measurement*. Also, the basic *regulated variable* is taken to be the generated torque τ, which is however *unmeasurable* since it depends on the variables $\dot{q}_r$. Notice that the motor speed is related to torque via a simple LTI system (9.7). The choice of torque control is rationalized in terms of passive operators in Section 3.

The torque control problem is therefore formulated as follows:

Definition 9.1 (Output feedback tracking problem.) *Consider the $n_e + 2$ dimensional machine model (9.6)-(9.8) with state vector $[\dot{q}_e^\top, q_m, \dot{q}_m]^\top$, inputs $u \in \mathbb{R}^{n_s}$, regulated output τ, measurable outputs $\dot{q}_s, q_m, \dot{q}_m$ and smooth disturbance $\tau_L \in \mathcal{L}_\infty$. Find conditions on $D, R, \mu(q_m), \eta(q_m)$ such that, for all continuously differentiable reference output functions $\tau_* \in \mathcal{L}_\infty$ with known derivative $\dot{\tau}_* \in \mathcal{L}_\infty$, global torque tracking with internal stability is achieved, i.e. $\lim_{t \to \infty} |\tau - \tau_*| = 0$ with all internal signals bounded. Further, for underactuated machines, asymptotic flux amplitude tracking will be required, that is, for a given twice differentiable positive and bounded function $\beta_*(t) \geq \delta > 0$, with known and bounded $\dot{\beta}_*(t)$, $\ddot{\beta}_*(t)$, $\lim_{t \to \infty} |\, \|\lambda_r\| - \beta_*(t) \,| = 0$ must hold.*

2.3   Remarks to the model

Remark 9.2 (Lumped system and Euler-Lagrange formulation.) Magnetic and electric fields are distributed phenomena, which are naturally modeled with partial differential equations to give a boundary value problem. This is usually done during the construction phase, when effects of different materials and geometric shapes are to be studied. Even in the case of two-dimensional fields, this results in problems which must be solved numerically, for instance by finite-element analysis.

A distributed model is too complicated for control purposes, and a lumped model is generally considered to give satisfactory results. In this chapter it is assumed that a lumped model of the system in terms of inductance and resistance matrices is already available. The lumped parameters are generally derived as functions of material constants, turns and span of distributed windings, air-gap parameters and approximations, by integration of surface-current densities over rotor and stator periphery [140, 179]. From a lumped description, the dynamical equations of motion are derived using the Euler-Lagrange procedure.

As discussed in [285] (see also Chapter 2), it could be argued that insight into the physical process is *lost* when the variational approach to modeling is used instead of basic force laws, even though the model equations are equivalent. Against this, it could be argued that physical insight is *gained* because coupling terms between various subsystems are derived in an analytically formal way, due to the generality of the method. This property has previously been exploited in numerous examples of controller designs for purely mechanical systems in Chapters 2–5, (see also [189] for some recent developments), and it is also the main argument behind its use here. In particular, the coupling between electrical and mechanical subsystems is highlighted, and storage and dissipation functions for subsequent PBC design are easily obtained. These properties are obscured if the EL equations are translated into a state-space formulation of the dynamical equations.

Remark 9.3 (Voltage balance equations.) As pointed out in Chapter 2 we can alternatively represent the electrical part of an electromechanical system in terms of fluxes (9.1) and currents (which are the generalized electrical momenta and velocities of the Hamiltonian formalism [285]) as

$$\dot{\lambda} + R_e \dot{q}_e = M_e u \tag{9.9}$$

This is a voltage balance equation. Particularly useful for further developments is the following relationship between rotor fluxes and rotor currents for motors where the rotor windings are short circuited (e.g. induction motors)

$$\dot{\lambda}_r + R_r \dot{q}_r = 0 \tag{9.10}$$

where $\lambda \triangleq [\lambda_s^{\mathsf{T}}, \lambda_r^{\mathsf{T}}]^{\mathsf{T}}$ with $R_s = r_s I_{n_s} \in \mathbb{R}^{n_s \times n_s}$, $R_r = r_r I_{n_r} \in \mathbb{R}^{n_r \times n_r}$. The importance of (9.10) is that it defines a dynamic relationship $\ddot{q}_e = f(q_e, \dot{q}_e, q_m, \dot{q}_m)$ that is *invariant* with respect to the control action that will have to taken into account when defining a "desired behavior" for the machine. This relationship will play a fundamental role in the explicit derivation of the PBC as explained in Sections 4 and 5.

Remark 9.4 (Ignorable coordinates.) An interesting property of the machine and other magnetic field devices in which currents are of main interest, is that the electrical charges are *ignorable* [285], (also known as cyclic in mechanics [12]). That is, the Lagrangian of the system does not contain q_e, although it contains the corresponding currents $\dot{q}_e$. It must be pointed out that when choosing the form (2.6) of Chapter 2 of the EL equations, it is crucial that the currents be expressed in their natural frames, where electrical charges can be obtained by integration of currents, to avoid the introduction of quasi coordinates. See also [285].

Remark 9.5 (Mechanical commutation.) For a class of machines with mechanical commutation, the relation between the physically applied port currents and the rotor currents introduces *non-holonomic constraints* [285], and the dynamic equations cannot be obtained directly from (2.6) with the given Lagrangian. Instead quasi coordinates could be introduced, and the dynamic equations derived by use of the Boltzmann-Hamel [285] or Gaponov [191] form of the EL equations. These procedures are however quite involved, and the dynamic equation for this class of machines is therefore usually not derived from variational principles, but by the use of basic force laws as Faraday's law, Ohm's law and Euler's law. As pointed out in Section 1, motors with mechanical commutation are of less interest for nonlinear control design, and will not be considered in this text.

Remark 9.6 (Parameters.) It is well known that the lumped parameters are not constant, but that they change due to temperature variations, skin effects, current displacement and magnetic hysteresis and saturation. In some cases, like for the switched reluctance machine, magnetic nonlinearities must be taken into account in the modeling procedure for satisfactory performance. In other cases the assumption of linear magnetics leads to controller designs which works satisfactory, at least if the machine is forced to operate at flux levels below the saturation limit. A controller design based on a model including magnetic saturation is however highly desirable for performance improvements. For instance, while older uniform air-gap induction motors often were robustly designed to operate under linear magnetic conditions (constant inductances), there is now an increasing interest of operating these under magnetic saturation (current/flux dependent inductances), saving cost and weight.

The EL procedure is based on energy properties of the system, and the dynamic equations for machines with nonlinear magnetics can also be derived by this procedure. However, it is believed that the simpler problems of constant parameters and linear magnetics must first be rigorously solved using passivity-based methods, before the formal framework can be used to incorporate and solve the more complex problems stated above. Hence, magnetic nonlinearities, time-varying resistances, or state dependent inductances will not be considered in this text. However, some robustness results to uncertain parameters, as well as tuning and adaptation rules for the controller gains are presented in Chapter **11**. For the induction motors case, work along the line of extending the PBC to handle nonlinear magnetics has been reported in [95].

Remark 9.7 (Complex versus real notation.) In this book a *real* representation of the machine's model is used, meaning that rotations of vectors will be presented by matrix exponentials, allowing for easy manipulation of the equations. This is a common approach which have been widely used for design of nonlinear controllers, at least within the control theory community. The complex notation, which is closely related to the transfer function language, has been widely used for analysis of stationary operation and presentation of two-axis theory using space phasors, but has not become the common approach for the application of recent nonlinear control theory to electric machines. Of course, the two notations are equivalent[10], and nonlinear analysis could be done in either of the two settings. Recent results from nonlinear control and analysis of induction motors reported in [214] and [176], have taken advantages of using a complex formulation of the induction motor model because of its compactness.

Remark 9.8 (Load torque.) The model in (9.5) is sufficiently general to include mechanical friction, windage, and pump or compressor loads. The scaling of the hyperbolic tangent function is done to mimic the signum function, replacing the discontinuity in the friction model with a curve of finite slope. This is done to avoid introduction of additional problems with differential equations having discontinuous right hand sides in the stability analysis. However, this model does not provide a true

---

[10]To state it more mathematically, this stems from the fact that the metric space of complex numbers is *homeomorphic* to the metric space of matrices on the form $\alpha I_2 + \beta \mathcal{J}$, where α, $\beta \in \mathbb{R}$ and $\mathcal{J}$ is a skew-symmetric matrix, i.e. there exists a continuous and invertible mapping from one space to the other, with a continuous inverse.

stiction mode, something which may be important if the period of the stick-slip limit cycle is long [11]. The issue of adaptive friction compensation is addressed in Chapter 5. Also, in Chapter **12** we address the far more challenging problem of controlling robot manipulators with AC drives.

2.4 Examples

In [157] several examples of electric machines modeled by (9.6), (9.7) are given. In this section two examples of *fully actuated* machines, i.e. machines with $n_s = n_e$ and $M_e = I_{n_e}$, are presented.

Example 9.9 (PM synchronous motor.) The 3ϕ *PM synchronous motor* [140] has $n_e = 3$ and the parameters

$$
D_e(q_m) = \begin{bmatrix} L_{ls} + L_A - L_B \cos 2n_p q_m & -\frac{1}{2}L_A - L_B \cos 2(n_p q_m - \frac{\pi}{3}) \\ -\frac{1}{2}L_A - L_B \cos 2(n_p q_m - \frac{\pi}{3}) & L_{ls} + L_A - L_B \cos 2(n_p q_m - \frac{2\pi}{3}) \\ -\frac{1}{2}L_A - L_B \cos 2(n_p q_m + \frac{\pi}{3}) & -\frac{1}{2}L_A - L_B \cos 2(n_p q_m + \pi) \end{bmatrix}
$$

$$
\begin{matrix} -\frac{1}{2}L_A - L_B \cos 2(n_p q_m + \frac{\pi}{3}) \\ -\frac{1}{2}L_A - L_B \cos 2(n_p q_m + \pi) \\ L_{ls} + L_A - L_B \cos 2(n_p q_m + \frac{2\pi}{3}) \end{matrix} \quad (9.11)
$$

$$
\mu(q_m) = \lambda_m \begin{bmatrix} \sin n_p q_m \\ \sin(n_p q_m - \frac{2\pi}{3}) \\ \sin(n_p q_m + \frac{2\pi}{3}) \end{bmatrix} \quad (9.12)
$$

where L_{ls}, L_A, L_B are inductance parameters, n_p is the number of pole pairs and λ_m is the amplitude of the flux linkage established by the permanent magnet. □

Example 9.10 (PM stepper motor.) The 2ϕ *PM stepper motor* has $n_e = 2$ and the following parameters [295]

$$
D_e = \begin{bmatrix} L & 0 \\ 0 & L \end{bmatrix} \quad (9.13)
$$

$$
\mu(q_m) = \frac{K_m}{N_r} \begin{bmatrix} \cos(N_r q_m) \\ \sin(N_r q_m) \end{bmatrix} \quad (9.14)
$$

L is the self-inductance of each winding, and K_m is the torque constant. N_r is the number of rotor teeth of same polarity. In this case the torque has a term due to the interaction between the permanent magnet and the magnetic material in the stator (detent torque), and $\eta(q_m) = -K_D \sin(4N_r q_m)$, $K_D \approx 5 - 10\%$ of $K_m i_0$, where i_0 is a rated current. □

3   A passivity-based approach for controller design

We have seen already in the simple levitating ball example of Chapter **8** that for electromechanical systems the PBC approach can be applied in at least two different ways, which lead to different controllers. In the first, more direct form, a PBC is designed for the whole electromechanical system using as storage function the total energy of the full system. Another way to design PBC is to first decompose the system into its electrical and mechanical dynamics. Then, observing that the mechanical dynamics can be treated as a "passive disturbance", we design a PBC only for the electrical subsystem using as storage function the electrical total energy. One important advantage of the second feedback–decomposition approach is that it leads to simpler controllers, which as will be shown below do not require observers. For these reasons we follow in this chapter the second route to solve the torque tracking problem posed above. In Section **10.4** we illustrate how the first approach can be applied to the induction machine, with the addition of an observer.

3.1   Passive subsystems feedback decomposition

In this subsection it is shown that the dynamic model of the generalized electric machine satisfies the conditions of Proposition 2.10 in Chapter **2**, and consequently it can be decomposed into the feedback interconnection of two passive subsystems. This result formalizes our approach that concentrates in the electrical dynamics and treats the mechanical subsystem as a "passive disturbance".

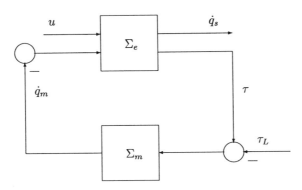

Figure 9.2: Passive subsystems decomposition of generalized electric machine.

Proposition 9.11 (Passive subsystems decomposition.) *The system (9.6)-(9.8) can be represented as the negative feedback interconnection of two passive subsystems*

(see Fig. 9.2)

$$\Sigma_e : \mathcal{L}_{2e}^{n_s+1} \to \mathcal{L}_{2e}^{n_s+1} \quad : \quad \begin{bmatrix} u \\ -\dot{q}_m \end{bmatrix} \mapsto \begin{bmatrix} \dot{q}_s \\ \tau \end{bmatrix}$$

$$\Sigma_m : \mathcal{L}_{2e} \to \mathcal{L}_{2e} \quad : \quad (\tau - \tau_L) \mapsto \dot{q}_m$$

$\square$

Proof. The proof is a corollary of Proposition 2.10, noting that the Lagrangian of the electric machine (9.4) can be decomposed into $\mathcal{L}(q, \dot{q}) = \mathcal{L}_e(\dot{q}_e, q_m) + \mathcal{L}_m(\dot{q}_m)$ with $\mathcal{L}_e(\dot{q}_e, q_m)$, $\mathcal{L}_m(\dot{q}_m)$ as defined in (9.2), (9.3), respectively.

■

3.2 Design procedure

The rationale of the design stems from the passive subsystems decomposition given above and, "disregarding" the mechanical dynamics, attempts to control the generated torque τ by imposing a desired value to the currents $\dot{q}_e$. There are therefore three natural steps to follow:

1. Apply the passive subsystems decomposition of Section 3.1 to the machine, to view Σ_e as the "system to be controlled", and Σ_m as a *passive disturbance*. To ensure the latter does not "destroy" the stability of the loop *damping* must be injected to Σ_e to strengthen its passivity property to *strict* passivity.

2. Define a set of *"attainable" currents* $\dot{q}_{ed}$, i.e. currents for which it is possible to find a control law that ensures $\lim_{t\to\infty} \|\dot{q}_e - \dot{q}_{ed}\| = 0$. To this end, *the energy* of the closed loop must be *shaped* to match a desired energy (storage) function, which is chosen here as $H_{ed} \triangleq \frac{1}{2} \dot{\tilde{q}}_e^\top D_e(q_m) \dot{\tilde{q}}_e$, with current error defined as

$$\dot{\tilde{q}}_e \quad \overset{\triangle}{=} \quad \dot{q}_e - \dot{q}_{ed} \tag{9.15}$$

3. Among the attainable currents choose $\dot{q}_{ed}$ to deliver the desired reference torque τ_*, that is, such that if $\dot{q}_e \equiv \dot{q}_{ed}$ then $\tau \equiv \tau_*$. Finally, give conditions under which $\lim_{t\to\infty} \|\dot{q}_e - \dot{q}_{ed}\| = 0$ implies $\lim_{t\to\infty} |\tau - \tau_*| = 0$ with internal stability.

4 A globally stable torque tracking controller

In Section 2 we presented the model of a generalized electric AC machine and formulated the torque tracking problem. In this section we identify a subclass of this model for which we can design a globally stable torque tracking PBC using the feedback–decomposition approach advanced in Section 3.1. As explained in the previous chapters the PBC is first expressed in an implicit form involving the controller states and its derivatives, the control signal and the external references. In a second stage we must express these equations explicitly to get an implementable form of the controller. It is at this stage that restrictions on the admissible machines will be imposed. From the analytic point of view these conditions boil down to an invertibility assumption. Interestingly enough this assumption is satisfied if the machine satisfies a decoupling condition –called Blondel–Park (BP) transformability– which is well-known in the electric machines literature. We proceed along the BP transformability route, which is more familiar to practitioners, in the first derivation of the controller, and then give an interpretation of this condition in terms of the geometric concepts of relative degree and zero dynamics, which are well–known to control theorists.

4.1   Strict passifiability via damping injection

The first step of our design procedure, the decomposition of the model into passive subsystems Σ_e, Σ_m, can be carried out as described in Section 3.1. In this section it is explained how *damping* is *injected* to Σ_e such that the mapping from control input to measurable output is *output strictly passive*.

Proposition 9.12 *Consider the electrical subsystem Σ_e described by (9.6). Assume*

A 9.1 The rotor resistance is positive, that is[11] $R_r = r_r I_{n_r} > 0$.

A 9.2 The $n_r \times n_r$–dimensional $(2,2)$ block of the matrix $W_1(q_m) = \frac{dD_e(q_m)}{dq_m}$ is zero, i.e.

$$W_1(q_m) \;=\; \left[\begin{array}{cc} (W_1(q_m))_{11} & (W_1(q_m))_{12} \\ (W_1(q_m))_{21} & 0_{n_r \times n_r} \end{array} \right]$$

A 9.3 The non-actuated rotor components of the vector $\mu(q_m)$ are independent of q_m, that is

$$W_2(q_m) = \frac{d\mu(q_m)}{dq_m} \triangleq \left[\begin{array}{c} W_{2s}(q_m) \\ 0 \end{array} \right], \; W_{2s}(q_m) \in \mathbb{R}^{n_s}$$

Under these conditions there is an output feedback *of the form*

$$u \;=\; v + W_{2s}(q_m)\dot{q}_m - K_1(q_m, \dot{q}_m)\dot{q}_s \tag{9.16}$$

such that the mapping $v \mapsto \dot{q}_s$ is output strictly passive *for all $q_m, \dot{q}_m \in \mathcal{L}_{2e}$.*   □

---

[11]This condition is obviously always satisfied in physical machines. It is stated here as an "assumption" to underscore the importance of having a good estimate of R_r, as discussed in Remark 9.15.

Proof. The dynamics of Σ_e is described by (9.6), which is repeated here for ease of reference

$$D_e(q_m)\ddot{q}_e + W_1(q_m)\dot{q}_m\dot{q}_e + W_2(q_m)\dot{q}_m + R_e\dot{q}_e = M_e u$$

Closing the loop with (9.16) results in

$$D_e(q_m)\ddot{q}_e + C_e(q_m,\dot{q}_m)\dot{q}_e + R_{es}(q_m,\dot{q}_m)\dot{q}_e = M_e v \qquad (9.17)$$

where the matrices $C_e(q_m,\dot{q}_m)$ and $R_{es}(q_m,\dot{q}_m)$ have been defined as

$$C_e(q_m,\dot{q}_m) \triangleq \frac{1}{2}W_1(q_m)\dot{q}_m$$

$$R_{es}(q_m,\dot{q}_m) \triangleq R_e + \frac{1}{2}W_1(q_m)\dot{q}_m + \begin{bmatrix} K_1(q_m,\dot{q}_m) & 0 \\ 0 & 0 \end{bmatrix} \qquad (9.18)$$

Taking the time derivative of electric part of the total energy of Σ_e, that is $H_e = \frac{1}{2}\dot{q}_e^\top D_e(q_m)\dot{q}_e$, along the trajectories of Σ_e gives

$$\dot{H}_e = \dot{q}_s^\top v - \dot{q}_e^\top R_{es}(q_m,\dot{q}_m)\dot{q}_e$$

Now, let

$$K_1(q_m,\dot{q}_m) = K_1^\top(q_m,\dot{q}_m)$$
$$> \sup_{q_m,\dot{q}_m} \{\frac{\dot{q}_m^2}{4}(W_1(q_m))_{12}R_r^{-1}(W_1(q_m))_{12}^\top - \frac{1}{2}(W_1(q_m))_{11}\dot{q}_m\} \qquad (9.19)$$

Then, by use of standard matrix results (see Section 4 in Appendix **D**), the symmetric matrix $R_{es}(q_m,\dot{q}_m)$ can be shown to be uniformly positive definite in the sense that

$$\inf_{q_m,\dot{q}_m} \lambda(R_{es}(q_m,\dot{q}_m)) \geq \alpha > 0 \qquad (9.20)$$

Integration of $\dot{H}_e$ completes the proof. ■

A Remarks to conditions for damping injection

Remark 9.13 Notice that strict passivity is achieved, via the *nonlinear damping term* $K_1(q_m,\dot{q}_m)\dot{q}_s$, which recovers the positivity of the "damping" matrix $R_{es}(q_m,\dot{q}_m)$.

Remark 9.14 In the case of full actuation, i.e. $n_s = n_e$, the damping matrix can be a full matrix, and the required positivity of $R_{es}(q_m,\dot{q}_m)$ is guaranteed if (see (9.18))

$$K_1(q_m,\dot{q}_m) = K_1^\top(q_m,\dot{q}_m) > -R_e - \frac{1}{2}W_1(q_m)\dot{q}_m \qquad (9.21)$$

Remark 9.15 Assumption **A9.1** is a reasonable condition of *damping of the unactuated dynamics* which is satisfied in all electric machines. The problem is with condition (9.19), which shows that to overcome the imprecise knowledge of the rotor resistances, *high gains* will have to be injected into the loop.

Remark 9.16 A9.2 is a *decoupled dynamics condition* equivalent to requiring that the contribution to the magnetic coenergy of the terms quadratic in $\dot{q}_r$ must be independent of q_m. Physically, this translates into the condition that if there are rotor windings, then the rotor flux induced by the rotor currents must be independent of the rotor position. This means that the stator must have uniform reluctance properties (non-salient and of uniform magnetic material). This assumption is satisfied by many machines, e.g. classical Park [292] and polyphase machines. [288]

Remark 9.17 Since the torque (9.8) consists of one component due to the currents, and the other of purely magnetical origin, and since there is no control on the fields from the permanent magnets, it is reasonable to expect that the effect on Σ_e of the flux linkages due to the permanent magnets must be eliminated. This explains the need for Assumption **A9.3**. Physically, this assumption also means that if the machine has rotor windings, then the stator must have uniform reluctance properties i.e., if the machine has permanent magnets, then they can only be placed on the rotor. As can be seen from (9.16), the term from the permanent magnets must be cancelled out. The need for this cancellation is a drawback of the scheme. However, the term is generally a vector with periodic functions of a measurable quantity (position), and proportional to a constant which can be precisely identified.

4.2 Current tracking via energy-shaping

The "attainable" currents $\dot{q}_{ed}$ in the second step of the design procedure are now to be defined.

Proposition 9.18 *If in (9.16) v, $\dot{q}_{ed} \in \mathcal{L}_\infty^{n_e}$ satisfy*

$$M_e v = D_e(q_m)\ddot{q}_{ed} + C_e(q_m, \dot{q}_m)\dot{q}_{ed} + R_{es}(q_m, \dot{q}_m)\dot{q}_{ed} \qquad (9.22)$$

then (see (9.15) for the definition of the current error) $\tilde{\dot{q}}_e \to 0$ as $t \to \infty$ independently of $q_m, \dot{q}_m$ and the choice of $\dot{q}_{ed}$. Furthermore, when $\dot{q}_{ed}$ is bounded then $\dot{q}_m$, $\dot{q}_e$ and τ are also bounded. In addition, boundedness of $\ddot{q}_{ed}$ ensures that $\ddot{q}_e$ and v are also bounded. □

Proof. Rewriting (9.17) in terms of the error signals gives

$$D_e(q_m)\ddot{\tilde{q}}_e + C_e(q_m, \dot{q}_m)\dot{\tilde{q}}_e + R_{es}(q_m, \dot{q}_m)\dot{\tilde{q}}_e = \psi \qquad (9.23)$$

with

$$\psi \triangleq M_e v - [D_e(q_m)\ddot{q}_{ed} + C_e(q_m, \dot{q}_m)\dot{q}_{ed} + R_{es}(q_m, \dot{q}_m)\dot{q}_{ed}] \qquad (9.24)$$

Then (9.22) implies $\psi \equiv 0$, and the dynamics of the system is fully described by

$$D_e(q_m)\ddot{\tilde{q}}_e + C_e(q_m, \dot{q}_m)\dot{\tilde{q}}_e + R_{es}(q_m, \dot{q}_m)\dot{\tilde{q}}_e = 0 \tag{9.25}$$
$$D_m\ddot{q}_m + R_m\dot{q}_m = \tau - \tau_L \tag{9.26}$$

These equations are locally Lipschitz in state, and under the assumptions on the desired torque and load torque, they are continuous in t, so there exists $t_1 > 0$ such that in the time interval $[0, t_1)$ the solutions exists and are unique.

Taking the time derivative of the desired energy function $H_{ed} = \frac{1}{2}\dot{\tilde{q}}_e^\top D_e(q_m)\dot{\tilde{q}}_e$ along the trajectories of (9.23) results in

$$\dot{H}_{ed} = -\dot{\tilde{q}}_e^\top R_{es}(q_m, \dot{q}_m)\dot{\tilde{q}}_e, \ \forall t \in [0, t_1) \tag{9.27}$$

It follows from (9.27) and the proof of Proposition 9.12 that

$$\dot{H}_{ed} = -\dot{\tilde{q}}_e^\top R_{es}(q_m, \dot{q}_m)\dot{\tilde{q}}_e \leq -\alpha\|\dot{\tilde{q}}_e\|^2, \ \forall t \in [0, t_1)$$

and it can be concluded that

$$\|\dot{\tilde{q}}_e(t)\| \leq m_e\|\dot{\tilde{q}}_e(0)\|e^{-\rho_e t}, \ \forall t \in [0, t_1) \tag{9.28}$$

holds with $m_e = \sqrt{\frac{\overline{\lambda}(D_e(q_m))}{\underline{\lambda}(D_e(q_m))}}$ and $\rho_e = \frac{\alpha}{\overline{\lambda}(D_e(q_m))}$, where α is defined in (9.20). These two constants are independent of t_1.

From this, and since $\dot{q}_{ed} \in \mathcal{L}_\infty^{n_e}$, it can be deduced that $\dot{q}_e$ is bounded on the open interval $[0, t_1)$. Now, it must be proved that it remains bounded also on the *closed* interval $[0, t_1]$. To this end, notice that the right hand side of (9.26) is also bounded on $[0, t_1)$, thus its solution can not grow faster than an exponential, and consequently $\dot{q}_m, q_m$ remain bounded $\forall t \in [0, t_1]$. This in its turn ensures the boundedness of R_{es}, and consequently $\|\dot{\tilde{q}}_e\|$ cannot escape to infinity in this time interval.

Since m_e, ρ_e and α are independent of t_1, it is possible to repeat this argument for a new initial condition, to define solutions on the time interval $[t_1, 2t_1]$. This shows that it is possible to extend this procedure to prove existence of solutions for the whole real axis, and the system cannot have finite escape time. It follows from (9.28) that

$$\lim_{t \to \infty} \dot{\tilde{q}}_e = 0$$

and it can be concluded that $\dot{q}_e$ is bounded, which implies that τ is also bounded. From this, and the definition of the load torque (9.5), it follows that $\dot{q}_m$ remains bounded. Now, it follows from (9.25) and (9.24) that $\ddot{q}_e$ and v will be bounded if $\ddot{q}_{ed}$ is also bounded. ∎

4.3   From current tracking to torque tracking

It is now convenient to take a brief respite and recapitulate the previous derivations. The first step of the design procedure described in Section 3.2 was carried out in Section 4.1, where an inner control loop was designed to ensure that (9.17) defines a strictly passive mapping $v \mapsto \dot{q}_s$. In Section 4.2 the second step was carried out, that is, a relationship between the control signals v and $\dot{q}_{ed}$ (9.22) which implies $\lim_{t \to \infty} \|\dot{q}_e - \dot{q}_{ed}\| = 0$, was established. The third step, to which this section is devoted, demands the definition of $\dot{q}_{ed}$ from the "attainable" set that delivers the desired torque, and the establishment of conditions under which current tracking implies torque tracking. Notice that these steps are not straightforward since D_e, C_e, R_{es} in (9.22) and τ in (9.8), depend on q_m and $\dot{q}_m$, thus some additional conditions on the couplings between the subsystems must be satisfied. These conditions are expressed in terms of restrictions on the parameters $D_e(q_m), R_e(q_m, \dot{q}_m), \mu(q_m), \eta(q_m)$ of the generalized electric machine model.

A   Desired current behavior

Motivated by (9.8) it is proposed to define $\dot{q}_{ed}$ such that for the given reference torque τ_*, the equation

$$\tau_* = \frac{1}{2}\dot{q}_{ed}^\top W_1(q_m)\dot{q}_{ed} + W_2(q_m)\dot{q}_{ed} + \eta(q_m) \tag{9.29}$$

holds. This gives, using (9.8)

$$\tau - \tau_* = \frac{1}{2}\dot{\tilde{q}}_e^\top W_1(q_m)\dot{\tilde{q}}_e + \dot{\tilde{q}}_e^\top W_1(q_m)\dot{q}_{ed} + W_2^\top(q_m)\dot{\tilde{q}}_e$$

with the error signal $\dot{\tilde{q}}_e \triangleq \dot{q}_e - \dot{q}_{ed}$. Since $W_1(q_m)$ and $W_2(q_m)$ are bounded, it follows that asymptotic *torque tracking* will be achieved if $\lim_{t \to \infty} \|\dot{q}_e - \dot{q}_{ed}\| = 0$, with $\dot{q}_{ed} \in \mathcal{L}_\infty^{n_e}$, can be ensured.

- The PBC is implicitly characterized by (9.22) and (9.29). To attain the torque tracking objective we must now solve these equations explicitly and furthermore ensure that the resulting $\dot{q}_{ed}$ is bounded.

Towards this end, first notice that in the case of *fully actuated* machines $M_e = I_{n_e}$, thus there are no restrictions on the set of "attainable" currents. That is, in this case, *for any* given $\ddot{q}_{ed}, \dot{q}_{ed}$ we simply plug in these functions in (9.22) to calculate v (hence making $\psi \equiv 0$). The only remaining question is then the solution of (9.29) for a given $\tau_*, \dot{\tau}_*$. In Section 6 we show the procedure for the synchronous and the PM stepper motors.

Now, in the more difficult case of *underactuated* machines there are not enough control actions to directly set $\psi \equiv 0$ for any given $\dot{q}_{ed}$. It will be shown in the sequel that the well known BP transformation conditions –which are fundamental in the analysis of rotating machines [157, 179]– will allow us to get an explicit expression for (9.22) and (9.29).

B   Decoupling conditions

The following definition is in order:

Definition 9.19 (BP transformable machine.) *The machine (9.6)–(9.8) is BP transformable if there exists a coordinate transformation for the current of the form*

$$\dot{z}_e(q_m, \dot{q}_e) \;=\; P(q_m)\dot{q}_e \;=\; P_1 \; e^{-U q_m} \dot{q}_e \tag{9.30}$$

such that the dynamics of Σ_e in these coordinates are independent of q_m (but still dependent on $\dot{q}_m$). P_1 is any nonsingular constant matrix. If furthermore the matrix U is of the form

$$U = \begin{bmatrix} 0 & 0 \\ 0 & U_{22} \end{bmatrix} = -U^{\top} = \begin{bmatrix} 0 & 0 \\ 0 & -U_{22}^{-1} \end{bmatrix} \in SS(n_e)$$

then the machine is strongly BP transformable.

¿From the structure of the matrix U above it can be seen that in strongly BP transformable machines we can remove the dependence on q_m of Σ_e by rotating *only the rotor variables*. As will become clear later, this condition is needed when the rotor circuits are not actuated, as in the induction motor case.

In the fundamental paper [157] necessary and sufficient conditions for BP transformability are given. Since the definition of the BP transformation given above is slightly different from the one given in [157], and for the sake of self–containment, a simplified version of their theorem is given below. A proof can be found in Section 3 of Appendix **D**.

Proposition 9.20 *If there exists a constant matrix $U \in \mathbb{R}^{n_e \times n_e}$ such that*

$$UD_e(q_m) - D_e(q_m)U \;=\; W_1(q_m) \tag{9.31}$$

$$R_e U \;=\; U R_e \tag{9.32}$$

$$UW_2(q_m) \;=\; \frac{dW_2(q_m)}{dq_m} \tag{9.33}$$

then the machine (9.6)–(9.8) is BP transformable. In this case the dynamics of Σ_e (see (9.6)) in the coordinates $\dot{z}_e$ is described by

$$D_e(0)P_1^{-1}\ddot{z}_e + UD_e(0)P_1^{-1}\dot{q}_m\dot{z}_e + W_2(0)\dot{q}_m + R_e P_1^{-1}\dot{z}_e \;=\; e^{-U q_m} M_e u = M_e u' \tag{9.34}$$

while the dynamics of Σ_m (see (9.7)) is described by

$$D_m \ddot{q}_m + R_m \dot{q}_m = \tau - \tau_L \tag{9.35}$$
$$\tau = \dot{z}_e^\top P_1^{-\top} U D_e(0) P_1^{-1} \dot{z}_e + W_2^\top(0) P_1^{-1} \dot{z}_e + \eta(q_m)$$

□

Example 9.21 (Park's transformation.) For a class of electric machines with inductance matrix as in (9.11), the BP transformation to the $dq0$-frame is given as [140]

$$P(q_m) = p_1 \begin{bmatrix} \cos(n_p q_m) & \cos(n_p q_m - \frac{2\pi}{3}) & \cos(n_p q_m + \frac{2\pi}{3}) \\ \sin(n_p q_m) & \sin(n_p q_m - \frac{2\pi}{3}) & \sin(n_p q_m + \frac{2\pi}{3}) \\ \frac{1}{2} & \frac{1}{2} & \frac{1}{2} \end{bmatrix} \tag{9.36}$$

where $p_1 = 2/3$ (or $\sqrt{2/3}$) is a constant. This transformation can also be written as

$$P(q_m) = P_1 \, e^{-U q_m}$$

where

$$U = \frac{n_p}{\sqrt{3}} \begin{bmatrix} 0 & -1 & 1 \\ 1 & 0 & -1 \\ -1 & 1 & 0 \end{bmatrix} \qquad P_1 = p_1 \begin{bmatrix} 1 & -\frac{1}{2} & -\frac{1}{2} \\ 0 & -\frac{\sqrt{3}}{2} & \frac{\sqrt{3}}{2} \\ \frac{1}{2} & \frac{1}{2} & \frac{1}{2} \end{bmatrix}$$

and U satisfies (9.31)-(9.33). The inverse transformation for $p_1 = 2/3$ is given as

$$P^{-1}(q_m) = e^{U q_m} P_1^{-1} = \begin{bmatrix} \cos(n_p q_m) & \sin(n_p q_m) & 1 \\ \cos(n_p q_m - \frac{2\pi}{3}) & \sin(n_p q_m - \frac{2\pi}{3}) & 1 \\ \cos(n_p q_m + \frac{2\pi}{3}) & \sin(n_p q_m + \frac{2\pi}{3}) & 1 \end{bmatrix} \tag{9.37}$$

Several slightly different forms can be found in the literature, [80]. This stems from the choice of the constant factor p_1, which is sometimes chosen to preserve power in the transformed phases ($p_1 = 2/3$), or with the objective of making P_1 orthogonal ($p_1 = \sqrt{\frac{2}{3}}$). □

C   Remarks to the BP transformation

Remark 9.22 For the purpose of the present work, the key feature of BP transformable machines is that it allows us to get, via a kind of system inversion, an explicit expression for the PBC (9.22) and (9.29). Notice also that, for constant speed, the electrical subsystem in (9.34) is linear and time-invariant when u' is taken as the new input. This fundamental property has been exploited in the literature to determine stability properties in stationary operation [279].

Remark 9.23 The underlying fundamental assumption for the machine to be BP transformable, is that the windings are sinusoidally distributed [292], giving a sinusoidal air-gap magnetomotive force (MMF) and sinusoidally varying elements in the inductance matrix $D_e(q_m)$. For a practical machine, this means that the first harmonic in a Fourier approximation of the MMF must give a sufficiently close approximation of the real MMF. Examples of machines in which higher order harmonics must be taken into account, are the square wave brushless DC motors in [184], and machines with significant saliency in the air gap [263]. The squirrel-cage induction machine is an example of a machine where the squirrel-cage rotor with non-sinusoidally distributed MMF is replaced by an equivalent fictitious sinusoidally wound rotor for analytical purposes, without introducing detrimental effects to controller design.

Remark 9.24 It is interesting to remark that the BP transformation can not be derived from a *canonical transformation* [97] $z = Z(q)$ of the generalized coordinates and momenta. This fact is presented in Section 3 of Appendix **D**.

Remark 9.25 Since the matrix U is real and skew-symmetric, it follows that $\mathbf{e}^{-Uq_m}$ is an orthogonal transformation, and the transformation $P(q_m)$ is bounded.

Remark 9.26 It is worth to point out that the passivity properties, being input-output properties, are invariant under a change of coordinates on the tangent bundle of the configuration manifold, hence they are preserved for the transformed systems Σ_e and Σ_m. This can be proved by evaluating the time derivate of

$$H(\dot{z}_e, q_m) \;=\; \frac{1}{2}\dot{z}_e^\top P_1^{-\top} D_e(0) P_1^{-1}\dot{z}_e + \mathcal{V}(q_m)$$

along the trajectories of (9.34), and from the fact that the transfer function between $\tau - \tau_L$ and $\dot{q}_m$ is positive real, and the mapping is passive. Thus it is possible to design PBC also for the transformed system as done in [206] for the case of induction motors. See also [77] for further discussion in this respect.

4.4 PBC for electric machines

The property of BP transformability will now be related with the problem of realizability of the implicit PBC (9.22), (9.29). For the sake of clarity we discuss separately underactuated ($n_s < n_e$) and fully actuated machines ($n_s = n_e$).

A   Underactuated machines, $n_s < n_e$

Proposition 9.27 (Desired currents for underactuated machines.) *Assume that the machine (9.6)–(9.8) is strongly BP transformable, $\mu(q_m) = \eta(q_m) \equiv 0$, the (2,1) block of $D_e(q_m)$ is nonsingular, and that $\beta_*(t)$ is a bounded strictly positive twice* *differentiable function with known first and second order derivatives. Under these conditions, the following definition of $\dot{q}_{ed}$ satisfies (9.22) and (9.29) for any given τ_**

$$\dot{q}_{ed} = \begin{bmatrix} \dot{q}_{sd} \\ \dot{q}_{rd} \end{bmatrix} = \begin{bmatrix} (D_e(q_m))_{21}^{-1} \left[I_{n_r} + (D_e(q_m))_{22} \left\{ \frac{1}{\beta_*^2(t)} \tau_* U_{22}^{-1} + \frac{\dot{\beta}_*(t)}{\beta_*(t)} R_r^{-1} \right\} \right] \lambda_{rd} \\ -\left[\frac{1}{\beta_*^2(t)} \tau_* U_{22}^{-1} + \frac{\dot{\beta}_*(t)}{\beta_*(t)} R_r^{-1} \right] \lambda_{rd} \end{bmatrix}$$

$$(9.38)$$

where λ_{rd} is the solution of the differential equation

$$\dot{\lambda}_{rd} = \frac{1}{\beta_*^2(t)} \left[\tau_* R_r U_{22}^{-1} + \dot{\beta}_*(t)\beta_*(t) \right] \lambda_{rd} \tag{9.39}$$

with initial conditions such that $\|\lambda_{rd}(0)\| = \beta_*(0)$. Furthermore,

$$\|\lambda_{rd}(t)\| = \beta_*(t), \ \forall t \geq 0$$

$\square$

Proof.   The last statement of the proposition follows immediately by taking the time derivate of $\frac{1}{2}\|\lambda_{rd}\|^2$, substituting (9.39), and using the fact that strong BP transformability implies $R_r U_{22}^{-1} + (R_r U_{22}^{-1})^\top = 0$.

To simplify the notation of the rest of the proof, it is convenient to introduce the *desired flux* $\lambda_d = [\lambda_{sd}^\top, \lambda_{rd}^\top]^\top$ as

$$\lambda_d \triangleq D_e(q_m)\dot{q}_{ed} \tag{9.40}$$

Some simple calculations using (9.24), (9.18) and (9.40) show that[12]

$$\psi_r \equiv 0 \ \Leftrightarrow \ \dot{\lambda}_{rd} + R_r \dot{q}_{rd} = 0 \tag{9.41}$$

Now, notice that *BP transformability* of the machine implies that (9.29) can be rewritten as (see (9.31))

$$\tau_* = \dot{q}_{ed}^\top U D_e(q_m)\dot{q}_{ed}$$

which, in terms of the desired fluxes and currents looks like

$$\tau_* = \dot{q}_{ed}^\top U \lambda_d$$

If further the machine is *strongly* BP transformable then

$$\begin{aligned} \tau_* &= \dot{q}_{rd}^\top U_{22} \lambda_{rd} \\ &= -\dot{\lambda}_{rd}^\top R_r^{-1} U_{22} \lambda_{rd} \end{aligned} \tag{9.42}$$

---

[12]See (9.1) and Remark 9.2 in Section 2.3 for the physical motivation behind this choice of relations between desired fluxes and currents.

where $\psi_r \equiv 0$ and (9.41) has been used in the last equation. From this it can be seen that for (9.29) to hold, λ_{rd} must be defined such that (9.42) always holds. It is straightforward to verify that this is the case when λ_{rd} is defined as in (9.39), using $U_{22}^{-\top} = -U_{22}^{-1}$ (see Definition 9.19), the symmetry of R_r and the fact that $\|\lambda_{rd}(t)\| = \beta_*(t)$.

The proof is completed using (9.41) to obtain $\dot{q}_{rd}$ and the definition of λ_d to calculate $\dot{q}_{sd}$.

∎

The main result for underactuated machines is contained in the proposition below.

Proposition 9.28 (PBC for underactuated machines.) *Consider the machine model (9.6)–(9.8). Assume that the machine is strongly BP transformable (Definition 9.19), $\mu(q_m) = 0$, $\eta(q_m) = 0$, $(D_e(q_m))_{21}$ is nonsingular and* **A9.1**–**A9.3** *of Proposition 9.12 hold. Under these conditions, there exists a dynamic output feedback controller that ensures global asymptotic torque tracking with internal stability. Furthermore, for all $\beta_*(t)$ (strictly positive bounded twice differentiable with known bounded first and second order derivatives) the rotor flux λ_r satisfies $\lim_{t\to\infty} |\,\|\lambda_r\| - \beta_*(t)\,| = 0$.*

□

Proof. The control law is obtained from (9.16) and (9.24), setting $\psi_s \equiv 0$. To this end, the definition of $\dot{q}_{ed}$ in Proposition 9.27 is used. Notice that $\ddot{q}_{ed}$ is bounded, and can be computed from the available measurements provided $\dot{\tau}_*$ is known.

Convergence of $\dot{\tilde{q}}_e \to 0$ follows from the arguments of Section 4.2. Boundedness of $\dot{q}_{ed}$ follows from (9.38) and the boundedness of λ_{rd} and τ_*. This establishes asymptotic torque tracking.

Electrical rotor flux norm tracking is a consequence of the convergence of the currents to their desired values and $\beta_*(t) = \|\lambda_{rd}\|$, since

$$\lambda_{rd} - \lambda_r = (D_e(q_m))_{21}(\dot{q}_{sd} - \dot{q}_s) + (D_e(q_m))_{22}(\dot{q}_{rd} - \dot{q}_r)$$

and $D_e(q_m)$ is bounded.

∎

B Fully actuated machines, $n_s = n_e$

For fully actuated machines $M_e = I_{n_e}$, and as previously explained, $\psi \equiv 0$ can be obtained by a suitable selection of v for given $\dot{q}_{ed}$ and $\ddot{q}_{ed}$. The main difficulty is to find $\dot{q}_{ed}$ such that (9.29) is satisfied. This is done by choosing the desired currents from the BP transformed torque equation, since the matrices relating the transformed currents and the torque are no longer dependent on q_m, which considerably simplifies the choice.

Proposition 9.29 (PBC for fully actuated machines.) *Consider the machine model (9.6)–(9.7) and (9.8). Assume the machine is BP transformable (Definition 9.19), and* **A9.1–A9.3** *of Proposition 9.12 hold. Let the desired currents and their derivatives be defined as*

$$\dot{q}_{ed} = \mathbf{e}^{\,Uq_m} P_1^{-1} \dot{z}_{ed}$$

$$\ddot{q}_{ed} = U \, \mathbf{e}^{\,Uq_m} P_1^{-1} \dot{q}_m \dot{z}_{ed} + \mathbf{e}^{\,Uq_m} P_1^{-1} \ddot{z}_{ed}$$

where $\dot{z}_{ed}$ is chosen to satisfy

$$\tau_* - \eta = \dot{z}_{ed}^\top P_1^{-\top} U D_e(0) P_1^{-1} \dot{z}_{ed} + W_2^\top(0) P_1^{-1} \dot{z}_{ed} \qquad (9.43)$$

with $\dot{z}_{ed}, \ddot{z}_{ed} \in \mathcal{L}_\infty^{n_e}$. Under these conditions, use of the dynamic output feedback controller defined in (9.16), with

$$v = D_e(q_m)\ddot{q}_{ed} + C_e(q_m, \dot{q}_m)\dot{q}_{ed} + R_{es}(q_m, \dot{q}_m)\dot{q}_{ed} \qquad (9.44)$$

will ensure global asymptotic torque tracking with internal stability. □

Proof.   The expression for the torque in the transformed system is, according to (9.35)

$$\tau - \eta = \dot{z}_e^\top P_1^{-\top} U D_e(0) P_1^{-1} \dot{z}_e + W_2^\top(0) P_1^{-1} \dot{z}_e$$

Setting $\tilde{z}_e = \dot{z}_e - \dot{z}_{ed}$ and using (9.43) gives

$$\tau - \tau_* = \tilde{z}_e^\top P_1^{-\top} U D_e(0) P_1^{-1} \tilde{z}_e + 2\tilde{z}_e^\top P_1^{-\top} U D_e(0) P^{-1} \dot{z}_{ed} + W_2^\top(0) P_1^{-1} \tilde{z}_e$$

Since $\tilde{z}_e = P(q_m)\tilde{\dot{q}}_e$, and $P(q_m)$ is a bounded transformation, it follows that

$$\lim_{t\to\infty} \tilde{\dot{q}}_e = 0 \qquad \Leftrightarrow \qquad \lim_{t\to\infty} \tilde{z}_e = 0$$

$$\Downarrow$$

$$\lim_{t\to\infty} \tau = \tau_*$$

It is clear that $\dot{z}_{ed} \in \mathcal{L}_\infty^{n_e} \Leftrightarrow \dot{q}_{ed} \in \mathcal{L}_\infty^{n_e}$, and $\ddot{z}_{ed} \in \mathcal{L}_\infty^{n_e}, \dot{q}_m \in \mathcal{L}_\infty \Rightarrow \ddot{q}_{ed} \in \mathcal{L}_\infty^{n_e}$. Defining v as in (9.44) gives $\psi = 0$, and the arguments of Section 4.2 hold. ■

C   Remarks to the controllers

Remark 9.30 As can be seen from (9.38) there is a need for the derivative of the torque reference for the computation of $\ddot{q}_{ed}$. This may in its turn lead to the need of $\ddot{q}_m$, if the speed $\dot{q}_m$ is used in an outer loop controller. It will be explained in Section **10**.3 and Chapter **12** how this need for acceleration measurement can be avoided by use of linear filters or nonlinear observers.

Remark 9.31 Notice that the assumption in Proposition 9.27 of the (2,1) block of D_e being nonsingular implies that the number of actuated and nonactuated windings must be equal. This is the case for typical induction motors, where this matrix is a nonsingular rotation matrix.

Remark 9.32 Equation (9.39) has a solution[13] of the form

$$\lambda_{rd}(t) \;=\; \frac{\beta_*(t)}{\beta_*(0)}\, \mathrm{e}^{\,U_{22}^{-1}\rho_d(t)}\lambda_{rd}(0), \;\; \dot{\rho}_d(t) = \frac{r_r}{\beta_*^2(t)}\tau_*(t), \rho_d(0) = 0$$

This gives an interpretation of the desired flux in terms of its rotation angle, whose speed (*the desired slip*) is related to the desired torque.

Remark 9.33 During the derivation of the model and controller, it has been assumed that a VSI has been used to generate the inputs to the actuated windings. If the inverter used is a CSI or a VSI with fast current control, it follows that the inputs to the actuated windings will be the currents $\dot{q}_s \equiv \dot{q}_{sd}$, where $\dot{q}_{sd}$ is the vector of desired currents for the actuated windings, as defined in the previous sections. Given their great practical importance these so–called current–fed machines will be studied in great detail in Chapter **11**.

Remark 9.34 The quadratic form in (9.43) is in general not easy to solve for the components of $\dot{z}_{ed}$. Examples of solutions for certain machines are given in Section 6.

Remark 9.35 For 3ϕ machines the currents of the transformed system are usually denoted with subscripts d (direct-axis component), q (quadrature-axis component) and 0 (zero-sequence component). For machines in which the symmetrical windings has an isolated neutral, the zero sequence of the transformed currents is exactly zero, which defines a natural choice for desired value of this current.

Remark 9.36 An interesting connection between the PBC described above and the field-oriented approach, that we briefly review in Section **10**.5, may be established as follows. The transformed torque equation is generally given as

$$\tau \;=\; c\{\lambda_d \dot{z}_q - \lambda_q \dot{z}_d\}$$

where c is a constant, and λ_d, λ_q are d and q components of the transformed flux vector. If $\lambda_q \dot{z}_d$ can be made equal to zero and λ_d is constant, it will be possible to control the torque by specifying $\dot{z}_q$. This is the basic idea of the field-oriented approach. Notice that in this case the angle of the transformation is known, and there is no need to estimate it.

---

[13]Recall that if the matrices $A(t)$ and $\int_0^t A(s)ds$ commute, then the differential equation $\dot{x}(t) = A(t)x(t)$, $x(0) = x_0$ $t \geq 0$, has the solution $x(t) = \mathrm{e}^{\int_0^t A(s)ds}x_0$, see pp. 595–596 in [114].

5   PBC of underactuated electrical machines revisited

In this section we revisit the PBC derived above for underactuated electrical machines from the perspective of systems invertibility, a concept which is well understood in the geometric formulation of nonlinear control theory [110, 197]. In particular, we show that the BP transformability assumption needed for the realizability of the PBC controller is akin –but not equivalent– to an invertibility condition. For ease of presentation we will restrict ourselves here to the case when there are no permanent magnets, hence $\mu(q_m) = \eta(q_m) = \mathcal{V}(q_m) = 0$. Furthermore, we choose β_* to be constant.

5.1   Realization of the PBC via BP transformability

For the purposes of this section it is convenient to express the electrical dynamics Σ_e of the machine (9.6) and the torque (9.8) in terms of the fluxes defined as (9.1). Hence, recalling (9.9) of Remark 9.3 we can write (9.6) and (9.8) as

$$\dot{\lambda} + R_e D_e^{-1}(q_m)\lambda = M_e u \tag{9.45}$$
$$\tau = \frac{1}{2}\lambda^\top C_1(q_m)\lambda$$

where we have defined

$$C_1(q_m) \triangleq D_e^{-1}(q_m)W_1(q_m)D_e^{-1}(q_m)$$

The PBC of Proposition 9.28 can be written in the form

$$M_e u = \dot{\lambda}_d + R_e D_e^{-1}(q_m)\lambda_d + \begin{bmatrix} K_1(q_m, \dot{q}_m) & 0 \\ 0 & 0 \end{bmatrix} D_e^{-1}(q_m)\tilde{\lambda} \tag{9.46}$$
$$\tau_d = \frac{1}{2}\lambda_d^\top C_1(q_m)\lambda_d \tag{9.47}$$

where $\tilde{\lambda} \triangleq \lambda - \lambda_d$.

The first equation of the PBC (9.46) is a "copy" of the motor dynamics (9.45) with an additional damping injection term needed to get the strict passivity. The second equation (9.47) is a constraint that "clamps" the controller dynamics. As explained in Section 4.3, the motivation for this constraint stems from the fact that the output to be controlled, that is τ, can be written in terms of the errors as

$$
\begin{aligned}
\tau &= \frac{1}{2}\lambda^\top C_1(q_m)\lambda \\
&= \frac{1}{2}(\tilde{\lambda}+\lambda_d)^\top C_1(q_m)(\tilde{\lambda}+\lambda_d) \\
&= \frac{1}{2}\lambda_d^\top C_1(q_m)\lambda_d + \frac{1}{2}\tilde{\lambda}^\top C_1(q_m)(\tilde{\lambda}+2\lambda_d) \\
&= \tau_* + \frac{1}{2}\tilde{\lambda}^\top C_1(q_m)(\tilde{\lambda}+2\lambda_d)
\end{aligned}
$$

where we have replaced (9.47) to get the last identity. Noting that $C_1(q_m)$ is bounded, and recalling that $\tilde{\lambda} \to 0$, we see from the expression above that $\tau \to \tau_*$ provided λ_d is bounded.

The PBC (9.46), (9.47) is given in *implicit* form, to get an *explicit* realization some assumptions on the system Σ_e are needed, a stage that we called "from current tracking to torque tracking" in Section 4.3 above. To solve the problem we imposed the condition of BP transformability of the motor (Definition 9.19), which is a condition on $D_e(q_m)$ and R_r that we use to reduce the expression for τ_* to the form (9.42). Applied to the machine itself we get

$$
\tau = -\dot{\lambda}_r^\top R_r^{-1} U_{22}\lambda_r
$$

where U_{22} is skew–symmetric. This is a key property since it allows us to explicitly *solve* this equation as

$$
\dot{\lambda}_r = \frac{\tau}{\|\lambda_r\|^2}R_r U_{22}^{-1}\lambda_r \tag{9.48}
$$

(Recall that the BP transformability condition ensures, via (9.32), that R_r and U_{22} commute). Applying these derivations to the controller equations (9.46), (9.47), we obtain the controller dynamics (9.39), which for the case $\beta_* = 0$ reduces to

$$
\dot{\lambda}_{rd} = \frac{\tau_*}{\beta_*^2}R_r U_{22}^{-1}\lambda_{rd}
$$

where we recall that β_* and τ_* are the reference values for $\|\lambda_r\|$ and τ, respectively. In this case, boundedness of λ_{rd} is guaranteed, because $\|\lambda_{rd}(t)\| = \beta_*$. The calculation of the controller equations proceeds then as follows. Once we have the explicit expressions for λ_{rd} we can calculate $\dot{q}_{rd}$ from the last $n_e/2$ equations of (9.46). From here, and $\lambda_d = D_e(q_m)\dot{q}_{ed}$, we can calculate λ_{sd} and, upon *differentiation*, obtain $\dot{\lambda}_d$ which we replace in the remaining controller equations (9.46) to get u.

In summary what we needed to solve the controller equations for the machines was an *invertibility* assumption. Indeed, the "output" equation $\tau = h_1(\lambda, q_m)$ was first expressed as $\tau = h_2(\lambda_r, \dot{\lambda}_r)$, and then we inverted this function to get $\lambda_r = h_3(\lambda_r, \tau)$, where $h_i(\cdot)$, $i = 1, 2, 3$ are suitably defined functions.

Remark 9.37 (Periodic zero dynamics.) The equation (9.48) shows that for this class of machines, if we choose as outputs τ and $\|\lambda_r\|$, the zero dynamics are *periodic*. See also Remark 9.32. We will comment further on this property for the case of induction machines in the next chapter.

5.2 A geometric perspective

The invertibility condition above is akin to the condition of full rank of the decoupling matrix in standard geometric control [110]. Actually we can approach the problem of realization of our PBC from that perspective. First, we need to complete an output vector

$$y = h(\lambda, q_m) \;=\; [\tau, \|\lambda_r\|, h_3(\lambda, q_m), \cdots, h_{n_s}(\lambda, q_m)]^\top \tag{9.49}$$

with respect to which the system (9.46) has a well–defined relative degree $(r_1, \cdots, r_{n_s})$. Then, the system must be transformed to the normal form

$$\begin{aligned}
\dot{\zeta} &= q(\zeta, Y^{(k)}, q_m) + p(\zeta, Y^{(k)}, q_m)u \\
y_i^{(r_i)} &= a_i(\zeta, Y^{(k)}, q_m) + b_i^\top(\zeta, Y^{(k)}, q_m)u
\end{aligned}$$

where $Y^{(k)}$ denotes a vector containing as many derivatives of the output as needed. The relative degree condition ensures that the matrix

$$B(\zeta, Y^{(k)}, q_m) \stackrel{\triangle}{=} \left[\begin{array}{c} b_1^\top(\zeta, Y^{(k)}, q_m) \\ \vdots \\ b_{n_s}^\top(\zeta, Y^{(k)}, q_m) \end{array} \right]$$

is, at least locally invertible. The PBC will then be defined as

$$\begin{aligned}
\dot{\zeta}_d &= q(\zeta_d, Y_*^{(k)}, q_m) + p(\zeta_d, Y_*^{(k)}, q_m)u \\
u &= B^{-1}(\zeta, Y_*^{(k)}, q_m)\left(\left[\begin{array}{c} y_{1*}^{(r_1)} \\ \vdots \\ y_{n_s*}^{(r_{n_s})} \end{array} \right] - \left[\begin{array}{c} a_1^\top(\zeta_d, Y_*^{(k)}, q_m) \\ \vdots \\ a_{n_s}^\top(\zeta_d, Y_*^{(k)}, q_m) \end{array} \right] \right) + \\
&\quad + \left[\begin{array}{cc} K_1(q_m, \dot{q}_m) & 0 \end{array} \right] D_e^{-1}(q_m)\tilde{\lambda}
\end{aligned}$$

for which we have fixed the output and its derivatives to the reference values $y^{(i)} = y_*^{(i)}$, carried the inversion from the normal form, and added the damping injection.

The requirement of boundedness of the controller state ζ_d will then be related with the stability of the zero dynamics for the given reference outputs. Even though this approach is more systematic, it requires the calculation of the normal form, a step which is usually far from obvious. Furthermore, it is not clear how to define the additional outputs to obtain a square system. It is interesting to note that we have been able to avoid these computations invoking the practically reasonable assumption of BP transformability. One interesting application of this geometric derivation of PBC could be for the extension of these schemes to machines which are *not BP-transformable.*

6 Examples

In this section controllers are derived for some common fully actuated machines. Controller design for underactuated machines will be the issue of subsequent chapters, where the squirrel-cage induction motor will be studied.

Example 9.38 (Synchronous motors.) In the last years, synchronous motors, and in particular permanent magnet motors have become attractive alternatives to induction motors in the low to medium power range [37]. These machines are generally more expensive than induction motors, but have higher efficiency due to the fact that the rotor losses are negligible. This results in reduced size and cooling problems as compared to induction motors. As low price high-energy permanent magnets become available, the market for these machines will increase even more.

The controller given by (9.16), (9.44) with currents satisfying (9.43), can be applied to this type of motor as follows.

Using the transformation given in (9.36)-(9.37) with $p_1 = 2/3$, and the model given in (9.11)-(9.12), the torque can be expressed in new coordinates $\dot{z}_e = [\dot{z}_d, \dot{z}_q, \dot{z}_0]^\top$ as

$$\tau = \frac{3n_p}{2}\{(L_d - L_q)\dot{z}_d\dot{z}_q + \lambda_m\dot{z}_q\}$$

where $L_d = L_{ls} + \frac{3}{2}(L_A + L_B)$, $L_q = L_{ls} + \frac{3}{2}(L_A - L_B)$.

The desired currents are chosen as

$$\dot{z}_{ed} = \begin{bmatrix} 0 \\ \frac{2\tau_*}{3n_p\lambda_m} \\ 0 \end{bmatrix}$$

from which it follows that $\dot{z}_{ed}, \ddot{z}_{ed} \in \mathcal{L}_\infty^3$, whenever $\tau_*, \dot{\tau}_* \in \mathcal{L}_\infty$.

To satisfy (9.21), taking the uncertainty of the resistances into account, K_1 is chosen as

$$K_1 = -\frac{1}{2}W_1(q_m)\dot{q}_m + kI_3, \ k > 0$$

The input is then given from (9.44) and (9.16).

This approach can also be extended to synchronous reluctance motors[14] with the same inductance matrix as in (9.11). In these machines there are no permanent magnets or windings in the rotor, hence $\lambda_m = 0$ and the torque is given as [101]

$$\tau = \frac{3n_p}{2}(L_d - L_q)\dot{z}_d\dot{z}_q$$

from which it follows that one of the desired currents should be constant, and the other proportional to desired torque.

Also, if the synchronous machine has a field winding on the rotor instead of permanent magnets, λ_m will be proportional to the current in the field winding, which is usually chosen to be constant or varying according to a field weakening objective. The choice of the other desired currents could be done as previously explained. □

Example 9.39 (PM stepper motor.) As another example, it will be shown how to apply the proposed controller (9.16), (9.44) with currents satisfying (9.43), to a PM stepper motor.

With the model given in (9.13)-(9.14), the transformation to the dq-frame is [157]

$$P(q_m) = \mathbf{e}^{-Uq_m} = \begin{bmatrix} \cos(N_r q_m) & \sin(N_r q_m) \\ -\sin(N_r q_m) & \cos(N_r q_m) \end{bmatrix}, \quad U = \begin{bmatrix} 0 & -N_r \\ N_r & 0 \end{bmatrix}$$

where U satisfies (9.31)-(9.33). This transformation is orthogonal, and $P^{-1}(q_m) = P^T(q_m)$.

The torque expressed in new coordinates $\dot{z}_e = [\dot{z}_d, \dot{z}_q]^T$ is

$$\tau = K_m\dot{z}_q - K_D\sin(4N_r q_m)$$

and the desired currents can be chosen as

$$\dot{z}_{ed} = \begin{bmatrix} 0 \\ \frac{1}{K_m}\{\tau_* + K_D\sin(4N_r q_m)\} \end{bmatrix}$$

Notice that (9.21) is satisfied for the choice $K_1 = kI_2 > 0$, where $k > 0$, and $\tau_* \in \mathcal{L}_\infty$ implies $\dot{z}_{ed} \in \mathcal{L}_\infty^2$, and boundedness of $\ddot{q}_{ed}$ follows from the boundedness of $\dot{\tau}_*$ and $\dot{q}_m$. The input is then given from (9.44) and (9.16).

It is worth to point out that the controller in [33] can be obtained from the passivity-based approach if it is applied to the full system, without dividing the system into electrical and mechanical parts.

As previously pointed out, the underlying assumption of BP-transformability is the sinusoidally distribution of the MMF. It can be discussed whether this is a good approximation in the case of stepper motors, with concentrated windings, significant air gap saliency and often hybrid rotor constructions. The BP transformation above has however been used in several papers, among them [295] and [25].

□

---

[14]This motor has been proposed as an alternative to other AC machines, see [156].

7 Conclusions

7.1 Summary

In this chapter the output feedback global tracking problem for a generalized electric machine model has been studied. A passivity-based method was used to design the controller in three steps. First, the dynamics of the machine was decomposed as the feedback interconnection of two passive subsystems –electrical and mechanical–. Then, a nonlinear damping was injected to make the electrical subsystem strictly passive. Finally, an energy-shaping controller was designed to make the currents converge exponentially to desired functions, such that the desired torque is generated. The main contribution is the establishment of physically interpretable conditions on the model, such that the method can be successfully applied. To further relax these conditions, it is believed that passivity ideas must be combined with the powerful new dynamic extension techniques for stabilization of nonlinear systems. Some research along these lines for the robotics problem has been reported in [40].

The passivity-based approach gives control schemes which provide global stability results for the closed-loop system. There is no need for observers since unmeasurable states are not used, hence the robustness problems associated with observer-based designs are avoided (e.g. numerical problems from open-loop integrations, unknown parameters). Further, PBC do not introduce singularities, and the need for special precautions to be taken at for example start-up is obviated. The performance of the scheme, as measured with the exponential convergence rate of desired currents (and consequently outputs) to their desired values, can be explicitly derived for each machine using the results in Sections 4.1 and 4.2. It follows that the rate of convergence is restricted by the convergence rate of the unactuated dynamics, i.e. the resistance of the unactuated windings. This is a consequence of our inability to add damping into this dynamics, since the involved states are unmeasurable.

It must be pointed out that the energy properties of the system are invariant under a change of coordinates, and this gives the possibility of controller implementation in a general dq-frame, chosen from the objectives of minimizing computational burden and increasing numerical robustness. This also allows for implementations without measurement of rotor position, if the stator fixed frame is chosen for model representation.

To establish a relationship of the controller in this work to existing schemes, it should be noticed that this control input consists of a nonlinear damping term added to the reference dynamics. Henceforth, it can be classified as an indirect vector control scheme, which is the most widely used implementation of FOC (especially well suited for operation close to zero speed [158]). In particular, for speed control of the induction motor, it is shown in Section 11.3 that PBC exactly reduces to the indirect field-oriented controller under some simplifying assumptions, namely speed regulation with use of a current-fed converter (or high-gain current control), for which the additional problem of stator dynamics is not present. In practice, the assumptions of constant and known parameters will not hold. Resistances will for instance vary due to temperature changes and the skin effect at high frequencies, and inductances will change when magnetic saturation occurs. Some answer to this challenging questions will be given, for the case of induction motors, in Chapters **10–11**.

7.2   Open issues

It is worthwhile to point out that no systematic comparison of all the control approaches discussed in Section 1 has been done for motor control. Intuitively, the schemes seem to have much in common, but to the best of the authors' knowledge, no comparing analysis from the implementation of more than two or three schemes on the same experimental setup has yet been reported. Also, the schemes are not general in the sense that estimation and control are not based on a compact and general model. Equations of dynamics are commonly specialized to a particular machine before any results are derived. A comparison of the above results in a general setting, would be a highly interesting and challenging task.

The problem of combined parameter and state estimation is still an unresolved issue in general nonlinear control theory, and a lot of effort has been devoted to this problem of interlacing controller design with the design of adaptive observers. As an example, the main driving force within the field of induction motor research is to provide a satisfactory solution to the problem of rotor resistance adaptation. Adaptive observers have been designed, and the first results from interlacing these results with output feedback controller design, have been reported for the backstepping scheme and the adaptive feedback linearizing scheme. This problem is still open for the other schemes.

Another issue to be solved in a general setting, is the problem of incorporating additional nonlinearities arising from inherently nonlinear magnetics and actuator saturation. Some results along this line using nonlinear control theory have been reported in [31] for a feedback linearizing controller and in [95] for the PBC, but this is still an active area of research.

The problem of digital implementation of controllers for nonlinear systems is yet another problem. Usually nonlinear analysis starts with a continuous model, and stability results are established for the total continuous system. The controllers are however invariably implemented on digital processors. Today this is commonly done by use of an emulation technique with some *ad hoc* discretization (e.g. ZOH) , under an assumption of "sufficiently short" sampling period. This assumption can be recast in terms of a bound on sampling period relative time constants of the system in the linear case. For nonlinear systems there is no such equivalent, and the performance issues of the resulting systems with nonlinear controllers which have been discretized by use of *ad hoc* methods, are yet not fully understood. As an example, it is not obvious that exact cancelation of dynamics and stable zero dynamics can be achieved with a discretized feedback linearizing controller, when directly applied to a continuous system. Promising first results towards the rigorous design of discrete-time controllers for electric machines, have been been reported for the CSI induction motor in [215] (see Section 11.7 for a presentation of this scheme), and for the synchronous motor in [94]. It is expected that the solution of this problem will lead to the design of discrete nonlinear controllers at *converter level*. This means that the discrete nature of switched converters will be taken advantage of in the controller design, removing the PWM block between controller and converter, and instead directly specifying the switching pattern by the discrete-time controller. Hysteresis controllers for direct control of torque and flux, which specify the converter switching via lookup tables, are now becoming alternatives to field-oriented controllers for industrial use [119]. This principle should be focused on also from the viewpoint of nonlinear control theory, and some interesting results have been reported in [167]

Torque ripple due to current harmonics must be addressed to reduce mechanical vibrations and to obtain higher power efficiency. It is expected that the solution of the above points regarding nonlinear magnetics, unknown parameters and converter switching will significantly reduce this problem.

There have not been reported many results where both speed and flux observers have been interlaced with controller designs, and rigorously analyzed using nonlinear theory. Observers for flux and speed are generally designed separately in an *ad hoc* way, even though a high quality estimate of speed is needed for good flux estimates, and vice versa. This is a problem to be solved for both higher dynamic performance and power efficiency [32]. A natural extension of this problem, is to the design of nonlinear controllers which can be implemented without rotational sensors. This is a challenging nonlinear estimation problem for which no rigorous solution has yet been derived.

Chapter 10

Voltage–fed induction motors

The induction[1] motor, and especially the squirrel-cage induction motor, has tradition-ally been the workhorse of industry, due to its mechanical robustness and relatively low cost. In a wide range of servo applications with high-performance requirements it has now, due to advances in control theory and power electronics, replaced DC and synchronous drives. As a continuation of our studies on torque–control of the generalized machine in Chapter **9**, we address here the problem of passivity–based *speed/position* control of this particularly important machine. The two phase squirrel-cage induction motor model[2] is first given in Section 1. Various equivalent represen-tations, often encountered in the literature are also presented. The speed/position control problem is then formulated in Section 2.

In Section 3 we add to the torque–tracking PBC of Proposition 9.28 an outer-loop speed–position controller for generation of the desired torque signal. A nice feature of this scheme is that it has the nested–loop configuration that is standard in applications. Furthermore, we show in Chapter **11** that, under some suitable modeling assumptions, the PBC reduces to the well–known FOC. Other extensions of the scheme like adaptation and integral action, as well as an important remark for its practical implementation, are also presented.

We have in previous chapters discussed how different PBCs can be designed, based on the choice of desired storage functions for mechanical (Chapter **3**) and electromechanical (Chapter **8**) systems. In Section 4 we illustrate this particularly nice feature for the induction motor. In this case the energy–shaping stage is applied to the full machine model, –instead of just the electrical subsystem–, using as a storage function the total energy of the motor. Although conceptually simpler, this controller requires, in contrast with the previous scheme, the implementation of an observer.

---

[1]The induction motor was invented by Tesla ca. 1887, and a first analysis of its dynamic behaviour was presented some years later in [259].

[2]This is the most commonly used model for control purposes. The derivation of the 2ϕ equations from the 3ϕ-model can be found in any standard textbook on electric machines, e.g. [140, 275].

For both controllers we can prove global asymptotic speed/position tracking in a nested–loop configuration. While for the first PBC we give a complete proof, for the sake of brevity we present only the torque tracking version of the second PBC. In Section 6 we present experimental results for both PBCs.

As pointed out in Chapter **9**, a classical and widely used control strategy for electrical machines in general, and in particular for induction motors, is the principle of field–oriented control (FOC). To underscore the connections and differences with our PBC, we present in some detail this scheme in Section 5. We also discuss in this section the theoretically interesting feedback–linearization strategy. Connections between these three strategies are further elaborated for current–fed machines in Section **11.2.3**.

1 Induction motor model

1.1 Dynamic equations

Under the same assumptions about the physical construction of the machine as in Section **9.2**, the standard two phase $\alpha\beta$–model[3] of an n_p pole pair squirrel-cage induction motor with uniform air-gap has $n_e = 4$, $n_s = n_r = 2$ and electrical parameters

$$D_e(q_m) = \begin{bmatrix} L_s I_2 & L_{sr}\mathrm{e}^{\mathcal{J} n_p q_m} \\ L_{sr}\mathrm{e}^{-\mathcal{J} n_p q_m} & L_r I_2 \end{bmatrix}, \; \eta(q_m) = 0, \mu(q_m) = 0$$

$$R_e = \begin{bmatrix} R_s I_2 & 0 \\ 0 & R_r I_2 \end{bmatrix}, \; M_e = \begin{bmatrix} I_2 \\ 0 \end{bmatrix}, \; \mathcal{J} = \begin{bmatrix} 0 & -1 \\ 1 & 0 \end{bmatrix} = -\mathcal{J}^\top$$

$$\mathrm{e}^{\mathcal{J} n_p q_m} = \begin{bmatrix} \cos(n_p q_m) & -\sin(n_p q_m) \\ \sin(n_p q_m) & \cos(n_p q_m) \end{bmatrix}, \; \mathrm{e}^{-\mathcal{J} n_p q_m} = (\mathrm{e}^{\mathcal{J} n_p q_m})^\top$$

$L_s, L_r, L_{sr} > 0$ are the stator, rotor and mutual inductance, $R_s, R_r > 0$ are stator and rotor resistances. I_2 is the 2×2 identity matrix.

The dynamic equations are derived by direct application of the Euler-Lagrange equations (2.6) as in Section **9.2** with the Lagrangian from (9.4). This results in

$$D_e(q_m)\ddot{q}_e + W_1(q_m)\dot{q}_m\dot{q}_e + R_e\dot{q}_e = M_e u \tag{10.1}$$

$$D_m\ddot{q}_m + R_m\dot{q}_m = \tau(\dot{q}_e, q_m) - \tau_L \tag{10.2}$$

$$\tau(\dot{q}_e, q_m) = \frac{1}{2}\dot{q}_e^\top W_1(q_m)\dot{q}_e \tag{10.3}$$

where

$$W_1(q_m) = \frac{dD_e(q_m)}{dq_m} = \begin{bmatrix} 0 & n_p L_{sr}\mathcal{J}\mathrm{e}^{\mathcal{J} n_p q_m} \\ -n_p L_{sr}\mathcal{J}\mathrm{e}^{-\mathcal{J} n_p q_m} & 0 \end{bmatrix} \tag{10.4}$$

---

[3]In this model the axes for the stator have a fixed position while those corresponding to the rotor are rotating at the rotor (electrical) angular speed.

$\dot{q}_e \overset{\triangle}{=} [\dot{q}_s^\top, \dot{q}_r^\top]^\top = [\dot{q}_{s1}, \dot{q}_{s2}, \dot{q}_{r1}, \dot{q}_{r2}]^\top$ is the current vector with stator and rotor components, $\dot{q}_m$ is the rotor angular velocity. $D_m > 0$ is the rotor inertia. The control signals $u = [u_1, u_2]^\top$ are the stator voltages, τ_L is the external load torque, and $R_m \geq 0$ is the mechanical viscous damping constant.

The flux vector $\lambda \overset{\triangle}{=} [\lambda_s^\top, \lambda_r^\top]^\top = [\lambda_{s1}, \lambda_{s2}, \lambda_{r1}, \lambda_{r2}]^\top$ is related to the current vector $\dot{q}_e$ via

$$\lambda = D_e(q_m)\dot{q}_e \tag{10.5}$$

from which the second of these vector equations

$$\lambda_r = L_{sr}\mathbf{e}^{-\mathcal{J}n_p q_m}\dot{q}_s + L_r\dot{q}_r \tag{10.6}$$

is of particular interest for use in the subsequent analysis.

Also, notice that due to the short circuited windings of the squirrel-cage rotor, the second of the equations in (10.1) is given by

$$\dot{\lambda}_r + R_r\dot{q}_r = 0 \tag{10.7}$$

1.2 Some control properties of the model

A Input–output properties

The model of the induction machine derived above is a particular case of the generalized machine model (9.6)–(9.8) of Chapter **9**. It therefore has the passivity properties which were established in Proposition 9.11 of Section **9.3**. In particular, the model (10.1)–(10.2), may be rewritten in "Newton's second law form" as

$$\underbrace{\mathcal{D}(q)\ddot{q}}_{\text{mass} \times \text{acceleration}} = \underbrace{\begin{bmatrix} -W_1(q_m)\dot{q}_m\dot{q}_e \\ \frac{1}{2}\dot{q}_e^\top W_1(q_m)\dot{q}_e \end{bmatrix} - \mathcal{R}\dot{q} + \mathcal{M}u + \xi}_{\text{sum of forces}}$$

where $\mathcal{D}(q) = \text{diag}\{D_e(q_m), D_m\}$, $\mathcal{R} = \text{diag}\{R_e, R_m\}$, $\dot{q} = [\dot{q}_e^\top, \dot{q}_m]^\top$, $\mathcal{M} = [M_e^\top, 0]^\top$, $\xi = [0, -\tau_L]^\top$.

The second term on the right–hand side of the equation above, corresponds to the dissipation forces, while the last two terms on the right–hand side constitute the external forces. The *cornerstone* of the passivity-based design philosophy is to reveal the *workless forces*, in this case the first term on the right–hand side. This is easily established with the systems total energy

$$\mathcal{T}(q, \dot{q}) = \frac{1}{2}\dot{q}^\top \mathcal{D}(q)\dot{q}$$

which has a rate of change (*the systems work*)

$$\dot{\mathcal{T}} = \dot{q}^\top(-\mathcal{R}\dot{q} + \mathcal{M}u + \xi)$$

Workless forces do not affect the systems energy balance, which results from the integration of the equation above

$$\underbrace{\mathcal{T}(t) - \mathcal{T}(0)}_{\text{stored energy}} = \underbrace{-\int_0^t \dot{q}^\top\mathcal{R}\dot{q}ds}_{\text{dissipated}} + \underbrace{\int_0^t \dot{q}^\top(\mathcal{M}u + \xi)ds}_{\text{supplied}}$$

As a result of this fact, the effect of these forces can, roughly speaking, be disregarded in the stability analysis.

The energy balance equation also proves that the mapping $[u^\top, -\tau_L]^\top \mapsto [\dot{q}_s^\top, \dot{q}_m]^\top$ is passive, with storage function $\mathcal{T}(q, \dot{q})$. Furthermore, as shown in Section 9.3, the motor model can be decomposed as the feedback interconnection of two passive operators with storage functions $\mathcal{T}_e(q_m, \dot{q}_e)$ and $\mathcal{T}_m(\dot{q}_m)$, respectively. These passivity properties, and their corresponding storage functions, will be the basis for the two PBCs to be presented in this chapter.

B Geometric properties

We now exhibit the "invertibility" property of the induction motor model which was presented for a general strictly BP transformable machine in Section 9.5. We recall that this property is essential for obtaining an explicit expression of the torque–tracking PBC.

From (10.3) and (10.4) we see that the torque can be written as

$$\tau = \frac{1}{2}\dot{q}_e^\top W_1(q_m)\dot{q}_e = n_p L_{sr}\dot{q}_s^\top \mathcal{J}e^{\mathcal{J}n_p q_m}\dot{q}_r \qquad (10.8)$$

where the fact that $\mathcal{J}$ and $e^{\mathcal{J}n_p q_m}$ commute ($\mathcal{J}e^{\mathcal{J}n_p q_m} = e^{\mathcal{J}n_p q_m}\mathcal{J}$), and the skew-symmetry of $\mathcal{J}$ ($\mathcal{J}^\top = -\mathcal{J} \Rightarrow x^\top \mathcal{J}x = 0, \ \forall \ x \in \mathbb{R}^2$) has been used. Now, solving (10.6) for $\dot{q}_r$

$$\dot{q}_r = \frac{1}{L_r}\left(\lambda_r - L_{sr}e^{-\mathcal{J}n_p q_m}\dot{q}_s\right) \qquad (10.9)$$

and substituting it into (10.8) gives

$$\tau = n_p\frac{L_{sr}}{L_r}\dot{q}_s^\top \mathcal{J}e^{\mathcal{J}n_p q_m}\lambda_r \qquad (10.10)$$

Finally, (10.6) can be solved for $\dot{q}_s$

$$\dot{q}_s = \frac{1}{L_{sr}}e^{\mathcal{J}n_p q_m}\left(\lambda_r - L_r\dot{q}_r\right)$$

and then substituted into (10.10) to give

$$\tau = -n_p \dot{q}_r^{\mathsf{T}} \mathcal{J} \lambda_r = \frac{n_p}{R_r} \dot{\lambda}_r^{\mathsf{T}} \mathcal{J} \lambda_r \tag{10.11}$$

where $\dot{q}_r = -\frac{1}{R_r} \dot{\lambda}_r$ from (10.7) has been used. This is the key expression that we used in Section **9.5** to "invert" the systems dynamics. In this case (9.48) takes the form

$$\dot{\lambda}_r = \frac{\tau}{\|\lambda_r\|} \frac{R_r}{n_p} \mathcal{J} \lambda_r$$

As pointed out in Remark 9.37, the equation above shows that the zero dynamics of the motor with outputs τ and $\|\lambda_r\|$ are periodic. This fact becomes clearer if we evaluate the angular speed of the rotor flux vector relative the rotor fixed frame (*the slip speed*) as

$$\begin{aligned}
\dot{\rho} &= \frac{d}{dt} \arctan(\frac{\lambda_{r2}}{\lambda_{r1}}) = \frac{1}{1 + (\frac{\lambda_{r2}}{\lambda_{r1}})^2} \frac{\dot{\lambda}_{r2}\lambda_{r1} - \lambda_{r2}\dot{\lambda}_{r1}}{\lambda_{r1}^2} \\
&= \frac{1}{\|\lambda_r\|^2} \dot{\lambda}_r^{\mathsf{T}} \mathcal{J} \lambda_r \\
&= \frac{R_r}{n_p \|\lambda_r\|^2} \tau \tag{10.12}
\end{aligned}$$

From this equation we that if τ and $\|\lambda_r\|$ are fixed to constant values, the rotor flux rotates at a constant speed. This expression also shows that torque can be controlled by controlling rotor flux norm and slip speed.

1.3 Coordinate transformations

In order to highlight some aspects of the machine, the model is sometimes presented in a particular set of coordinates[4]. On the other hand, the use of different motor representations has been a source of confusion. In this section we follow [140] and introduce a general transformation from which we can derive most of the models considered in the control literature.

The induction motor model, previously presented in its natural frames of reference, will now be presented in a frame of reference rotating at an arbitrary speed $\omega_a(t)$. In this model, the natural machine variables (current, voltages, flux linkages) associated with stator and rotor windings, are substituted with dq-variables[5] (direct

---

[4]The importance of coordinate changes was probably first underscored by Copernicus, who pointed out that the planetary motions are better understood from the sun's perspective [57]. A change of coordinates is, of course, also the underlying principle of FOC.

[5]There is no consensus in the electric drives community concerning the names of the different representations, here we follow the one adopted by [140].

and quadrature variables) associated with fictitious windings. Although this model could be written in terms of new current variables, it is common practice to present the model in terms of fictitious stator currents i_{dq} and rotor flux linkages λ_{dq}

$$\lambda_{dq} \triangleq \begin{bmatrix} \lambda_d \\ \lambda_q \end{bmatrix} = e^{-\mathcal{J}(\theta_a - n_p q_m)} \lambda_r, \quad i_{dq} \triangleq \begin{bmatrix} i_d \\ i_q \end{bmatrix} = e^{-\mathcal{J}\theta_a} \dot{q}_s$$

$$u_{dq} \triangleq \begin{bmatrix} u_d \\ u_q \end{bmatrix} = e^{-\mathcal{J}\theta_a} u \tag{10.13}$$

where θ_a is the solution of $\dot{\theta}_a = \omega_a$, $\theta_a(0) = 0$, with ω_a a function to be defined later, depending on each particular choice of reference frame.

Substituting the expression for $\dot{q}_r$ from (10.9) into (10.7) and multiplying by $\frac{L_r}{R_r}$ gives

$$T_r \dot{\lambda}_r + \lambda_r = L_{sr} e^{-\mathcal{J} n_p q_m} \dot{q}_s \tag{10.14}$$

where $T_r \triangleq \frac{L_r}{R_r}$ is the time constant of the rotor dynamics.

Noting that $\lambda_r = e^{\mathcal{J}(\theta_a - n_p q_m)} \lambda_{dq}$, computing its derivate, substituting it into (10.14), rearranging terms, multiplying from the left by $e^{-\mathcal{J}(\theta_a - n_p q_m)}$ and using $i_{dq} = e^{-\mathcal{J}\theta_a} \dot{q}_s$, finally gives

$$T_r \dot{\lambda}_{dq} + T_r(\omega_a - n_p \dot{q}_m) \mathcal{J} \lambda_{dq} + \lambda_{dq} = L_{sr} i_{dq} \tag{10.15}$$

To express the upper two stator equations of (10.1) in this new reference frame, it is started by expressing (10.9) in terms of i_{dq} and λ_{dq}. This expression and the relation $\dot{q}_s = e^{\mathcal{J}\theta_a} i_{dq}$ are then substituted into the stator part of (10.1), written as $\frac{d}{dt}\lambda_s + R_s \dot{q}_s = u$, or equivalently as

$$\frac{d}{dt}\left(L_s \dot{q}_s + L_{sr} e^{\mathcal{J} n_p q_m} \dot{q}_r\right) + R_s \dot{q}_s = u$$

Differentiating, multiplying from the left by $e^{-\mathcal{J}\theta_a}$, substituting for $\frac{d}{dt}\lambda_{dq}$ from (10.15) and rearranging terms, the stator equations can be written

$$\frac{d}{dt} i_{dq} + [\omega_a \mathcal{J} + \gamma I_2] i_{dq} + \frac{L_{sr}}{\sigma L_s L_r}[n_p \dot{q}_m \mathcal{J} - \frac{1}{T_r} I_2] \lambda_{dq} = \frac{1}{\sigma L_s} u_{dq} \tag{10.16}$$

$\sigma = 1 - \frac{L_{sr}^2}{L_s L_r} > 0$ is the total leakage factor of the motor, and $\gamma = \frac{R_s}{L_s \sigma} + \frac{L_{sr}^2}{L_s \sigma L_r T_r}$.

To express the torque in terms of dq-variables, substitution of $\lambda_r = e^{\mathcal{J}(\theta_a - n_p q_m)} \lambda_{dq}$ and $\dot{q}_s = e^{\mathcal{J}\theta_a} i_{dq}$ into (10.10) gives

$$\tau = n_p \frac{L_{sr}}{L_r} i_{dq}^\top \mathcal{J} \lambda_{dq} = n_p \frac{L_{sr}}{L_r}(\lambda_d i_q - \lambda_q i_d) \tag{10.17}$$

Definition of ω_a	Reference frame
0	Stator-fixed frame (ab-frame)
$n_p \dot{q}_m$	Rotor-fixed frame
$n_p \dot{q}_m + \frac{L_{sr}}{T_r} \frac{i_q}{\lambda_d}$ or $n_p \dot{q}_m + \frac{d}{dt} \arctan \left(\frac{\lambda_{r2}}{\lambda_{r1}} \right)$	Rotor-flux frame

Table 10.1: Definitions of $\dot{\theta}_a = \omega_a$.

Some of the most commonly used definitions for the reference angle θ_a of this rotation, are given in Table 10.1.

A choice of reference frame which is of particular interest is the stator fixed frame, where rotor variables are associated with fictitious stationary windings [258]. The model is of special interest because it is widely used in the many implementations of controllers based on *backstepping* [142] and *feedback linearization* [172]. This model is denoted the ab-model and can be derived from (10.15) and (10.16) by setting $\omega_a = 0$, $\theta_a = 0$. Since this model is usually written in state space form instead of the second order Euler-Lagrange form in (10.1)-(10.2), it is of interest to rewrite it as

$$\dot{\lambda}_{dq} = n_p \dot{q}_m \mathcal{J} \lambda_{dq} - \frac{1}{T_r} \lambda_{dq} + \frac{L_{sr}}{T_r} i_{dq}$$

$$\frac{d}{dt} i_{dq} = -\gamma i_{dq} - \frac{L_{sr}}{\sigma L_s L_r} [n_p \dot{q}_m \mathcal{J} - \frac{1}{T_r} I_2] \lambda_{dq} + \frac{1}{\sigma L_s} u_{dq}$$

This model has been used in many works in the control literature, where instead of subscripts d and q, a and b are used. For ease of reference we write the equations above out in detail as

$$\dot{\hat{i}}_a = \frac{L_{sr} R_r}{L_s \sigma L_r^2} \lambda_a + \frac{n_p L_{sr}}{L_s \sigma L_r} \dot{q}_m \lambda_b - \gamma i_a + \frac{1}{L_s \sigma} u_1 \qquad (10.18)$$

$$\dot{\hat{i}}_b = \frac{L_{sr} R_r}{L_s \sigma L_r^2} \lambda_b - \frac{n_p L_{sr}}{L_s \sigma L_r} \dot{q}_m \lambda_a - \gamma i_b + \frac{1}{L_s \sigma} u_2$$

$$\dot{\lambda}_a = -\frac{R_r}{L_r} \lambda_a - n_p \dot{q}_m \lambda_b + \frac{R_r L_{sr}}{L_r} i_a$$

$$\dot{\lambda}_b = -\frac{R_r}{L_r} \lambda_b + n_p \dot{q}_m \lambda_a + \frac{R_r L_{sr}}{L_r} i_b$$

$$\ddot{q}_m = \frac{1}{D_m} (\tau - \tau_L)$$

$$\tau = \frac{n_p L_{sr}}{L_r} (\lambda_a i_b - \lambda_b i_a) \qquad (10.19)$$

where $\lambda_{ab} = [\lambda_a, \lambda_b]^\top$ is the rotor flux, $i_{ab} = [i_a, i_b]^\top = \dot{q}_s$ is the stator current, and it has been assumed that $R_m = 0$. Defining the state vector $x \triangleq [\dot{q}_m, \lambda_{ab}^\top, i_{ab}^\top]^\top$, the model can be rewritten in the classical state–space form

$$\dot{x} = f(x) + Gu \qquad (10.20)$$

with an obvious definition of the vector field $f(x)$ and the constant matrix G with columns g_1, g_2.

The state space ab-model of the induction motor was used in this work only for implementing a simulation model of the induction motor. It is well suited for this purpose, since it has no rotational transformations. It is also written down for the purpose of comparison with the second order Euler-Lagrange model which has been used in this work, since it is believed that this second order model is rather uncommon within the motor control literature. While the EL model has a structure suitable for PBC design, the model in (10.20) is a set of mathematical equations which are well suited for controller design and analysis using tools from geometric control theory.

1.4   Remarks to the model

Remark 10.1 (Squirrel–cage rotor.) For analytical purposes it is common to substitute the squirrel– (single)–cage rotor, which has a uniform conductor distribution, with an equivalent fictitious rotor with the same number of phases as the stator, and sinusoidally distributed conductors. This implies that in the analysis, only the first order harmonic of the rotor MMF is accounted for. Experimental results indicate that analysis and controller designs based on this simplified model will also be valid for the real machine. In cases of deep bar or double–cage rotors, care should however be taken when modeling these with sinusoidally distributed windings [275, 285].

Remark 10.2 (PBC with the dq–model.) It is interesting to remark that in our first works on PBC of induction machines, i.e. [204, 206], we used the model (10.15)–(10.17). We treated ω_a as an additional *input* which we then later integrated to get the change of coordinates (10.13). In control terms: for a control system $\dot{x} = f(x, u)$ we introduced a change of coordinates $z = \Phi(x, \theta_a)$, with θ_a a signal to be defined. Then we designed the PBC based on the system

$$\dot{z} = \frac{\partial \Phi}{\partial x} f + \frac{\partial \Phi}{\partial \theta_a} \dot{\theta}_a = g(z, \theta_a, u, \omega_a)$$

with the "inputs" u and ω_a, and added an integrator to recover the original control signals. This is an interesting idea which, to the best of our knowledge, has not been explored elsewhere.

Remark 10.3 (Flux control.) The rotor flux norm needs to be controlled for system optimization (e.g. power efficiency, torque maximization) during changing operating conditions and under inverter limits [29, 92].

Remark 10.4 (Measured variables.) A common instrumentation of a standard high-performance industrial 3ϕ induction motor drive is the use of two current transducers and one rotational transducer. Currents are often measured using Hall-effect sensors with magnetic compensation, and can give high precision measurements with high bandwidth (DC to 100 kHz) and isolation from measured currents. With knowledge of the switching instants of the inverter, the currents in each of the phases can also be computed from measurement of the DC-side current with a shunt resistor. This allows for current measurement without expensive sensors. Velocity measurement can be expensive, and estimation of rotational speed from position measurement with a high resolution digital incremental encoder can give significantly better results than often noisy analog measurements with DC tachometers [159]. In some rare cases both velocity and position transducers are used, but the most common approach is to use an encoder, and estimate speed by simple numerical differentiation, or by the use of speed observers driven by reference- or estimated torque and updated from discrete position measurements [158].

Since rotational transducers and their associated digital or analogue circuits give extra costs and reduce the mechanical robustness of the total system, there has been an increasing interest in schemes without rotational sensors. In some of these cases, speed is estimated by exploiting the influence of the rotational voltages in the dynamic equations. These methods are parameter sensitive with typically low performance at speeds close to zero. Recently, promising results from sensorless speed and even position estimation have been obtained by modifying rotor slots (introducing magnetic saliencies), and injecting balanced high frequency voltage signals at the terminals [112]. By signal processing of measured voltages and currents in combination with a closed-loop observer, sensorless control is achieved. This field is an area of active research, and some successful implementations have already been reported in the recent survey [224]. The rigorous solution of this *nonlinear estimation problem* is yet to be derived, and drives without rotational sensors suffer from a performance degradation, which hampers their use in high performance speed or position control applications.

Additional voltage transducers are also used, not only for some control schemes without rotational sensors, but also in experimental setups for parameter identification [185]. Signal filtering is then needed, especially with PWM converters. In experimental laboratory setups and in some industrial applications (e.g. ships), torque and input/output power are also sometimes measured using current and voltage transducers on the DC-link, and strain gauge rosettes on the motor shaft.

Many of the nonlinear control schemes are derived under the assumption of the full state being measured. This is rarely the case, since the rotor currents or the fluxes, are not directly available for measurement. Measurement of currents in the squirrel-cage rotor is very difficult. Flux sensors (Hall-sensors, extra sensing coils) require expensive modifications of standard motors, and are not robust to mechanical vibrations and other conditions encountered in rough industrial environments. Also, due to space harmonics, it is very difficult to get good flux estimates by interpolating measurements from a few point sensors. Hence, flux measurement is impractical and against the benefits of the squirrel-cage motor [153]. It is therefore only used in experimental setups, or special purpose installations.

1.5   Concluding remarks

In this section the dynamic equations for an n_p pole-pair squirrel-cage induction motor with smooth air-gap have been presented. The second order model structure follows naturally from the Euler-Lagrange approach for modeling, and is well suited for passivity-based analysis.

As pointed out in Section 9.2.3, there are parameter variations due to heating and magnetic saturation, and these have not been been specified in the model given in this chapter. It is well known that the rotor resistance R_r can vary significantly[6], and ability to compensate for this variation, or at least to guarantee stability despite variations, is of utmost importance for any control design.

For comparison purposes, the more commonly encountered stator fixed ab-model has also been derived. This model is used extensively for analysis with geometric tools, and in a large number of implementations of controllers based on other designs than the passivity-based. For implementation purposes, this choice of model is sometimes preferred since measured quantities and controls need not be rotated to other frames [143]. It is however important to be aware of bandwidth considerations for controllers in such implementations. While currents will be constant in a dq-frame at stationary conditions (constant flux and speed), they will be oscillating with a speed dependent frequency in the stationary frame. The bandwidth must consequently be higher for current controllers if they are implemented in the stationary frame.

It is worthwhile to point out the generality of the model presented in this chapter. It could be interpreted as an equivalent model of the usual 3ϕ machine, or stemming from a reduction of phases in the more general polyphase machine through transformations like those presented in [285]. Notice that depending on the transformation used to go from 3ϕ to 2ϕ, there is a factor appearing in the torque equation, depending on the form (power invariant, non-power invariant, see [275] or [140]) of the transformation used.

2   Problem formulation

We will now formulate the control problem to be solved in this chapter. For ease of presentation we solve the speed control problem, since the extension to position control is straightforward, see Remark 10.8.

---

[6]A change $0.75R_r^N \leq R_r \leq 1.5R_r^N$ is not unreasonable [153].

Definition 10.5 (Speed and rotor flux norm tracking problem.) *Consider the six–dimensional induction motor model (10.1), (10.2) with state vector* $[\dot{q}_e^\top, q_m, \dot{q}_m]^\top$, *inputs the stator voltages* $u \in \mathbb{R}^2$, *regulated outputs speed* $\dot{q}_m$ *and rotor flux norm* $\|\lambda_r\|$. *Assume:*

A 10.1 Stator currents $\dot{q}_s$, rotor speed $\dot{q}_m$ and position q_m are available for measurement.

A 10.2 All motor parameters are exactly known.

A 10.3 The load torque $\tau_L(t)$ is a known[7] bounded function with known bounded first order derivate, such that $|\tau_L(t)| \leq c_1 < \infty$, $\forall t \in [0, \infty)$.

A 10.4 The desired rotor speed $\dot{q}_{m*}(t)$ is a bounded and twice differentiable function with known bounded first and second order derivatives, such that $|\ddot{q}_{m*}(t)| \leq c_2 < \infty$, $\forall t \in [0, \infty)$.

A 10.5 The desired rotor flux norm $\beta_*(t)$ is a strictly positive bounded twice differentiable function with known bounded first and second order derivatives, such that $0 < \delta_1 \leq \beta_* \leq \bar{\beta}_* < \infty$, $\dot{\beta}_* \leq \bar{\dot{\beta}}_* < \infty$ and $\ddot{\beta}_* \leq \bar{\ddot{\beta}}_* < \infty$.

Under these conditions, design a PBC that ensures global asymptotic speed and rotor flux norm tracking, that is,

$$\lim_{t\to\infty} |\dot{q}_m - \dot{q}_{m*}(t)| = 0, \; \lim_{t\to\infty} | \, \|\lambda_r\| - \beta_*(t)| = 0$$

with all internal signals uniformly bounded.

3   A nested–loop PBC

In Chapter **9** we adopted a feedback decomposition approach for the design of a torque tracking PBC. In this section we solve the speed-position tracking problem posed above by adding an outer–loop controller to this torque tracking PBC. This leads to the nested–loop (i.q. cascaded) scheme depicted in Fig. 10.1, where $\mathcal{C}_{il}$ is the inner–loop torque tracking PBC and $\mathcal{C}_{ol}$ is an outer–loop speed controller, which generates the desired torque[8] τ_d. We will show in this section that $\mathcal{C}_{ol}$ may be taken

---

[7]This assumption is made for simplicity, the result can be extended for the case of unknown linearly parameterized load torque as discussed in Remark 10.9.

[8]In the case where the torque is explicitly given as an external reference signal, as in the torque tracking problem, we denote it τ_*. If it can be considered as the output from a controller (which takes for instance the speed reference $\dot{q}_{m*}$ as an external input), we use τ_d.

as an LTI system that asymptotically stabilizes the mechanical dynamics. The main technical obstacle for its design stems from the fact that C_{il} requires the knowledge of $\dot{\tau}_d$, and this in its turn implies measurement of $\ddot{q}_m$. To overcome this obstacle we proceed, as done in Section **3**.2.4.B.2 for the robotics problem, and replace $\ddot{q}_m$ by its approximate differentiation, while preserving the global stabilization property.

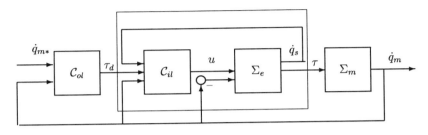

Figure 10.1: Nested-loop control configuration.

A very interesting property of the resulting scheme, which is further elaborated in Chapter **11**, is that if the inverter can be modeled as a current source and the desired speed and rotor flux norm are constant, the controller exactly reduces to the well known indirect FOC, hence providing a solid theoretical foundation to this popular control strategy.

Even though the torque–tracking PBC can be directly obtained from the derivations in Section **9**.4, we will present it in detail for three reasons. First, to underscore the additional difficulties associated with the control of the mechanical subsystem. Second, we show that the damping injection term which was instrumental for the strict passivation in Section **9**.4.1, can be *obviated* if we choose a different storage function. It is however required if we use instead the usual desired electrical energy

$$H_{ed} = \frac{1}{2}\dot{\tilde{q}}_e^{\top} D_e(q_m)\dot{\tilde{q}}_e \tag{10.21}$$

where $\dot{\tilde{q}}_e \overset{\triangle}{=} \dot{q}_e - \dot{q}_{ed}$, and $\dot{q}_{ed}$ is generated by the PBC.

Third, to provide further insight into the philosophy of PBC and highlight its connection with other control approaches, we present the design this time from the "asymptotic inversion" perspective taken in Section **9**.5. Comments to the relation with the general results from Section **9**.4 will be given at appropriate points.

The rest of this section is organized as follows. First, in Section 3.1, we review the derivations of Section **9**.5, as specialized to the induction machine. The main result of this section, a globally defined observer-less speed and rotor flux norm tracking controller, is then given, followed by its proof. It is also shown how this controller can be robustified with an integral term in the stator currents, and how adaptation of stator parameters can be added. A negative result concerning rotor resistance adaptation is then presented in Section 3.6. It is shown that there exists a fundamental structural obstacle in the closed–loop equations for the application of the existing adaptive control theory based on passivity of the error system (or equivalently, on decoupled Lyapunov equations.)

Motivated by practical considerations, a dq-implementation of the scheme is derived in Section 3.7. Examples of how to define the desired rotor flux norm for minimization of power losses and to mimic field weakening are also included, followed by simulation results in Section 3.9. For ease of comparison with the various schemes, all experimental results, including the ones of the nested–loop PBC, have been grouped at the end of this chapter in Section 6.

3.1 A systems "inversion" perspective of the torque tracking PBC

In this section we derive the torque tracking PBC from the perspective of systems "inversion" studied in Section **9.5**. As discussed in that section, this derivation is easier to understand if the flux vector is used in the calculations instead of the current vector. Therefore, using (10.5), we rewrite (10.1), (10.3) as

$$\dot{\lambda} + R_e D_e^{-1}(q_m)\lambda = M_e u \tag{10.22}$$

$$\tau = \frac{n_p}{R_r}\dot{\lambda}_r^\top \mathcal{J}\lambda_r \tag{10.23}$$

where, for ease of reference, we have repeated (10.11).

The PBC of Proposition 9.28 is a "copy" of the electrical dynamics of the motor (10.22) with an additional damping injection term introduced to get strict passivity of the mapping $v \mapsto \dot{q}_s$, see Section **9.4**. In its *implicit* form, this yields

$$M_e u = \dot{\lambda}_d + R_e D_e^{-1}(q_m)\lambda_d + \begin{bmatrix} K_1(\dot{q}_m) & 0 \\ 0 & 0 \end{bmatrix} D_e^{-1}(q_m)\tilde{\lambda} \tag{10.24}$$

$$\tau_* = \frac{n_p}{R_r}\dot{\lambda}_{rd}^\top \mathcal{J}\lambda_{rd} \tag{10.25}$$

where $\tilde{\lambda} \triangleq \lambda - \lambda_d$ and $\lambda_d = \begin{bmatrix} \lambda_{sd}^\top, \lambda_{rd}^\top \end{bmatrix}^\top$. Notice that the damping injection is independent of q_m.

An *explicit* realization is obtained –as explained in Section **9.5**–, by "inversion of" (10.25)

$$\dot{\lambda}_{rd} = \frac{1}{\beta_*^2(t)}\left(\frac{R_r}{n_p}\tau_*\mathcal{J} + \dot{\beta}_*(t)\beta_*(t)I_2\right)\lambda_{rd}, \ \lambda_{rd}(0) = \begin{bmatrix} \beta_*(0) \\ 0 \end{bmatrix}$$

which can actually be solved as

$$\lambda_{rd} = \mathbf{e}^{\,\mathcal{J}\rho_d} \left[\begin{array}{c} \beta_*(t) \\ 0 \end{array} \right] \qquad (10.26)$$

$$\dot{\rho}_d = \frac{R_r}{n_p \beta_*^2(t)} \tau_*, \ \rho_d(0) = 0 \qquad (10.27)$$

The description of the controller is completed by replacement of λ_{rd} and $\dot{\lambda}_{rd}$ in the last two equations of (10.24) to get λ_{sd}. After differentiation we get $\dot{\lambda}_{sd}$ which can be replaced in the first two equations of (10.24) to get[9]

$$u = \dot{\lambda}_{sd} + \left[\begin{array}{cc} I_2 & 0 \end{array} \right] \left(R_s D_e^{-1}(q_m)\lambda_d + \left[\begin{array}{cc} K_1(\dot{q}_m) & 0 \\ 0 & 0 \end{array} \right] D_e^{-1}(q_m)\tilde{\lambda} \right)$$

From the expression above we can see a difficulty for the implementation of the nested–loop scheme of Fig. 10.1. Namely that the control law depends on $\dot{\lambda}_{sd}$, which in its turn will depend on $\dot{\tau}_*$. Notice that the latter was assumed to be known in Proposition 9.27. On the other hand, the signal τ_* will now be generated by an outer–loop controller $\mathcal{C}_{ol}$, which will generally depend on $\dot{q}_m$. We will see in the Proposition 10.6 how to overcome this obstacle by the use of a linear filter.

Let us now analyze the stability of the closed–loop. The error equation for the fluxes is obtained from (10.22) and (10.24) as

$$\dot{\tilde{\lambda}} + \left(R_e + \left[\begin{array}{cc} K_1(\dot{q}_m) & 0 \\ 0 & 0 \end{array} \right] \right) D_e^{-1}(q_m)\tilde{\lambda} = 0$$

Global convergence of $\dot{\tilde{q}}_e$ (and consequently of[10] $\tilde{\lambda}$) to zero was established in Proposition 9.28 (for a suitably defined damping injection term $K_1(q_m, \dot{q}_m)$, see (9.19)) using the storage function (10.21). With this storage function it is possible to prove that the closed–loop equations define an output strictly passive mapping $v \mapsto \dot{\tilde{q}}_s$, where v is some external signal added to the control input. It is important to stress the fact that $\dot{\tilde{q}}_s$ is known since $\dot{q}_s$ is *measurable*.

We will show now that it is also possible to prove convergence, even *without* the damping injection. To this end, consider the storage function[11]

$$H_\lambda = \frac{1}{2}\tilde{\lambda}^\top R_e^{-1}\tilde{\lambda} \geq 0$$

whose derivative satisfies

$$\dot{H}_\lambda = -\tilde{\lambda}^\top D_e^{-1}(q_m)\tilde{\lambda} \leq -\alpha H_\lambda$$

---

[9]An explicit state space description is given in Proposition 10.6.

[10]Recall that $\tilde{\lambda} = D_e(q_m)\dot{\tilde{q}}_e$.

[11]This function was used in [176] to give an "implicit observer" interpretation of the PBC controller.

for some $\alpha > 0$. Hence, $\tilde{\lambda} \to 0$ exponentially fast. It is easy to see that the closed–loop equations define an output strictly passive mapping $v \mapsto \tilde{\lambda}_s$, with storage function H_λ. Notice, however, that this time the output is *not measurable*. We will see in Chapter **12** that this feature of measurable signals is essential for handling the case when the load torque is due to some nonlinear dynamics that we capture in Σ_m, like in robots with AC drives. (It is also required for the implementation of hybrid PBC for direct–torque control as presented in [167].) Even though the analysis above clearly leads to a simpler controller for the problem at hand, it is for its later extension to the case with nonlinear load torques, that we present the controller *with* a damping injection term and the proof based on the storage function (10.21).

To illustrate the second difficulty in the stability analysis of the nested–loop scheme, let us turn our attention to the torque tracking error $\tilde{\tau} \stackrel{\triangle}{=} \tau - \tau_*$. After some simple operations from (10.23) and (10.25) we get

$$
\begin{aligned}
\tilde{\tau} &= \frac{n_p}{R_r} \left\{ \dot{\tilde{\lambda}}_r^\top \mathcal{J} \tilde{\lambda}_r + \dot{\tilde{\lambda}}_r^\top \mathcal{J} \lambda_{rd} + \dot{\lambda}_{rd} \mathcal{J} \tilde{\lambda}_r \right\} \\
&= n_p \left\{ -\dot{\tilde{q}}_r^\top \mathcal{J} \lambda_r + \frac{1}{R_r} \dot{\lambda}_{rd} \mathcal{J} \tilde{\lambda}_r \right\}
\end{aligned}
$$

where we have used $\dot{\tilde{q}}_r = -\frac{1}{R_r} \dot{\tilde{\lambda}}_r$, and the definition of $\tilde{\lambda}_r$ to get the last identity. In the torque control problem we assumed that the external reference τ_* and its derivative $\dot{\tau}_*$ were bounded, and convergence of $\tilde{\tau} \to 0$ was proved as follows. First, λ_{rd} and $\dot{\lambda}_{rd}$ are bounded by construction, see (10.26). Then, we have shown above that $\tilde{\lambda} \to 0$ (exp.), consequently also $\dot{\tilde{q}}_r \to 0$ and λ_r is bounded. From here we conclude that $\tilde{\tau} \to 0$. In position–speed control τ_* and $\dot{\tau}_*$ are not *a priori* bounded, since they will be generated by $\mathcal{C}_{ol}$. Therefore, $\mathcal{C}_{ol}$ must be chosen with care and a new argument should be invoked to complete the proof. Proposition 10.6 below shows that $\mathcal{C}_{ol}$ can be taken as a linear filter.

A   Connection with system inversion

Before closing this section, we will view the PBC from the geometric perspective discussed in Section **9**.5. The purpose of the exercise is to show that if we complete in a suitable manner the "output" vector (9.49), then the standard inversion algorithm of geometric control, –as applied to the *reference signals*–, will give us the PBC controller above. This fact is important for at least two reasons, first because the inversion algorithm is a general *systematic* procedure, while the procedure used above is somewhat *ad hoc*. Second, it provides a clear connection with feedback linearization where the same inversion algorithm is invoked, but now applied to the *output signals*.

To this end, we propose to complete the "output" vector for the system (10.22) as $y = [\tau, \ y_2]^\top$, where

$$y_2 \overset{\triangle}{=} -n_p \lambda_r^\top \begin{bmatrix} 0 & I_2 \end{bmatrix} D_e^{-1}(q_m)\lambda = \frac{n_p}{R_r}\lambda_r^\top \dot{\lambda}_r$$

Recall that τ can be written in the form (10.23).

Following the inversion algorithm we evaluate the decoupling matrix by taking the time derivative of y

$$\dot{y} = \frac{n_p}{R_r} \begin{bmatrix} \lambda_r^\top \mathcal{J} \\ \lambda_r^\top \end{bmatrix} \left(R_e D_e^{-1}(q_m) \right)_{21} u + m(\lambda) \overset{\triangle}{=} G(\lambda_r, q_m)u + m(\lambda)$$

where $m(\lambda)$ is some function of λ. It can be shown that

$$\left(R_e D_e^{-1}(q_m) \right)_{21} = \frac{R_r L_{sr}}{L_s L_r \sigma} e^{\mathcal{J} n_p q_m}$$

which is globally invertible. On the other hand

$$\begin{bmatrix} \lambda_r^\top \mathcal{J} \\ \lambda_r^\top \end{bmatrix}^{-1} = \frac{1}{\|\lambda_r\|^2} \begin{bmatrix} -\mathcal{J}\lambda_r & \lambda_r \end{bmatrix}$$

Consequently, the decoupling matrix $G(\lambda_r, q_m)$ is nonsingular everywhere except when $\|\lambda_r\| = 0$. This implies that the system (10.22) with output y has relative degree $\{1, \ 1\}$. A feedback linearizing controller is chosen as

$$u = G(\lambda_r, q_m)^{-1}[\dot{y}_* - m(\lambda) + v]$$

where v is an additional stabilizing controller. On the other hand, the control signal for the PBC derived above can be obtained by "evaluating the inversion for the reference signals", that is

$$u = G(\lambda_{rd}, q_m)^{-1}[\dot{y}_* - m(\lambda_d)] + u_{di}$$

where u_{di} is the damping injection term and λ_{rd} is obtained from (10.26) and (10.27).

Roughly speaking, we can summarize the discussion above as follows:

While input–output linearization implements a *right inverse* of the system $\Sigma(\lambda, q_m)$, that is

$$u_{FL} \quad = \quad \Sigma^{-1}(\lambda, q_m)(y_* + v),$$

the PBC implements a *left inverse*

$$u_{PBC} \quad = \quad \Sigma^{-1}(\lambda_d, q_m)y_* + u_{di}$$

This interpretation is depicted in Fig. 10.2 below. Notice that, except for the damping injection, the PBC is open–loop in λ. However, the loop is closed with q_m.

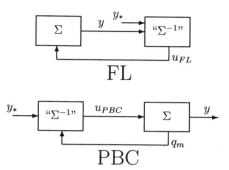

Figure 10.2: Connection with system inversion.

3.2　Observer-less PBC for induction motors

The main result of this section can now be formulated:

Proposition 10.6 (Speed and rotor flux norm tracking.) *The nonlinear dynamic output feedback nested–loop controller*

$$u = \underbrace{L_s\ddot{q}_{sd} + L_{sr}e^{\mathcal{J}n_p q_m}\ddot{q}_{rd} + n_p L_{sr}\mathcal{J}e^{\mathcal{J}n_p q_m}\dot{q}_m\dot{q}_{rd}}_{\dot{\lambda}_{sd}} + R_s\dot{q}_{sd} - K_1(\dot{q}_m)\dot{\tilde{q}}_s \quad (10.28)$$

with

$$\dot{q}_{ed} = \begin{bmatrix} \dot{q}_{sd} \\ \dot{q}_{rd} \end{bmatrix} = \begin{bmatrix} \frac{1}{L_{sr}}\left[(1 + \frac{L_r\dot{\beta}_*}{R_r\beta_*})I_2 + \frac{L_r}{n_p\beta_*^2}\tau_d\mathcal{J}\right]e^{\mathcal{J}n_p q_m}\lambda_{rd} \\ -\left(\frac{\tau_d}{n_p\beta_*^2}\mathcal{J} + \frac{\dot{\beta}_*}{R_r\beta_*}I_2\right)\lambda_{rd} \end{bmatrix} \quad (10.29)$$

where

$$\tilde{\dot{q}}_e = \begin{bmatrix} \tilde{\dot{q}}_s \\ \tilde{\dot{q}}_r \end{bmatrix} \doteq \begin{bmatrix} \dot{q}_s - \dot{q}_{sd} \\ \dot{q}_r - \dot{q}_{rd} \end{bmatrix}$$

$$K_1(\dot{q}_m) \triangleq \frac{n_p^2 L_{sr}^2}{4\epsilon}\dot{q}_m^2 + k_1, \ 0 < \epsilon \leq R_r, \ k_1 \geq 0$$

$$\tau_d = D_m\ddot{q}_{m*} - z + \tau_L \quad (10.30)$$

and controller state equations

$$\dot{\lambda}_{rd} = \left(\frac{R_r}{n_p\beta_*^2}\tau_d\mathcal{J} + \frac{\dot{\beta}_*}{\beta_*}I_2\right)\lambda_{rd}, \ \lambda_{rd}(0) = \begin{bmatrix} \beta_*(0) \\ 0 \end{bmatrix} \quad (10.31)$$

$$\dot{z} = -az + b\tilde{\dot{q}}_m, \ z(0) = \tilde{\dot{q}}_m(0) \quad (10.32)$$

with $\dot{\tilde{q}}_m \triangleq \dot{q}_m - \dot{q}_{m*}$ and $a, b > 0$, provides a solution to the speed and rotor flux norm tracking problem of Definition 10.5. That is, when placed in closed–loop with (10.1)–(10.3) it ensures

$$\lim_{t\to\infty} \dot{\tilde{q}}_m = 0, \ \lim_{t\to\infty} | \, \|\lambda_r\| - \beta_*(t)| = 0$$

for all initial conditions and with all internal signals uniformly bounded. □

Proof. First, notice that (10.1) can be rewritten as

$$D_e(q_m)\ddot{q}_e + C_e(q_m, \dot{q}_m)\dot{q}_e + R(q_m, \dot{q}_m)\dot{q}_e \ = \ M_e u \tag{10.33}$$

where

$$C_e(q_m, \dot{q}_m) \triangleq \begin{bmatrix} 0 & n_p L_{sr} \mathcal{J} e^{\mathcal{J} n_p q_m} \\ 0 & 0 \end{bmatrix} \dot{q}_m$$

$$R(q_m, \dot{q}_m) \triangleq \begin{bmatrix} R_s I_2 & 0 \\ -n_p L_{sr} \mathcal{J} e^{-\mathcal{J} n_p q_m} \dot{q}_m & R_r I_2 \end{bmatrix}$$

By substitution of (10.28) into (10.33) and using (10.7), it follows that the closed loop system is fully described by

$$D_e(q_m)\ddot{\tilde{q}}_e + C_e(q_m, \dot{q}_m)\dot{\tilde{q}}_e + [R(q_m, \dot{q}_m) + \mathcal{K}(\dot{q}_m)]\,\dot{\tilde{q}}_e = 0 \tag{10.34}$$

$$D_m\ddot{\tilde{q}}_m = -z + \tau(\dot{q}_e, q_m) - \tau_d \tag{10.35}$$

$$\dot{z} = -az + b\dot{\tilde{q}}_m \tag{10.36}$$

$$\dot{\lambda}_{rd} = \left(\tfrac{R_r}{n_p \beta_*^2} \tau_d \mathcal{J} \lambda_{rd} + \tfrac{\dot{\beta}_*}{\beta_*} \lambda_{rd} \right) \tag{10.37}$$

where $\mathcal{K}(\dot{q}_m) \triangleq \mathrm{diag}\{K_1(\dot{q}_m)I_2, 0\}$. For later convenience, (10.35) and (10.36) are rewritten as

$$\begin{bmatrix} \ddot{\tilde{q}}_m \\ \dot{z} \end{bmatrix} \ = \ \begin{bmatrix} 0 & -\tfrac{1}{D_m} \\ b & -a \end{bmatrix} \begin{bmatrix} \dot{\tilde{q}}_m \\ z \end{bmatrix} + \begin{bmatrix} \tfrac{1}{D_m} \\ 0 \end{bmatrix} (\tau - \tau_d)$$

$$\Updownarrow$$

$$\dot{x} \ = \ \mathcal{A}x + \mathcal{B}(\tau - \tau_d) \tag{10.38}$$

The matrix $\mathcal{A}$ is Hurwitz for all positive values of a and b.

Under Assumptions **A10.3–A10.5** the system (10.34)-(10.37) is locally Lipschitz in the state $[\dot{\tilde{q}}_e^\top, \dot{\tilde{q}}_m, z, \lambda_{rd}^\top]^\top$ and continuous in t. This condition ensures that there exists a time interval $[0, T)$ where the solutions exist and are unique. First, $\|\lambda_{rd}\| = \beta_*(t)$, $\forall t \in [0, T)$, and it is consequently bounded. Now, consider the quadratic function (10.21) whose time derivative along the solutions of (10.34) for all $t \in [0, T)$, is given by

$$\dot{H}_{ed} \ = \ -\dot{\tilde{q}}_e^\top [R(q_m, \dot{q}_m) + \mathcal{K}(\dot{q}_m)]_{\mathrm{sy}} \, \dot{\tilde{q}}_e \tag{10.39}$$

where the skew symmetry property

$$\dot{D}_e(q_m) = C_e(q_m, \dot{q}_m) + C_e^\top(q_m, \dot{q}_m)$$

has been used, and

$$[R(q_m, \dot{q}_m) + \mathcal{K}(\dot{q}_m)]_{\text{SY}} = \begin{bmatrix} (R_s + K_1(\dot{q}_m)) I_2 & \frac{1}{2} n_p L_{sr} \mathcal{J} e^{\mathcal{J} n_p q_m} \dot{q}_m \\ -\frac{1}{2} n_p L_{sr} \mathcal{J} e^{-\mathcal{J} n_p q_m} \dot{q}_m & R_r I_2 \end{bmatrix}$$

is the symmetric part of $R(q_m, \dot{q}_m) + \mathcal{K}(\dot{q}_m)$. This matrix will in the rest of this chapter be denoted[12] $R_{es}(q_m, \dot{q}_m)$. The matrix is strictly positive definite, uniformly in $\dot{q}_m$, namely

$$R_{es}(q_m, \dot{q}_m) \geq \delta I_4 > 0 \tag{10.40}$$

where I_4 is the 4×4 identity matrix. This can be proved by using standard results from matrix theory and the facts that $R_r > 0$ and $\mathcal{J} e^{-\mathcal{J} n_p q_m} = e^{-\mathcal{J} n_p q_m} \mathcal{J}$, which leads to (10.40) holding if and only if

$$R_s + K_1(\dot{q}_m) - \frac{n_p^2 L_{sr}^2}{4(R_r - \delta)} \dot{q}_m^2 \geq \delta \tag{10.41}$$

See Section 4.3 of Appendix **D** for a detailed derivation of this requirement, which is fulfilled with the chosen definition of $K_1(\dot{q}_m)$.

Therefore, from the above and (10.39) it follows that

$$\dot{H}_{ed} = -\dot{\tilde{q}}_e^\top R_{es}(q_m, \dot{q}_m) \dot{\tilde{q}}_e \leq -\inf_{q_m, \dot{q}_m} \underline{\lambda}(R_{es}) \|\dot{\tilde{q}}_e\|^2$$

and it can be derived that for some constants $m_e > 0$ and $\rho_e > 0$ independent of T

$$\|\dot{\tilde{q}}_e(t)\| \leq m_e \|\dot{\tilde{q}}_e(0)\| e^{-\rho_e t}, \ \forall t \in [0, T) \tag{10.42}$$

Notice that, unless $\dot{q}_m$ escapes to infinity in finite time, $\lim_{t \to \infty} \dot{\tilde{q}}_e = 0$. Thus, it must be proved first that this is not the case by showing that the input $(\tau - \tau_d)$ to the linear filter (10.35)-(10.36) is linearly bounded by the filter state. To this end, notice that the desired torque τ_d can be written as

$$\tau_d = \frac{1}{2} \dot{q}_{ed}^\top W_1(q_m) \dot{q}_{ed}$$

hence it follows that

$$\tau - \tau_d = \frac{1}{2} \dot{\tilde{q}}_e^\top W_1(q_m) \dot{\tilde{q}}_e + \dot{\tilde{q}}_e^\top W_1(q_m) \dot{q}_{ed} \tag{10.43}$$

---

[12]Notice that the symmetric matrix above is exactly the matrix $R_{es}(q_m, \dot{q}_m)$ defined in (9.18).

and the following bound holds

$$|\tau - \tau_d| \leq \frac{n_p L_{sr}}{2} \left(\|\dot{\tilde{q}}_e\|^2 + 2\|\dot{\tilde{q}}_e\| \|\dot{q}_{ed}\| \right) \tag{10.44}$$

On the other hand, writing the desired currents in (10.29) as

$$\dot{q}_{ed} = \begin{bmatrix} \frac{1}{L_{sr}} \left(\frac{L_r}{n_p \beta_\star^2} (D_m \ddot{q}_{m*} + \tau_L) \mathcal{J} + (1 + \frac{L_r \dot{\beta}_\star}{R_r \beta_\star}) I_2 \right) e^{\mathcal{J} n_p q_m} \lambda_{rd} \\ - \left(\frac{1}{n_p \beta_\star^2} (D_m \ddot{q}_{m*} + \tau_L) \mathcal{J} + \frac{\dot{\beta}_\star}{R_r \beta_\star} I_2 \right) \lambda_{rd} \end{bmatrix}$$

$$+ \begin{bmatrix} -\frac{L_r}{L_{sr} n_p \beta_\star^2} \mathcal{J} e^{\mathcal{J} n_p q_m} \lambda_{rd} \\ \frac{1}{n_p \beta_\star^2} \mathcal{J} \lambda_{rd} \end{bmatrix} z$$

$$\triangleq \alpha_1(t) + \alpha_2(t) z \tag{10.45}$$

and noting that

$$\|\alpha_1(t)\| \leq$$

$$\sqrt{(\frac{1}{L_{sr}^2} \| \frac{L_r}{n_p \beta_\star^2} (D_m \ddot{q}_{m*} + \tau_L) \mathcal{J} + (1 + \frac{L_r \dot{\beta}_\star}{R_r \beta_\star}) I_2 \|^2 + \| \frac{1}{n_p \beta_\star^2} (D_m \ddot{q}_{m*} + \tau_L) \mathcal{J} + \frac{\dot{\beta}_\star}{R_r \beta_\star} I_2 \|^2)} \| \lambda_{rd} \|$$

$$\leq \bar{\alpha}_1 < \infty$$

$$\|\alpha_2(t)\| \leq \sqrt{((\frac{L_r}{L_{sr} n_p \beta_\star^2})^2 + \frac{1}{n_p^2 \beta_\star^4})} \| \lambda_{rd} \| \leq \bar{\alpha}_2 < \infty$$

yields $\|\dot{q}_{ed}\| \leq \bar{\alpha}_1 + \bar{\alpha}_2 |z|$. Replacing this bound, together with $\|\dot{\tilde{q}}_e\| \leq m_e \|\dot{\tilde{q}}_e(0)\|$, $\forall t \in [0, T)$ in (10.44), it follows that

$$|\tau - \tau_d| \leq \frac{n_p L_{sr}}{2} (m_e^2 \|\dot{\tilde{q}}_e(0)\|^2 + 2m_e \|\dot{\tilde{q}}_e(0)\| \bar{\alpha}_1) + n_p L_{sr} m_e \|\dot{\tilde{q}}_e(0)\| \bar{\alpha}_2 |z|, \ \forall t \in [0, T)$$

This last inequality proves, via Gronwall's inequality, that x, and consequently $\dot{q}_m$, can not grow faster than an exponential in the time interval $[0, T)$. Moreover, since all the constants in the above bound are independent of T, this argument can be repeated to extend the time interval of existence of solutions to the whole real axis.

Having proved that (10.42) holds as $t \to \infty$, it must be proved that this implies that $\lim_{t \to \infty} x = 0$, with $\dot{q}_{ed}$ bounded.

Inserting (10.43) into (10.38), and using (10.45) to express $\dot{q}_{ed}$, results in the system

$$\dot{x} = \mathcal{A}x + \begin{bmatrix} 0 & \frac{1}{D_m} \dot{\tilde{q}}_e^\top W_1(q_m) \alpha_2(t) \\ 0 & 0 \end{bmatrix} \begin{bmatrix} \dot{\tilde{q}}_m \\ z \end{bmatrix}$$

$$+ \begin{bmatrix} \frac{1}{2D_m} \{ \dot{\tilde{q}}_e^\top W_1(q_m) \dot{\tilde{q}}_e + 2\dot{\tilde{q}}_e^\top W_1(q_m) \alpha_1(t) \} \\ 0 \end{bmatrix}$$

$$\Updownarrow$$

$$\dot{x} = \mathcal{A}x + B(t)x + c(t) \tag{10.46}$$

Calculation of norms gives

$$\|B(t)\| \le \frac{1}{D_m} n_p L_{sr} \bar{\alpha}_2 m_e \|\dot{\tilde{q}}_e(0)\| e^{-\rho_e t} \tag{10.47}$$

$$\|c(t)\| \le \frac{n_p L_{sr}}{2D_m} \{ m_e \|\dot{\tilde{q}}_e(0)\| e^{-\rho_e t} + 2\bar{\alpha}_1 \} m_e \|\dot{\tilde{q}}_e(0)\| e^{-\rho_e t}$$

From this (see Example 5.9 in [127]) it can be concluded that the system $\dot{x} = [\mathcal{A} + B(t)]x$ is globally exponentially stable, and since $\|c(t)\| \to 0$ as $t \to \infty$, it follows that $x \to 0$. Further, from (10.45) it can be established that $\dot{q}_{ed}$ is bounded.

The proof of asymptotic rotor flux norm tracking follows from (see (10.6) and (10.7))

$$\lambda_r - \lambda_{rd} = L_{sr} e^{-\mathcal{J} n_p q_m} (\dot{q}_s - \dot{q}_{sd}) + L_r (\dot{q}_r - \dot{q}_{rd})$$
$$\dot{\lambda}_r - \dot{\lambda}_{rd} = -R_r (\dot{q}_r - \dot{q}_{rd})$$

and convergence of current errors to zero.

■

3.3 Remarks to the controller

Remark 10.7 (Relation with Proposition 9.28.) Notice that the inner–loop PBC above follow directly from Proposition 9.28 with $U_{22} = -n_p \mathcal{J}$. In the case of a two-dimensional flux vector, the initial condition on rotor flux norm can easily be given in terms of one of its components.

Remark 10.8 (Position control.) It is easy to see that choosing the desired torque in the controller above as

$$\tau_d = D_m \ddot{q}_{m*} - z - f\tilde{q}_m + \tau_L \tag{10.48}$$

yields global asymptotic *position* tracking for all positive values of a, b, f. In this case, the error equation (10.38) has

$$x = \begin{bmatrix} \tilde{q}_m \\ \dot{\tilde{q}}_m \\ z \end{bmatrix}, \quad \mathcal{A} \triangleq \begin{bmatrix} 0 & 1 & 0 \\ -\frac{f}{D_m} & 0 & -\frac{1}{D_m} \\ 0 & b & -a \end{bmatrix}, \quad \mathcal{B} \triangleq \begin{bmatrix} 0 \\ \frac{1}{D_m} \\ 0 \end{bmatrix}$$

The matrix $\mathcal{A}$ is Hurwitz for all positive values of a, b, f, and the proof of global asymptotic rotor flux norm and position tracking follows *verbatim* from the proof of the main result above.

Remark 10.9 (Adaptation of load torque.) We can extend the result in Proposition 10.6 to the case of unknown but linearly parameterized load

$$\tau_L = \theta^\top \phi(q_m, \dot{q}_m)$$

where $\theta \in \mathbb{R}^q$ is a vector of unknown constant parameters, and $\phi(q_m, \dot{q}_m)$ is a measurable regressor. We replace the last equation of (10.30) by the certainty equivalence law

$$\tau_d = D_m \ddot{q}_{m*} - z + \hat{\tau}_L$$

where $\hat{\tau}_L = \hat{\theta}^\top \phi(q_m, \dot{q}_m)$, with $\hat{\theta}$ an on–line estimate of θ to be defined below. Notice that the proof of convergence of $\dot{\tilde{q}}_e$ is not affected by this change, with only equation (10.38) being replaced by

$$\dot{x} = \mathcal{A}x + \mathcal{B}(\tau - \tau_d + \tilde{\tau}_L)$$

where $\tilde{\tau}_L \stackrel{\triangle}{=} \hat{\tau}_L - \tau_L$. This driving term appears again in the last part of the proof in (10.46), which should be replaced by

$$\dot{x} = \mathcal{A}x + \mathcal{B}\tilde{\tau}_L + B(t)x + c(t)$$

Thus we are confronted with the well-known problem of designing an estimator for a full–state measurable LTI system with exponentially decaying additive disturbances. The use of the estimator

$$\dot{\hat{\theta}} = -\gamma x^\top P \mathcal{B} \phi(q_m, \dot{q}_m), \ \gamma > 0$$

with P a symmetric positive definite matrix satisfying $\mathcal{A}^\top P + P \mathcal{A} < 0$ will solve this problem.

Remark 10.10 (Damping injection to Σ_m.) To overcome the problem of measurement of acceleration it was assumed in [76] that $R_m > 0$, and the speed control strategy

$$\tau_d = D_m \ddot{q}_{m*} + R_m \dot{q}_{m*} + \tau_L$$

was proposed. Two drawbacks of this scheme are that it is open loop in the speed tracking error, and that its convergence rate is limited by the mechanical time constant $\frac{D_m}{R_m}$. Defining the desired torque τ_d as in (10.30) with z from (10.32), allows for effectively feeding back the speed tracking error without acceleration measurement.[13] Even though, the convergence rate of the mechanical subsystem is independent of the natural mechanical damping, it is however restricted by the convergence of the electrical subsystem. This, in its turn, is limited by the rotor resistance, as seen from (10.39).

Remark 10.11 (Damping injection to Σ_e.) In Chapter 8 we carried out, in a simple magnetic levitation system, a simulation study of the effect of damping injection to Σ_e in PBC. We showed that without damping injection the convergence rate was inadmissibly slow. On the other hand, in Section 6 of this chapter we will present

---

[13]This result was first reported in [209].

experimental evidence showing that the damping injection term introduces undesirable high gains that excite the unmodeled dynamics, induces actuator saturation and amplify the noise. Also, in Chapter **12** we will present simulation studies for robots with AC drives where a similar undesirable behaviour is observed for a backstepping–based design, whose stability analysis heavily relies on this kind of nonlinear damping injection terms. Further research is clearly needed to enhance our understanding of these nonlinear phenomena to be able to trade–off between speed of response and high–gain injection into the control loop.

3.4 Integral action in stator currents

It is common in applications to add an integral loop around the stator current errors to the input voltages. The experimental evidence presented in Section 6 shows that this indeed robustifies the PBC by compensating for unmodeled dynamics. It is interesting to note that the global tracking result above still holds for this case, as shown in the proposition below.

Proposition 10.12 *The result in Proposition 10.6 is still valid if the integral term*

$$u_I = -K_{Is} \int_0^t \dot{\tilde{q}}_s \, dt, \; K_{Is} \geq 0$$

is added to the control, with K_{Is} a positive semidefinite matrix. That is, if we set $u = u_{PBC} + u_I$, where u_{PBC} is defined in (10.28). □

Proof. To study the stability of this new system, a term $\frac{1}{2} \left[\int_0^t \dot{\tilde{q}}_s \, dt \right]^\top K_{Is} \left[\int_0^t \dot{\tilde{q}}_s \, dt \right]$ is added to H_{ed}. Computing the derivate of this new H_{ed} with the new closed loop system (see (10.34)),

$$D_e(q_m)\ddot{\tilde{q}}_e + C_e(q_m, \dot{q}_m)\dot{\tilde{q}}_e + [R(q_m, \dot{q}_m) + \mathcal{K}(\dot{q}_m)] \dot{\tilde{q}}_e = \psi \tag{10.49}$$

with

$$\psi = \begin{bmatrix} -K_{Is} \int_0^t \dot{\tilde{q}}_s \, dt \\ 0 \end{bmatrix}$$

results in

$$\dot{V}_1 = -\dot{\tilde{q}}_e^\top R_{es}\dot{\tilde{q}}_e - K_{Is}\dot{\tilde{q}}_s^\top \int_0^t \dot{\tilde{q}}_s \, dt + \frac{d}{dt}\left(\frac{1}{2}K_{Is} \left[\int_0^t \dot{\tilde{q}}_s \, dt \right]^\top \left[\int_0^t \dot{\tilde{q}}_s \, dt \right] \right)$$

$$= -\dot{\tilde{q}}_e^\top R_{es}\dot{\tilde{q}}_e$$

It follows that $\dot{\tilde{q}}_e$ and the integral term will be bounded for a closed time interval. This is enough to complete the previous details proving that there is no finite escape time, and it follows that $\dot{\tilde{q}}_e \in \mathcal{L}_\infty^4 \cap \mathcal{L}_2^4$ and $\int_0^t \dot{\tilde{q}}_s \, dt \in \mathcal{L}_\infty^2$. It is however not enough to claim convergence of current errors to zero. For this, the additional requirement of $\ddot{\tilde{q}}_e \in \mathcal{L}_\infty^4$ will be sufficient. From (10.49) it can be seen that this is the case if the speed is bounded.

Notice from (10.47) that $\|B(t)\| = |\frac{1}{D_m} \dot{\tilde{q}}_e^\top W_1(q_m)\alpha_2(t)| \leq \frac{1}{D_m} n_p L_{sr} \bar{\alpha}_2 \|\dot{\tilde{q}}_e\|$. Now, since $\dot{\tilde{q}}_e \in \mathcal{L}_2^4$, it follows that

$$\int_0^\infty \|B(t)\|^2 \, dt \; < \; \infty$$

and by use of standard results on stability of linear time-varying systems (see [127], pp. 227,257), it follows that the origin is a globally exponentially stable equilibrium of the linear part $\dot{x} = [\mathcal{A} + B(t)] x$ of the system (10.46)

$$\dot{x} \; = \; [\mathcal{A} + B(t)] x + c(t)$$

Since $\dot{\tilde{q}}_e \in \mathcal{L}_\infty^4$ implies that the perturbation term $c(t)$ is bounded, it follows by use of total stability arguments that x and consequently the speed $\dot{q}_m$ will be bounded. With bounded speed $\dot{q}_m$ it follows as previously explained that the current errors converge to zero, which again implies convergence of the speed tracking error to zero since $c(t) \to 0$.

Thus, global asymptotic speed and rotor flux tracking tracking with all internal states bounded can be proved even when the nested–loop PBC is robustified with integral action in stator currents.

∎

3.5 Adaptation of stator parameters

The controller in (10.28) can also be modified to compensate for unknown stator resistance and inductance, by using estimates $\hat{L}_s$, $\hat{R}_s$ instead of real parameters in the controller. This results in a closed loop system (assuming for simplicity there is no integral action) with

$$D_e(q_m)\ddot{\tilde{q}}_e + C_e(q_m, \dot{q}_m)\dot{\tilde{q}}_e + [R(q_m, \dot{q}_m) + \mathcal{K}(\dot{q}_m)]\, \tilde{q}_e \; = \; \psi$$
$$= \; \begin{bmatrix} \left(\hat{L}_s - L_s\right) \ddot{q}_{sd} + \left(\hat{R}_s - R_s\right) \dot{q}_{sd} \\ 0 \end{bmatrix}$$

To prove the same stability results as before using a Lyapunov approach, extra terms $\frac{\gamma_{L_s}}{2}\left(\hat{L}_s - L_s\right)^2$ and $\frac{\gamma_{R_s}}{2}\left(\hat{R}_s - R_s\right)^2$ with γ_{L_s}, $\gamma_{R_s} > 0$ must be added to V_1. Calculation of $\dot{V}_1$ then gives

$$\dot{V}_1 = -\dot{\tilde{q}}_e^\top Res \dot{\tilde{q}}_e + \dot{\tilde{q}}_s^\top \left\{ \left(\hat{L}_s - L_s \right) \ddot{q}_{sd} + \left(\hat{R}_s - R_s \right) \dot{q}_{sd} \right\}$$
$$+ \frac{d}{dt} \left(\frac{\gamma_{L_s}}{2} \left(\hat{L}_s - L_s \right)^2 + \frac{\gamma_{R_s}}{2} \left(\hat{R}_s - R_s \right)^2 \right)$$

Assuming that L_s and R_s are constant, and choosing parameter adaptation laws to be

$$\dot{\hat{L}}_s = -\frac{1}{\gamma_{L_s}} \dot{\tilde{q}}_s^\top \ddot{q}_{sd}$$

$$\dot{\hat{R}}_s = -\frac{1}{\gamma_{R_s}} \dot{\tilde{q}}_s^\top \dot{q}_{sd}$$

results in

$$\dot{V}_1 = -\dot{\tilde{q}}_e^\top Res \dot{\tilde{q}}_e$$

The rest of the proof follows as explained in the previous section. This approach will not be pursued here, since adaptation of only these parameters is of less interest, especially if integral action in stator currents is added.

3.6 A fundamental obstacle for rotor resistance adaptation

We have shown already in previous chapters that it is relatively easy to incorporate adaptation features to PBC. This stems from the following two facts, first, that PBC is based on EL descriptions which preserve linearity (and minimality) on the parameters. Second, that PBC enforces a passive operator in closed–loop, which is a fundamental property for adaptation, since as pointed out in Section 3.4.2, the standard (gradient or least squares) estimators themselves define also passive operators. Unfortunately, the error equations that result of the nested–loop PBC of induction motors exhibit a fundamental obstacle for adaptation of the practically important parameter R_r. (See Remark 9.6 for a discussion on parameter uncertainty.) Namely, that the operator required to be passive for a successful adaptation is not passive *with respect to* the classical quadratic storage function. It is important to underline the qualifier "with respect to", because this does not rule out the possibility that it is passive for some other storage function.

Let us now proceed to establish this negative result. We assume that the only uncertain parameter is R_r and propose an adaptive implementation of the nested–loop PBC (10.28)–(10.32) with a time–varying estimate $\hat{R}_r$ replacing R_r. For simplicity let us consider the case of $\beta_* = $ const. In this case R_r appears only in (10.31), which in the adaptive case takes the form

$$\dot{\lambda}_{rd} = \frac{\hat{R}_r}{n_p \beta_*^2} \tau_d \mathcal{J} \lambda_{rd}$$

Going through the calculations and defining the parameter error $\tilde{R}_r \triangleq \hat{R}_r - R_r$ we see that this error propagates to the current error equations (10.34) as

$$D_e(q_m)\ddot{\tilde{q}}_e + C_e(q_m, \dot{q}_m)\dot{\tilde{q}}_e + R_{es}(q_m, \dot{q}_m)\dot{\tilde{q}}_e = \begin{bmatrix} 0 \\ \dot{q}_{rd} \end{bmatrix} \tilde{R}_r \qquad (10.50)$$

¿From the derivations above we see that (10.50) defines an output strictly passive operator $\dot{q}_{rd}\tilde{R}_r \mapsto \dot{\tilde{q}}_r$ with storage function H_{ed}. As first pointed out in [206], this property allows us to design a globally stable adaptation law, in the case of *full state measurement*, as

$$\dot{\hat{R}}_r = -\gamma \dot{q}_{rd}^\top \dot{\tilde{q}}_r \qquad (10.51)$$

with the adaptation gain $\gamma > 0$. The convergence proof can be easily established with a passivity argument as in Section 3.4.2, or with the Lyapunov function

$$V = H_{ed} + \frac{1}{2\gamma}\tilde{R}_r^2$$

which gives $\dot{V} = -\dot{\tilde{q}}_e^\top R_{es}(q_m, \dot{q}_m)\dot{\tilde{q}}_e \leq -\alpha\|\dot{\tilde{q}}_e\|^2$, for some $\alpha > 0$. Unfortunately, the only computable current errors are $\tilde{q}_s$, and (10.51) is therefore not implementable. The question is then whether we can define *measurable* 2–dimensional vector functions Φ_1, Φ_2 so that, along the dynamics (10.50) the operator $\Phi_1\tilde{R}_r \mapsto \Phi_2$ is strictly passive. In this way, the update law $\dot{\hat{R}}_r = -\gamma\Phi_1^\top\Phi_2$ would solve the problem.

To help us investigate this question, it is convenient to work in a coordinate frame where the desired storage function (and for that matter all the electrical equations) is independent of q_m. As shown in Section 1 this is the case if the ab-coordinates (10.20) are used.[14] The electrical error equations can be expressed in these coordinates as

$$\begin{aligned} \dot{e}_e &= A(\dot{q}_m)e_e + B\Phi_1(\tau_d, \lambda_{rd})\tilde{R}_r \\ \Phi_2 &= \dot{\tilde{q}}_s = \begin{bmatrix} I_2 \\ 0 \end{bmatrix} e_e \end{aligned}$$

where $e_e \triangleq [\dot{\tilde{q}}_s^\top, \tilde{\lambda}_{ab}^\top]^\top$ and we have defined

$$A(\dot{q}_m) \triangleq \begin{bmatrix} -\frac{L_s L_{sr}}{L_{r\sigma}}K_1(\dot{q}_m)I_2 & -\frac{L_s L_{sr}}{L_{r\sigma}}\eta(\dot{q}_m) \\ \frac{L_{sr}R_r}{L_r}I_2 & \eta(\dot{q}_m) \end{bmatrix}, \quad B \triangleq \begin{bmatrix} \frac{L_s}{L_{r\sigma}}I_2 \\ -I_2 \end{bmatrix}$$

with $\eta(\dot{q}_m) \triangleq \dot{q}_m\mathcal{J} - \frac{R_r}{L_r}I_2$ and a regressor vector

$$\Phi_1(\tau_d, \lambda_{rd}) \triangleq \frac{\tau_d}{\beta_*^2 L_{sr}^2}\mathcal{J}\lambda_{rd}$$

---

[14]See also [76] for a derivation of the torque–tracking PBC in this reference frame.

To investigate the passivity properties of the operator $\Phi_1(\tau_d, \lambda_{rd})\tilde{R}_r \mapsto \Phi_2$ we follow the standard procedure in adaptive control and look for a constant matrix $P = P^\top > 0$ so that $W = \frac{1}{2}e_e^\top P e_e$ with $P = P^\top > 0$ qualifies as a storage function. In other words, we must satisfy the conditions of the Kalman–Yakubovich–Popov lemma (see Section 6 in Appendix A)

$$PA(\dot{q}_m) + A^\top(\dot{q}_m)P \; < \; 0 \tag{10.52}$$

$$PB \; = \; \begin{bmatrix} I_2 \\ 0 \end{bmatrix} \tag{10.53}$$

Notice that (10.52) is satisfied with the choice $P = \begin{bmatrix} \frac{L_{r\sigma}}{L_s}I_2 & 0 \\ 0 & I_2 \end{bmatrix}$, which is the one corresponding to H_{ed} in the ab coordinates.

Now, it is easy to see that the matrix $A(\dot{q}_m)$ can be factored as

$$A(\dot{q}_m) = \begin{bmatrix} \times & -B\eta(q_m, \dot{q}_m) \end{bmatrix}$$

Consequently, in view of the condition (10.53) we have that $(PA)_{22} = 0$, and the condition (10.52) cannot be satisfied (with strict inequality). Therefore, we conclude that there does not exist a quadratic storage function $W = \frac{1}{2}e_e^\top P e_e$ to prove the strict passivity of the operator $\Phi_1(\tau_d, \lambda_{rd})\tilde{R}_r \mapsto \Phi_2$.

3.7   A dq-implementation

For the VSI case, it is of interest to rotate the control in (10.28) to a dq-frame, not only for comparison with indirect FOC, but also for implementation purposes.

Using the results from Section 1.3, the equivalent of (10.28) can be derived in the reference frame of the desired rotor flux.

As previously pointed out, (10.28) consists of the *desired stator dynamics* and a nonlinear damping term. By using (10.13) in terms of also desired quantities[15] $i_{dq}^d = e^{-\mathcal{J}\theta_a}\dot{q}_{sd}$, $\lambda_{dq}^d = e^{-\mathcal{J}(\theta_a - n_p q_m)}\lambda_{rd}$, and comparing (10.28) with (10.16), it can be

---

[15]Superscript d will be used to denote desired quantities expressed in a dq-frame, e.g. desired stator currents in a dq-frame as i_{dq}^d.

seen that

$$u = L_s\ddot{q}_{sd} + L_{sr}e^{\mathcal{J}n_pq_m}\ddot{q}_{rd} + n_pL_{sr}\mathcal{J}e^{\mathcal{J}n_pq_m}\dot{q}_m\dot{q}_{rd} + R_s\dot{q}_{sd} - K_1(\dot{q}_m)\dot{\tilde{q}}_s$$

$$\Updownarrow$$

$$u = \sigma L_s e^{\mathcal{J}\theta_a}\left\{\frac{d}{dt}i_{dq}^d + [\omega_a\mathcal{J} + \gamma I_2]i_{dq}^d + \frac{L_{sr}}{\sigma L_s L_r}[n_p\dot{q}_m\mathcal{J} - \frac{1}{T_r}I_2]\lambda_{dq}^d\right\}$$
$$-K_1(\dot{q}_m)e^{\mathcal{J}\theta_a}\left(i_{dq} - i_{dq}^d\right)$$

$$\theta_a = \int_0^t \omega_a\,dt,\ \theta_a(0) = n_p\dot{q}_m(0)$$

$$\omega_a = n_p\dot{q}_m + \dot{\rho}_d$$

where $\dot{\rho}_d$ is defined by (10.27), and $i_{dq}^d = \frac{1}{L_{sr}}[\beta_* + \frac{L_r}{R_r}\dot{\beta}_*, \frac{L_r}{n_p\beta_*}\tau_d]^\top, \lambda_{dq}^d = [\beta_*, 0]^\top$.

3.8 Definitions of desired rotor flux norm

In the following two examples of how the desired rotor flux norm $\beta_*(t)$ can be designed, will be given. The first result is a direct adaptation of the results in [277] to the passivity-based controller for minimization of steady state losses. The other is an example of how the well known flux weakening approach can be mimicked.

Example 10.13 (Minimization of steady state losses.) To minimize power losses in the motor, the function

$$P_{\text{loss}} = \underbrace{u^\top\dot{q}_s}_{\text{supplied power}} - \underbrace{\tau\dot{q}_m}_{\text{mechanical output power}} \tag{10.54}$$

is considered.

The control u is first eliminated from P_{loss} by using (10.1) and (10.8), which gives

$$P_{\text{loss}} = \left(L_s - \frac{L_{sr}^2}{L_r}\right)\dot{q}_s^\top\ddot{q}_s + \left(R_s + R_r\frac{L_{sr}^2}{L_r^2}\right)\dot{q}_s^\top\dot{q}_s - \frac{L_{sr}R_r}{L_r}\dot{q}_s^\top e^{\mathcal{J}n_pq_m}\lambda_r \tag{10.55}$$

Detailed derivations of this and the following expressions are given in Section 5 of Appendix **D**.

It is assumed that the stator currents $\dot{q}_s$ have converged to their desired values, given in (10.29), and all occurrences of $\dot{q}_s$ and $\ddot{q}_s$ can be substituted by their reference values. This gives

$$
\begin{aligned}
P_{\text{loss}} = {} & \frac{L_r^2L_s - L_{sr}^2L_r}{L_{sr}^2R_r^2}\ddot{\beta}_*\dot{\beta}_* + \frac{L_sL_r - L_{sr}^2}{R_r}\ddot{\beta}_*\beta_* + \frac{L_sR_r + 2L_rR_s}{L_{sr}^2R_r}\dot{\beta}_*\beta_* \\
& + \frac{R_rL_rL_s + L_r^2R_s}{L_{sr}^2R_r}\dot{\beta}_*^2 + \frac{L_rL_{sr}^2 - L_r^2L_s}{n_p^2L_{sr}^2}\frac{\dot{\beta}_*}{\beta_*^3}\tau_d^2 + \frac{R_s}{L_{sr}^2}\beta_*^2 \\
& + \left[\frac{L_r^2L_s - L_rL_{sr}^2}{n_p^2L_{sr}^2}\dot{\tau}_d\tau_d + \frac{L_r^2R_s + R_rL_{sr}^2}{n_p^2L_{sr}^2}\tau_d^2\right]\frac{1}{\beta_*^2}
\end{aligned}
\tag{10.56}
$$

The minimization of the above criterion with respect to a general time-varying strictly positive function[16] $\beta_*(\tau_d, \dot{\tau}_d)$, is a nontrivial dynamic optimization problem.

As a first approach, only the task of minimizing the above expression at stationary conditions with constant torque is considered here. In this case $\beta_*(t)$ can also be considered as a constant, which gives

$$P_{\text{loss}} = \frac{R_s}{L_{sr}^2}\beta_*^2 + \frac{L_r^2 R_s + R_r L_{sr}^2}{n_p^2 L_{sr}^2}\frac{1}{\beta_*^2}\tau_d^2 \qquad (10.57)$$

By evaluating and setting $\partial P_{\text{loss}}/\partial(\beta_*^2) = 0$, the only extremum is found to be

$$(\beta_*^{\text{opt}})^2 = \sqrt{\frac{L_r^2}{n_p^2} + \frac{R_r}{R_s}\frac{L_{sr}^2}{n_p^2}}\,|\tau_d|$$

Evaluation of the second order partial derivative of P_{loss} with respect to β_*^2 gives

$$\frac{\partial^2 P_{\text{loss}}}{\partial^2(\beta_*^2)} = 2\frac{R_s L_r^2 + R_r L_{sr}^2}{n_p^2 L_{sr}^2}\frac{1}{\beta_*^6}\tau_d^2 > 0, \,\forall \beta_* \geq \delta > 0$$

This implies that the function (10.57) is convex in β_*^2, and $(\beta_*^{\text{opt}})^2$ is a global minimum.

It must be pointed out that the loss model is very simple, with no core losses, current/voltage limits or effects from nonlinear magnetics included. Despite this, the model indicates that to minimize power losses, the flux norm reference should be proportional to the square root of the desired torque, at least as long as the resulting reference is below the maximum allowable value. $\qquad\square$

Example 10.14 (Flux weakening.) In this example it is explained how the controller can be used to mimic the well known *flux weakening* approach for operation in the constant power region above nominal speed.

To get the desired smoothness properties of the norm reference, a linear second order filter is introduced,

$$\begin{bmatrix} \dot{x}_1(t) \\ \dot{x}_2(t) \end{bmatrix} = \begin{bmatrix} 0 & 1 \\ -\omega_n^2 & -2\zeta\omega_n \end{bmatrix}\begin{bmatrix} x_1(t) \\ x_2(t) \end{bmatrix} + \begin{bmatrix} 0 \\ \omega_n^2 \end{bmatrix}\beta_{\text{ref}}(t) \qquad (10.58)$$

$$\beta_*(t) = x_1(t), \; \dot{\beta}_*(t) = x_2(t)$$

with $[x_1(0), x_2(0)]^\top = [\beta_{\text{ref}}(0), 0]^\top$, $\zeta = 1$ and $\omega_n > 0$.

The input signal to the filter (10.58) is given by

$$\beta_{\text{ref}}(t) = \begin{cases} \beta_*^N, & -\dot{q}_m^N \leq \dot{q}_{m*} \leq \dot{q}_m^N \\ \beta_*^N\frac{\dot{q}_m^N}{|\dot{q}_{m*}|}, & |\dot{q}_{m*}| > \dot{q}_m^N \end{cases}$$

---

[16]Notice that the derivative of desired torque is also needed for dynamic optimization.

with β_*^N the desired constant nominal value for the rotor flux norm and $\dot{q}_m^N$ the base speed.

With the given value for the damping constant ζ, it is ensured that the output of the filter, β_*, will always be positive, and singularity points in the controller are avoided.

The flux reference has been defined in a feed-forward way from reference speed, instead of by using a memoryless feedback from actual speed. This is due to the additional technicalities involved in stability analysis when an extra feedback loop is used.

A possible drawback of this approach is that speed control has to be tight to avoid unwanted saturation and low performance when operating at speeds close to nominal speed and during transients. This implies that the acceleration constraint of the system must be taken into consideration when defining the reference for speed tracking.   □

3.9   Simulation results

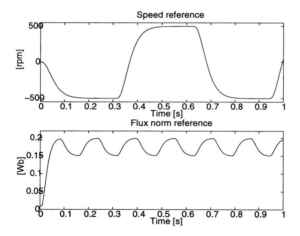

Figure 10.3: References for speed and flux norm.

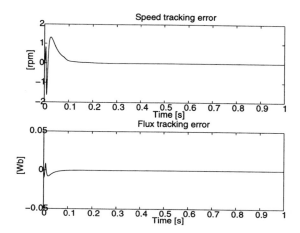

Figure 10.4: Tracking errors for speed and flux norm.

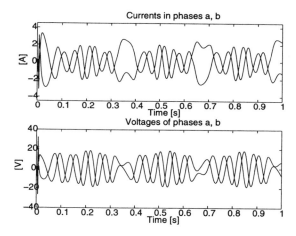

Figure 10.5: 3ϕ Currents and voltages i_a, i_b, u_a, u_b.

To verify the qualitative behavior of the scheme in Proposition 10.6, a simulation study with a SIMULINK™implementation of the induction motor ab-model was done. This is a stiff system with fast electrical dynamics, and relatively slow mechanical dynamics, for which the choice of integration method has to be carefully considered. The Gear-method with small step-size was found to give satisfactory numerical accuracy. The same model parameters (see Section 6.1) as for the experimental setup were used, and simulations were done under the assumptions of ideal conditions (zero load torque, known parameters, linear unsaturated amplifier, speed and position measurements). References were generated by use of step and square wave functions which were filtered using third order linear filters of the form

$$h(p) = \frac{1}{(\frac{p}{\omega_0} + 1)} \frac{1}{(\frac{p^2}{\omega_0^2} + 2\zeta\frac{p}{\omega_0} + 1)}$$

The necessary first and second order derivatives were obtained from the state space realizations of the filters.

Filtering of flux and speed references was found to be of high importance for avoiding current and voltage saturations, and the values $\zeta = 1$, $\omega_0 = 50$, 100 rad/s were chosen for speed and flux norm reference filtering, respectively.

The following parameters were used: $a = 1000$, $b = 300$, $k_1 = 30$, $\epsilon = 0.45R_r$. Integral action was not used in the controller, and all initial conditions were set to zero.

References, tracking errors and voltages/currents are shown in Figs. 10.3–10.5. As can be seen from these figures, there is no interaction between flux and speed control. This is consistent with the previous analysis. The 3ϕ voltages[17] and currents are well within limits of the system. The linear term k_1 in the gain K_1 was found to be significant for satisfactory behavior around zero speed, when the speed dependent term is close to zero.

4   A PBC with total energy–shaping

Instrumental for the development of the nested–loop controller of the previous section were the decomposition of the motor dynamics into the feedback interconnection of passive subsystems, and the possibility to make the mapping $v \mapsto \tilde{q}_s$ output strictly passive via output feedback. In this way we could apply the PBC methodology solely to the electrical subsystem and obviate the need of an (explicit) observer. In this section we take a different approach were we, instead of shaping only the energy of the electrical dynamics, design a PBC based on the total energy of the system, mechanical and electrical. The design philosophy is in principle simpler, however, the price to be paid is the need for an observer. This observer-based approach has been pursued in [77, 204, 206]. An additional assumption that we require here is that the friction coefficient R_m is not zero. This restriction is needed only to complete the stability proof and does not restrict the convergence rate of the scheme, which is limited, –as in the nested–loop PBC (see Remark 10.18)–, by the open–loop electrical dynamics, in this case via the observer rate. We have seen in the previous section

---

[17]Only two of the 3ϕ quantities are shown, since the third is a linear combination of the first two. See Section 6.1 for the transformations from two-phase quantities.

how to select the reference torque τ_* in a torque tracking scheme to achieve speed or position control. Consequently, we will concentrate here on the torque tracking problem.

Similar to the trajectory tracking problems in robotics of Chapter 4, for the design of a PBC the workless forces must be factored in a suitable way. More specifically, they must be linear in $\dot{q}$. In the case of the induction machine an additional restriction on this factorization is imposed by the lack of full state measurement. The output feedback torque tracking problem is solved with an *interlaced controller-observer* design which follows the passivity-based approach. A proof of *global tracking* is given under the assumption of known motor parameters. As the previous nested-loop PBC, the new control law does not require measurement of rotor variables, is always well defined and does not rely on (intrinsically non-robust) nonlinear dynamics cancelation. To simplify the presentation we solve first the problem assuming full state measurement, and then we focus on the output feedback case.

4.1   Factorization of workless forces

To carry out the PBC design, a suitable *linear factorization* of the workless forces discussed in Section 1 into a form

$$\left[\begin{array}{c} W_1(q_m)\dot{q}_m\dot{q}_e \\ -\frac{1}{2}\dot{q}_e^\top W_1(q_m)\dot{q}_e \end{array} \right] \;=\; \mathcal{C}(q,\dot{q})\dot{q}$$

must be found. Specifically, $\mathcal{C}(q,\dot{q})$ will be required to be such that:

i) $\dot{\mathcal{D}}(q) = \mathcal{C}(q,\dot{q}) + \mathcal{C}^\top(q,\dot{q})$;

ii) The third and fourth rows of $\mathcal{C}(q,\dot{q})$ are *independent* of $\dot{q}_e$.

The first condition is equivalent to the one imposed in Chapter 4 for the robotics problem. It stems from the fact that, even though passivity ensures that

$$\dot{q}^\top[\dot{\mathcal{D}}(q) - 2\mathcal{C}(q,\dot{q})]\dot{q} = 0$$

for all factorizations $\mathcal{C}(q,\dot{q})\dot{q}$, for the energy shaping stage we need the stronger condition

$$z^\top[\dot{\mathcal{D}}(q) - 2\mathcal{C}(q,\dot{q})]z = 0, \quad \forall z \in \mathbb{R}^5$$

which is ensured by the skew-symmetry property i) above. The second condition is related with the underactuated nature of the machine, and will be clarified below.

Using (10.4) and the transposed of the last expression in (10.8), it is clear that the objectives can be achieved with the choice

$$\mathcal{C}(q,\dot{q}) \;\triangleq\; \left[\begin{array}{ccc} 0 & 0 & f(q_m,\dot{q}_r) \\ -n_p L_{sr}\mathcal{J}\mathbf{e}^{-\mathcal{J}n_p q_m}\dot{q}_m & 0 & 0 \\ -f^\top(q_m,\dot{q}_r) & 0 & 0 \end{array} \right]$$

with

$$f(q_m, \dot{q}_r) \stackrel{\triangle}{=} n_p L_{sr} \mathcal{J} e^{\mathcal{J} n_p q_m} \dot{q}_r \tag{10.59}$$

This factorization leads to the following compact model representation:

$$\mathcal{D}(q)\ddot{q} + \mathcal{C}(q, \dot{q})\dot{q} + \mathcal{R}\dot{q} = \mathcal{M}u + \xi \tag{10.60}$$

4.2 Problem formulation

The following problem will be solved by the use of an observer-based PBC in this section:

Definition 10.15 (Torque and flux tracking with unknown load.) *Consider the induction motor model (10.60) with outputs torque τ and rotor flux norm $\|\lambda_r\|$ to be controlled. Assume:*

A 10.6 The load torque τ_L is an unknown constant.

A 10.7 Stator currents $\dot{q}_s$, rotor speed $\dot{q}_m$ and position q_m are available for measurement.

A 10.8 All motor parameters are exactly known, and the viscous mechanical damping constant is *nonzero*, i.e. $R_m > 0$.

    Let the desired torque $\tau_*(t)$ be a bounded and differentiable function with known bounded first order derivative, and the desired rotor flux norm be a strictly positive bounded and twice differentiable function $\beta_*(t)$ with known bounded first and second order derivatives. Under these conditions, design a control law that will ensure internal stability and asymptotic torque and rotor flux norm tracking, that is, the closed loop system must give

$$\lim_{t \to \infty} |\tau - \tau_*(t)| = 0, \quad \lim_{t \to \infty} |\ \|\lambda_r\| - \beta_*(t)| = 0 \tag{10.61}$$

from all initial conditions and with all signals uniformly bounded.

4.3 Ideal case with full state feedback

For the sake of clarity of presentation, the problem will first be solved under the *temporary* assumption of full state measurement (measurable rotor currents) and known load torque. This is referred to as *the ideal case*. It will then be explained in the next section how the controller can be modified to remove this assumption.

Following the PBC approach we want to shape the motor total energy $\mathcal{T}(q, \dot{q}) = \frac{1}{2} \dot{q}^\top \mathcal{D}(q) \dot{q}$ to the incremental form $\mathcal{T}_d(q, \dot{q}) = \frac{1}{2} \tilde{\dot{q}}^\top \mathcal{D}(q) \tilde{\dot{q}}$ where we defined the error signals $\tilde{\dot{q}} \triangleq \dot{q} - \dot{q}_d = \left[\tilde{\dot{q}}_e^\top, \tilde{\dot{q}}_m \right]^\top$, with $\dot{q}_d$ the vector of desired currents and *internal desired* rotor speed, $\dot{q}_d \triangleq [\dot{q}_{ed}^\top, \dot{q}_{md}]^\top$. The introduction of the *internal speed reference* $\dot{q}_{md}$ is specific to this observer-based approach, and it should not be confused with an *external* speed reference, which we will denote with $\dot{q}_{m*}$.

Equation (10.60) can then be rewritten as

$$\mathcal{D}(q) \tilde{\ddot{q}} + \mathcal{C}(q, \dot{q}) \tilde{\dot{q}} + [\mathcal{R} + \mathcal{K}] \tilde{\dot{q}} = \psi$$

where $\mathcal{K}$ is a positive semidefinite matrix (to be defined below) that injects the required damping in the output feedback case. It can be set to zero when the state is measurable. The right-hand side in the equation above is defined as

$$\psi \triangleq -\mathcal{D}(q) \ddot{q}_d - \mathcal{C}(q, \dot{q}) \dot{q}_d - \mathcal{R} \dot{q}_d + \mathcal{K} \tilde{\dot{q}} + \mathcal{M} u + \xi$$

Setting this perturbation term to zero is part of the implicit definition of the PBC, and results in

$$L_s \ddot{q}_{sd} + L_{sr} e^{\mathcal{J} n_p q_m} \ddot{q}_{rd} + f(q_m, \dot{q}_r) \dot{q}_{md} + R_s \dot{q}_{sd} = u \tag{10.62}$$
$$\lambda_{rd} + R_r \dot{q}_{rd} = 0 \tag{10.63}$$
$$D_m \ddot{q}_{md} - f^\top(q_m, \dot{q}_r) \dot{q}_{sd} + R_m \dot{q}_{md} = -\tau_L \tag{10.64}$$

with

$$\lambda_{rd} \triangleq L_{sr} e^{-\mathcal{J} n_p q_m} \dot{q}_{sd} + L_r \dot{q}_{rd} \tag{10.65}$$

The second equation is the "clamped dynamics" condition that ensures the desired torque is delivered. This is defined by (9.47) and can consequently be solved like in the previous controller, that is, by setting

$$\lambda_{rd} = \beta_* \begin{bmatrix} \cos(\rho_d) \\ \sin(\rho_d) \end{bmatrix} = e^{\mathcal{J} \rho_d} \begin{bmatrix} \beta_* \\ 0 \end{bmatrix}, \quad \dot{\rho}_d = \frac{R_r}{n_p \beta_*^2} \tau_* , \quad \rho_d(0) = 0 \tag{10.66}$$

There are two changes to the previous observer-less controller. First, from (10.59) we observe that for the implementation of the control signal (10.62), the rotor currents $\dot{q}_r$ must be measured. Second, the fifth equation of $\psi \equiv 0$ defines an additional controller dynamics (10.64).

The above derivations are summarized in the following proposition:

Proposition 10.16 (PBC with full state measurement.) *Consider the induction motor model (10.60) in closed loop with (10.62), (10.64), (10.59) where $\dot{q}_{sd}$, $\ddot{q}_{sd}$, $\dot{q}_{rd}$, $\ddot{q}_{rd}$ are calculated from (10.65) and (10.63) using (10.66). Then, for all initial conditions equation (10.61) holds with all signals uniformly bounded.* $\qquad\square$

4.4   Observer-based PBC for induction motors

The main result of this section, a nonlinear *observer-based* PBC, is presented in the proposition below.

Proposition 10.17 (Observer-based PBC for induction motors.) *Consider the induction motor model (10.60) with outputs to be controlled torque τ and rotor flux norm $\|\lambda_r\|$, and Assumptions* **A10.6-A10.8.** *Let the control law be defined as*

$$u = L_s \ddot{q}_{sd} + L_{sr} e^{\mathcal{J} n_p q_m} \ddot{q}_{rd} + n_p L_{sr} \mathcal{J} e^{\mathcal{J} n_p q_m} \dot{q}_r \dot{q}_{md} + R_s \dot{q}_{sd} - K_1(\dot{q}_{md})\dot{\tilde{q}}_s \quad (10.67)$$

where

$$\dot{q}_{rd} = -e^{\mathcal{J}\rho_d} \begin{bmatrix} \frac{\dot{\beta}_*}{R_r} \\ \frac{\tau_*}{n_p \beta_*} \end{bmatrix}$$

$$\dot{q}_{sd} = \frac{1}{L_{sr}} e^{\mathcal{J}(n_p q_m + \rho_d)} \begin{bmatrix} \beta + \frac{L_r}{R_r}\dot{\beta}_* \\ \frac{L_r}{n_p \beta_*}\tau_* \end{bmatrix}$$

and with controller dynamics

$$\dot{\rho}_d = \frac{R_r}{n_p \beta_*^2}\tau_* \; , \rho_d(0) = 0 \quad (10.68)$$

$$\ddot{q}_{md} = \frac{1}{D_m}\left(-n_p L_{sr}\dot{q}_r^\top \mathcal{J} e^{-\mathcal{J} n_p q_m} \dot{q}_{sd} - R_m \dot{q}_{md} - \hat{\tau}_L + K_2(\dot{q}_d)\dot{\tilde{q}}_m\right), \quad (10.69)$$

with $\dot{q}_{md}(0) = \dot{q}_m(0)$. The gains $K_1(\dot{q}_{md})$ and $K_2(\dot{q}_d)$ are given as

$$K_1(\dot{q}_{md}) \;\triangleq\; \frac{n_p^2 L_{sr}^2}{4\epsilon_1}\dot{q}_{md}^2 + k_1, \; k_1 \geq 0 \quad (10.70)$$

$$K_2(\dot{q}_d) \;\triangleq\; \frac{n_p^2 L_{sr}^2}{4\epsilon_1}\left(\dot{q}_{1d}^2 + \dot{q}_{2d}^2\right) + k_2, \; 0 < \epsilon_1 < \frac{R_r}{2}, \; k_2 > 0 \quad (10.71)$$

while the state estimator and load adaptation law are

$$D_e(q_m)\ddot{\hat{q}}_e + W_1(q_m)\dot{q}_m\dot{\hat{q}}_e + R_e\dot{\hat{q}}_e \;=\; M_e u - L(q_m, \dot{q}_m)\dot{e}_e \quad (10.72)$$

$$\dot{\hat{\tau}}_L \;=\; -\gamma_{\tau_L}\dot{\tilde{q}}_m, \; \gamma_{\tau_L} > 0 \quad (10.73)$$

with $\dot{e}_e \triangleq \dot{\hat{q}}_e - \dot{q}_e$ the observation error and

$$L(q_m, \dot{q}_m) = \begin{bmatrix} 0 & 0 \\ n_p L_{sr}\mathcal{J} e^{-\mathcal{J} n_p q_m} & 0 \end{bmatrix}\dot{q}_m \quad (10.74)$$

Under these conditions, the closed loop system achieves global torque and rotor flux norm tracking with all signals uniformly bounded. $\square$

Proof. Since the control law above uses the estimated instead of the real states, contrary to the ideal case, the error equation in this case takes the following form

$$\mathcal{D}(q)\ddot{\tilde{q}} + \mathcal{C}(q,\dot{q})\dot{\tilde{q}} + [\mathcal{R} + \mathcal{K}(\dot{q}_d)]\,\dot{\tilde{q}} \;=\; \mathcal{S}(q_m,\dot{q}_d)\dot{e}_e + \tilde{\xi} \tag{10.75}$$

where $\mathcal{K}(\dot{q}_d) = \text{diag}\{K_1(\dot{q}_{md})I_2, 0, 0, K_2(\dot{q}_d)\}$, $\tilde{\xi} \triangleq [0,0,0,0,\tilde{\tau}_L]^\top = [0,0,0,0,(\hat{\tau}_L - \tau_L)]^\top$ and

$$\mathcal{S}(q_m,\dot{q}_d) \;=\; \begin{bmatrix} 0 & n_p L_{sr} \mathcal{J} e^{\mathcal{J} n_p q_m} \dot{q}_{md} \\ 0 & 0 \\ 0 & -n_p L_{sr} \dot{q}_{sd}^\top \mathcal{J} e^{\mathcal{J} n_p q_m} \end{bmatrix} \in \mathbb{R}^{5\times 4}$$

On the other hand, from (10.72) and (10.1) the observation error $\dot{e}_e$ satisfies the following equation

$$D_e(q_m)\ddot{e}_e + [W_1(q_m)\dot{q}_m + L(q_m,\dot{q}_m)]\,\dot{e}_e + R_e\dot{e}_e \;=\; 0 \tag{10.76}$$

Now, consider the composite Lyapunov function candidate

$$V \;=\; \frac{1}{2}\dot{\tilde{q}}^\top \mathcal{D}(q)\dot{\tilde{q}} + \frac{1}{2}\dot{e}_e^\top D_e(q_m)\dot{e}_e + \frac{1}{2\gamma_{\tau_L}}\tilde{\tau}_L^2$$

whose derivative, taking into account the skew-symmetry of

$$\dot{\mathcal{D}}(q) - 2\mathcal{C}(q,\dot{q})$$

and

$$\dot{D}_e(q_m) - 2\,[W_1(q_m)\dot{q}_m + L(q_m,\dot{q}_m)]$$

yields

$$\dot{V} \;=\; -\dot{\tilde{q}}^\top [\mathcal{R} + \mathcal{K}(\dot{q}_d)]\,\dot{\tilde{q}} + \dot{\tilde{q}}^\top \mathcal{S}(q_m,\dot{q}_d)\dot{e}_e - \dot{e}_e^\top R_e \dot{e}_e + \frac{1}{\gamma_{\tau_L}}\tilde{\tau}_L\dot{\tilde{\tau}}_L + \dot{\tilde{q}}^\top \tilde{\xi}$$

Use of (10.73) in the equation above and defining $z \triangleq [\dot{\tilde{q}}^\top, \dot{e}_e^\top]^\top$, results in the following quadratic function

$$\dot{V} \;=\; -z^\top \mathcal{M} z$$

with

$$\mathcal{M} \;=\; \begin{bmatrix} \mathcal{R} + \mathcal{K}(\dot{q}_d) & -\frac{1}{2}\mathcal{S}(q_m,\dot{q}_d) \\ -\frac{1}{2}\mathcal{S}^\top(q_m,\dot{q}_d) & R_e \end{bmatrix}$$

Checking that (10.70) and (10.71) ensures strictly positive definiteness of $\mathcal{M}$ (see Section 4.2 of Appendix D), i.e.

$$\mathcal{M} \;\geq\; \delta I_9 > 0 \tag{10.77}$$

it can be concluded that $\dot{\tilde{q}} \in \mathcal{L}_\infty^5, \dot{e}_e \in \mathcal{L}_\infty^4$ and $\hat{\tau}_L \in \mathcal{L}_\infty$ hold (i.e. boundedness) and that $\dot{e}_e$ and $\dot{\tilde{q}}$ are also square integrable (i.e. belongs to $\mathcal{L}_2^n$, $n = 4, 5$). Since $\dot{q}_{sd}, \dot{q}_{rd}$ are bounded by construction, then $\dot{q}_e$ is bounded, which together with $\dot{e}_e$ in its turn implies that $\dot{\hat{q}}_e$ is bounded. The fact that $\dot{q}_{md}$ is bounded follows from boundedness of $\dot{\hat{q}}_r$ and (10.69) with $R_m > 0$, and implies $\dot{q}_m$ bounded since $\dot{\tilde{q}}_m$ is bounded. From (10.75) and (10.76) it now follows that $\ddot{\tilde{q}}, \ddot{e}_e$ are bounded. Since $\dot{\tilde{q}}, \dot{e}_e$ are bounded signals with bounded derivatives, they are also uniformly continuous. Together with square integrability, this implies convergence of current errors to zero. Hence, rotor flux norm and torque tracking can be concluded, with all internal signals uniformly bounded. ∎

4.5   Remarks to the controller

Remark 10.18 (Observer structure.) Under the assumptions that $u \in \mathcal{L}_\infty^2$ and $\dot{q} \in \mathcal{L}_\infty^5$, global exponential convergence of the estimated currents to their real values can be proved, using only the part of V quadratic in $\dot{e}_e$ together with (10.76). To get this result, a speed dependent term (10.74) proportional to the observation error in stator currents is used to update the estimates in (10.72). This is common to several reduced order observers for which global stability results exists, see [280]. The choice of $L(q_m, \dot{q}_m)$ follows naturally from the Euler-Lagrange structure of the model, aiming at getting $\dot{D}_e(q_m) - 2\left[W_1(q_m)\dot{q}_m + L(q_m, \dot{q}_m)\right]$ skew-symmetric. Unfortunately the convergence rate of the estimation errors depends on the minimum resistance, i.e. $\lambda(R_e)$.

Notice also that even global exponential convergence of current estimation errors to zero is not enough to claim stability of the total system with estimated states in the controller. Nonlinear damping terms must be introduced in the controller equations to ensure global stability when estimated instead of real states are used. An interesting task would be to incorporate other globally valid and exponentially convergent observers into this scheme. This would give the possibility of having a convergence rate which does not depend on the resistances, and may lead to a performance improvement.

Remark 10.19 (Comparison with observer-less case.) Comparing (10.28) with (10.67), it can be seen that the difference is in the third term on the right hand side, where $\hat{q}_r \dot{q}_{md}$ is used instead of $\dot{q}_{rd} \dot{q}_m$ as in (10.28), and in the nonlinear damping term, where the internal reference speed $\dot{q}_{md}$ is used instead of the real speed $\dot{q}_m$. This internal reference speed is defined by (10.69), and it depends on the estimated rotor currents and the desired stator currents. After estimates and real currents have converged to their desired values, this speed will indeed be the actual rotor speed. To extend the torque tracking objective above to classical speed/position tracking problems, an outer loop is needed to define the desired torque τ_*. Since the derivatives of the desired currents are needed in the control (10.67), $\dot{\tau}_*$ must also be known.

The speed/position control loop in Section 3.2 is an example of how the outer torque generating loop can be defined.

Notice that the observed rotor currents and the derivatives of the desired currents are used only in the case of a voltage input (10.67), and hence they are not needed if a CSI or a VSI with current control is used. In these cases the controls will be the stator currents $\dot{q}_s \equiv \dot{q}_{sd}$, and the controller above exactly reduces to the controller of Proposition 11.8 in Chapter **11**.

4.6   A *dq*-implementation

For the purpose of implementation, it is of interest to formulate the controller in an arbitrary rotating frame of reference. This can be done by using the results from Section 1.3, giving

$$\lambda_{dq} \triangleq \begin{bmatrix} \lambda_d \\ \lambda_q \end{bmatrix} = e^{-\mathcal{J}(\theta_a - n_p q_m)} \lambda_r, \ i_{dq} \triangleq \begin{bmatrix} i_d \\ i_q \end{bmatrix} = e^{-\mathcal{J}\theta_a} \dot{q}_s$$

$$u_{dq} \triangleq \begin{bmatrix} u_d \\ u_q \end{bmatrix} = e^{-\mathcal{J}\theta_a} u, \ \dot{q}_r^{dq} \triangleq e^{-\mathcal{J}(\theta_a - n_p q_m)} \dot{q}_r$$

θ_a is the solution of $\dot{\theta}_a = \omega_a$, $\theta_a(0) = 0$, with ω_a the angular speed of rotation for the reference frame relative to the stator fixed frame. The real rotor currents rotated to a dq-frame, have been denoted $\dot{q}_r^{dq}$.

Using these definitions and following the same procedure as for the derivation of (10.15)–(10.16), it follows that the observer in (10.72) and (10.74) can be rewritten

$$\sigma L_s \left\{ \frac{d}{dt}\hat{i}_{dq} + [\omega_a \mathcal{J} + \gamma I_2]\hat{i}_{dq} + \frac{L_{sr}}{\sigma L_s L_r}[n_p \dot{q}_m \mathcal{J} - \frac{1}{T_r}I_2]\hat{\lambda}_{dq} \right\} = u_{dq} = e^{-\mathcal{J}\theta_a} u$$

$$T_r \dot{\hat{\lambda}}_{dq} + T_r(\omega_a - n_p \dot{q}_m)\mathcal{J}\hat{\lambda}_{dq} + \hat{\lambda}_{dq} = L_{sr}\hat{i}_{dq} - T_r n_p L_{sr} \dot{q}_m \mathcal{J}\left(\hat{i}_{dq} - i_{dq}\right) \quad (10.78)$$

Next, a relation between $\dot{\hat{q}}_r^{dq}$ and the other estimated and measured dq-quantities must be derived. Notice from (10.72) and (10.74) that

$$\dot{\hat{\lambda}}_r + R_r \dot{\hat{q}}_r = -n_p L_{sr} \dot{q}_m \mathcal{J} e^{-\mathcal{J} n_p q_m}\left(\dot{\hat{q}}_s - \dot{q}_s\right)$$

Differentiation of $\hat{\lambda}_r = e^{\mathcal{J}(\theta_a - n_p q_m)}\hat{\lambda}_{dq}$ and substitution of this expression in the equation above, gives after a rearrangement of terms

$$\dot{\hat{q}}_r^{dq} = e^{-\mathcal{J}(\theta_a - n_p q_m)}\dot{\hat{q}}_r = -\frac{1}{R_r}\left[(\omega_a - n_p \dot{q}_m)\mathcal{J}\hat{\lambda}_{dq} + \dot{\hat{\lambda}}_{dq} + n_p L_{sr}\mathcal{J}\dot{q}_m\left(\hat{i}_{dq} - i_{dq}\right)\right]$$

Substitution of the first two terms in the bracket with terms from (10.78) results in

$$\dot{\hat{q}}_r^{dq} = -\frac{1}{R_r}\left[\frac{L_{sr}}{T_r}\hat{i}_{dq} - \frac{1}{T_r}\hat{\lambda}_{dq}\right] \quad (10.79)$$

To express the control (10.67) in terms of dq-quantities, notice that if the desired stator and rotor currents are written in terms of rotated desired quantities[18] i_{dq}^d, $\dot{q}_{rd}^{dq}$ as $\dot{q}_{sd} = e^{\mathcal{J}\theta_a} i_{dq}^d$ and $\dot{q}_{rd} = e^{\mathcal{J}(\theta_a - n_p q_m)} \dot{q}_{rd}^{dq}$, it follows that

$$\ddot{q}_{sd} = \frac{d}{dt}\left[e^{\mathcal{J}\theta_a} i_{dq}^d\right] = e^{\mathcal{J}\theta_a}\left[\omega_a \mathcal{J} i_{dq}^d + \frac{d}{dt} i_{dq}^d\right] \tag{10.80}$$

$$\ddot{q}_{rd} = \frac{d}{dt}\left[e^{\mathcal{J}(\theta_a - n_p q_m)} \dot{q}_{rd}^{dq}\right] = e^{\mathcal{J}(\theta_a - n_p q_m)}\left[(\omega_a - n_p \dot{q}_m)\mathcal{J}\dot{q}_{rd}^{dq} + \frac{d}{dt}\dot{q}_{rd}^{dq}\right]$$

From the above and $\dot{\hat{q}}_r = e^{\mathcal{J}(\theta_a - n_p q_m)}\dot{\hat{q}}_r^{dq}$, it follows that

$$n_p L_{sr}\mathcal{J}e^{\mathcal{J}n_p q_m}\dot{\hat{q}}_r \dot{q}_{md} = n_p L_{sr}\mathcal{J}e^{\mathcal{J}\theta_a}\dot{\hat{q}}_r^{dq} \dot{q}_{md}$$

$$-n_p L_{sr}\dot{\hat{q}}_r^\top \mathcal{J}e^{-\mathcal{J}n_p q_m}\dot{q}_{sd} = -n_p L_{sr}\left(\dot{\hat{q}}_r^{dq}\right)^\top \mathcal{J}i_{dq}^d \tag{10.81}$$

Substitution of the above expressions into (10.67) gives the control in any reference frame. Especially, with the choice

$$\omega_a = n_p \dot{q}_m + \dot{\rho}_d$$

where $\dot{\rho}_d$ is defined in (10.68), (10.67) can be rewritten using (10.80)–(10.81) as

$$u = e^{\mathcal{J}(n_p q_m + \rho_d)}\Big[L_s\left\{\frac{d}{dt} i_{dq}^d + (n_p \dot{q}_m + \dot{\rho}_d)\mathcal{J}i_{dq}^d\right\} + L_{sr}\left\{\frac{d}{dt}\dot{q}_{rd}^{dq} + \dot{\rho}_d\mathcal{J}\dot{q}_{rd}^{dq}\right\}$$

$$+ n_p L_{sr}\mathcal{J}\dot{\hat{q}}_r^{dq}\dot{q}_{md} + R_s i_{dq}^d - K_1(\dot{q}_{md})\left(i_{dq} - i_{dq}^d\right)\Big]$$

where $\dot{\hat{q}}_r^{dq}$ is given in (10.79), and $\dot{q}_{md}$ is defined by

$$\ddot{q}_{md} = \frac{1}{D_m}\left(-n_p L_{sr}\left(\dot{\hat{q}}_r^{dq}\right)^\top \mathcal{J}i_{dq}^d - R_m \dot{q}_{md} - \hat{\tau}_L + K_2(\dot{q}_d)\dot{\hat{q}}_m\right),$$

$$\dot{q}_{md}(0) = \dot{q}_m(0)$$

The desired stator and rotor currents in the dq-frame are in this case given as (see Proposition 10.17)

$$\dot{q}_{rd}^{dq} = e^{-\mathcal{J}\rho_d}\dot{q}_{rd} = -\begin{bmatrix} \frac{\dot{\beta}_*}{R_r} \\ \frac{\tau_*}{n_p \beta_*} \end{bmatrix}$$

$$i_{dq}^d = \frac{1}{L_{sr}}\begin{bmatrix} \beta_* + \frac{L_r}{R_r}\dot{\beta}_* \\ \frac{L_r}{n_p \beta_*}\tau_* \end{bmatrix}$$

The nonlinear gains K_1 and K_2 can be calculated directly from (10.70) and (10.71), noting that the squared norm of $\dot{q}_{sd}$ is equal to the squared norm of i_{dq}^d.

---

[18]Desired stator currents $\dot{q}_{sd}$ expressed in a dq-frame are denoted $i_{dq}^d = \left[i_d^d, i_q^d\right]^\top$.

4.7   Simulation results

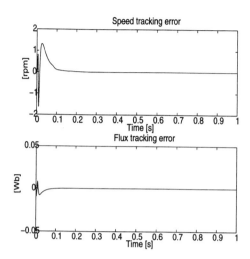

Figure 10.6: Tracking errors for flux norm and speed. Observer-based controller.

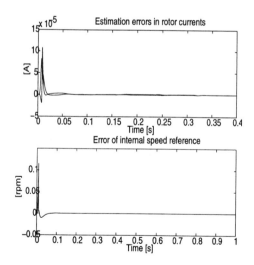

Figure 10.7: Components of estimation error $\dot{\hat{q}}_r - \dot{q}_r$ and error between the real speed $\dot{q}_m$ and the internal speed $\dot{q}_{md}$ from (10.69). Observer-based controller.

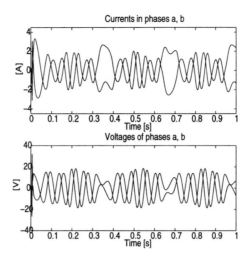

Figure 10.8: 3ϕ Voltages and currents u_a, u_b, i_a, i_b. Observer-based controller.

Simulations with the same reference trajectories (see Fig. 10.3), filter values and under exactly the same conditions as in Section 3.9 were performed with the observer-based controller in Proposition 10.17. The outer speed control loop was chosen as in (10.30) to generate τ_* , with filtering of the speed error to allow for speed tracking without acceleration measurement. Control parameters were chosen as: $a = 1000$, $b = 300$, $k_1 = 30$, $k_2 = 2$, $\epsilon_1 = 0.45R_r$, $\gamma_{\tau_L} = 0$. All initial conditions were set to zero. Results from the simulations are shown in Figs. 10.6–10.8.

As it can be seen from Fig. 10.6, there is no interaction between speed and flux norm control, as predicted from the analysis. Also for this controller filtering of references was found to be of high importance for avoidance of saturations in currents and voltages, and the parameter k_1 in the additional damping term was important for improving tracking performance for reference speeds close to zero. For small damping R_m, the parameter k_2 was also important for the quality of speed tracking.

Comparison with Figs. 10.4-10.5 from the observer-less case shows that there is little difference between the two approaches under ideal conditions. This can be explained as follows: After a short initial transient the internal speed $\dot{q}_{md}$ and the rotor current estimates converge to their real values (and consequently also their desired values), and then the two controllers behave exactly similar.

4.8   Concluding remarks

In this section an observer-based solution to the output feedback torque and rotor flux tracking problem for an induction motor model was proposed. The controller is an outgrowth of the work in [77, 204, 206]. More specifically, it is an extension of the work presented in the last paper to include the important case of rotor flux norm tracking. In this approach the design of controller and observer is interlaced, and global stability results for the total system can be proved, provided that nonlinear damping terms are added to the controller to compensate for estimation errors.

It must be pointed out that the observer used in this approach is simple, with only updating from current error terms in the rotor equations. It would in general be advantageous to have updating also in stator equations for making the observer more robust. The inclusion of error terms with adjustable gains will also give the possibility of specifying a convergence rate less sensitive to resistance parameters. These are possible future extensions of this observer-based result.

However, the observer is only needed when a voltage input is used, since the controller proposed in this section reduces to the controller of Proposition 11.8 (see Chapter 11) for a motor with a CSI or a VSI with current control. Consequently, the robustness results from [68] (see Section 11.5) with respect to uncertainty in rotor flux time constant T_r hold for this controller too in the case of current inputs.

5   Field-oriented control and feedback linearization

The concept of field-oriented control (FOC) was introduced by the German researchers Hasse and Blaschke in the beginning of the seventies. It was based on the space vector description of the dynamic model, formulated in the fifties by the researchers Lyon, Kovacs and Racz, for the analysis of transients in AC-machines. The ideas presented in [100] for vector control of a PWM inverter fed induction motor, showed a remarkable improvement in response as compared to previous scalar methods, which were based on steady state linear models. The ideas did not seem to be of general interest, and their realization was quite complex. Some years later a general theory for vector control of AC machinery, based on deep understanding of the physics of the systems and a nonlinear model, was presented in [24]. This theory is now the *de facto* standard for high-performance control of AC machinery, in good part due to the strong influence by the pioneering work of Professor Leonhard. The method consists of a nonlinear change of coordinates together with a nonlinear asymptotically decoupling state feedback. Seen from the viewpoint of modern control theory, FOC was one of the earliest implementations of ideas which are now considered to belong to the more general field of geometric control theory. The fact that it gives superior performance as compared to methods based on classical linear theory, has motivated and paved the way for other applications of nonlinear control theory.

In the rest of this section the rationale behind *rotor-flux-oriented* vector control[19], and the arguments commonly used against this approach are explained. In particular, it is argued that better performance can be achieved with a controller that *exactly* (and not just asymptotically) linearizes and decouples the motor dynamics. A controller that achieves this objective is presented.

There are generally two types of FOCs, and in this presentation, we adopt the following definitions of *direct* and *indirect* FOC:

Definition 10.20 (Direct and indirect FOC.) Direct *FOC refers to an implementation where the full state of the system is* measured *(Hall-effect sensors, search coils, tapped stator windings), while* indirect *FOC refers to an implementation where only stator currents, position and speed are measured.*

There is some inconsistency in the literature regarding these definitions, and sometimes direct FOC (also called flux feedback control) refers to the case where flux is measured or estimated, while approaches using the reference values are denoted indirect (or flux feed-forward control). See, e.g. pp. 124–125 in [275]. We prefer to take the clear distinction of state or output–feedback given above.

It must also be pointed out that there is no clear definition in the literature of what exactly an estimator or a reference signal is. Our PBC has an implicit estimator, in the sense the $\tilde{\lambda} \to 0$, with the reference λ_d interpreted as an estimate.

5.1   Rationale of field–oriented control

The simplicity of controlling a DC motor is due to mechanical commutation, which ensures that the main flux from the field winding in the stator is always orthogonal to the magneto-motive force created by the current in the armature winding. It follows that the torque, which is proportional to the vector cross product of flux and current, will be proportional to armature current when flux is kept constant. Hence, it can easily be controlled to give high dynamic performance by use of two linear loops, controlling flux and armature current. In AC machinery flux and MMF distributions rotate with different speeds, resulting in varying relative angle. This motivates the use of rotating reference frames to analyze the dynamics.

Looking at the torque equation for the rotated model of an induction motor in (10.17),

$$\tau = n_p \frac{L_{sr}}{L_r}(\lambda_d i_q - \lambda_q i_d)$$

---

[19]There are several implementations of FOC, depending on the frame of reference chosen for model representation, e.g. rotor-, magnetizing flux- and stator-oriented. The rotor-oriented implementation is what now commonly is called *field-oriented control* [275].

it can be seen that if one of the components of λ_{dq} is held constant, while the other is zero, torque can be controlled by controlling one of the rotated stator currents, analogous to the DC motor. To achieve this objective a reference frame in which *the rotor flux vector is aligned with one of the axis*, should first be chosen. Thus, the angle of the rotor flux vector must be known.

Assuming that the d-axis of the reference frame has been aligned with the rotor flux vector, it can be derived from (10.12) using (10.17), that the rotor flux speed (relative to the rotor frame) can be written in terms of dq-components of flux and currents as

$$\dot{\rho} = \frac{L_{sr}}{T_r} \frac{i_q}{\lambda_d}$$

Indeed, if it is assumed that $|\lambda_d| \geq \epsilon > 0$, and ω_a is chosen to be the rotational speed of flux relative to the fixed stator frame,

$$\omega_a = n_p \dot{q}_m + \frac{L_{sr}}{T_r} \frac{i_q}{\lambda_d} \tag{10.82}$$

the second of the equations in (10.15) gives

$$\dot{\lambda}_q = -\frac{1}{T_r} \lambda_q$$

Hence λ_q converges exponentially to zero.

The first equation in (10.15) becomes

$$T_r \dot{\lambda}_d + \lambda_d = L_{sr} i_d + \epsilon_t \tag{10.83}$$

with ϵ_t and exponentially decaying term. Thus, λ_d can (in principle) be controlled by i_d. For instance, if $i_d = \frac{\beta_*}{L_{sr}}$ with $\beta_* > 0$ a constant, it follows that $\lim_{t\to\infty} \lambda_d = \beta_*$.

Assuming that λ_d is held constant, while λ_q is still zero, it follows that

$$\tau = n_p \frac{L_{sr}}{L_r} \lambda_d i_q$$

and torque can be controlled by the q-component of the rotated stator currents. This is the essence of FOC.

The remaining problem is to control the rotated stator currents i_{dq}. This can either be done by high-gain current control, or if this does not give satisfactory performance, by exact cancelation of parts of the dynamics in (10.16). High-gain current control leads, invoking a singular–perturbation argument, to a simpler model, the so–called current–fed machine. This case is treated in greater detail in Chapter **11**.

Otherwise, with the nonlinear decoupling input

$$u_{dq} = \sigma L_s \left(\omega_a \mathcal{J} i_{dq} + \frac{L_{sr}}{\sigma L_s L_r} \left[n_p \dot{q}_m \mathcal{J} - \frac{1}{T_r} I_2 \right] \lambda_{dq} + v_{dq} \right) \tag{10.84}$$

equation (10.16) can be written as

$$\frac{d}{dt}i_{dq} = -\gamma i_{dq} + v_{dq}$$

The inputs v_{dq} can now easily be defined to force i_d and i_q to their desired values. Usually v_{dq} consist of nested PI-loops as in [153]

$$v_d = H_1(p)\left(i_d^d - i_d\right) \qquad (10.85)$$
$$v_q = H_2(p)\left(i_q^d - i_q\right) \qquad (10.86)$$

where $H_i(p)$, $i = 1, 2$ are PI-controllers, and the desired stator currents i_d^d, i_q^d could be defined as

$$i_d^d = \frac{\beta_*}{L_{sr}} \qquad (10.87)$$
$$i_q^d = \frac{L_r}{n_p L_{sr}\beta_*}\tau_d \qquad (10.88)$$

with τ_d generated by an outer speed PI-controller

$$\tau_d = H_3(p)(\dot{q}_{m*} - \dot{q}_m) \qquad (10.89)$$

Instead of the feed-forward definition of i_d^d, it could be defined using feedback as

$$i_d^d = H_4(p)(\beta_* - \lambda_d)$$

with $H_4(p)$ a PI-controller.

Assuming rotor flux is available for measurement, the voltage input for direct FOC may then be implemented as

$$u = e^{\mathcal{J}\theta_a}u_{dq}$$
$$= e^{\mathcal{J}\theta_a}\sigma L_s \left(\omega_a \mathcal{J}i_{dq} + \frac{L_{sr}}{\sigma L_s L_r}\left[n_p\dot{q}_m\mathcal{J} - \frac{1}{T_r}I_2\right]\lambda_{dq} + v_{dq}\right) \qquad (10.90)$$

It is also possible to implement high-gain current control by neglecting all terms except v_{dq} in the equation above.

Notice that to compute ω_a from (10.82), the norm of the rotor flux must be strictly greater than zero. This assumption does not hold at startup, giving a controller singularity at this point. To avoid the unwanted controller blow-up for small values of rotor flux norm measurements or estimates, some heuristics is added to the control scheme to make it work, for instance exciting i_d before i_q or setting λ_d used in the controller equal to a constant value, when the measured/estimated value is smaller than a certain limit.

5.2 State estimation or reference values

In the above derivations it has been assumed that all states including rotor flux norm and angle could be measured. In general this assumption does not hold, as explained in Remark 10.4. This problem has been a longstanding research topic, and generally there are two ways to solve it. The first one is to *estimate* the rotor flux angle and amplitude, while the other is to use reference values for these two quantities.

As an example of the first method, rotor flux can be estimated in open loop from stator current measurements using the first equation of (10.15), and its angle can be found by integrating (10.82) with the estimated value of λ_d as

$$\hat{\omega}_a = n_p \dot{q}_m + \frac{L_{sr}}{T_r} \frac{\hat{i}_q}{\hat{\lambda}_d} \tag{10.91}$$

where

$$\hat{\lambda}_d = \frac{L_{sr}}{T_r p + 1} \hat{i}_d \tag{10.92}$$

The estimated currents $\hat{i}_{dq}$ are computed from measured currents using the estimated angle $\hat{\theta}_a$ and the rotation defined in (10.13). This simple estimation scheme has been used for high-performance control of an induction motor in [30]. See [280] for other solutions to the estimation problem.

To implement indirect FOC, the same feed-forward way of defining the desired currents as in (10.87) and (10.88) is used, but now the rotor flux speed in (10.82) is also computed using reference values

$$\omega_a = n_p \dot{q}_m + \frac{L_{sr}}{T_r} \frac{i_q^d}{\beta_*}$$

and θ_a is found by integration as before.

The indirect approach has become the most popular implementation of FOC since it does not require flux sensors or a flux model, avoiding the need for estimation [275]. Also, its performance at low rotor speed is generally better than for direct schemes, for which there are estimation problems present when the ohmic losses become dominant in the stator equation, and signal integration is problematic [158]. However, all of these schemes are highly sensitive to parameters, especially to changes in the rotor time constant T_r. When this parameter is uncertain, asymptotic decoupling of flux norm and torque is lost, resulting in second order dynamic flux and torque interactions.

It is often argued that the indirect approach is much more sensitive to parameter uncertainty than other direct approaches. This issue has recently been addressed in [68], where it is shown that the first requirement of the system, *global stability*, is preserved for the indirect approach with as much as a 200% error in rotor resistance estimate. This is a remarkably strong result, and to the best of the authors' knowledge, there is no such result for the direct approach, where controllers based on a "nonlinear separation principle" are used. See Section **11**.5 for a presentation of these robustness results.

Motivated by the successful use of the voltage decoupling terms in (10.84) in combination with estimated states, there have been several attempts to implement equivalents by using reference values instead of the real the states, see pp. 152-158 in [275]. The first problem with these approaches is that the term $\sigma L_s \frac{d}{dt} i_{dq}^d$ with the derivative of the stator current references i_{dq}^d, is either neglected, resulting in unwanted torque overshoot, or has to be computed in an *ad hoc* way by numerical differentiation. The other problem is that these schemes are only suited for constant rotor flux norm. The indirect scheme proposed in Proposition 10.6 (see also Section 3.7) is an alternative method which do not possess these problems, and which also gives a theoretical explanation of why these reference value "decoupling" approaches actually work.

5.3   Shortcomings of FOC

There are mainly two arguments used against the various implementations of FOC:

i) It does not give full decoupling

The decoupling between flux and torque (or speed) control is only asymptotic [172]. There is only decoupling when the flux has converged to its *constant* value, giving also a decoupled and linear speed dynamics. This means that simultaneously tracking of both flux amplitude and torque/speed/position using FOC most likely will cause problems, especially if the flux reference is not slowly varying. For instance, as pointed out in [263], operation in the flux-weakening regime will excite the coupling between flux and speed. This gives undesired speed fluctuations, and could possibly cause instability. The problem is due to the rationale behind FOC, which is to make it behave like a DC motor, where torque is proportional to current when flux is constant. However, this does not give a decoupling between speed/torque and flux norm, which are the outputs a tracking controller must be designed for. For the reasons above it has been natural to operate the machine at maximum constant flux level below rated speed, something which restricts the possibility of power efficient operation.

The problem of full (dynamical) decoupling of rotor flux norm and torque (exact input-output linearization), still under the assumption of full state feedback, was solved in [143]. The basic idea for the dq-implementation (under the assumption of ideal field orientation, $\lambda_q = 0$) is to choose λ_d and the torque

$$\tau = n_p \frac{L_{sr}}{L_r} \lambda_d i_q$$

as controlled variables, instead of λ_d and i_q. For a motor with current control loops as previously explained, decoupling can be achieved directly by defining the current reference for i_q as

$$i_q^d = \frac{L_r}{n_p L_{sr}} \frac{\tau_*}{\lambda_d} \tag{10.93}$$

with τ_* an external torque reference.

Some additional insight into the decoupling problem can be gained if the new nonlinear decoupling terms in the q-direction are found by directly evaluating the dynamic equation for the torque, giving

$$
\begin{aligned}
\dot{\tau} = {} & n_p \frac{L_{sr}}{L_r} \left[\dot{\lambda}_d i_q + \lambda_d \frac{d}{dt} i_q \right] - \left(\frac{1}{T_r} + \gamma \right) \tau + \frac{L_{sr}}{T_r} \frac{i_d}{\lambda_d} \tau - \omega_a \frac{n_p L_{sr}}{L_r} \lambda_d i_d \\
& - \frac{L_{sr}^2 n_p^2}{\sigma L_s L_r^2} \dot{q}_m \lambda_d^2 + \frac{n_p L_{sr}}{\sigma L_r L_s} \lambda_d u_q
\end{aligned}
$$

where (10.16), (10.83), $i_q = \frac{L_r}{n_p L_{sr}} \frac{\tau}{\lambda_d}$, and ω_a as defined in (10.82) have been used.

From the above it follows that the choice

$$u_q = \frac{\sigma L_r L_s}{n_p L_{sr} \lambda_d} \left[v_q - \frac{L_{sr}}{T_r} \frac{i_d}{\lambda_d} \tau + \omega_a \frac{n_p L_{sr}}{L_r} \lambda_d i_d + \frac{L_{sr} n_p}{\sigma L_s L_r} \dot{q}_m \lambda_d^2 \right]$$

gives

$$\dot{\tau} = - \left(\frac{1}{T_r} + \gamma \right) \tau + v_q$$

and τ can be controlled with a linear inner PI-controller $H(p)$

$$v_q = H(p)(\tau_d - \tau)$$

where τ_d could be chosen as in (10.89). The voltage decoupling term in the d-axis is still given by the first of the equations in (10.84).

With another choice of outputs, the decoupling problem is not as easy to solve as above, and tools adopted from differential geometry must be used for its solution. This approach will be explained in the next section.

ii) It is based on "a nonlinear separation principle"

It is well known that for linear time-invariant systems, the problem of stabilizability with only partial state measurement can be solved by use of an observer, at least if the system is both stabilizable and detectable. The stability of the total system is only dependent on the observer stability, and the stability of the feedback control system when full state measurement is assumed. This is the so-called (deterministic) *separation principle.*

Motivated by the successful linear controller designs, this principle has been carried over to controller designs for nonlinear and non autonomous systems, giving a *"nonlinear separation principle"*. Generally no theoretical stability analysis are given for systems resulting from such designs, and the performance is only verified through simulations or experiments. It can be showed that even for very simple nonlinear systems, with an exponentially convergent state observer[20] and known parameters, the approach of separate observer-controller design can lead to explosive instability in terms of *finite escape time*, when estimated states are used as if they were real states, even if the controller with real states would ensure global exponential stability of the total system [142]. Analysis of separate observer-controller design using linearization techniques leads to local stability results, even under the assumptions of known constant parameters. With only local results, it is very difficult to predict what will happen for certain choices of controller parameters and initial conditions under real operation, and the designer is left with a trial and error approach to controller design, making tuning an often time consuming and difficult task.

It must also be pointed out than when observers are used for implementation, asymptotic properties are added to the schemes, i.e. there is only exact decoupling after the estimates have converged to their real values.

Even if some heuristics are implemented to avoid instability, estimation errors can lead to severe effects on system performance and power efficiency, especially during high speed transients. This is why much of the research in nonlinear control of induction motors has aimed at interlacing design of controller and observer, with additional terms in the controller or the observer to counteract for estimation errors, and obtaining global stability results. These schemes have the benefit of guaranteeing stability (under given assumptions), and hence giving *a priori* information about which modifications can reasonably be expected to work and which ones will probably not. In the cases where exponential stability results are established for a nominal system, robustness result for bounded perturbations can also be established [127].

It must be pointed out that even if the robustness and stability issues of the many implementations of FOC are not rigorously established, this scheme has through years of practical experience been developed to a level giving a performance which is difficult to compete with for other nonlinear approaches, at least for nonlinear designs implemented directly from theoretical desk designs. Years of experimental work has resulted in modifications of FOC based on experience and intuition, and problems of parameter uncertainty, flux saturation and other unmodeled dynamics can be compensated for by several *ad hoc* methods. Even if these methods are based on "nonlinear certainty equivalence", using estimated parameters as if they were real parameters in the controller, and proposed without theoretical justification, the most important aspect from an application oriented view is that they work and improve performance. The rigorous analysis of the resulting schemes is left as challenging

---

[20]This is generally the best convergence that can be expected.

problems for the academics.

5.4   Feedback linearization

As a solution to the decoupling problem in standard FOC, an *input-output* linearization from voltage inputs to speed and square of rotor flux norm in the case of the
stator fixed *ab*-model, was derived in [143] (see also [172]), using the new powerful
tools adopted from differential geometry.

The linearization procedure starts with (10.20) and the definition of a set of new
outputs $y_i : \mathbb{R}^5 \mapsto \mathbb{R}$, $i = 1, 2, 3$, which are functions of the states $x = [\dot{q}_m, \lambda_{ab}^\mathsf{T}, i_{ab}^\mathsf{T}]^\mathsf{T}$

$$
\begin{aligned}
y_1(x) &= \dot{q}_m \\
y_2(x) &= L_f y_1(x) = \frac{n_p L_{sr}}{D_m L_r}(\lambda_a i_b - \lambda_b i_a) - \frac{\tau_L}{D_m} \\
y_3(x) &= \|\lambda_{ab}\|^2 \\
y_4(x) &= L_f y_3(x) = -\frac{2R_r}{L_r}\|\lambda_{ab}\|^2 + \frac{2R_r L_{sr}}{L_r}\lambda_{ab}^\mathsf{T} i_{ab} \\
y_5(x) &= \arctan(\frac{\lambda_b}{\lambda_a})
\end{aligned}
$$

In the equations above, $L_f y_i(x) \stackrel{\triangle}{=} (\partial y_i/\partial x)^\mathsf{T} f(x)$ is the directional derivative (or
the *Lie derivative*[21]) of the function $y_i(x)$ along the vector field $f(x)$, and satisfies
$L_f(L_f^{m-1} y_i(x)) = L_f^m y_i(x)$.

The definition of new outputs is not defined for $\|\lambda_{ab}\| = 0$.

Differentiation of the equations above under the assumption that the load torque
τ_L is constant, results[22] in

$$
\begin{aligned}
\dot{y}_1(x) &= y_2(x) \\
\dot{y}_2(x) &= L_f^2 y_1(x) + L_{g_1} L_f y_1(x) u_1 + L_{g_2} L_f y_1(x) u_2 \\
\dot{y}_3(x) &= y_4(x) \\
\dot{y}_4(x) &= L_f^2 y_3(x) + L_{g_1} L_f y_3(x) u_1 + L_{g_2} L_f y_3(x) u_2 \\
\dot{y}_5(x) &= L_f y_5(x)
\end{aligned}
$$

from which it can be seen that if the real inputs are given as

$$
\begin{bmatrix} u_1 \\ u_2 \end{bmatrix} = B^{-1}(x) \begin{bmatrix} -L_f^2 y_1(x) + v_1 \\ -L_f^2 y_3(x) + v_2 \end{bmatrix} \tag{10.94}
$$

---

[21]Named after the famous Norwegian mathematician Marius Sophus Lie (1842–1899).
[22]Notice that the terms $L_{g_1} y_i(x)$, $L_{g_2} y_i(x)$, $i = 1, 3, 5$ are zero.

with auxiliary inputs v_1, v_2, and the *decoupling matrix*

$$
\begin{aligned}
B(x) &= \begin{bmatrix} L_{g_1} L_f y_1(x) & L_{g_2} L_f y_1(x) \\ L_{g_1} L_f y_3(x) & L_{g_2} L_f y_3(x) \end{bmatrix} \\
&= \begin{bmatrix} -\dfrac{n_p L_{sr}}{D_m L_r \sigma L_s} \lambda_b & \dfrac{n_p L_{sr}}{D_m L_r \sigma L_s} \lambda_a \\ \dfrac{2 L_{sr} R_r}{\sigma L_s L_r} \lambda_a & \dfrac{2 L_{sr} R_r}{\sigma L_s L_r} \lambda_b \end{bmatrix}
\end{aligned} \tag{10.95}
$$

the first four equations of the resulting system will be

$$
\begin{aligned}
\dot{y}_1(x) &= y_2(x) \\
\dot{y}_2(x) &= v_1 \\
\dot{y}_3(x) &= y_4(x) \\
\dot{y}_4(x) &= v_2
\end{aligned}
$$

which are two second-order linear and totally decoupled systems for speed $y_1(x)$ and square of rotor flux norm $y_3(x)$. These two outputs can now be controlled independently with v_1 and v_2 outputs from simple linear controllers.

The fifth differential equation of the system represents the dynamics of the flux angle

$$
\dot{y}_5(x) = n_p \dot{q}_m + \frac{R_r}{n_p} \frac{1}{\|\lambda_{ab}\|^2} (D_m \ddot{q}_m + \tau_L)
$$

which is *unobservable* with the choice of outputs and new controls in this approach [172].

From (10.95) it is clear that the above scheme has a singularity for zero rotor flux norm. Also, for the implementation of this scheme, measurement of the rotor flux is required, and with the use of observers, the points from the previous section regarding use of a "nonlinear separation principle" apply to this scheme as well.

It can also be seen from (10.94) that the need for exact cancelation of parts of the dynamics is essential for the implementation of this scheme. The terms $-L_f^2 y_i(x)$, $i = 1, 3$, which depend on the full state vector $x = [\dot{q}_m, \lambda_{ab}^\top, i_{ab}^\top]^\top$, must be cancelled with the control signals u_1 and u_2.

It was also shown in [143] that the implementation complexity of this scheme is not much greater than for FOC. Experimental results from various implementations (stator fixed frame of reference, decoupling control for square of rotor flux norm and speed or torque) of this scheme have been presented in [129, 274], and [30] (*dq*-frame implementation with high-gain current control).

In Section **11.2.3** we carry out a detailed comparison between PBC, FOC and feedback linearization for the case of current–fed machines.

6 Experimental results

6.1 Experimental setup

In this section the experimental setup used in the testing of different induction motor controllers is described. The setup was built from scratch with basic ideas for implementation taken from [235], but modified to fit specific needs, and avoid reported problems with noise and signal transmission. Additional equipment for load torque generation and torque measurement were built from standard components. For the software interface, an integrated system from dSPACE was chosen. This choice allowed for fast prototype implementation without extensive C or assembly language coding. The motor used in the experiments was a 4-pole 3ϕ squirrel-cage induction motor with a voltage source switched converter and a standard PWM scheme, and the controllers were implemented on a DSP. A more detailed description of the induction motor part of the setup can be found in [192].

A   Hardware description

The controller was implemented on a DS1102 controller board from dSPACE. This board has a 40 MHz TI320C31 32/32 bit floating-point DSP and a 25 MHz TI320P14 32/16 bit micro controller DSP. In addition there are 4 A/D-converters which are used for current (12 bit resolution, 3 μs conversion time) and torque[23] (16 bit resolution, 10 μs conversion time) measurements, 16 bit digital I/O of which 6 bits were used for pulse-width modulation and 1 bit for a converter enable signal, and two 24 bit encoder interfaces, of which one was used for position measurement. The micro controller computed the switching signals for the symmetric carrier-based PWM of the three phases from reference values transferred to it from the main processor at the end of each sampling interval. The three signals for the upper transistors in the converter legs were converted to optical signals on an interface card before they were transmitted through optical fibers to the converter, where complementary switching signals and blanking time of the converter (2 μs) were generated in hardware using an IXYS IXDP630 digital deadtime generator for 3ϕ PWM control.

A 1.1 kW Lust FU2235 voltage source converter with a DC-link voltage of $U_{\mathrm{DC}} = 300$V was used. The converter could give maximum line currents of $I_{max} = 6.8$ A. This converter was connected to a 4-pole ($n_p = 2$) Lust ASH-11-10I63-00 400 W squirrel-cage induction motor, as shown in Fig. 10.12 on page 369. The following nominal two phase parameters of the motor were given in the data sheet: $L_s^N = 99.0$ mH, $L_{sr}^N = 92.3$ mH, $L_r^N = 97.1$ mH, $R_s^N = 1.8$ Ω, $R_r^N = 2.2$ Ω, $D_m = 2.8$ kgcm^2, $\dot{q}_m^N = 3000$ rpm, $\beta^N = 0.2$ Wb, $\tau^N = 1.5$ Nm, $R_m \approx 0.005$ Nms/rad. A Lust BC1200

---

[23]The torque was measured only for illustration purposes. It was not used for feedback in any of the experiments presented in this section.

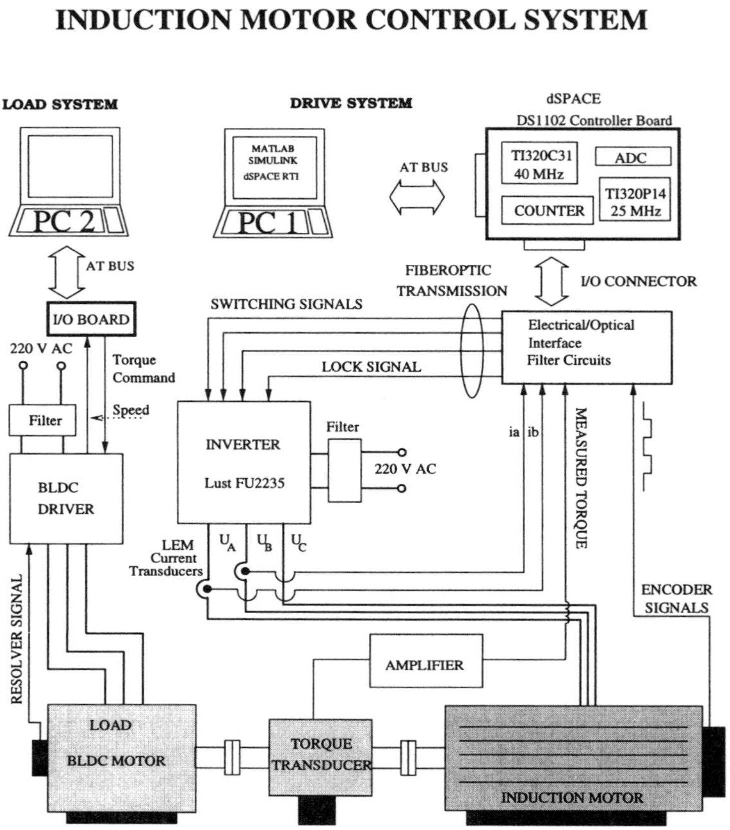

INDUCTION MOTOR CONTROL SYSTEM

Figure 10.9: Block diagram of experimental setup.

brake chopper was connected to the DC link for power dissipation.

To allow for optical transmission of the three switching signals and one enable signal from the DSP board to the converter, the standard microprocessor board for voltage/frequency control was removed from the inverter, and replaced by a specially designed interface card. Over-current protection was implemented both in software and hardware.

Position was measured using an incremental encoder with 4096 lines, and a quadruple counter was used, giving a position measurement resolution of $\frac{2\pi}{4\cdot4096}$ rad.

The currents were measured with LEM LA 25-NP current transducers and filtered with first order analog anti-aliasing filters, having a cut-off frequency of 1.1 kHz, before they were converted by the 12 bit A/D converters. An offset correction of the current measurements was done at startup, when the inverter switches were disabled and the currents in the motor windings were zero.

As a load for the induction motor, a current controlled BSM 80A250 brushless DC-motor from Baldor with a BSC1105 driver was used. This motor has a moment of inertia equal to 2.13 kgcm^2, and is capable of producing a nominal torque of 3.2 Nm. A separate PC with a PCL-711B I/O board and external electronics was used to control the load.

Torque was measured using a HBM T1 50 Nm strain gauge torque transducer with an AE101 amplifier, and the signal was filtered with a first order low-pass filter. This torque transducer has a moment of inertia equal to 0.6 kgcm^2.

An overview of the total system is given in Fig. 10.9.

B   Software description

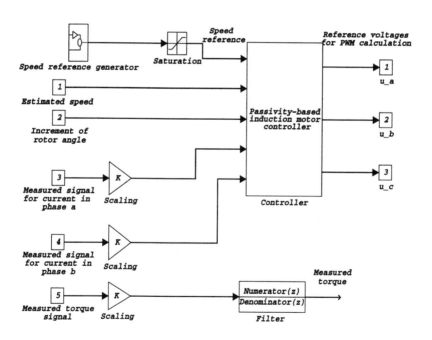

SIMULINK Block Diagram for C-Code Generation

Figure 10.10: Main block diagram for C-code generation from SIMULINK™.

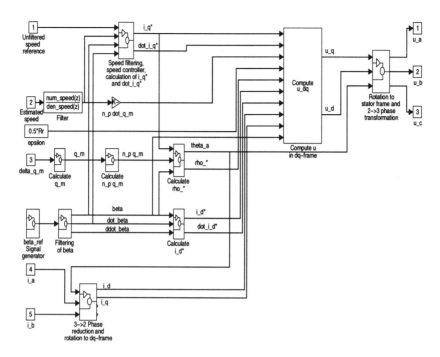

Figure 10.11: Example of SIMULINK™block diagram for controller.

The DSP board was installed in a 80486/66MHz PC with MATLAB™/SIMULINK™, and a RTI C code generator was used, which converted the controller graphically described in SIMULINK to C code which could be compiled and run on the DSP. For logging of signals and online tuning of parameters, the programs TRACE and COCKPIT[24] were used to communicate with the DSP board.

Computation of speed, position increment, current and torque measurements and the communication between the main processor and the micro controller, was implemented in external C-code which was linked with the code generated from the block diagram. These routines also provided the mapping between software inputs/outputs specified in the block diagram, and hardware addresses on the board. The code for the PWM running on the micro controller was written in assembly and down-loaded to the slave-processor during startup. In Fig. 10.10 the main block diagram used for code generation is shown, and in Fig. 10.11 an example of the diagram for the controller block is shown.

---

[24]MATLAB™and SIMULINK™are registered trademarks of The MathWorks Inc. RTI, COCKPIT and TRACE are registered trademarks of dSPACE GmbH.

The controller was implemented with multirate computation, where the PWM calculation was run relatively fast on the micro controller at a sampling period of $T_{\text{PWM}} = 100$ μs independent of control algorithm, with a slower computation of reference voltages and speed control at a period of T_{sampl} seconds. These sampling periods depended on the implemented control algorithm.

C Controller discretization and speed estimation

In the derivations below a base sample period of T [s] for controller implementation is assumed, and $y(k)$, $k \in \mathbb{Z}_+$ is used to denote a sample of the signal $y(t)$ at time $t = kT$. Discretization of controller equations were done using the ZOH (Zero-Order-Hold) approximation of an integration. In this approach the discrete equivalents of transfer functions for linear continuous systems are derived by using the formula

$$h_{ZOH}(z) = (1 - z^{-1})\mathcal{Z}\left\{\frac{h(p)}{p}\right\}$$

where z^{-1} is the delay operator, and $\mathcal{Z}$ denotes the z-transform. This gives [87]

Linear filter: $\dfrac{y}{u}(p) = \dfrac{1}{T_1 p + 1}$

$\downarrow$

$$y(k) = e^{-\frac{T}{T_1}}y(k-1) + (1 - e^{-\frac{T}{T_1}})u(k-1)$$

Integration: $\dfrac{y}{u}(p) = \dfrac{1}{p}$

$\downarrow$

$$y(k) = y(k-1) + Tu(k-1)$$

To avoid integral wind-up, conditional integration was implemented in all PI-controllers. In this approach, the integral term is held constant when the output of the controller exceeds an adjustable limit.

Since the induction motor used in this work had no speed transducer, speed had to be estimated from discrete time position measurement. This was done using the simple backward difference approximation [158]

$$\hat{\dot{q}}_m = \frac{q_m(k) - q_m(k-3)}{3T}$$

This average speed is a rather rough first estimate with a resolution (assuming quadruple counter) of

$$\dot{q}_m^{\text{res}} = \frac{\frac{2\pi}{4N\text{pos}}}{3T}$$

N_{pos} is the encoder resolution in pulses per rotation. For $T = 300$ μs and $N_{\mathrm{pos}} = 4096$, the resolution is $\dot{q}_m^{\mathrm{res}} \approx 0.43$ rad/s (≈ 4.1 rpm). The backward difference estimate of speed was smoothened using a discrete implementation of the linear filter

$$h(p) = \frac{1}{3Tp + 1}$$

before it was used for control purposes.

D   Phase transformations

To transform the measured 3ϕ line currents I_a and I_b to equivalent 2ϕ phase currents $\dot{q}_s$ used in the controller calculations, the linear transformation

$$\dot{q}_s = \left[\begin{array}{c} \dot{q}_{s1} \\ \dot{q}_{s2} \end{array} \right] = \sqrt{\frac{2}{3}} \left[\begin{array}{cc} \frac{3}{2} & 0 \\ \frac{\sqrt{3}}{2} & \sqrt{3} \end{array} \right] \left[\begin{array}{c} \frac{1}{\sqrt{3}} I_a \\ \frac{1}{\sqrt{3}} I_b \end{array} \right]$$

was used. The 3ϕ-voltages u_a, u_b, u_c used for the PWM calculation were computed from the voltages u_1, u_2 calculated by the controller as

$$\left[\begin{array}{c} u_a \\ u_b \\ u_c \end{array} \right] = \sqrt{\frac{2}{3}} \left[\begin{array}{cc} 1 & 0 \\ -\frac{1}{2} & \frac{\sqrt{3}}{2} \\ -\frac{1}{2} & -\frac{\sqrt{3}}{2} \end{array} \right] \left[\begin{array}{c} u_1 \\ u_2 \end{array} \right]$$

Note that $u_c = -(u_a + u_b)$, hence only u_a and u_b are needed.

The form of the transformation used here is the so-called *power-invariant* form of the transformation, see [275].

E   Pulse-width modulation

A standard symmetric carrier-based PWM was chosen for generation of the switching signals of the converter.

In this method the switching signals for the transistors in the three bridge legs of the converter (see Fig. 10.12) are generated to be symmetric to the middle of the switching interval, and the on-time for each upper switch t_{ON}^i, $i \in \{a, b, c\}$ is computed from

$$t_{\mathrm{ON}}^i = \left\{ \begin{array}{ll} 0 & u_i < -\frac{U_{\mathrm{DC}}}{2} \\ (\frac{1}{2} + \frac{u_i}{U_{\mathrm{DC}}}) T_{\mathrm{PWM}} & -\frac{U_{\mathrm{DC}}}{2} \leq u_i \leq \frac{U_{\mathrm{DC}}}{2} \\ T_{\mathrm{PWM}} & u_i > \frac{U_{\mathrm{DC}}}{2} \end{array} \right.$$

where u_i is the constant value of the reference at the beginning of the switching interval which starts at time $t = kT_{\mathrm{PWM}}$, $k \in \mathbb{Z}_+$.

To get the desired symmetry of the digital switching signals, the slave DSP generates two signals for each upper switching signal. These are then passed through an external XOR gate. Both inputs to the XOR gate are logical high at the beginning of each switching interval, $t = kT_{PWM}$. The first input is set low at time $t = kT_{\text{PWM}} + T_{\text{PWM}}/2 - t^i_{\text{ON}}/2$, and the other at time $t = kT_{\text{PWM}} + T_{\text{PWM}}/2 + t^i_{\text{ON}}/2$.

The three TTL voltage outputs from the XOR gates were then converted to current signals suited for generating optical signals using external electronics, and transmitted to the converter. Complementary signals for switching of the lower transistors and blanking time were generated by an IXYS IXDP630 PWM controller in the converter.

INVERTER-MOTOR CONFIGURATION

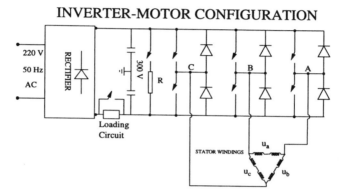

Figure 10.12: Inverter-motor configuration.

6.2 Outline of experiments

Several series of experiments were carried out with the equipment described above and different controllers, and to limit the number of figures, only plots from what was considered as illustrative experiments are shown. Unless something else is explicitly stated, the load torque used in the experiments was only due to friction. However, the converter of the load was turned on during all experiments. Even if the reference for the load torque[25] was set to zero, this gave some additional high frequency oscillations of small amplitude in the system. Also, to limit the number of plots, measured currents and reference voltages for PWM have only been included for a few of the reported experiments. In those cases where saturation was experienced, this is commented explicitly. The references were generated by linear first order filtering of step/square-wave signals from real-time implementations of the SIMULINK™ signal

---

[25]The control of the load torque was feed-forward in currents (i.e. torque) for the brushless DC motor.

generator. Higher order derivatives of references for position/speed and flux amplitude $\beta_*(t)$ were obtained from state space representations of linear filters, similar to those reported in the simulation part of Section 3.9. For this reason, in the cases where both a desired quantity and an estimated or measured quantity are shown in the same plot, the desired value can be distinguished from the other as the smoothest curve.

It was aimed at showing that flux norm and speed can be independently controlled, and for this reason either the speed- or flux reference was held constant during each experiment, while the other reference was a varying function. The unfiltered time-varying references were chosen to be square-waves, with a maximum amplitude of the flux reference equal to the nominal flux level of the motor.

The outline of the rest of this section is as follows. In Section 6.3 some of the results from an implementation of the observer-less scheme described in Section 3.2 are reported. Section 6.4 contains experimental results with the observer-based controller from Section 4.4. In Section 6.5 results from an implementation of the rotor-flux-oriented control scheme in Section 5, are given for the purpose of comparison with the passivity-based controllers. Finally, concluding remarks to the experimental work are given in Section 6.6.

6.3   Observer-less control

The behavior of the controller presented in Section 3.2, was first investigated in a series of experiments. For later convenience (10.28) is rewritten here in terms of its different components

$$
u = \underbrace{L_s \ddot{q}_{sd} + L_{sr} e^{\mathcal{J} n_p q_m} \ddot{q}_{rd} + n_p L_{sr} \mathcal{J} e^{\mathcal{J} n_p q_m} \dot{q}_m \dot{q}_{rd} + R_s \dot{q}_{sd}}_{\text{desired dynamics}}
$$
$$
- \underbrace{(\frac{n_p^2 L_{sr}^2}{4\epsilon} \dot{q}_m^2 + k_1) \tilde{\dot{q}}_s}_{\text{nonlinear damping term}} - \underbrace{K_{Is} \int_0^t \tilde{\dot{q}}_s dt}_{\text{integral term}} \tag{10.96}
$$

where a possible integral term in stator currents have been added.

Since flux measurement was not implemented in the setup, a flux observer had to be run in parallel with the controller for the purpose of verifying flux tracking. For later comparison with an implementation of FOC, the estimation scheme in (10.91)–(10.92) on p. 357 was chosen. To avoid the singularity in the rotor flux speed estimation for zero flux estimate, it was necessary to substitute $\hat{\lambda}_d$ in the division (see (10.91)) by a small constant $c = 0.001$ whenever $\hat{\lambda}_d \leq c$.

To the desired torque defined in (10.30) and (10.32) (or (10.48) for position control), an integral term was added to compensate for unknown load torque, giving

$$\tau_d = D_m\ddot{q}_{m*} - z - f(q_m - q_{m*}) + \hat{\tau}_L \tag{10.97}$$
$$\dot{z} = -az + b(\dot{q}_m - \dot{q}_{m*}), \ a, b > 0, \ z(0) = \dot{q}_m(0) - \dot{q}_{m*}(0)$$
$$\dot{\hat{\tau}}_L = -\gamma_{\tau_L}e, \ \gamma_{\tau_L} \geq 0$$

where q_{m*}, $\dot{q}_{m*}$ are the rotor position and speed reference, and $f \equiv 0$ for speed tracking. The error term in the integral action was set to $e = q_m - q_{m*}$ for position tracking, and $e = \dot{q}_m - \dot{q}_{m*}$ in the case of speed tracking.

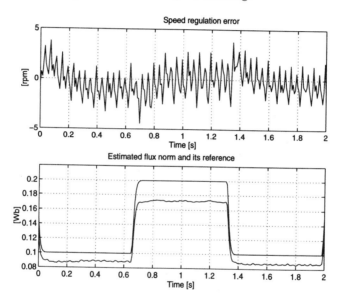

Figure 10.13: Speed regulation/flux tracking without integral action in currents.

The controller was first tested without integral action in stator currents, and a typical response is given in Fig. 10.13. As can be seen from the figure, speed regulation is satisfactory, except from some high-frequency ripple, due to a combination of unknown parameters, PWM, load torque and unmodeled dynamics. There is however an error in flux amplitude, which can be explained as follows: In (10.28) there is only proportional action in stator currents, even if the gain is a nonlinear function of speed. There will always be some unmodeled dynamics in the system, in addition to the introduced discretization effects[26] and parameter uncertainty. For this reason real currents deviates from their desired values. In the speed controller there is an outer

---

[26]It can be shown that discretization introduces coupling terms in the dynamic equations which are proportional to sampling period and speed [275]. See also Section **11.7**.

loop integral action, which forces the speed error to zero, despite error in q-axis current. There is no such feedback in flux control, which is feed-forward. Consequently the error in d-axis current gives an error in flux tracking.

For the reasons above, the integral term in (10.96) had to be used for satisfactory performance, in addition to the other terms. The following controller parameters were used in the rest of the experiments reported here: $\epsilon = 0.5R_r$, $k_1 = 30$, $K_{Is} = 0.3$, $a = 1000$, $b = 300$, $\gamma_{\tau_L} = 3.85$.

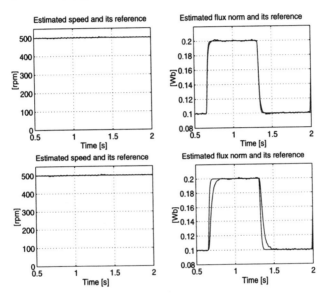

Figure 10.14: Speed regulation/flux tracking *with* (upper two figures), and *without* $\dot{\beta}_*$ in the controller.

In Fig. 10.14 the importance of the $\dot{\beta}_*$ term in desired currents is shown. This feed-forward term is significant for high-performance flux tracking.

As can be seen from Fig. 10.15, it was difficult to get good low speed tracking performance. This is due to the resolution of speed estimation together with friction (especially stiction) in the load. Since the speed estimation gives the average speed between the sampling intervals, it is difficult to detect the sign transition precisely, and this gives problems with compensation of stiction terms.

In Fig. 10.16 an example of load torque rejection is shown, after a step in load torque of approximately 0.8 Nm. The controller compensates fast for the disturbances, and no steady state error is present.

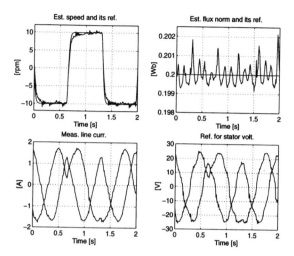

Figure 10.15: Speed tracking/flux regulation at low speed (±10 rpm).

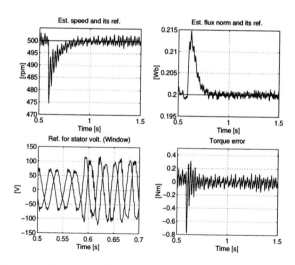

Figure 10.16: Speed regulation with load torque disturbance, $\tau_L \approx 0.8$ Nm for $t \geq 0.58$ s. $\dot{q}_{m*} = 500$ rpm. Error between desired torque and measured torque is shown in lower right plot.

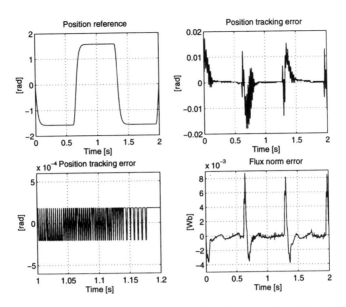

Figure 10.17: Position control. Passivity-based controller. $\beta_* = 0.2$ Wb.

Fig. 10.17 shows an example of position tracking for filtered steps in position reference of $\pm\pi$ rad. The maximum error is approximately 1°. The steady state error is only restricted by the resolution of the position measurement system. In a real implementation, the digital jittering would be eliminated by a dead-zone. The controller parameters $\epsilon = 0.5R_r$, $k_1 = 30$, $K_{Is} = 0.3$, $a = 1000$, $b = 95$, $\gamma_{\tau_L} = 70$, $f = 41$ were used in this experiment.

It is well known that for small and medium size motors with relatively high sampling frequencies for control calculations and PWM, high-performance control can be achieved with only high-gain current control [186]. This was also experienced in this experimental work, were it was found that the desired dynamics in the controller (see (10.96)) had relatively low influence on the performance for the chosen sample period of $T = 300$ μs, when integral action was also used. However, for a sample period twice this value, the influence of the desired terms was significant, as can be seen from Fig. 10.18.

It must be pointed out that in all the reported experiments there are interactions between flux and torque control, resulting in small flux norm peaks during transients. For a real system with unmodeled dynamics and with parameters taken from the data sheet, perfect control can hardly be expected. The peaks are not detrimental for system operation, and result in negligible speed transients. Also, for high speed operation (more than 2000 rpm), the nonlinear damping term in (10.96) became

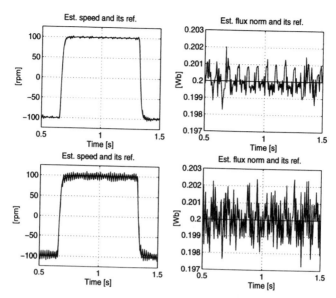

Figure 10.18: Effect of desired dynamics in controller for high sampling period, $T_{sampl} = 600\ \mu s$. Lower two plots are result from an experiment with only integral action and the nonlinear damping term in the controller.

large, and this resulted in amplification of noise from measured currents, which again gave saturation in voltages. For this reason the nonlinear damping term had to be disconnected for high speeds. With integral action in stator currents, the effect of this term was found to be negligible.

6.4 Observer-based control

The controller in Section 4.4 was implemented, with the desired torque as in (10.97). Since this controller exactly reduces to the observer-less controller if stator currents are controlled by high-gain, it was aimed at testing how well it worked without integral action in stator currents. An extensive simulation study was carried out with its discretized version, which work well under ideal conditions. The controller was then tested experimentally.

As can be seen from Fig. 10.19, except from some ripple, the speed regulation is quite good, but flux regulation is far from good. This was general for all the experiments, and can be explained with both missing integral action in stator currents, and only current error terms in the updating of the rotor quantities of the observer.

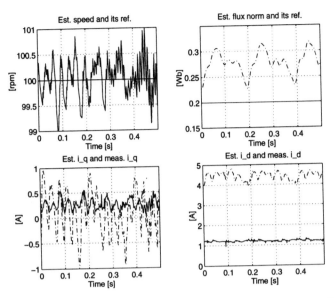

Figure 10.19: Speed and flux regulation. Observer-based controller. Estimates of electrical quantities denoted by '-.'.

The effect of the first point has been explained in the previous section. Since there is no updating in the stator terms of (10.72), the estimated stator currents drift off from the measured values due to uncertainty in parameters, noise and other unmodeled dynamics. This again introduces errors in the estimated rotor currents, which are used in the controller. It was possible to reconstruct a similar behavior under simulations, when the parameters used in the observer deviated from the real parameters.

6.5   Comparison with FOC

For the purpose of comparing the observer-less passivity-based controller with another scheme, the rotor-flux-oriented controller from Section 5 was also implemented. More specifically, the controller in (10.90) was used. Estimated rotor flux amplitude $\hat{\lambda}_d$ and angle $\hat{\theta}_a$ were computed from (10.92) and (10.91) (see p.357). The PI-controllers given in (10.85)-(10.86), together with the current references from (10.87), (10.93), were used to give the voltage references v_{dq}. For speed control the desired torque was defined as in (10.89). The controller parameters were the same as in the previous section for the passivity-based controller, and the parameters $K_{1P} = 30.2$, $K_{1I} = 0.3$ (PI speed controller), $K_{2P} = 0.3$, $K_{2I} = 3.8$ (PI current controllers) were used in the implementation of the FOC scheme.

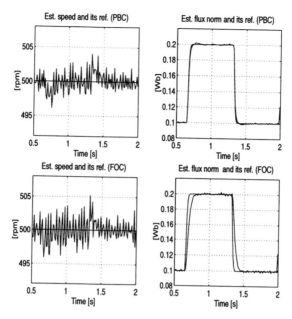

Figure 10.20: Comparison of the passivity-based controller (PBC) with an implementation of FOC. Speed regulation/flux tracking.

Figs. 10.20–10.21 are representative for the comparison between the two schemes. In both cases the controller parameters were tuned such that currents saturated during transients. The FOC scheme generally gave slower responses, and somewhat higher maximum tracking errors. This scheme was also more difficult to tune than the passivity-based scheme. In the FOC controller implementation it was advantageous to use saturation limits corresponding to the systems constraints both in reference currents and voltages.

To investigate the robustness of the schemes, an artificial change in rotor resistance was introduced by using a value different from the nominal value in the controllers. Both controllers were then tuned to give a "best performance" in terms of transients and overshoot. The FOC scheme was experienced to give a more oscillatory behavior than the passivity-based for different values of R_r. Examples are shown in Figs. 10.22–10.23 for $R_r = 1.5R_r^N$.

The execution time[27] was approximately 150 μs for the passivity-based scheme (with the flux observer running in parallel), and 145 μs for the FOC scheme.

---

[27] Only the execution time was logged, and no signal generators were implemented.

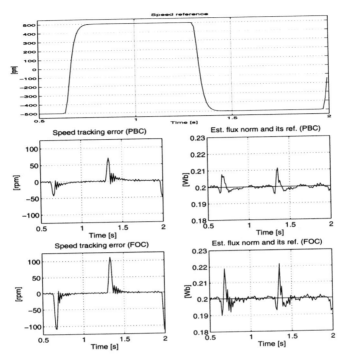

Figure 10.21: Comparison with FOC. Speed tracking/flux regulation.

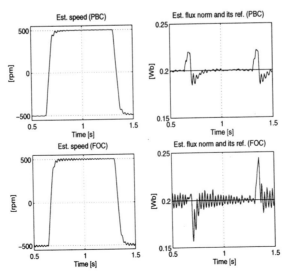

Figure 10.22: Comparison with FOC. Speed tracking/flux regulation. $R_r = 1.5R_r^N$.

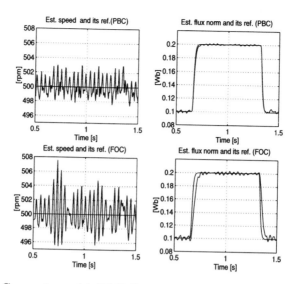

Figure 10.23: Comparison with FOC. Speed regulation/flux tracking. $R_r = 1.5R_r^N$

6.6 Concluding remarks

The experimental testing of the proposed controllers can be summarized as follows:

The observer-less passivity-based controller was found to give the best dynamic performance and robustness to unmodeled dynamics, as compared to a (direct) FOC scheme and a passivity-based controller with observer. The observer-less controller was easy to tune, and could basically be down-loaded with parameters taken from the simulations, without the need for extensive tuning. Significantly more tuning was necessary to make the FOC scheme give results comparable to the performance of the passivity-based controller.

For the passivity-based controller, the feed-forward term from the derivative $\dot{\beta}_*$ of the flux reference improved flux tracking significantly. The effect of this simple modification of the passivity-based controller to allow for global flux tracking is an interesting result, especially when it is related to conventional implementations of indirect FOC schemes, which only can handle flux regulation.

It must be pointed out that high-gain current control was necessary for satisfactory performance of all the controllers which were tested. The nonlinear damping terms which were introduced to prove stability, could not compensate for the unmodeled dynamics of the motor and the converter, and integral action was needed. These damping terms also have certain disadvantages with respect to current noise amplification. Together with the terms in the voltage controller stemming from the reference dynamics, they can be removed when integral action current control is implemented, at least when sampling frequency is high.

For the case of observer-based control along the direction of the result in Section 4.4, it should be focused on including an observer which is more robust to unmodeled dynamics into the scheme. This could for instance be a scheme which also has updating from stator current error terms in the stator equations, as in [280]. An observer-based controller is of interest for future extensions along the line of adaptive observers, and it is believed that it will be possible to include other observers with the desired robustness properties into the scheme, provided they give exponentially convergent estimates.

The aspects of discrete implementation and speed estimation also have to be carefully considered for performance improvements. In this experimental work standard *ad hoc* schemes for controller discretization, speed estimation, and generation of switching signals have been used. This is a drawback of the implementation, since there is no theoretical justification for such an approach.

Other experimental results from the application of the observer-less passivity-based controller to induction motors have recently been reported in [132] (comparison with the scheme of [169]) showing similar promising results. See also Chapter 11 for experimental results with current–fed induction motors.

Chapter 11

Current–fed induction motors

"Experience does not err, it is only your judgment that errs in promising itself results which are not caused by your experiments."

Leonardo da Vinci.

In Chapter **3** we proved that PBC of mechanical systems reduces, in regulation tasks with full state feedback, to the classical PD controller used in most robotic applications. Furthermore, when velocities are not available for measurement the PBC methodology suggests to replace the velocities by their approximate derivatives, which is also a standard procedure in applications. This "downward compatibility" of PBC with current engineering practice is a remarkable feature whose importance can hardly be overestimated. On one hand, it provides a solid system–theoretic foundation to popular control strategies which enhances their understanding and paves the way for subsequent improvements. On the other hand, viewing the new controllers as "upgrades" of the existing ones, it facilitates the transfer of these developments to practitioners. In this chapter we will show that, under some simplifying assumptions on the machine model, the PBC for electrical machines presented in previous chapters also has a "downward compatibility" property with the industry standard field-oriented controller (FOC).

FOC has already been introduced for voltage–fed induction machines in the previous chapter. It was pointed out that field orientation, in one of its many forms, is an established control method for high dynamic performance AC drives. In particular, for induction motors *indirect FOC* is a simple and highly reliable scheme which has become the *de facto* industry standard. In spite of its widespread popularity, the stability and robustness properties of FOC schemes are not theoretically well understood. An approximate analysis (based on steady–state behaviour, time–scale assumptions, and linearizations, e.g., [36, 153]) can be combined with the designer expertise to commission the controller in simple applications. However, to meet large bandwidth requirements, or other tight specifications, this *ad–hoc* commissioning stage may be time–consuming and expensive, if at all possible. To simplify the off–line tuning of FOC, and eventually come to terms with its achievable performance, a better theoretical understanding of the dynamic behaviour of FOC is unquestionably needed. Such an analysis is unfortunately stymied by the fact that, as we have seen before, the dynamic behaviour of the closed loop is described by a complex nonlinear relationship. However, as we will see in this chapter, for current–fed machines it is possible to define some suitable coordinate changes[1] that reveal useful passivity properties that can be profitably exploited to overcome this obstacle.

Realizing the practical importance of FOC, and motivated by the need to clarify its theoretical underpinnings we analyze in this chapter the indirect FOC for current–fed induction motors in detail. We start with a presentation of the model of the current–fed induction motor in Section 1, and to explain the rationale of indirect FOC we find it convenient to recall in the next section first the simpler concept of *direct* FOC, which was presented in Section 5 of the previous chapter for the voltage-fed induction motor. We underscore also in this section the links of direct FOC and the theoretically interesting idea of feedback linearization, which for current–fed machines are more transparent. Then, we show in Section 3 that the observer–less PBC developed for induction motors in Section **10**.3.2 *exactly reduces* to the well-known indirect FOC in speed regulation applications with constant load torque, for current–fed machines. A corollary of this result is a rigorous proof of the *global* asymptotic stability of FOC, first reported in [214].

In Section 5 we address the problem of robustness of FOC with respect to uncertainty in the rotor time constant, and we establish the stronger property of global *exponential* stability. In Section 6 we give some simple rules for the *off–line tuning* of PI gains, which will ensure *robust* stability. Not surprisingly, instrumental for this analysis is a suitable decomposition of the system into a passive feedback interconnection.

A new *discrete-time* FOC for current–fed induction motors (first reported in [215]) is presented in Section 7. This controller ensures global asymptotic speed regulation and rotor flux norm tracking provided a condition relating the sampling rate with the *controller parameters* is satisfied. In analogy with the continuous time controller design, this condition disappears as the sampling rate goes to zero.

Many of the sections have experimental results included for illustration of the theoretical results. In particular, we present in this Section 4 a detailed experimental comparison of FOC and a feedback linearizing scheme.

Concluding remarks to this chapter are given in Section 8.

Remark 11.1 (Notation.) Throughout the chapter we will be interested in the ro-

---

[1]The importance of coordinate changes was probably first underscored by Copernicus who pointed out that the planetary motions are better understood from the sun's perspective [57]. An ingenious change of coordinates is, of course, also the underlying principle of FOC.

bustness –with respect to parameter uncertainty– of the different controllers. Whenever we want to stress this point, we will use $(\hat{\cdot})$ to denote a *fixed estimate* of a parameter $(\cdot)$. Also, with the objective of simplifying the notation and the presentation, in some sections where the value of certain parameters is of no relevance for the discussion, they are set equal to unity.

1   Model of the current–fed induction motor

In Chapter 10.1 we have shown that the induction motor in the fixed stator frame is described by the state equations (10.18)–(10.19), which we repeat here for ease of reference

$$
\dot{\hat{i}}_a = \frac{L_{sr}R_r}{L_s\sigma L_r^2}\lambda_a + \frac{n_pL_{sr}}{L_s\sigma L_r}\dot{q}_m\lambda_b - \gamma i_a + \frac{1}{L_s\sigma}u_1
$$

$$
\dot{\hat{i}}_b = \frac{L_{sr}R_r}{L_s\sigma L_r^2}\lambda_b - \frac{n_pL_{sr}}{L_s\sigma L_r}\dot{q}_m\lambda_a - \gamma i_b + \frac{1}{L_s\sigma}u_2
$$

$$
\dot{\lambda}_a = -\frac{R_r}{L_r}\lambda_a - n_p\dot{q}_m\lambda_b + \frac{R_rL_{sr}}{L_r}i_a
$$

$$
\dot{\lambda}_b = -\frac{R_r}{L_r}\lambda_b + n_p\dot{q}_m\lambda_a + \frac{R_rL_{sr}}{L_r}i_b
$$

$$
\ddot{q}_m = \frac{1}{D_m}(\tau - \tau_L)
$$

$$
\tau = \frac{n_pL_{sr}}{L_r}(\lambda_a i_b - \lambda_b i_a)
$$

where $\lambda_{ab} = [\lambda_a, \lambda_b]^\top$ is the rotor flux vector, $i_{ab} = [i_a, i_b]^\top = \dot{q}_s$ is the stator current vector, and $u_{ab} = [u_1, u_2]^\top$ is the vector of stator voltages. R_s, R_r $[\Omega]$ are stator and rotor resistances, L_s, L_r $[H]$ are the inductances of the stator and rotor windings and L_{sr} $[H]$ is the mutual inductance. $\sigma = 1 - \frac{L_{sr}^2}{L_sL_r} > 0$ is the total leakage factor of the motor, and $\gamma = \frac{R_s}{L_s\sigma} + \frac{L_{sr}^2}{L_s\sigma L_rT_r}$, with the rotor time constant defined as $T_r \stackrel{\triangle}{=} L_r/R_r$ [s], D_m [kgm^2] is the rotor inertia and $\dot{q}_m$ [rad/s] the rotor speed.

In many practical applications high–gain current loops (sometimes with PI actions) of the form

$$
u = \frac{1}{\epsilon}(i_{ab}^d - i_{ab})
$$

are used to force i_{ab} to track their corresponding references i_{ab}^d, where ϵ is a small positive number. It is reasonable then to consider the singularly perturbed reduced model obtained by setting $\epsilon \to 0$, that is

$$
\begin{aligned}
\dot{\lambda}_{ab} &= -\frac{R_r}{L_r}\lambda_{ab} + n_p\dot{q}_m\mathcal{J}\lambda_{ab} + \frac{R_rL_{sr}}{L_r}i_{ab} \\
D_m\ddot{q}_m &= \tau - \tau_L \\
\tau &= \frac{n_pL_{sr}}{L_r}i_{ab}^\top\mathcal{J}\lambda_{ab}
\end{aligned}
\tag{11.1}
$$

with the skew-symmetric matrix

$$J = \begin{bmatrix} 0 & -1 \\ 1 & 0 \end{bmatrix}$$

The underlying assumption of this model is that the stator currents are exactly equal to their references, i.e. $i_{ab} \equiv i_{ab}^d$. To further simplify the equations, we introduce the (globally defined) change of coordinates

$$v = e^{-Jn_pq_m}i_{ab}, \quad \lambda_r = e^{-Jn_pq_m}\lambda_{ab} \tag{11.2}$$

with the *rotation matrix*

$$e^{-Jn_pq_m} = \begin{bmatrix} \cos(n_pq_m) & \sin(n_pq_m) \\ -\sin(n_pq_m) & \cos(n_pq_m) \end{bmatrix}, \quad (e^{-Jn_pq_m})^{-1} = (e^{-Jn_pq_m})^\top = e^{Jn_pq_m}$$

Hence $v = [v_1, v_2]^\top$, $\lambda_r = [\lambda_{r1}, \lambda_{r2}]^\top$ are quantities expressed in a frame rotating with the (electrical) speed of the rotor. In the new state coordinates $[\lambda_r^\top, q_m, \dot{q}_m]^\top$, and with the *new control inputs* v, we have the following *bilinear* model

$$\begin{aligned} T_r\dot{\lambda}_r &= -\lambda_r + L_{sr}v \\ D_m\ddot{q}_m &= \tau - \tau_L \\ \tau &= \frac{n_pL_{sr}}{L_r}v^\top J\lambda_r \end{aligned} \tag{11.3}$$

Throughout the remaining of the chapter, we will assume that the behaviour of the (so–called) current–fed induction motor is captured by the dynamical model (11.3). As we will show below this apparently innocuous system can exhibit an amazingly complex behaviour and poses a significant challenge for control system design.

Unless otherwise stated we will concentrate in the sequel on the problem of *speed control* and assume that load torque τ_L is *constant*. The modifications needed for position control and–or adaptation for a linearly parameterized τ_L are analogous to the ones explained for voltage–fed machines, henceforth will be omitted.

Remark 11.2 (Validity of model.) In current-fed machines we assume that the (rotated) stator currents $v(t)$ are equal to possibly discontinuous references calculated by the controller. In real motors, stator currents must be continuous so that they cannot follow discontinuous references exactly. However, the stator currents can follow the references well if the stator inductances are sufficiently small. On the other hand, the assumption of perfect current control is based on sufficiently high ceiling voltage and switching frequency of the inverter, for the control bandwidth to be adequate. For instance, in the case of high power drives with thyristors switching at only a few hundred Hz, perfect current control can not be assumed, and the interactions of the stator voltage equations must be taken into account for controller design.

Remark 11.3 (Relation with the asymptotic model.) It is important to underscore the fact that λ_r is a vector quantity. This model should not be confused with the machine model used in decoupling control, e.g., (7.75), (7.82), (2.78) of [36], which describes the *asymptotic* behaviour of the motor in closed-loop with an ideal direct FOC.

Remark 11.4 (Relation with the non–holonomic integrator.) The model coincides (up to the presence of the term λ_r in the first equation of (11.3) and the load torque) with the celebrated *non–holonomic integrator* of Brockett for which a vast amount of research has been devoted in the last years. Interesting connections (extensions and simplifications) between the time–varying controllers used for this system and the PBC presented in this text are explored in [73].

2   Field orientation and feedback linearization

2.1   Direct field–oriented control

Defining the rotor flux amplitude $\beta = \|\lambda_r\|$, and the rotor flux angle $\rho = \arctan(\lambda_{r2}/\lambda_{r1})$, and introducing an additional input change of coordinates

$$i_{dq} \triangleq \begin{bmatrix} i_d \\ i_q \end{bmatrix} = \mathbf{e}^{-\mathcal{J}\rho}v$$

we can rewrite the model (11.3) in polar coordinates as

$$
\begin{aligned}
T_r \dot{\beta} &= -\beta + L_{sr} i_d \\
\dot{\rho} &= \frac{R_r}{n_p \beta^2}\tau \\
\tau &= \frac{n_p L_{sr}}{L_r} i_q \beta
\end{aligned}
\tag{11.4}
$$

Notice that β is the output of a linear filter with "input" i_d, while τ is simply the product of the second "control input" i_q and β. This two facts, together with the natural *time scale* separation of the electrical and mechanical dynamics, motivates the classical direct FOC (see also Section **10**.5) where i_d is chosen to regulate β to its reference value β_*, while i_q is used to drive the torque to some desired reference τ_d, as follows

$$
i_{dq} = \frac{1}{\hat{L}_{sr}} \left[\begin{array}{c} \beta_* \\ \underbrace{\frac{\hat{L}_r}{n_p \beta_*} C(p)\dot{\tilde{q}}_m}_{\tau_d} \end{array} \right] \quad \left(= \mathbf{e}^{-\mathcal{J}(\rho + n_p q_m)} i_{ab} \right)
\tag{11.5}
$$

where $\dot{\tilde{q}}_m \triangleq \dot{q}_m - \dot{q}_{m*}$, β_*, $\dot{q}_{m*}$ are the *desired* rotor flux magnitude and rotor speed, respectively, $(\hat{\cdot})$ denotes an estimate, $C(p) = -(K_P + K_I \frac{1}{p})$, $p \triangleq \frac{d}{dt}$, and K_P, $K_I > 0$ are *tuning gains*.

It is important to remark that for the implementation of the *actual control signals*[2]

$$i_{ab} = \frac{1}{\hat{L}_{sr}} e^{\mathcal{J}(\rho + n_p q_m)} \begin{bmatrix} \beta_* \\ \frac{\hat{L}_r}{n_p \beta_*} \underbrace{C(p)\dot{\tilde{q}}_m}_{\tau_d} \end{bmatrix}$$

the flux angle ρ must be measured or estimated. The unavailability of this signal, coupled with the high sensitivity (with respect to the highly uncertain parameter T_r) of its estimate, are the main drawbacks of this approach. See Section **10**.5 for additional comments.

2.2   Indirect field-oriented control

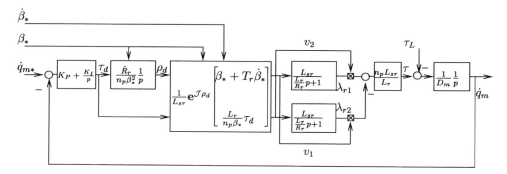

Figure 11.1: Current–fed induction motor with indirect FOC. Motor model presented in a frame of reference rotating with rotor (electrical) speed.

To remove the need of the rotor flux angle ρ from the direct FOC above we can think of replacing it by it's "asymptotic certainty equivalent" estimate, that is the value to which it will converge if β and τ behave as desired. Applying this rationale to the second equation of (11.4), and calling this signal ρ_d, yields

$$\dot{\rho}_d = \frac{\hat{R}_r}{n_p \beta_*^2} \tau_d \tag{11.6}$$

$$\tau_d = C(p)\dot{\tilde{q}}_m = -(K_P + K_I \frac{1}{p})\dot{\tilde{q}}_m$$

where we have kept the same definition of τ_d as above. We can now use this estimate instead of the actual ρ in the direct FOC (11.5) to get

$$i_{dq} = \frac{1}{\hat{L}_{sr}} \begin{bmatrix} \beta_* + \hat{T}_r \dot{\beta}_* \\ \frac{\hat{L}_r}{n_p \beta_*} C(p)\dot{\tilde{q}}_m \end{bmatrix} \left(= e^{-\mathcal{J}(\rho_d + n_p q_m)} i_{ab} \right) \tag{11.7}$$

---

[2]Recall that we assume current–fed operation.

where the term $\hat{T}_r \dot{\beta}_*$ in the i_d-component of the currents is introduced to achieve flux norm tracking and is motivated by the first equation in (11.4). The resulting control law, that is,

$$i_{ab} = \frac{1}{\hat{L}_{sr}} e^{\mathcal{J}(\rho_d + n_p q_m)} \left[\begin{array}{c} \beta_* + \hat{T}_r \dot{\beta}_* \\ \frac{\hat{L}_r}{n_p \beta_*} C(p) \dot{\tilde{q}}_m \end{array} \right]$$

will be called in the sequel *indirect FOC*. This scheme may be found, for instance, in [190]. It is often presented in the literature without the flux tracking term $\hat{T}_r \dot{\beta}_*$.

The total closed-loop system consisting of the induction motor (11.3) and the controller (11.6)–(11.7) is shown in Fig. 11.1.

2.3 Observer–based feedback–linearizing control

The observer-based feedback–linearizing controller (OBFL) of [169][3] may be viewed as a variation of direct FOC where, in order to achieve a linear system in closed loop, some terms are added to the control law to cancel the motor nonlinearities. A similar philosophy was adopted in [143] for the case of the full motor dynamics, see also [172] where some adaptation terms are added to the basic scheme of [143]. Although the objective of feedback linearization is quite luring, the resulting schemes suffer from serious drawbacks, from both theoretical and practical viewpoints. First, they invariably require the explicit implementation of an observer. This, besides increasing the computational burden, makes the stability analysis extremely difficult. For instance, it is well known that for nonlinear systems the certainty equivalence principle fails. Also, it widely recognized that at this stage, our understanding of nonlinear observers is quite rudimentary. It suffices to say that in spite of many years of research, to the best of our knowledge, a complete stability analysis for exact linearization schemes in the full motor model case is conspicuous by its absence. However, it should be pointed out that for the simplified model of current-fed machines this problem is elegantly solved in [169]. Second, since these schemes are based on nonlinearity cancelations, it is expected that potential instability due to parameter mismatch will arise. One such instability mechanism for the OBFL, which appears even in the state feedback case, is identified in this work, and observed in the experiments.

For later comparison with direct FOC and to exhibit the instability mechanism, we will distinguish between the state–feedback and the output feedback cases in the presentation of the OBFL.

---

[3]Called in that paper observer-based adaptive controller.

A   State feedback

First, we notice from the first equation of (11.4), that β^2 satisfies the differential equation

$$\frac{T_r}{2}\frac{d}{dt}\beta^2 = -\beta^2 + L_{sr}\beta i_d \tag{11.8}$$

Hence, assuming that the state is measurable and that the *parameters are known*, we can choose

$$i_d = \frac{1}{L_{sr}}[\beta - \frac{T_r k_\lambda}{2\beta}(\beta^2 - \beta_*^2)] \tag{11.9}$$

with $k_\lambda > 0$ a *design parameter*, to obtain a linear dynamics for β^2 as

$$\frac{d}{dt}\beta^2 = -k_\lambda(\beta^2 - \beta_*^2)$$

A feedback linearizing and decoupling control is completed with a suitable definition of i_q, for instance if τ_L is assumed known,[4] we can choose

$$i_{ab} = \frac{1}{\hat{L}_{sr}}e^{\mathcal{J}(n_p q_m + \rho)}\left[\begin{array}{c} \beta - \frac{\hat{T}_r k_\lambda}{2\beta}(\beta^2 - \beta_*^2) \\ \frac{1}{\beta}\frac{\hat{L}_r}{n_p}(\tau_L - \hat{D}_m K_P \dot{\tilde{q}}_m) \end{array}\right] \tag{11.10}$$

The connection between direct FOC and OBFL is easily established comparing (11.10) with (11.5). There are two essential differences, the utilization of the actual β instead of β_* throughout, and the inclusion of the second and third terms in the first component of the vector. These modifications are introduced to achieve the exact linearization and decoupling, which are only asymptotic in direct FOC. This is exactly the opposite to the approach taken in PBC, where the actual rotor quantities are replaced by their desired values. Of course, the main contribution of [169] resides in the ingenious construction of the observer to achieve global stability, as explained below.

We observe that

$$e^{\mathcal{J}(n_p q_m + \rho)} = \beta\left[\begin{array}{cc} \lambda_a & -\lambda_b \\ \lambda_b & \lambda_a \end{array}\right] \tag{11.11}$$

which will be used to explain the rationale of the OBFL below.

---

[4]Instead of the expression for i_q above we could have chosen, as in FOC, a PI speed loop, that is $i_q = \frac{1}{\beta}C(p)\dot{\tilde{q}}_m$. This idea underlies the work of [129], and has recently been pursued by [30]. Other linearizing schemes are presented in [58, 186], see also Section **10.5**.

B   Output feedback

In this case estimates of the rotor flux and the load torque are provided by the *nonlinear observer*

$$\dot{\hat{\lambda}}_{ab} = -\frac{1}{\hat{T}_r}\hat{\lambda}_{ab} + n_p\dot{q}_m\mathcal{J}\hat{\lambda}_{ab} + \frac{\hat{L}_{sr}}{\hat{T}_r}i_{ab} - \frac{n_p\hat{L}_{sr}}{\hat{L}_r\hat{D}_m}(\delta_0\dot{q}_m + \delta_1\dot{q}_m^3)\mathcal{J}i_{ab}$$

$$\dot{\hat{\tau}}_L = -\frac{K_I}{\hat{D}_m}(\delta_0\dot{q}_m + \delta_1\dot{q}_m^3)$$

with δ_0, $\delta_1 > 0$ *design parameters*. Notice that if we set these parameters to zero the observer is just a copy of the rotor flux dynamics (11.1).

The control is a modified version of (11.10), and is given as

$$i_{ab} = \frac{1}{\hat{L}_{sr}}D^{-1}(\hat{\lambda}_{ab}, \dot{q}_m)\left[\begin{array}{c} \|\hat{\lambda}_{ab}\|^2 - \frac{k_\lambda\hat{T}_r}{2}(\|\hat{\lambda}_{ab}\|^2 - \beta_*^2) \\ \frac{\hat{L}_r}{n_p}(\hat{\tau}_L - \hat{D}_mK_P\dot{q}_m) \end{array}\right]$$

where the elements of the matrix $D(\hat{\lambda}_{ab}, \dot{q}_m)$ are given by:

$$D_{11}(\hat{\lambda}_{ab}, \dot{q}_m) \triangleq \hat{\lambda}_a - \frac{n_p}{\hat{R}_r\hat{D}_m}(\delta_0\dot{q}_m + \delta_1\dot{q}_m^3)\hat{\lambda}_b$$

$$D_{12}(\hat{\lambda}_{ab}, \dot{q}_m) \triangleq \hat{\lambda}_b + \frac{n_p}{\hat{R}_r\hat{D}_m}(\delta_0\dot{q}_m + \delta_1\dot{q}_m^3)\hat{\lambda}_a$$

$$D_{21}(\hat{\lambda}_{ab}, \dot{q}_m) \triangleq -\hat{\lambda}_b$$

$$D_{22}(\hat{\lambda}_{ab}, \dot{q}_m) \triangleq \hat{\lambda}_a$$

Notice that we have added to the rotation matrix (11.11) some additional terms, hence the control *is not* a certainty-equivalent version of (11.10). It is interesting to note that, as shown in [173], to ensure invertibility of the matrix $D(\hat{\lambda}_{ab}, \dot{q}_m)$, it is enough to choose the *initial conditions* of the observer bounded away from zero, which imposes no constraint in a practical application.

C   An instability mechanism of the OBFL

In the following we will show that, in the face of parameter uncertainty, the OBFL scheme may become unstable. We have decided to present here the state feedback case to underscore the fact that the instability is not due to the observation error transient, but to the nonlinearity cancelation term. The same calculations can be carried out, leading to the same result, for the observer feedback case.

Proposition 11.5 *Consider the current-fed motor model (11.1) in closed loop with the OBFL, which in the state feedback case simplifies to (Eq. (4) in [169])*

$$i_{ab} = \left[\begin{array}{cc} \lambda_a & \lambda_b \\ -\lambda_b & \lambda_a \end{array}\right]^{-1}\left[\begin{array}{c} \frac{1}{\hat{L}_{sr}}(\|\lambda_{ab}\|^2 - \frac{k_\lambda\hat{T}_r}{2}\tilde{\Psi}) \\ \frac{\hat{L}_r}{n_p\hat{L}_{sr}}(\hat{\tau}_L - \hat{D}_mK_P\dot{q}_m) \end{array}\right]$$

where $\tilde{\Psi} \overset{\triangle}{=} \beta^2 - \beta_*^2$, and we have assumed $\|\lambda_{ab}\| \neq 0$. In this case, the system becomes unstable *if the mutual inductance L_{sr} is underestimated and the gain k_λ is small.* More precisely, the flux grows unbounded if

$$\frac{\hat{L}_{sr}}{L_{sr}} < 1 - \hat{T}_r \frac{k_\lambda}{2} \tag{11.12}$$

□

Proof.   The proof consists in showing that the term introduced in to cancel the rotor flux norm, i.e., $\frac{1}{\hat{L}_{sr}}\|\lambda_{ab}\|^2$, induces an unstable behaviour on the flux norm error when (11.12) holds.

The first term of the control is given as

$$i_d = \frac{1}{\hat{L}_{sr}}[\beta - \frac{\hat{T}_r k_\lambda}{2\beta}\tilde{\Psi}]$$

That is, the estimated parameter version of (11.9). Replacing the equation above in (11.8) and doing some simple calculations we see that

$$\dot{\tilde{\Psi}} = C_1\tilde{\Psi} + \frac{2}{T_r}(\frac{L_{sr}}{\hat{L}_{sr}} - 1)\beta_*^2 \tag{11.13}$$

where

$$C_1 \overset{\triangle}{=} \frac{2}{T_r}\{-1 + \frac{L_{sr}}{\hat{L}_{sr}}(1 - \hat{T}_r\frac{k_\lambda}{2})\}$$

This completes the proof.   ■

2.4   Remarks to OBFL and FOC

Remark 11.6 (High-gain control.) It is clear from (11.12) that the instability mentioned above can be easily avoided by choosing k_λ sufficiently large, namely $k_\lambda > \frac{2}{T_r}$ suffices. This will have the additional beneficial effect of reducing the steady–state error in (11.13). Experimental and simulation evidence have shown, however, that large values of k_λ induce a resonant behaviour. This is illustrated in the simulation of Fig. 11.2. The simulation was performed in SIMULINK using the motor parameters of Table 11.1, see page 396. A step signal was applied for the speed reference, $\dot{q}_{m*} = 50$ rad/s, at $t = 0$ s and we observed the speed and amplitude of the stator current vector as k_λ was varied. Fig. 11.2(a) corresponds to the ideal case without parameter uncertainty. Figs. 11.2(b)–11.2(f) show the effect of varying k_λ as in the practically reasonable case when L_{sr} and L_r are underestimated 20% and R_r is overestimated 50%. Notice that instability is triggered when k_λ is small (≤ 9), as predicted by the proposition, but also when it is chosen to large (≥ 100). Further studies are needed to understand the latter instability mechanism, but it is clear that the use of high gains to avoid instability is not a sensible approach.

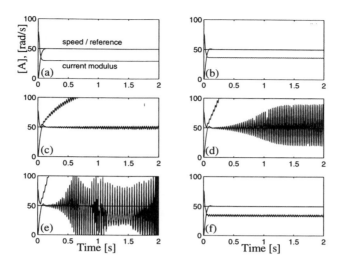

Figure 11.2: Simulation illustrating instability of feedback linearizing control. (a) Ideal case with $k_\lambda = 50$. (b)–(f) parameters perturbed, and $k_\lambda = 50$, $k_\lambda = 9$, $k_\lambda = 7$, $k_\lambda = 5$ and $k_\lambda = 100$.

Remark 11.7 (Modification of FOC which gives robust flux tracking.) We bring to the readers attention the (apparently not well known) fact that a simple modification to direct FOC allows us to achieve also flux norm tracking, see e.g. [190]. This is an important feature that is sometimes invoked to claim superiority of exact linearization schemes over direct FOC. To this end, in (11.5), we choose

$$i_d = \frac{1}{\hat{L}_{sr}}[\beta_* + \hat{T}_r \dot{\beta}_* - k_\lambda \tilde{\beta}]$$

with $\tilde{\beta} \triangleq \beta - \beta_*$. This gives from (11.4) the closed loop equation

$$T_r \dot{\beta} = C_2 \beta + C_3 \beta_* + C_4 \dot{\beta}_* \tag{11.14}$$

where C_3, C_4 are some constants, and

$$C_2 \triangleq -1 - \frac{L_{sr}}{\hat{L}_{sr}} k_\lambda < 0$$

Notice that (11.14), in contrast to (11.13), is always stable, hence avoiding the instability mechanism of OBFL. When the parameters are exactly known we have

$$T_r \dot{\tilde{\beta}} = -(k_\lambda + 1)\tilde{\beta}$$

which ensures the desired tracking with arbitrary rate of convergence.

3 Passivity–based control of current–fed machines

The main message that we should retain from Section 1 is that under assumptions that can be satisfied in many practical applications, the stator currents can be taken as control inputs for the induction motor. In other words, in some applications the inverter can be modeled as an ideal current source. In this section we will prove that, under this condition, our PBC for voltage–fed machines considerably simplifies and actually *reduces* to the indirect FOC presented above.

3.1 PBC is downward compatible with FOC

In Section 10.3.2 we derived an observer–less PBC for speed and rotor flux norm tracking with the full voltage–fed machine model. This controller has the following fundamental *downward compatibility* property:

Proposition 11.8 *In the case of current–fed machines with constant speed reference the PBC (10.28)–(10.32) developed for the voltage–fed machine model exactly reduces to the indirect FOC (11.7).*   □

Proof.   For current–fed machines the control signal is actually $\dot{q}_{sd}$, that is the first row of (10.29) and we do not need to calculate u as in (10.28). Consequently $\dot{\tau}_d$, which was required because of the presence of the term $\ddot{q}_{ed}$, is no longer needed for the implementation. Hence, we can remove the filtered speed error and replace it directly by the speed error. In this way the controller reduces to

$$\dot{q}_s \equiv \dot{q}_{sd} \;=\; \frac{1}{L_{sr}} \left[\left(1 + T_r \frac{\dot{\beta}_*}{\beta_*} \right) I_2 + \frac{L_r}{n_p \beta_*^2} \tau_d \mathcal{J} \right] \mathrm{e}^{\mathcal{J} n_p q_m} \lambda_{rd} \qquad (11.15)$$

$$\dot{\lambda}_{rd} \;=\; \left(\frac{R_r}{n_p \beta_*^2} \tau_d \mathcal{J} + \frac{\dot{\beta}_*}{\beta_*} \right) \lambda_{rd}, \; \lambda_{rd}(0) = \left[\begin{array}{c} \beta_*(0) \\ 0 \end{array} \right] \qquad (11.16)$$

$$\tau_d \;=\; D_m \ddot{q}_{m*} - a \dot{\tilde{q}}_m + \tau_L \qquad (11.17)$$

where $\tilde{q}_m \overset{\triangle}{=} \dot{q}_m - \dot{q}_{m*}$ is the speed error and β_* is the time–varying flux reference. Now, notice that the controller states in (11.16) can be exactly integrated as

$$\lambda_{rd} \;=\; \beta_* \left[\begin{array}{c} \cos(\rho_d) \\ \sin(\rho_d) \end{array} \right] = \mathrm{e}^{\mathcal{J}\rho_d} \left[\begin{array}{c} \beta_* \\ 0 \end{array} \right]$$

where ρ_d is the solution of

$$\dot{\rho}_d \;=\; \frac{R_r}{n_p \beta_*^2} \tau_d, \; \rho_d(0) = 0 \qquad (11.18)$$

By use of these expressions in (11.15) we get

$$\dot{q}_s = \frac{1}{L_{sr}} e^{\mathcal{J}(n_p q_m + \rho_d)} \begin{bmatrix} \beta_* + T_r \dot{\beta}_* \\ \frac{L_r}{n_p \beta_*} \tau_d \end{bmatrix} \tag{11.19}$$

On the other hand, assuming that the desired speed is constant ($\ddot{q}_{m*} = 0$) and replacing the exact load torque cancelation by an integral action we get from (11.17)

$$\tau_d = -(a + \frac{K_I}{p})\dot{\tilde{q}}_m, \ K_I > 0 \tag{11.20}$$

That is, a PI action around the speed error.

Let us now express the controller equations in the stator fixed coordinates introduced in Section 1. To this end, recall that $\dot{q}_s = \dot{q}_{sd} = i_{ab}$, hence from (11.19)

$$i_{ab} = \frac{1}{L_{sr}} e^{\mathcal{J}(n_p q_m + \rho_d)} \begin{bmatrix} \beta_* + T_r \dot{\beta}_* \\ \frac{L_r}{n_p \beta_*} \tau_d \end{bmatrix} \tag{11.21}$$

which is exactly the indirect FOC in (11.7). ∎

3.2 Stability of indirect FOC for known parameters

The proposition below follows as a corollary of Proposition 10.6. To highlight some issues that will become important in the next sections we give a proof here only for speed and flux regulation. The proof for the tracking case follows *verbatim* from the proof in Section **10.3.2**, when the stator currents are used as inputs.

Proposition 11.9 *Consider the current–fed induction motor model (11.3) in closed–loop with the controller (11.18), (11.20), (11.21), where τ_L, $\dot{q}_{m*}$ and β_* are constant, and the parameters L_{sr}, L_r and R_r are known. Under these conditions, the controller ensures* global speed *and rotor flux norm regulation. That is,*

$$\lim_{t \to \infty} |\dot{q}_m - \dot{q}_{m*}| = 0, \ \lim_{t \to \infty} | \ \|\lambda_r\| - \beta_*| = 0$$

holds for all initial conditions and with all signals uniformly bounded. □

Proof.

It follows from (11.15), (11.16) and (11.2) that the PBC can be written as

$$v = \frac{1}{L_{sr}}(\lambda_{rd} + T_r \dot{\lambda}_{rd}) \tag{11.22}$$

Defining $\tilde{\lambda}_r \stackrel{\Delta}{=} \lambda_r - \lambda_{rd}$ and using (11.3) we see that $\dot{\tilde{\lambda}}_r = -\frac{1}{T_r}\tilde{\lambda}_r$, hence $\tilde{\lambda}_r$ converges (exponentially fast) to zero.

Finally, with some lengthy but straightforward calculations, we can get

$$
\begin{aligned}
D_m \ddot{\tilde{q}}_m = D_m \ddot{q}_m &= \frac{n_p L_{sr}}{L_r} v^\top \mathcal{J} (\tilde{\lambda} + \lambda_{rd}) - \tau_L \\
&= \frac{n_p}{L_r} \lambda_{rd}^\top \mathcal{J} \tilde{\lambda} - \tau_L + (1 + \frac{1}{\beta_*^2} \lambda_{rd}^\top \tilde{\lambda}_r) \tau_d \\
&= -K_P (1 + \frac{1}{\beta_*^2} \lambda_{rd}^\top \tilde{\lambda}) \dot{\tilde{q}}_m - K_I (1 + \frac{1}{\beta_*^2} \lambda_{rd}^\top \tilde{\lambda}) \frac{1}{p} \dot{\tilde{q}}_m + \frac{n_p}{L_r} \lambda_{rd}^\top \mathcal{J} \tilde{\lambda}_r - \tau_L
\end{aligned}
$$

Noting that $\|\lambda_{rd}(t)\| = \beta_*$, we have that $\frac{1}{\beta_*^2} \lambda_{rd}^\top \tilde{\lambda}_r$ and $\lambda_{rd}^\top \mathcal{J} \tilde{\lambda}_r$ are exponentially decaying terms. Asymptotic stability of $\dot{\tilde{q}}_m \to 0$ follows invoking standard arguments of LTV systems with exponentially convergent coefficients. ∎

In Section 5 we will give an alternative proof of *exponential stability* based on construction of a strict (quadratic) Lyapunov function.

4 Experimental comparison of PBC and feedback linearization

Several experimental studies have been carried out to illustrate the theoretical results described above. Some of these results are presented in this section. Other related works, which have been reported within the industrial electronics community may be found in [48, 96].

We present here an experimental comparison[5] between PBC and the OBFL of [169], which was presented in Section 2.3. As shown in the previous section, the PBC controller is obtained as a particular case, for the approximate model used for current–fed machines, of the globally stable and globally defined output feedback scheme of Section 10.3.2, which was developed for the full machine model. One important feature of this controller is that there is no need to reconstruct the state, i.e., to estimate the rotor flux. This translates into a considerable reduction in the computational complexity, a feature that, – due to cost and numerical robustness considerations–, can hardly be overestimated in an application of this nature.

The experiments were carried out on a two-pole squirrel-cage current-fed induction motor of 1.2 kW and a microcomputer-based control system developed for an industrial application by [183]. In contrast with other experimental studies, where sophisticated special purpose equipment is installed, we used in these experiments standard low–cost hardware (e.g., Motorola 68000 microprocessor) readily available

---

[5] *Caveat emptor*   In a comparative experimental study it is difficult to distinguish the relative merits (or demerits) of a technique from the talents (and prejudices) of the designer. The best we can do to alleviate this difficulty is to put at the disposal of the interested reader our experimental facility.

for a practical application. An exhaustive set of experimental results may be found in [131]. We have decided to present here only some representative curves on speed regulation, load torque disturbance rejection, and robustness to rotor resistance variations.

The conclusions of our experimental comparison may be summarized as follows:

1. The high computational requirements of the OBFL forced us to double the sampling period achievable for PBC with obvious ensuing performance degradation.

2. Even at a lower sampling frequency (with respect to the fastest achievable one), PBC systematically achieved better speed transient performance, faster load torque disturbance rejection, and enhanced robustness *vis à vis* uncertainty in the rotor resistance.

3. Commissioning of PBC was also simpler, because the performance of OBFL is more sensitive to parameter uncertainty and its control effort was larger. The latter factor considerably limited the range of operation of the scheme.

Due to the extremely high computational requirements we were unable to test, with the installed equipment, backstepping-based designs. See [65] for some experimental evidence of this scheme.

4.1 Experimental setup

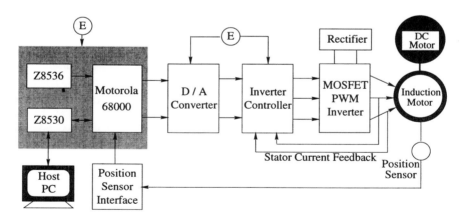

Figure 11.3: Experimental setup.

The experimental drive was assembled using commercially available products and circuits made in the laboratory. In Fig. 11.3 we show the experimental setup consisting

of the induction motor, a DC motor to simulate load torque, the main card with
a Motorola 68000 microprocessor, a timer and an input-output interface, the D/A
converter, a position sensor with interface card, a PWM inverter with control card,
current sensors, source of electric power, and a Macintosh host computer.

The parameters of the induction motor used in the experiments are given in Table
11.1.

Parameter	Notation	Value	Unit
Rated power	P	1.2	kW
Stator resistance	R_s	$7.8 \cdot 10^{-3}$	Ω
Rotor resistance	R_r	$8.0 \cdot 10^{-3}$	Ω
Mutural inductance	L_{sr}	0.193	mH
Stator inductance	L_s	0.25	mH
Rotor inductance	L_r	0.2	mH
Moment of inertia	D_m	$4.45 \cdot 10^{-4}$	kgm^2
Moment of inertia (with DC motor)	D_m	$4.62 \cdot 10^{-3}$	kgm^2
Pole pairs	n_p	2	

Table 11.1: Motor parameters.

A Hardware description

The main card has a 8 MHz Motorola 68000 (16 bits) microprocessor, a Zilog 8536
timer connected with the main processor for the management of the sampling time,
and an Zilog 8530 interface card used for the communication with the host computer.

The D/A converter card is inserted between the main card and the controller
for the inverter. The resolution of this converter is 12 bits and it includes four
D/A converters (one for a reference voltage and the others for commands). The
rotor position is transferred to the microprocessor from the position sensor, which is
an incremental encoder of type GESINC-1 with a resolution of 2000 lines (angular
resolution $360°/2000$). The power electronics which supplies the current to the motor
consist of a fast analogue current–loop controller card and a three-phase MOSFET
inverter.

The current control is performed in a classical manner. Two analogue current
reference signals are transmitted to the control card. The currents in two of the
phases are measured and compared with the reference signals. The regulation of the
third phase is carried out such that the mean value of the neutral point voltage is
equal to half the value of the feeding voltage. This non–redundant control approach

leads to good performance and utilization of the converter and allows for considering it as an ideal current source.

B Software description

The M68000 can be programmed in C and assembly language. Assembly code was used for the control routines which had to be executed in real time. The other routines (management of menus, data saving, etc.) were written in C. For communication with the processor, the application program *Versaterm* operated on a Macintosh was used as an interface between the user and the microprocessor.

The real-time application requires on-line solution of differential equations, trigonometric functions, square roots, integration, etc. A trigonometric function look-up table was used by the controller. To enhance precision, given the short time available for calculations, we used the second order Runge-Kutta method for solution of the differential equations.

Since the microprocessor can handle only fixed point numbers (of 8, 16 or 32 bits), we introduced a scaling proportional to the size of the numbers used in the controller. To avoid as much as possible a phenomenon of truncation during the calculation, an optimal multiplier was chosen for the control variables. For the multiplication of two variables, we developed multiplication algorithms of $(32\text{bits}) \times (32\text{bits}) \div 2^{16}$ and $(16\text{bits}) \times (32\text{bits}) \div 2^{16}$. An appropriate algorithm was chosen according to the size of each fixed-pointed constant or variable, and care was taken to avoid overflow when the result of the calculation exceeded 2^{32-1}.

In the implementation of OBFL, the terms containing $\delta_1 \dot{\tilde{q}}_m^3$ became zero in the fixed-point calculations even if we used the maximum multiplier ($\times 2^{16}$). The reason for this was that δ_1 had to be chosen very small to avoid oscillations, $\delta_1 \approx 10^{-12}$. Therefore we neglected these terms in the experiments[6]. In a simulation study [131] we established that the influence of this term is negligible (in current–fed applications).

Method	Assembly code		
	multiplier ($\times 2^{16}$) and multiplication algorithm (32bits $\times$ 32bits)	optimal multiplier ($\times 2^{13}$)	optimal multiplier ($\times 2^{13}$, $\times 2^{16}$)
PBC	2.5 ms	0.8 ms	1.5 ms
OBFL	6 ms	1.4 ms	2.4 ms

Table 11.2: Computation time.

In order to determine the smallest possible sampling time, we made a study of

---

[6]It is worth underscoring that setting $\delta_1 = 0$ does not change the stability properties of OBFL.

computing time in C and assembly code for different multipliers. The results are summarized in Table 11.2. Notice that the computing time of the assembly code for OBFL is two or three times larger than the corresponding time required for PBC.

To establish a good compromise between precision and speed we used the 2^{16} multiplier, with the multiplication algorithm of $(32\text{bits}) \times (32\text{bits}) \div 2^{16}$. To make a fair comparison between the schemes, we carried out the experiments with a sampling time of 7 ms for both controllers.

4.2 Selection of flux reference in experiments

The desired rotor flux magnitude is a primary design and application parameter which had to be selected equal for the two controllers. In general, high-performance servo applications require a low mean velocity, a continuous transient operation and a high peak torque with limited current amplitude amplifiers. In such applications, the selection of the rotor flux magnitude is usually based on the peak values of torque or acceleration which can be achieved by a given limited current [128]. Another important constraint for a design optimization is that magnetic saturation effects must be considered in the selection of the flux level.

There is a limit for the stator current which can be supplied by the inverter (the current obtained with the maximal amplitude for the given gain of the inverter). This limit is 100 A, but for safety reasons a lower value, $I_{\max} = 80$ A, was used. If the stator current magnitude is limited by the inverter to $I_{\max}$, then the torque becomes limited by the relative values of the flux-producing current, $i_d = \frac{\beta}{L_{sr}}$, and the torque-producing current,

$$i_q = \frac{L_r \tau}{L_{sr} n_p \beta}$$

which are attainable at $I_{\max}$, by the relation,

$$\sqrt{i_d^2 + i_q^2} = \sqrt{(\frac{\beta}{L_{sr}})^2 + (\frac{L_r \tau}{L_{sr} n_p \beta})^2} = I_{\max}$$

with use of the constant flux strategy. This limits the peak torque achievable with a limited current amplitude. In this case the optimal choice of rotor flux can be found by expressing the torque in terms of the stator current amplitude,

$$\tau = \frac{L_{sr}^2 n_p}{L_r} |I|^2 \frac{\sin(2\alpha)}{2}$$

where

$$\alpha = \tan^{-1}(\frac{i_q}{i_d}).$$

The maximum value of this expression corresponds to $\alpha = 45°$ or $i_d = i_q$ [128].

At $I_{\max}$, the magnetizing current I_{mr} can be selected as $I_{mr} = i_d = \frac{\beta}{L_{sr}} \simeq 56.57$ A using

$$\sqrt{i_d^2 + i_q^2} = \sqrt{(\frac{\beta}{L_{sr}})^2 + (\frac{L_r \tau}{L_{sr} n_p \beta})^2} = I_{\max}.$$

But the optimal choice of flux depends on the current selected in the optimization [128]. Thus the torque/amp can be maximized at the rated current of the inverter $I_0 = 40$ A. In this case, the magnetizing current can be selected as $I_{mr} \simeq 28.28$ A.

To evaluate this selection, we compared experimentally the speed tracking responses for different flux levels at low and high speeds [131]. As a consequence of this comparison, we used $I_{mr} = 30$ A, which gives $\beta_* = I_{mr} L_{sr} = 5.79 \cdot 10^{-3}$ Wb. We also limited the desired torque for PBC to $|\tau_d| < 0.82874$ Nm by the relation

$$I_{\max} = \sqrt{(\frac{\beta}{L_{sr}})^2 + (\frac{L_r \tau_d}{L_{sr} n_p \beta})^2}.$$

4.3   Speed tracking performance

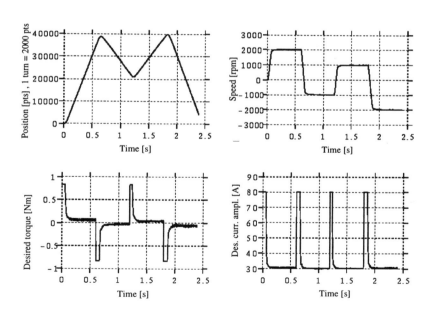

Figure 11.4: Results for a periodic square wave with amplitude changing between 2000 rpm and 1000 rpm. Sampling time 3 ms, $K_P = 0.012$, $K_I = 0.012$

Fig. 11.4 shows the response after starting the operation from zero initial conditions for the PBC, with a periodic square wave of large amplitude and a sampling time of 3 ms. The rotor speed converged very fast after the transient changes of speed reference. The desired torque and the desired current reached their maximum values during the transient. An integrator anti–windup technique [13] was implemented to avoid saturations. We can see from the graph of desired torque, that it did not converge to zero in steady state, because the integrator action was used to overcome friction. The desired current amplitude converged to the selected magnetizing current ($I_{mr} = 30$ A) plus the small torque drive current.

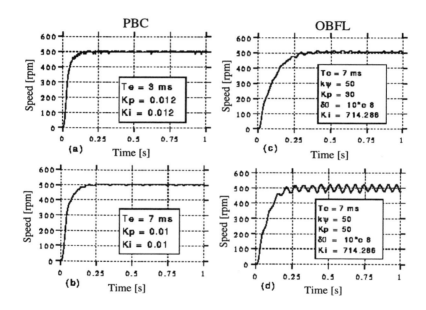

Figure 11.5: (a) and (b) show results with PBC for different sampling times. (c) and (d) show results with OBFL for different values of K_P.

As discussed in the previous section, because of the heavy computational burden of OBFL, we had to increase the sampling time to 7 ms. Also, and perhaps as a consequence of the increased sampling period, we were not able to tune properly the controller gains when we used a large change in the set points which saturated the control signals. Therefore we used a low speed reference of 500 rpm for the performance comparison of the two controllers. A larger regulation range for OBFL is possible with a slower reference, this at the cost of achievable bandwidth. As pointed out before, the initial condition for the observer in the OBFL must be chosen different from zero. We used $\hat{\lambda}_b(0) = 5.79 \cdot 10^{-3}$ equal to β_*. We also adjusted the gains of both controllers to have the most rapid response time without oscillations, which at the same time kept the current within the given amplitude limit of 80A.

As can be seen from Fig. 11.5 the response of PBC is faster than that of OBFL with the same sampling time of 7 ms. An increase of K_P of OBFL for a more rapid response caused oscillations and stator current saturation. It is also shown in [131] that an increase of δ_0 and K_I causes an overshoot in the speed tracking response, while a change of k_λ does not significantly affect the speed response.

4.4   Robustness and disturbance attenuation

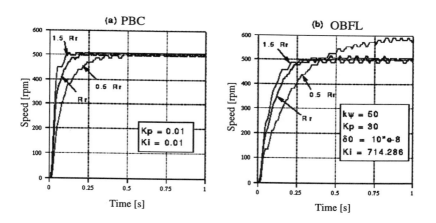

Figure 11.6: Variation of $\hat{R}_r (\pm 50\%)$. (a) show results of PBC, (b) show results of OBFL.

Fig. 11.6 shows the effects of $\pm 50\%$ errors in the rotor resistance estimate used in the two controllers with a speed reference of 500 rpm and for the same sampling time of 7 ms. We can see that PBC is unquestionably more robust to the change of this important parameter.

In Fig. 11.7, we can see the response of the two controllers for load torque compensation. We applied an unknown load torque (about 0.25 Nm) at $t = 3$ s using the DC motor which was connected to the shaft of the induction motor. The gain constants of the controllers had to be changed because the moment of inertia was changed to $D_m = 4.62 \cdot 10^{-3}$ kgm^2. The two controllers compensated well for the load torque disturbance within 2 s with an increase of about 15 A in the current amplitude, but the OBFL needed higher current amplitude than the PBC, both in the initial transient (50 A vs. 65 A), and at stationary conditions. Notice that the speed reference was reduced in this experiment to 250 rpm, which explains why the speed tracking of OBFL was improved as compared to the previous experiment.

We also observed experimentally the instability mechanism of OBFL presented in Section 2.3, under the same test conditions as in the simulation shown in Fig. 11.2.

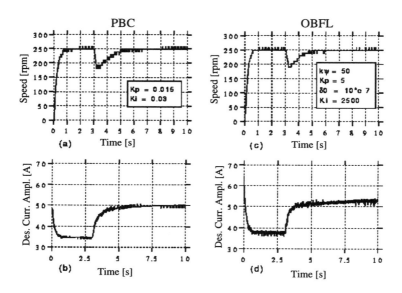

Figure 11.7: (a) and (b) show speed and desired current amplitude for the PBC. (c) and (d) show the same quantities for the OBFL.

4.5   Conclusion

We have presented in this section an experimental comparison of two controllers based on exact linearization and passivity ideas. It is our strong belief, and we hope the present study corroborates this point, that PBC outperforms schemes based on exact linearization, mainly because the control actions required to impose the linear dynamics typically will exceed the domain of validity of the models. This is clearly illustrated in the present experimental study where we were unable to achieve good performance of the OBFL for large and fast speed references. Of course, our experimental conclusions pertain only to the particular application at hand. Other experimental evidence of exact linearization controllers may be found in, e.g., [30, 186]. To the best of our knowledge the present work is the first attempt to compare two different novel nonlinear schemes on the same experimental facility. The recent paper [96] presents a similar study with analogous conclusions.

We conclude this section with the following remarks:

- We have illustrated that, in spite of a widespread erroneous belief, linearization (at least in the form which is presented here) does not allow us to impose arbitrary performances, even in the hypothetical case of no control constraints. In the present study the convergence rate of both schemes is limited by the rotor time constant, due to damping injection limitations in PBC, and to observer rate limits in OBFL.

- PBC will in general be simpler to implement than exact linearization controllers. This fundamental issue of computational complexity is, unfortunately, not fully appreciated in the recent literature of nonlinear control. It certainly goes beyond the arguments of availability of cheap and fast numerical processors, it pertains instead to poor numerical robustness of complicated arithmetic operations.

5 Robust stability of PBC

The stability proof presented in Section 3.2 for indirect FOC (see also Section **10**.3.2 for a proof in the case of the full model) critically depends on the cascade structure of the closed–loop equations, which is unfortunately lost when the rotor time constant is not exactly known. It is well known that this parameter is subject to significant changes during machine operation, hence the question of robust stability with respect to parameter uncertainty arises naturally.

In this section we recall the results of [68] and provide some answers to the question of robust stability. First, we present a result that states that, under very weak conditions, all signals in the closed–loop system are bounded. Then, we give necessary and sufficient conditions for uniqueness of the equilibrium point of the closed loop. Interestingly enough, both conditions above allow for a 200% error in the estimate of the rotor time–constant, a requirement which is not hard to satisfy in applications. Then, we give conditions on the motor and controller parameters, and the speed and rotor flux norm reference values that ensure either *global boundedness* of all solutions, or *(local or global) asymptotic stability or instability* of the equilibrium. The basis for these robustness results is the introduction of a new change of coordinates and the construction of a quadratic Lyapunov function from which we can establish the stronger property of *global exponential* stability.

The closed loop system that we study in this section is described by a fourth order nonlinear autonomous system that we repeat here for ease of reference. The *motor model* is given by

$$
\begin{aligned}
T_r \dot{\lambda}_r &= -\lambda_r + L_{sr} v \\
D_m \ddot{q}_m &= \tau - \tau_L \\
\tau &= \tfrac{n_p L_{sr}}{L_r} v^\top \mathcal{J} \lambda_r
\end{aligned}
\tag{11.23}
$$

and the control inputs are generated by the nonlinear dynamic output feedback PBC

$$
v = \frac{1}{L_{sr}} e^{\mathcal{J}\rho_d} \left[\begin{array}{c} \beta_* \\ \tfrac{L_r}{n_p \beta_*} \tau_d \end{array} \right]
\tag{11.24}
$$

$$
\dot{\rho}_d = \frac{\hat{R}_r}{n_p \beta_*^2} \tau_d, \; \rho_d(0) = 0
$$

$$
\tau_d = -(K_P + \frac{K_I}{p}) \dot{\tilde{q}}_m, \; K_I, \, K_P > 0
\tag{11.25}
$$

The flux norm and speed reference $\beta_*, \dot{q}_{m*}$, respectively, and the load torque τ_L are assumed to be constant.

To simplify the expressions below, and without loss of generality for the purposes of this study, all motor parameters will be set to unity except for the rotor resistance and the load torque, which are assumed to be constant but unknown. Setting all parameters except the rotor resistance to unity causes only small loss of generality, for two reasons: First, setting rotor moment of inertia and mutual inductance to unity changes only the loop gain by a factor, which can be compensated for by a scaling of the velocity. Second, the unknown parameter of importance to indirect field-oriented control is the rotor time constant, which is a function of both rotor resistance and rotor inductance (which can change due to nonlinear magnetics). Therefore the effect of an *unknown rotor time constant* can be investigated by considering the effect of an unknown rotor resistance only. To generalize the uniqueness and stability conditions derived in this section, R_r must be replaced by R_r/L_r and $\hat{R}_r$ by $\hat{R}_r/\hat{L}_r$.

5.1   Global boundedness

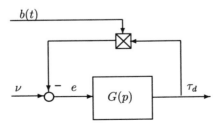

Figure 11.8: Input–output description of closed–loop system.

By the use of an input–output formulation of the problem we established in [68] the following result. Since the proof is quite technical we have omitted it here, and refer the interested reader to Section 6 of Appendix **D**.

Proposition 11.10 *The current–fed machine (11.23) in closed–loop with the PBC (11.24)–(11.25) may be described as the feedback interconnection of an LTI system with a time–varying bounded gain as*

$$\tau_d = G(p)e$$
$$e = \nu - b(t)\tau_d$$

see Fig. 11.8, where

$$G(p) = \frac{pK_P + K_I}{p^2 + (pK_P + K_I)\frac{R_r}{\hat{R}_r}}$$

with ν an external bounded signal, and $b(t) = b_\infty(t) + b_1(t)$, such that

$$|b_\infty(t)| \leq |\frac{\hat{R}_r - R_r}{\hat{R}_r}|$$

$$b_1(t) \in \mathcal{L}_1 .$$

$\square$

The corollary below follows from the application of the $\mathcal{L}_\infty$ small gain theorem and the evaluation of the $\mathcal{L}_1$ gain of $G(p)$.

Corollary 11.11 *Assume the impulse response of $G(p)$ is positive[7] and*

$$\hat{R}_r < 2R_r$$

Then the feedback interconnection of $G(p)$ and the $b(t)$-gain has finite $\mathcal{L}_\infty$ gain. That is, the PBC is bounded–input bounded–output stable.

5.2 Coordinate changes and uniqueness of equilibrium

To carry out the asymptotic stability analysis in the general case we find it convenient to work with a state space representation of the system, and introduce some coordinate transformations. First, let us define the coordinate transformation[8]

$$\vartheta = \begin{bmatrix} \vartheta_1 \\ \vartheta_2 \\ \vartheta_3 \\ \vartheta_4 \end{bmatrix} = \begin{bmatrix} \lambda_{rd}^\top \mathcal{J} \lambda_r \\ \lambda_{rd}^\top \lambda_r \\ \tau_d \\ \dot{\tilde{q}}_m \end{bmatrix} \tag{11.26}$$

---

[7]This assumption is made only to simplify the expressions, it can clearly be relaxed.

[8]Notice that this change of coordinates, although formally nonlinear because it involves a nonlinear function of the state variable ρ_d, may be considered as *linear* viewing $\lambda_{rd}(t)$ as a *bounded* function of time.

where

$$\lambda_{rd} = \mathbf{e}^{\mathcal{J}\rho_d} \begin{bmatrix} \beta_* \\ 0 \end{bmatrix}$$

This results in the following dynamic model

$$\dot{\vartheta} = \begin{bmatrix} -R_r & \hat{R}_r \frac{\vartheta_3}{\beta_*^2} & -R_r & 0 \\ -\hat{R}_r \frac{\vartheta_3}{\beta_*^2} & -R_r & 0 & 0 \\ -K_P & 0 & -K_P \frac{\vartheta_2}{\beta_*^2} & -K_I \\ 1 & 0 & \frac{\vartheta_2}{\beta_*^2} & 0 \end{bmatrix} \vartheta + \begin{bmatrix} 0 \\ R_r \beta_*^2 \\ K_P \tau_L \\ -\tau_L \end{bmatrix} \tag{11.27}$$

Now, we shift the equilibrium to the origin. To this end, we define the new coordinates $w \triangleq \vartheta - \bar{\vartheta}$ where $\bar{\vartheta} \in \mathbb{R}^4$ is an equilibrium of (11.27). Below we will show that, for all practical purposes, the equilibrium is unique. The transformed dynamic model becomes

$$\dot{w} = \begin{bmatrix} -R_r & \hat{R}_r \frac{\bar{\vartheta}_3+w_3}{\beta_*^2} & -R_r + \frac{\hat{R}_r}{\beta_*^2}\bar{\vartheta}_2 & 0 \\ -\hat{R}_r \frac{\bar{\vartheta}_3+w_3}{\beta_*^2} & -R_r & -\frac{\hat{R}_r}{\beta_*^2}\bar{\vartheta}_1 & 0 \\ -K_P & -\frac{K_P}{\beta_*^2}\bar{\vartheta}_3 & -K_P \frac{\bar{\vartheta}_2+w_2}{\beta_*^2} & -K_I \\ 1 & \frac{1}{\beta_*^2}\bar{\vartheta}_3 & \frac{\bar{\vartheta}_2+w_2}{\beta_*^2} & 0 \end{bmatrix} w \tag{11.28}$$

The equilibria of the model in (11.27) have the following property:

Proposition 11.12 *The equilibria of (11.27) are independent of K_P, K_I. Further, the equilibrium is unique for all values of τ_L if and only if $0 < \hat{R}_r \le 3R_r$.* $\square$

Proof. The equilibria of (11.27) are all solutions $\bar{\vartheta} \in \mathbb{R}^4$ to the equation

$$\begin{bmatrix} 0 \\ 0 \\ 0 \\ 0 \end{bmatrix} = \begin{bmatrix} -R_r & \hat{R}_r \frac{\bar{\vartheta}_3}{\beta_*^2} & -R_r & 0 \\ -\hat{R}_r \frac{\bar{\vartheta}_3}{\beta_*^2} & -R_r & 0 & 0 \\ -K_P & 0 & -K_P \frac{\bar{\vartheta}_2}{\beta_*^2} & -K_I \\ 1 & 0 & \frac{\bar{\vartheta}_2}{\beta_*^2} & 0 \end{bmatrix} \bar{\vartheta} + \begin{bmatrix} 0 \\ R_r \beta_*^2 \\ K_P \tau_L \\ -\tau_L \end{bmatrix}$$

For any equilibrium point we must have $\bar{\vartheta}_4 = 0$ because the PI-controller integrates ϑ_4. This simplifies the equilibrium equations to

$$\begin{bmatrix} 0 \\ 0 \\ 0 \end{bmatrix} = \begin{bmatrix} -R_r & \hat{R}_r \frac{\bar{\vartheta}_3}{\beta_*^2} & -R_r \\ -\hat{R}_r \frac{\bar{\vartheta}_3}{\beta_*^2} & -R_r & 0 \\ -K_P & 0 & -K_P \frac{\bar{\vartheta}_2}{\beta_*^2} \end{bmatrix} \begin{bmatrix} \bar{\vartheta}_1 \\ \bar{\vartheta}_2 \\ \bar{\vartheta}_3 \end{bmatrix} + \begin{bmatrix} 0 \\ R_r \beta_*^2 \\ K_P \tau_L \end{bmatrix}$$

From these equations, the following third order polynomial in $\bar{\vartheta}_3$ is derived.

$$R_r \hat{R}_r \bar{\vartheta}_3^3 - \hat{R}_r^2 \tau_L \bar{\vartheta}_3^2 + R_r \hat{R}_r \beta_*^4 \bar{\vartheta}_3 - R_r^2 \beta_*^4 \tau_L = 0 \tag{11.29}$$

If the equilibrium value of $\bar{\vartheta}_3$ is known, then $\bar{\vartheta}_1$ and $\bar{\vartheta}_2$ can be calculated (as functions of $\bar{\vartheta}_3$). Henceforth, we will concentrate on the solution of (11.29). In particular we will investigate the conditions under which the function $\tau_L = \tau_L(\bar{\vartheta}_3)$ is bijective, i.e., $\bar{\vartheta}_3$ is also a function of τ_L.

The expression for τ_L as a function of $\bar{\vartheta}_3$ is

$$\tau_L = \frac{R_r \hat{R}_r \bar{\vartheta}_3^3 + R_r \hat{R}_r \beta_*^4 \bar{\vartheta}_3}{R_r^2 \beta_*^4 + \hat{R}_r^2 \bar{\vartheta}_3^2}$$

Clearly, $\tau_L(\bar{\vartheta}_3)$ is continuous and surjective. Then it is a bijection if it is strictly monotone. The derivative of $\tau_L(\bar{\vartheta}_3)$ is

$$\frac{d\tau_L}{d\bar{\vartheta}_3} = \frac{R_r \hat{R}_r^3 \bar{\vartheta}_3^4 + \left(3R_r^3 \hat{R}_r \beta_*^4 - R_r \hat{R}_r^3 \beta_*^4\right) \bar{\vartheta}_3^2 + R_r^3 \hat{R}_r \beta_*^8}{\left(R_r^2 \beta_*^4 + \hat{R}_r^2 \bar{\vartheta}_3^2\right)^2} \qquad (11.30)$$

The denominator in this equation is always positive. Therefore, if the numerator is of constant sign, $\tau_L(\bar{\vartheta}_3)$ is bijective. The numerator of (11.30) is a polynomial in $\bar{\vartheta}_3^2$. This polynomial is of constant sign if its discriminant is less than or equal to zero, that is, if

$$9R_r^4 - 10R_r^2 \hat{R}_r^2 + \hat{R}_r^4 \le 0$$

The discriminant is a polynomial in $\hat{R}_r^2$ which is less than or equal to zero for $\hat{R}_r^2 \in [R_r^2, 9R_r^2]$.

Also, if $\hat{R}_r < \sqrt{3}R_r$ then all terms in the numerator of (11.30) are strictly positive. Then, τ_L is a monotone function of $\bar{\vartheta}_3$. If, on the other hand, $\hat{R}_r > 3R_r$ then values for $\bar{\vartheta}_3$ can be found where $\frac{d\tau_L}{d\bar{\vartheta}_3} < 0$, so that τ_L as a function of $\bar{\vartheta}_3$ is not monotone anymore and therefore not bijective. This concludes the proof. ■

As an example of the existence of multiple equilibria for certain ranges of τ_L, the roots of (11.29) will now be determined by application of the root locus technique to the more suitable form

$$1 - \tau_L \frac{\hat{R}_r}{R_r} \frac{\bar{\vartheta}_3^2 + \frac{R_r^2}{\hat{R}_r^2}\beta_*^4}{\bar{\vartheta}_3(\bar{\vartheta}_3^2 + \beta_*^4)} = 0$$

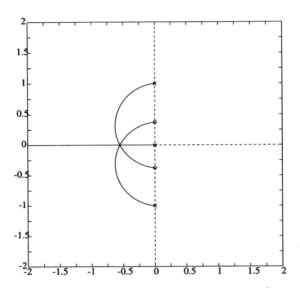

Figure 11.9: Root locus of the system equilibria for $\hat{R}_r = 3R_r$.

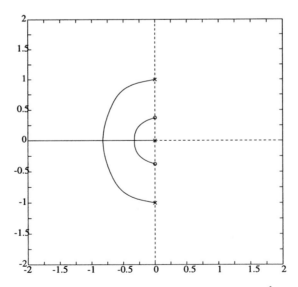

Figure 11.10: Root locus of the system equilibria for $\hat{R}_r > 3R_r$.

The uniqueness of equilibrium for $\hat{R}_r = 3R_r$ is evident from Fig. 11.9, since there are three *coinciding* real roots for one value of τ_L, while for any other value of τ_L there is only one real root. The non-uniqueness of the equilibria for $\hat{R}_r > 3R_r$ causes the locus of Fig. 11.10 to have three distinct real roots for a certain range of τ_L. If $\hat{R}_r < 3R_r$ then the two poles go directly to the zeros without crossing the real axis.

Before closing this section it is interesting to "pull out" the nonlinear terms of (11.28) as

$$
\dot{w} = \begin{bmatrix} -R_r & \hat{R}_r\frac{\bar{\vartheta}_3}{\beta_*^2} & -R_r + \frac{\hat{R}_r}{\beta_*^2}\bar{\vartheta}_2 & 0 \\ -\hat{R}_r\frac{\bar{\vartheta}_3}{\beta_*^2} & -R_r & -\frac{\hat{R}_r}{\beta_*^2}\bar{\vartheta}_1 & 0 \\ -K_P & -\frac{K_P}{\beta_*^2}\bar{\vartheta}_3 & -K_P\frac{\bar{\vartheta}_2}{\beta_*^2} & -K_I \\ 1 & \frac{1}{\beta_*^2}\bar{\vartheta}_3 & \frac{\bar{\vartheta}_2}{\beta_*^2} & 0 \end{bmatrix} w + \begin{bmatrix} \hat{R}_r w_2 \\ -\hat{R}_r w_1 \\ -K_P w_2 \\ w_2 \end{bmatrix} \frac{w_3}{\beta_*^2} \quad (11.31)
$$

Noting the presence in the right hand term of the scaling factor $\frac{w_3}{\beta_*^2}$, recalling that $w_3 = \tau_d - \bar{\tau}_d$, and refering to Fig. 11.1 (with $\dot{\beta}_* = 0$) on page 386, we see from (11.31) that, roughly speaking, the closed loop system behaves "almost linearly" if the PI speed loop is not too tight. That is, if $\frac{w_3}{\beta_*^2}$ is "small" and/or slowly time varying.

5.3 Local asymptotic stability

In this section we will study, via the *first Lyapunov method*, the local asymptotic stability of (11.28). Towards this end, we see that the systems first order approximation is simply the first right hand term of (11.31), whose characteristic polynomial is given by

$$
\begin{vmatrix} s + R_r & -\hat{R}_r\frac{\bar{\vartheta}_3}{\beta_*^2} & R_r - \frac{\hat{R}_r}{\beta_*^2}\bar{\vartheta}_2 & 0 \\ \hat{R}_r\frac{\bar{\vartheta}_3}{\beta_*^2} & s + R_r & \frac{\hat{R}_r}{\beta_*^2}\bar{\vartheta}_1 & 0 \\ K_P & \frac{K_P}{\beta_*^2}\bar{\vartheta}_3 & s + K_P\frac{\bar{\vartheta}_2}{\beta_*^2} & K_I \\ -1 & -\frac{1}{\beta_*^2}\bar{\vartheta}_3 & -\frac{\bar{\vartheta}_2}{\beta_*^2} & s \end{vmatrix} = 0
$$

Given the complexity of the expression above, (recall that $\bar{\vartheta}$ is itself a nonlinear function of the motor parameters), we are unable at this point to make a general statement concerning the stability of the roots of this polynomial. Consequently, we will only consider below some special cases. In particular we will show that, even with zero load torque, the equilibrium may become unstable.

A   Case of known R_r

As shown in Section 3.2, when $R_r = \hat{R}_r$ the equilibrium is GAS. However, to provide some tuning rules it is interesting to look at the behaviour of the roots of the linearized system, for instance as a function of the load torque. To this end, we write the characteristic polynomial in a τ_L-root locus form as

$$1 + \frac{\tau_L^2 R_r^2}{\beta_*^4} \frac{s^2 + K_P s + K_I}{(s + R_r)^2 (s^2 + K_P s + K_I)} = 0$$

The closed loop then has two poles at fixed positions determined by K_P, K_I and, as τ_L increases, the double pole at $s = -R_r$ moves along straight asymptotes in the left half plane.

B Case of zero load torque

Even though this case will be studied with greater detail in Section 6, where some tuning rules for the PI gains are derived, we present here this simple, but interesting, proposition.

Proposition 11.13 (Local stability for zero load torque.) *Assume* $\tau_L = 0$. *Then, the system (11.28) is locally asymptotically stable if either* $0 < \hat{R}_r \leq R_r + K_P$ *or* $K_P^2 > K_I$. *On the other hand, the equilibrium will be unstable if* $\hat{R}_r > R_r + K_P$ *and a large integral gain is used.* □

Proof. When $\tau_L = 0$ the characteristic equation reduces to

$$(s + R_r)^2 (s(s + K_P) + K_I) + K_P(s + R_r)(\hat{R}_r - R_r)s$$
$$+ (s + R_r)(\hat{R}_r - R_r)K_I = 0$$

The proof is completed noting that this equation has one root at $s = -R_r$, while from the Routh-Hurwitz criterion we know that the other roots are on the open the left hand plane if and only if

$$\hat{R}_r R_r K_P + \hat{R}_r K_P^2 > (\hat{R}_r - R_r - K_P)K_I$$

■

The proposition above shows that the system can be destabilized, in the sense of having unstable equilibria, if the rotor resistance is overestimated, the proportional gain is too small, and a large integral gain is used.

5.4 Global exponential stability

In this section we will investigate global exponential stability of the equilibrium using Lyapunov's second method. Namely, we will construct Lyapunov functions of the form

$$V(w) = \frac{1}{2} w^\top P w$$

where P is a positive definite symmetric constant matrix. To select P we first find positive semidefinite matrices P_i that lead to expressions without cubic terms in the derivative of V. Second, linear combinations of these positive semi-definite matrices are constructed that lead to a negative-definite $\dot{V}(w)$. Finally, the positive definiteness of P is checked.

To illustrate the procedure we first construct a Lyapunov function for the case where $R_r = \hat{R}_r$. Then, we treat the case when $\tau_L = 0$, $R_r \neq \hat{R}_r$, and derive a sufficient condition on $\hat{R}_r$ for GAS. The general case is then illustrated with a numerical example.

A Introducing positive semi-definite matrices to avoid cubic terms

To construct our Lyapunov-function candidate we consider for P linear combinations of the following positive semi-definite matrices P_i, $i = 1, \cdots, 4$.

$$
P_1 = \begin{bmatrix} 1 & 0 & 0 & 0 \\ 0 & 1 & 0 & 0 \\ 0 & 0 & 0 & 0 \\ 0 & 0 & 0 & 0 \end{bmatrix}, \quad
P_2 = \begin{bmatrix} \frac{1}{\hat{R}_r} & 0 & 0 & -1 \\ 0 & 0 & 0 & 0 \\ 0 & 0 & 0 & 0 \\ -1 & 0 & 0 & \hat{R}_r \end{bmatrix}
$$

$$
P_3 = \begin{bmatrix} 0 & 0 & 0 & 0 \\ 0 & 0 & 0 & 0 \\ 0 & 0 & 1 & K_P \\ 0 & 0 & K_P & K_P^2 \end{bmatrix}, \quad
P_4 = \begin{bmatrix} K_P^2 & 0 & K_P\hat{R}_r & 0 \\ 0 & 0 & 0 & 0 \\ K_P\hat{R}_r & 0 & \hat{R}_r^2 & 0 \\ 0 & 0 & 0 & 0 \end{bmatrix}
$$

The corresponding functions $V_i(w) = \frac{1}{2}w^\top P_i w$, $i = 1, \cdots, 4$ have derivatives

$$
\dot{V}_1(w) = w^\top P_1 \dot{w} = -R_r w_1^2 - \frac{R_r \beta_*^2 - \hat{R}_r \bar{\vartheta}_2}{\beta_*^2} w_1 w_3 - R_r w_2^2 - \frac{\hat{R}_r \bar{\vartheta}_1}{\beta_*^2} w_2 w_3
$$

$$
\dot{V}_2(w) = w^\top P_2 \dot{w} = -\frac{R_r + \hat{R}_r}{\hat{R}_r} w_1^2 - \frac{R_r}{\hat{R}_r} w_1 w_3 + (\hat{R}_r + R_r) w_1 w_4 + R_r w_3 w_4
$$

$$
\dot{V}_3(w) = w^\top P_3 \dot{w} = -K_I K_P w_4^2 - K_I w_3 w_4
$$

$$
\dot{V}_4(w) = w^\top P_4 \dot{w} = -K_P^2(R_r + \hat{R}_r) w_1^2 - \left(K_P \hat{R}_r (R_r + \hat{R}_r) + K_P^2 R_r \right) w_1 w_3
$$
$$
- K_P K_I \hat{R}_r w_1 w_4 - K_P R_r \hat{R}_r w_3^2 - K_I \hat{R}_r^2 w_3 w_4
$$

Since these derivatives have only quadratic terms, the derivative of $V(w) = w^\top P w$ will also have only quadratic terms if P is a linear combination of P_i, $i = 1, \cdots, 4$. As a consequence, the global negative definiteness of $\dot{V}(w)$ can be easily checked.

B   Lyapunov function for $\hat{R}_r = R_r$

For the nominal case $R_r = \hat{R}_r$, a Lyapunov function can be constructed that is valid for all τ_L as follows. Realize that for the nominal case, the equilibrium is

$$\begin{aligned}
\bar{\vartheta}_1 &= 0 \\
\bar{\vartheta}_2 &= \beta_*^2 \\
\bar{\vartheta}_3 &= \tau_L
\end{aligned}$$

Consider the matrix

$$P_a = P_3 + P_4$$

This choice of P results in the Lyapunov function candidate $V_a(w) = \frac{1}{2} w^\top P w$ with derivative

$$\begin{aligned}
\dot{V}_a(w) &= -2 K_P^2 R_r w_1^2 - R_r^2 K_P w_3^2 - K_I K_P w_4^2 \\
&\quad - \left(2 K_P R_r^2 + K_P^2 R_r \right) w_1 w_3 - K_P K_I R_r w_1 w_4 - (K_I + R_r^2 K_I) w_3 w_4
\end{aligned}$$

The cross-term in $w_3 w_4$ can be cancelled by adding a term in P_2 to P_a:

$$P_b = P_3 + P_4 + \frac{K_I + R_r^2 K_I}{R_r} P_2$$

which results in the candidate Lyapunov function $V_b(w)$ with derivative

$$\begin{aligned}
\dot{V}_b(w) &= -2 \left(K_P^2 R_r + \frac{K_I + R_r^2 K_I}{R_r} \right) w_1^2 - R_r^2 K_P w_3^2 - K_I K_P w_4^2 \\
&\quad - \left(2 K_P R_r^2 + K_P^2 R_r + \frac{K_I + R_r^2 K_I}{R_r} \right) w_1 w_3 \\
&\quad + \left(2(K_I + R_r^2 K_I) - K_P K_I R_r \right) w_1 w_4 \\
&= -a_1 w_1^2 - a_3 w_3^2 - a_4 w_4^2 - 2 b_{13} w_1 w_3 + 2 b_{14} w_1 w_4
\end{aligned}$$

This derivative can always be rendered negative definite by adding a component $(z_3 + z_4) P_1$ to the matrix P_b,

$$P_c = P_3 + P_4 + \frac{K_I + R_r^2 K_I}{R_r} P_2 + (z_3 + z_4) P_1$$

where the coefficients z_3, z_4 are chosen to compensate for the cross-terms as follows:

$$\begin{aligned}
z_3 &= \frac{1}{R_r} \frac{b_{13}^2}{a_3} \\
z_4 &= \frac{1}{R_r} \frac{b_{14}^2}{a_4}
\end{aligned}$$

so that the derivative of the Lyapunov function $V_c(w) = \frac{1}{2} w^\top P_c w$ becomes

$$
\begin{aligned}
\dot{V}_c(w) \;=\; & -a_1 w_1^2 \\
& -a_3 w_3^2 - a_4 w_4^2 - 2b_{13} w_1 w_3 + 2b_{14} w_1 w_4 \\
& -\frac{b_{13}^2}{a_3} w_1^2 - \frac{b_{14}^2}{a_4} w_1^2 \\
& -\left(\frac{b_{13}^2}{a_3} + \frac{b_{14}^2}{a_4} \right) w_2^2
\end{aligned}
$$

The function $V_c(w)$ is positive definite and its derivative is negative definite, therefore it is a strict Lyapunov function for $R_r = \hat{R}_r$.

C   Lyapunov functions for $\hat{R}_r \neq R_r$, $\tau_L = 0$

For the case $\tau_L = 0$ and $R_r \neq \hat{R}_r$, the cross-term $(R_r - \hat{R}_r) w_1 w_3$ appears in $\dot{V}_1(w)$. This constrains the construction of a Lyapunov-function. The following approach has been used to derive sufficient conditions for GAS.

Add P_3 and P_4 to obtain a $\dot{V}(w)$ with negative terms for w_3^2 and w_4^2.

$$
\begin{aligned}
\dot{V}_3(w) + \dot{V}_4(w) \;=\; & -K_I K_P w_4^2 - (K_I + \hat{R}_r^2 K_I) w_3 w_4 \\
& -K_P^2 (R_r + \hat{R}_r) w_1^2 - K_P(\hat{R}_r(R_r + \hat{R}_r) + K_P R_r) w_1 w_3 \\
& -K_P K_I \hat{R}_r w_1 w_4 - R_r \hat{R}_r K_P w_3^2
\end{aligned}
$$

Add an amount of P_2 such that the cross-term in $w_3 w_4$ is cancelled, which simplifies the quadratic term.

$$
\begin{aligned}
\dot{V}_3(w) + \dot{V}_4(w) + \frac{K_I(1 + \hat{R}_r^2)}{R_r} \dot{V}_2(w) \;=\; & -K_I K_P w_4^2 - R_r \hat{R}_r K_P w_3^2 \\
& -\left(K_P^2(R_r + \hat{R}_r) + \frac{K_I(1 + \hat{R}_r^2)}{R_r \hat{R}_r}(R_r + \hat{R}_r) \right) w_1^2 \\
& -\left(K_P(\hat{R}_r(R_r + \hat{R}_r) + K_P R_r) + K_I \frac{1 + \hat{R}_r^2}{\hat{R}_r} \right) w_1 w_3 \\
& -\left(K_P K_I \hat{R}_r - K_I \frac{1 + \hat{R}_r^2}{\hat{R}_r}(R_r + \hat{R}_r) \right) w_1 w_4 \\
=\; & -a_1 w_1^2 - 2b_{13} w_1 w_3 - 2b_{14} w_1 w_4 - a_3 w_3^2 - a_4 w_4^2
\end{aligned}
$$

To make this expression negative definite, one can add a term $\frac{z}{R_r} P_1$ so that the derivative of the total Lyapunov function becomes

$$
\begin{aligned}
\dot{V}(w) &= -a_1 w_1^2 - 2b_{13} w_1 w_3 - 2b_{14} w_1 w_4 - a_3 w_3^2 - a_4 w_4^2 \\
&\quad - z w_1^2 - 2z\xi w_1 w_3 - z w_2^2 \\
&= -w^\mathsf{T}
\begin{bmatrix}
z + a_1 & 0 & (z\xi + b_{13}) & b_{14} \\
0 & z & 0 & 0 \\
(z\xi + b_{13}) & 0 & a_3 & 0 \\
b_{14} & 0 & 0 & a_4
\end{bmatrix}
w \\
&= -w^\mathsf{T} Q w
\end{aligned}
$$

$\dot{V}(w)$ is thus negative if the matrix Q is positive. A necessary and sufficient condition for this matrix Q to be positive is that all its leading principal minors are positive:

$$ z + a_1 > 0 $$

$$
\begin{vmatrix}
z + a_1 & 0 \\
0 & z
\end{vmatrix} > 0
$$

$$
\begin{vmatrix}
z + a_1 & 0 & (z\xi + b_{13}) \\
0 & z & 0 \\
(z\xi + b_{13}) & 0 & a_3
\end{vmatrix} > 0
$$

$$
\begin{vmatrix}
z + a_1 & 0 & z\xi + b_{13} & b_{14} \\
0 & z & 0 & 0 \\
z\xi + b_{13} & 0 & a_3 & 0 \\
b_{14} & 0 & 0 & a_4
\end{vmatrix} > 0
$$

The first two conditions are satisfied. The third condition follows from the fourth which can be formulated as follows.

$$ (z + a_1)a_3 a_4 - (z\xi + b_{13})^2 a_4 - b_{14}^2 a_3 > 0 $$

The cross-factor ξ follows from $\hat{R}_r$ and R_r, while z can be chosen. The value for z where the expression is optimal is calculated from setting the derivative of the above expression to zero

$$ a_3 - 2\xi(z\xi + b_{13}) = 0 $$

$$ z = \frac{1}{\xi}\left(\frac{a_3}{2\xi} - b_{13}\right) $$

Replacing z, the condition for positive definiteness becomes:

$$ \frac{a_3 a_4}{4\xi^2} - \frac{a_4 b_{13}}{\xi} + a_1 a_4 - b_{14}^2 > 0 $$

This condition is always satisfied for sufficiently small ξ as a consequence of the positiveness of a_3, so that there are values of $\hat{R}_r$, with $\hat{R}_r \neq R_r$, where the system is GAS. The boundary values of ξ for which $\dot{V}(w)$ is still negative, are complicated functions of all parameters in the model.

D   Lyapunov functions for $\hat{R}_r \neq R_r$ and $\tau_L \neq 0$

The previous approach can also be applied to the case where $\tau_L \neq 0$, but the resulting sufficient conditions for GAS are not given here since they are rather complicated. Instead, a numerical example is given of a Lyapunov function that ensures GAS for the particular parameter values $K_P = 1$, $K_I = 0.1$, $\beta_* = 1$, $R_r = 2$, $\hat{R}_r = 1$ and all τ_L.

Consider the candidate Lyapunov function

$$V(w) \;=\; w^\top \left(\frac{1}{2} P_1 + 0.1 P_2 + P_3 + P_4 \right) w$$

This function is positive definite, and its derivative is

$$\dot{V}(w) \;=\; -w^\top Q(\xi_1, \xi_2) w$$

where

$$\xi_1 \;=\; \frac{R_r \beta_*^2 - \hat{R}_r \bar{\vartheta}_2}{2 R_r \beta_*^2}$$

$$\xi_2 \;=\; \frac{\hat{R}_r \bar{\vartheta}_1}{2 R_r \beta_*^2}$$

and $Q(\xi_1, \xi_2)$ is a constant symmetric matrix whose off-diagonal coefficients depend on ξ_1 and ξ_2. For $V(w)$ to be a Lyapunov function, $\dot{V}(w)$ must be negative definite, and $Q(\xi_1, \xi_2)$ must therefore be positive definite. For the particular parameter values of the numerical example, this positive definiteness can be proved using the property that ξ_1 and ξ_2 are bounded functions of $\bar{\vartheta}_3$

$$\xi_1 \;=\; \frac{1}{2} - \frac{\hat{R}_r}{2 R_r} + \frac{\hat{R}_r^2 \bar{\vartheta}_3^2}{2 R_r} \frac{\hat{R}_r - R_r}{R_r^2 \beta_*^4 + \hat{R}_r^2 \bar{\vartheta}_3^2}$$

$$\xi_2 \;=\; \frac{\hat{R}_r}{2} \frac{(\hat{R}_r - R_r)\beta_*^2}{R_r^2 \beta_*^4 + \hat{R}_r^2 \bar{\vartheta}_3^2} \bar{\vartheta}_3$$

Remark 11.14 We have gone with some detail in "by–hand" calculations above only for the sake of illustration of the ideas. It is clear that the procedure is much more efficient, and can be made systematic, with modern symbolic computation tools.

6   Off–line tuning of PBC

It is well known that the performance of indirect FOC critically depends on the tuning of the gains of the PI velocity loop, a task which is rendered difficult by the high uncertainty in the rotor time–constant. (Recall that we showed in Proposition 11.8 that indirect FOC is identical to PBC in current–fed machines). In this section we give some simple rules, first reported in [78], for the *off–line tuning* of the PI gains which will ensure *robust* stability. As is well–known, robust stability, as opposed to just stability, ensures better performance measures. We give then an algorithm that, for each setting of the PI gains, evaluates the maximum error of the rotor time–constant estimate for which *global stability* is guaranteed. In this way, without knowing the actual value of the rotor time constant, we can assess the performance of all PI settings before closing the loop. Not surprisingly, instrumental for our analysis is a suitable decomposition of the system into a *passive feedback interconnection*.

6.1   Problem formulation

It has been shown in Section 3.2 that the system (11.23)–(11.25) is globally asymptotically stable if $\hat{R}_r = R_r$. Furthermore, in Section 5 we proved that stability is actually *exponential* and showed that the system remains stable under large variations of the rotor resistance. The problem we address in this section is the selection of the parameters K_P and K_I which will ensure, not just stability of the closed–loop, but also a good *transient performance* in spite of the uncertainty in the rotor resistance R_r. To formulate mathematically this problem we make the following basic observation:

PI tunings which allow larger estimation errors in rotor resistance are more robust, hence their overall performance is better.

Thus the tuning problem can alternatively be formulated as:

Definition 11.15 (PI tuning problem for induction motors.) *Given the induction motor and controller equations (11.23)–(11.25) with controller parameters $\hat{R}_r$, K_P, K_I and β_*. Find a range of values of the motor resistance estimation error*

$$\tilde{R}_r \stackrel{\triangle}{=} \hat{R}_r - R_r$$

for which global stability of the closed–loop system is preserved. More precisely, we want to find an interval $[R_r^{min}, R_r^{max}] \in \mathbb{R}$ such that, if $R_r^{min} \leq R_r \leq R_r^{max}$, then the system (11.23)–(11.25) is globally stable.[9]

Even though some answers to this question may be found in Section 5, the procedure relies on the generation of Lyapunov functions, hence its computationally exhaustive and not very transparent to the user. In this section we give a very simple numerical algorithm that, for each setting of the controller gains, generates the

---

[9]It is clear that from the interval for R_r we can immediately obtain an interval for $\tilde{R}_r$ (which will contain zero) by simply subtracting $\hat{R}_r$.

required set of values for $\tilde{R}_r$. The *size* of this interval, which we will call in the sequel *the performance interval*, provides a robustness measure of the closed–loop system which guides the user in the choice of the PI gains.

Since we believe this result could be of interest for practitioners, and furthermore in the case we consider the expressions considerably simplify, we have decided to work out the details for the non–normalized motor model. However, we still limit our attention to uncertainty on R_r.

6.2   Change of coordinates

In order to solve the problem formulated above, we use the change of coordinates introduced in Section 5.2. We will show below that this representation reveals some new energy dissipation features of FOC which are instrumental for our theoretical developments.

Applying the transformation (11.26) to the closed–loop equations results in the following nonlinear dynamic model

$$
\dot{\vartheta} =
\begin{bmatrix}
-\frac{R_r}{L_r} & \frac{\hat{R}_r}{n_p\beta_*^2}\vartheta_3 & -\frac{R_r}{n_pL_{sr}} & 0 \\
-\frac{\hat{R}_r}{n_p\beta_*^2}\vartheta_3 & -\frac{R_r}{L_r} & 0 & 0 \\
-\frac{n_pK_PL_{sr}}{L_rD_m} & 0 & -\frac{K_PL_{sr}}{D_m\beta_*^2}\vartheta_2 & -K_I \\
\frac{n_pL_{sr}}{D_mL_r} & 0 & \frac{L_{sr}}{D_m\beta_*^2}\vartheta_2 & 0
\end{bmatrix}
\vartheta +
\begin{bmatrix}
0 \\ \frac{R_r\beta_*^2}{L_rL_{sr}} \\ 0 \\ 0
\end{bmatrix}
+
\begin{bmatrix}
0 \\ 0 \\ \frac{K_P}{D_m} \\ -\frac{1}{D_m}
\end{bmatrix}
\tau_L
$$

where we have "pulled-out" the terms depending on τ_L to underscore the fact that it enters linearly in the equations. Henceforth, if we can prove exponential stability of the system with $\tau_L = 0$ then the system will remain stable (in the sense of global boundedness) even when $\tau_L \neq 0$. This fact will be invoked in our subsequent analysis.

In Section 5 it is shown that the equilibria of this system are defined by very complex algebraic relationships, and the system can actually have multiple equilibria. On the other hand, when $\tau_L = 0$, the equilibrium is unique and is given by

$$
[\bar{\vartheta}_1, \bar{\vartheta}_2, \bar{\vartheta}_3, \bar{\vartheta}_4]^\top = [0, \frac{\beta_*^2}{L_{sr}}, 0, 0]^\top
$$

Given this fact, and the stability consideration mentioned in the previous paragraph, we will treat the load torque as a disturbance and concentrate on the case $\tau_L = 0$.

Let us now shift the equilibrium to the origin by introducing the new coordinates $z_i = \vartheta_i - \bar{\vartheta}_i$, $i = 1, \cdots, 4$ to obtain

$$
\dot{z} =
\begin{bmatrix}
-\frac{R_r}{L_r} & \frac{\hat{R}_r}{n_p\beta_*^2}z_3 & \frac{1}{n_pL_{sr}}\left(\hat{R}_r - R_r\right) & 0 \\
-\frac{\hat{R}_r}{n_p\beta_*^2}z_3 & -\frac{R_r}{L_r} & 0 & 0 \\
-\frac{n_pK_PL_{sr}}{D_mL_r} & 0 & -\frac{K_PL_{sr}}{D_m\beta_*^2}\left(z_2 + \frac{\beta_*^2}{L_{sr}}\right) & -K_I \\
\frac{n_pL_{sr}}{D_mL_r} & 0 & \frac{L_{sr}}{D_m\beta_*^2}\left(z_2 + \frac{\beta_*^2}{L_{sr}}\right) & 0
\end{bmatrix}
z
\qquad (11.32)
$$

6.3   Local stability

In this section we present conditions that guarantee local stability of the closed–loop. To this end we recall the indirect Lyapunov method which states that (under some conditions verified here) a nonlinear system is locally stable *if and only if* its linear approximation is asymptotically stable, i.e. all the eigenvalues of the system matrix are in the open left hand plane. Thus we rewrite (11.32) "pulling out" its nonlinear terms as

$$
\dot{z} =
\begin{bmatrix}
-\frac{R_r}{L_r} & 0 & \frac{1}{n_p L_{sr}}\left(\hat{R}_r - R_r\right) & 0 \\
0 & -\frac{R_r}{L_r} & 0 & 0 \\
-\frac{n_p K_P L_{sr}}{D_m L_r} & 0 & -\frac{K_P}{D_m} & -K_I \\
\frac{n_p L_{sr}}{D_m L_r} & 0 & \frac{1}{D_m} & 0
\end{bmatrix}
z +
\begin{bmatrix}
\frac{\hat{R}_r}{n_p \beta_*^2} z_2 z_3 \\
-\frac{\hat{R}_r}{n_p \beta_*^2} z_1 z_3 \\
-\frac{K_P L_{sr}}{D_m \beta_*^2} z_2 z_3 \\
\frac{L_{sr}}{D_m \beta_*^2} z_2 z_3
\end{bmatrix}
$$

In compact notation we get $\dot{z} = A_L z + F(z)$. The systems first–order approximation is simply $\dot{z} = A_L z$, whose characteristic polynomial is

$$
\det(sI - A_L) = \left(s + \frac{R_r}{L_r}\right) g(s)
$$

with

$$
g(s) = s^3 + \left(\frac{K_P}{D_m} + \frac{R_r}{L_r}\right) s^2 + \left(\frac{K_P \hat{R}_r}{D_m L_r} + \frac{K_I}{D_m}\right) s + \frac{K_I \hat{R}_r}{D_m L_r}
$$

Thus, by applying the Routh-Hurwitz criterion, it can be shown that the conditions which must be satisfied for local stability are

i) $K_P > -\frac{D_m R_r}{L_r}$,

ii) $K_I > 0$

   and

iii) $c_1 \triangleq \dfrac{\left[\left(\frac{K_P}{D_m} + \frac{R_r}{L_r}\right)\left(\frac{K_P \hat{R}_r}{D_m L_r} + \frac{K_I}{D_m}\right) - \frac{K_I \hat{R}_r}{D_m L_r}\right]}{\frac{K_P}{D_m} + \frac{R_r}{L_r}} > 0$

   Noting that the first two conditions are trivially satisfied with positive values for the PI gains, the attention will be focused on the third one. It is easy to see that, because the denominator is always positive, this condition can be equivalently written in terms of the rotor resistance R_r as

$$
\textbf{iii)} \quad \Leftrightarrow \quad R_r > \frac{K_I \hat{R}_r}{\frac{K_P \hat{R}_r}{L_r} + K_I} - \frac{K_P L_r}{D_m} \tag{11.33}
$$

From the fact that

$$\hat{R}_r \geq \frac{K_I \hat{R}_r}{\frac{K_P \hat{R}_r}{L_r} + K_I}$$

for all K_P, K_I, L_r, we can now state the following proposition.

Proposition 11.16 *Consider the model of the current-fed induction motor in closed loop with the indirect FOC (11.23)–(11.25). If*

$$R_r > \hat{R}_r \qquad (11.34)$$

then the system is locally exponentially stable for $\tau_L = 0$. When $\tau_L \neq 0$ all trajectories enter (in finite time) a ball centered at the origin of radius $|\tau_L|$. $\square$

The importance of this result is that stability is preserved, for all PI gains, provided we *underestimate* the rotor resistance. However, our interest lies in obtaining tuning rules *independent* of R_r, hence we will study the case when

$$\frac{K_I \hat{R}_r}{\frac{K_P \hat{R}_r}{L_r} + K_I} \leq \frac{K_P L_r}{D_m}$$

i.e., when condition (11.33) holds *for all R_r*. This inequality can be equivalently written as

$$\hat{R}_r K_P^2 \geq K_I (D_m \hat{R}_r - K_P L_r)$$

from which, after some easy manipulations, the following result can be obtained:

Proposition 11.17 *Consider the model of the current-fed induction motor in closed loop with the indirect FOC (11.23)–(11.25). Then, the system (with $\tau_L = 0$) is locally exponentially stable for all R_r if and only if one of the conditions below hold*

Condition 1:

$$K_P \geq \frac{D_m \hat{R}_r}{L_r} \quad \text{and} \quad K_I > 0 \qquad (11.35)$$

Condition 2:

$$0 < K_P < \frac{D_m \hat{R}_r}{L_r} \quad \text{and} \quad 0 < K_I \leq \frac{K_P^2 \hat{R}_r}{D_m \hat{R}_r - K_P L_r} \qquad (11.36)$$

hold.

$\square$

It is interesting to note that the results presented in this section are independent of the parameter β_*. This will also be the case for the performance results given below.

6.4   A performance evaluation method based on passivity

In this section we define intervals of the rotor resistance for which global stability is preserved as a function of the PI gains. As mentioned above the size of these intervals gives a quantitative measure of the performance of the PI tuning. To this end, we find convenient to decompose the closed–loop system (11.32) as the feedback interconnection of two subsystems. One of the subsystems contains all the nonlinearities and turns out to be strictly *passive*. The second subsystem is LTI. Our motivation for the introduction of this decomposition is twofold: First, as shown in Appendix **A**, the negative feedback interconnection of two passive subsystems is still passive, and if one of them is strictly passive then the closed–loop system is stable. Second, for LTI systems there is a very simple analytic characterization of passivity in terms of positivity of the real part of its transfer function. Since this transfer function depends on the motor parameters and the PI gains, this positive realness test will provide us with the desired resistance intervals.

We can rewrite (11.32) as the feedback interconnection of the following two subsystems

$$
G_1 \; : \; u_1 \mapsto y_1 \left\{
\begin{array}{rcl}
\dot{z}_2 & = & -\frac{R_r}{L_r} z_2 + \frac{1}{\beta_*^2} z_3 u_1 \\
u_1 & = & -\frac{\hat{R}_r}{n_p} z_1 \\
y_1 & = & z_2 z_3
\end{array}
\right.
\tag{11.37}
$$

$$
G_2 \; : \; u_2 \mapsto y_2 \left\{
\begin{array}{rcl}
\dot{\nu} & = & A\nu + bu_2 \\
u_2 & = & z_2 z_3 \\
y_2 & = & c^\top \nu
\end{array}
\right.
\tag{11.38}
$$

with

$$
A = \begin{bmatrix}
-\frac{R_r}{L_r} & \frac{1}{n_p L_{sr}}\left(\hat{R}_r - R_r\right) & 0 \\
-\frac{n_p L_{sr} K_P}{D_m L_r} & -\frac{K_P}{D_m} & -K_I \\
\frac{n_p L_{sr}}{D_m L_r} & \frac{1}{D_m} & 0
\end{bmatrix}, \;
b = \frac{1}{\beta_*^2}\begin{bmatrix}
\frac{\hat{R}_r}{n_p} \\
-\frac{K_P L_{sr}}{D_m} \\
\frac{L_{sr}}{D_m}
\end{bmatrix}
$$

$$
c = \begin{bmatrix} \frac{\hat{R}_r}{n_p} \\ 0 \\ 0 \end{bmatrix}, \;
\nu = \begin{bmatrix} z_1 \\ z_3 \\ z_4 \end{bmatrix}
$$

and the obvious *interconnection structure*

$$
\begin{array}{rcl}
u_1 & = & -y_2 \\
u_2 & = & y_1
\end{array}
$$

This decomposition is shown in Fig. 11.11, where we have defined the transfer function $G_2(s) = c^\top (sI - A)^{-1}b$

$$
G_2(s) = \frac{L_r \hat{R}_r}{n_p^2 \beta_*^2} \frac{D_m \hat{R}_r s^2 + K_P R_r s + K_I R_r}{D_m L_r s^3 + (K_P L_r + D_m R_r)s^2 + (K_P \hat{R}_r + K_I L_r)s + K_I \hat{R}_r}
\tag{11.39}
$$

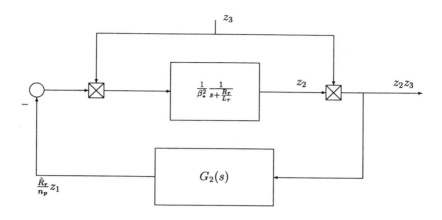

Figure 11.11: Decomposition of the closed–loop system.

The main features of this decomposition are: 1) We show below that the subsystem G_1, although nonlinear, is *passive*. 2) G_2 is an LTI relative degree one and minimum phase LTI system that can be made *strictly positive real* for suitable values of the motor and controller parameters.

We present now the main result of this section:

Proposition 11.18 *Consider the model of the current-fed induction motor in closed loop with the indirect FOC (11.23)–(11.25). Assume that the conditions for local stability ((11.34) or (11.35) or (11.36)) are satisfied, and that the following inequalities hold*

$$f_1 > 0 \quad and \quad f_2 > -2\sqrt{f_1 f_3} \tag{11.40}$$

where

$$f_1 \triangleq D_m^2 R_r \hat{R}_r + K_P D_m L_r (\hat{R}_r - R_r)$$
$$f_2 \triangleq K_P^2 R_r \hat{R}_r - K_I D_m (\hat{R}_r^2 + R_r^2)$$
$$f_3 \triangleq K_I^2 R_r \hat{R}_r$$

Then, the trivial equilibrium of the closed-loop system is GAS if $\tau_L = 0$. When $\tau_L \neq 0$ all trajectories enter (in finite time) a ball centered at the origin of radius $|\tau_L|$. $\square$

To establish the proof we need the following lemma.

Lemma 11.19 *The subsystem* G_1 : $u_1 \mapsto y_1$ *defined by (11.37) is output strictly passive. In particular, it satisfies the following inequality for all* $t \geq 0$

$$\frac{1}{\beta_*^2} \int_0^t u_1(\tau) y_1(\tau) d\tau = -\frac{\hat{R}_r}{n_p \beta_*^2} \int_0^t z_1(\tau) z_2(\tau) z_3(\tau) d\tau$$

$$\geq \frac{R_r}{L_r} \int_0^t z_2^2(\tau) d\tau + \alpha$$

with $\alpha \in \mathbb{R}$. $\qquad\qquad\qquad\qquad\qquad\qquad\qquad\qquad\qquad\qquad\qquad\qquad\qquad\qquad$ $\square$

Proof. Consider the function

$$V = \frac{1}{2} z_2^2 \geq 0$$

whose time derivative along the trajectories of G_1 is given by

$$\dot{V} = -\frac{R_r}{L_r} z_2^2 - \frac{\hat{R}_r}{n_p \beta_*^2} z_1 z_2 z_3$$

The proof is completed by integrating the above expression over the interval $[0, t]$, recalling that $V \geq 0$, and defining $\alpha \stackrel{\triangle}{=} -V(0)$.

$\qquad\qquad\qquad\qquad\qquad\qquad\qquad\qquad\qquad\qquad\qquad\qquad\qquad\qquad\qquad\qquad\qquad$ ■

Lemma 11.20 *Consider the transfer function (11.39). Then, the conditions of Proposition 11.18 above (i. e., stability and (11.40)) ensure that* $G_2(s)$ *is strictly positive real.* $\qquad\qquad\qquad\qquad\qquad\qquad\qquad\qquad\qquad\qquad\qquad\qquad\qquad\qquad\qquad\qquad$ $\square$

Proof. We will verify that $G_2(j\omega)$ satisfies condition **(ii')** of the Kalman–Yakubovich–Popov lemma, see Section 6 in Appendix **A**. First, notice that the real part of the transfer function (11.39) is given by

$$\Re\left\{G_2(j\omega)\right\} = k \frac{f_1 \omega^4 + f_2 \omega^2 + f_3}{[a_1 - a_2 \omega^2]^2 + [(a_3 - a_4 \omega^2)\omega]^2} \qquad (11.41)$$

with

$$k = \frac{L_r \hat{R}_r}{n_p^2 \beta_*^2}$$

$$a_1 = K_I \hat{R}_r$$

$$a_2 = K_P L_r + D_m R_r$$

$$a_3 = K_P \hat{R}_r + K_I L_r$$

$$a_4 = D_m L_r$$

The limit condition is clearly verified with $f_1 > 0$. Now, from (11.41) we see that the sign of this equation is only determined by the sign of its numerator. Moreover, viewing this polynomial as a function of the variable ω, two conditions must be satisfied in order to guarantee the positivity of the transfer function, namely: The coefficient f_1 must be positive and the polynomial

$$f(\omega) = f_1\omega^4 + f_2\omega^2 + f_3 \tag{11.42}$$

must not have real roots.

Finally, notice that (11.42) can be written as a polynomial of degree 2 over the variable $x = \omega^2$. The roots of this new polynomial are

$$\bar{x}_{1,2} = \frac{-f_2 \pm \sqrt{f_2^2 - 4f_1f_3}}{2f_1}$$

It can be seen that if $f_2^2 < 4f_1f_3$, then $\bar{x}_{1,2}$ are complex and therefore, the roots of the original polynomial, $\bar{\omega}_1, \cdots, \bar{\omega}_4$, are also complex. On the other hand, if $f_2^2 = 4f_1f_3$, the roots $\bar{x}_{1,2}$ are real of the form

$$\bar{x}_{1,2} = -\frac{f_2}{2f_1}$$

Hence, if $f_2 > 0$ then $\bar{\omega}_1, \cdots \bar{\omega}_4$ are complex, satisfying the condition for positivity. Finally, if $f_2^2 > 4f_1f_3$, then $\bar{x}_{1,2}$ are again real but in this case with the following structure

$$\bar{x}_1 = \frac{-f_2 + \sqrt{f_2^2 - 4f_1f_3}}{2f_1}$$

$$\bar{x}_2 = \frac{-f_2 - \sqrt{f_2^2 - 4f_1f_3}}{2f_1}$$

It is easy to see that, if $f_2 > 0$ then $\bar{x}_2 < 0$ and therefore $\bar{\omega}_1, \bar{\omega}_2$ are complex. In order to get $\bar{x}_1 < 0$, i.e. $\bar{\omega}_3, \bar{\omega}_4$ also complex, the condition that must be satisfied is $f_2 > \sqrt{f_2^2 - 4f_1f_3}$ which can be equivalently written as $0 > 4f_1f_3$. The proof of the proposition is then completed by noting that the condition $f_2 > -2\sqrt{f_1f_3}$ satisfies simultaneously the three required conditions to guarantee positivity of the polynomial (11.42). ∎

We can now present the proof of Proposition 11.18.

Proof. Consider the following function

$$V_1 = \underbrace{-\frac{\hat{R}_r}{n_p}\int_0^t z_1z_2z_3d\tau - \frac{R_r}{L_r}\int_0^t z_2^2d\tau - \alpha}_{\geq 0 \;\Leftarrow\; \text{Lemma 11.19}} + \frac{1}{2}\nu^\top P\nu \geq 0$$

whose time derivative along the trajectories of (11.38) is given by

$$\dot{V}_1 = -\frac{\hat{R}_r}{n_p}z_1 z_2 z_3 - \frac{R_r}{L_r}z_2^2 + \frac{1}{2}\nu^\top(PA + A^\top P)\nu + \nu^\top P b z_2 z_3$$

Invoking the strict positive realness of the transfer function (11.39), the above expression can be written as

$$
\begin{aligned}
\dot{V}_1 \;\; &= \tfrac{1}{2}\nu^\top Q\nu + y_2 z_2 z_3 \\
&= \qquad\qquad -\frac{R_r}{L_r}z_2^2 - \frac{1}{2}\nu^\top Q\nu \\
&\leq \qquad\qquad -\alpha\|z\|^2
\end{aligned}
$$

for some $\alpha > 0$. This proves that the whole state $z \in \mathcal{L}_2$, and furthermore that $\nu \in \mathcal{L}_\infty$. From (11.37) we conclude that $\dot{z}_2 \in \mathcal{L}_2$ also, hence z_2 tends to zero. Finally, from (11.38) and boundedness of z_3 we conclude that $\nu \to 0$ as well. $\blacksquare$

Standard convex optimization techniques can be used to obtain from the inequalities above the performance interval $[R_r^{\min}, R_r^{\max}]$. A very simple algorithm can, however, be derived as follows:

An algorithm for estimation of the performance interval

Step 1: Input data: Numerical values for the induction motor parameters, the rotor resistance estimate $\hat{R}_r > 0$ and the controller gains $K_P > 0$, $K_I > 0$.

Step 2: Set $R_r = \hat{R}_r$.

Step 3: Check conditions for local stability ((11.34) or (11.35) or (11.36)) and (11.40). If one of them is not satisfied then $R_r^{\max}$ (the maximum value that guarantees global stability) is found. Go to **Step 5**. If both conditions hold proceed with the following step.

Step 4: Increment the current values R_r by a small number $\delta > 0$, and go to **Step 3**.

Step 5: Set $R_r = \hat{R}_r$.

Step 6: Decrement the current value of R_r by a small number $\delta > 0$.

Step 7: Check conditions for local stability ((11.34) or (11.35) or (11.36)) and (11.40). If one of them is not satisfied then $R_r^{\min}$ (the minimum value that guarantees global stability) is found, and the seeking algorithm stops. If both conditions hold go to **Step 6**.

6.5   Illustrative examples

In this section some numerical and experimental results are presented. The objective is twofold: First, to illustrate the sharpness of the local stability–instability boundary predicted by the theory. Second, to validate our claim that the size of the performance intervals indeed provide a measure of the transient performance behaviour.

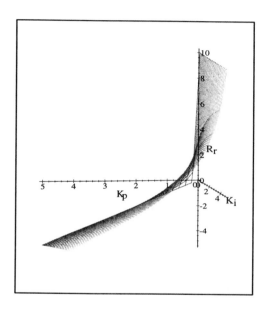

Figure 11.12: Manifold of local stability–instability.

A   Simulations

For clarity of presentation, we have in this section chosen an academic example where we set all parameters of the induction motor equal to unity except for R_r, which will be changed throughout the experiments. Also, we have set $\hat{R}_r = 10\ \Omega$. In Fig. 11.12 the stability–instability boundary predicted by condition (11.33) is illustrated in the R_r, K_P, K_I space. All values of the rotor resistance above this surface correspond to stable behaviours, while those below will yield an unstable closed–loop system. We have simulated the system with $K_P = 0.1$, $K_I = 1.0$ for which the critical value for R_r is $R_r = 4.9$. In Fig. 11.13 we show the time evolution of the speed for the stable $(R_r = 6)$, critical $(R_r = 4.9)$ and unstable $(R_r = 4)$ cases, with $\dot{q}_{m*} = 10$ rad/s and

$\dot{q}_m(0) = 10.1$ rad/s. Notice that we chose the initial condition of the speed very close to its reference to further underscore the stable–unstable behaviour.

We then evaluated, using the algorithm of Section 6.4, the performance interval for this setting to get $[R_r^{\min}, R_r^{\max}] = [9.1, 10.9]$. This is, of course, a very small robustness margin, thus we expect the performance to be below par. This is corroborated in the step response plot of Fig. 11.14(a) where the speed reference was changed from $\dot{q}_{m*} = 10$ rad/s to $\dot{q}_{m*} = 1$ rad/s. To improve transient performance we must retune the PI gains to enlarge the size of the performance interval. We set then $K_P = 5$, for which we get the bigger interval $[R_r^{\min}, R_r^{\max}] = [0.4, \infty]$. As expected, the transient response, shown in Fig. 11.14(b), is much better.

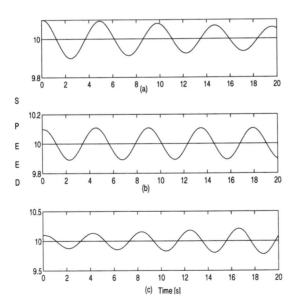

Figure 11.13: Simulation showing the stability–instability boundary. Speed in [rad/s] versus time.

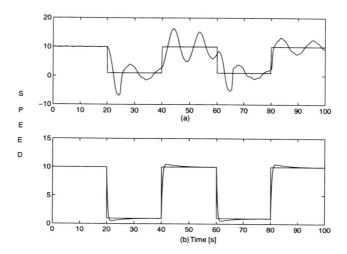

S
P
E

E
D

Figure 11.14: Simulation showing the improvement of performance. Speed in [rad/s] versus time.

B Experimental results

We have tried our tuning rules on the experimental setup of the Laboratoire de Génie Electrique de Paris. For the software interface an integrated system with an Intel 80486 50MHz microprocessor and a DSP32C digital signal processor allowed us to facilitate the assembly language coding. The squirrel–cage induction motor parameters can be found in Table 11.3.

The motor is driven by a pulse–width modulated (PWM) inverter which has a sampling period of $76.6 \cdot 10^{-6}$ s, and MOSFET bridges with current feedback loops as shown in Fig. 11.15. The measurement of currents is done using Hall–effect sensors, which have good accuracy (the linearity error is about 1%) and isolates the acquisition system electrically from the inverter. This reduces the measurement noise. The rotor position is measured by a high–resolution optical incremental encoder, with a sampling period of $5 \cdot 10^{-3}$ s. and the rotor speed is estimated from the position measurement.

Thus, from standard discretization considerations, the maximum sampling period is the estimation time of the speed $(1 \cdot 10^{-2}$ s).

Description	Notation	Value	Unit
Nominal power	P	1.1	kW
Power factor	$\cos(\phi)$	0.83	
Number of pole-pairs	n_p	2	
Maximum speed	$\dot{q}_m^{\max}$	1400	rpm
Maximum stator voltage		210	V
Maximum stator current		12	A
Nominal rotor flux norm	β^N	1.14	Wb
Nominal stator resistance	R_s	8	Ω
Nominal rotor resistance	R_r	3.6	Ω
Stator inductance	L_s	470	mH
Rotor inductance	L_r	470	mH
Mutual inductance	L_{sr}	0.44	mH
Total leakage factor($\sigma_r = \sigma_s = 0.068$)	σ	0.12	
Moment of inertia	D_m	0.01	kgm^2

Table 11.3: Motor parameters.

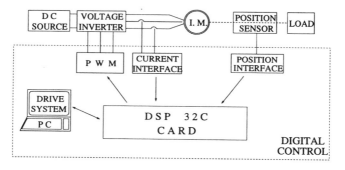

Figure 11.15: Block diagram of experimental setup.

First we illustrate the stability–instability boundary predicted by (11.33). To this end, we used the controller gains $K_P = 0.1, K_I = 7$ and fixed the rotor resistance estimate to $\hat{R}_r = 30\ \Omega$, which is clearly very far from its nominal value. The result is shown in Fig. 11.16 where we have taken, as in the simulation example, the initial speed very close to the reference, namely $\dot{q}_m(0) = 288$ rpm, and $\dot{q}_{m*} = 300$ rpm, respectively. It is clear that, for this particular setting, the bound (11.33) is only of theoretical interest.

We then fixed $\hat{R}_r = 3.6\ \Omega$ and evaluated the performance interval for $K_P = 0.1$ and various settings of K_I. For $K_I = 7$ we have $[R_r^{\min},\ R_r^{\max}] = [0.8, 3.9]\ \Omega$, which is a very narrow range. Decreasing K_I to $K_I = 0.5$ yields $[0.6, 9]$ Ω, and finally $K_I = 0.2$ gives the bound $[0.6, 15]$ Ω. The corresponding step responses, depicted in Fig. 11.17, corroborate our claim that transient performance is directly correlated with the size of the performance interval.

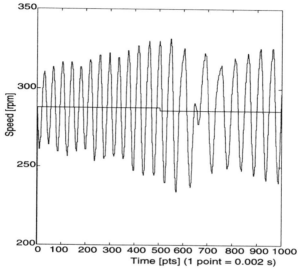

Figure 11.16: Experimental instability with $K_P = 0.1, K_I = 7, \hat{R}_r = 30$ Ω and $\dot{q}_m(0) = 288$ rpm.

7 Discrete–time implementation of PBC

The computation of an FOC scheme, which is a nonlinear dynamic feedback, is invariably carried out in discrete–time (with standard microprocessors or special purpose digital signal processors). Since both the machine model and the controller are nonlinear, the stability analysis of a discretized implementation of the controller is far from obvious. Furthermore, it is not even clear how such a discretization should be done. Some efforts in this direction may be found in [94]. In this section we present a new *discrete–time* FOC for current–fed induction motors (first reported in [215]) which ensures global asymptotic speed regulation and rotor flux norm tracking provided a condition relating the sampling rate with the *controller parameters* is satisfied. This condition disappears as the sampling period goes to zero. To ensure stability of the scheme we modify the the flux norm *reference*. If the latter is fixed, a trade–off must be established between the sampling period and the reference value.

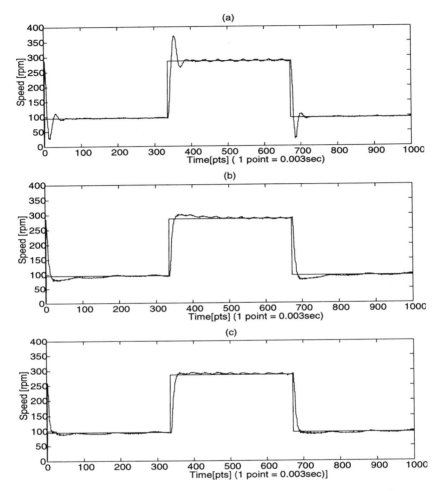

Figure 11.17: Experimental performance improvement with $K_P = 0.1$, $\dot{q}_m(0) = 288$ rpm and: (a) $K_I = 7$, (b)$K_I = 0.5$, (c)$K_I = 0.2$.

We show that the main, and far from obvious, effect of discretization is that the tasks of flux and torque regulation have to be shared between the two control channels, as opposed to a continuous–time implementation where they can be decoupled. A consequence of this coupled control approach is an enhancement in rotor flux tracking performance. One additional feature of this scheme is that, compared with the first difference approximation of the classical indirect FOC, the additional computational burden is negligible. Establishing the stability for a discretized FOC without this modification, remains a challenging open problem.

The performance of this new discrete controller is illustrated with some *experimental* results.

Remark 11.21 (Discrete time notation.) For a given (constant) sampling period $T \in \mathbb{R}_{\geq 0}$ [s] we will use the notation $x(k) = x(kT)$, $k \in \mathbb{Z}_+$, i.e., $x(k)$ denotes the samples of the continuous signal $x(t)$ at the time instants $t = kT$. Also, q will denote the shift operator, that is $q^{\pm i}x(k) = x(k \pm i)$, $i \in \mathbb{Z}_+$.

7.1 The exact discrete–time model of the induction motor

We will be concerned here with the, by now well–known, current–fed induction motor model (11.23) and PBC (11.24)–(11.25).

Before proceeding with the derivations of the discrete–time PBC it is important to underscore that the tasks of flux and torque regulation are decoupled between the first flux controlling component and the second torque controlling component. It will be shown below that to achieve a stable discrete–time implementation, this principle of decoupling in control signals will have to be abandoned. This is the main, and far from obvious, modification to FOC induced by the discretization. It is well known that discretization introduces coupling terms in the dynamic equations which are related to sampling period and speed [275]. This naturally leads to the question of how the discrete–time control components should be *coupled*, to avoid a degradation of performance because of the discretization. An answer to this question is given the next sections.

To obtain a stable discrete–time FOC the key observation made in [215] is that the motor equations (11.23) can be *exactly* discretized under the assumption of zero order hold control. In spite of the simplicity of the model, this is not an obvious fact. This is possible only due to the particular form of the bilinear term in the torque equation. Namely, we can integrate (11.23) to get

$$\dot{q}_m(k+1) = \dot{q}_m(k) + \frac{n_p L_{sr}}{D_m L_r} v(k)^\top \mathcal{J} \left(\int_{kT}^{(k+1)T} \lambda_r(s)ds \right) - \frac{T}{D_m}\tau_L$$

On the other hand, since $v(k)$ will be constant in the sampling interval, we can also integrate the solution of the first equation in (11.23) as

$$\int_{kT}^{(k+1)T} \lambda_r(s)ds = (1-a)T_r\lambda_r(k) + [T - (1-a)T_r]L_{sr}v(k), \quad a \stackrel{\triangle}{=} e^{-\frac{T}{T_r}}$$

which allows us, using the skew–symmetry of $\mathcal{J}$, to evaluate

$$v(k)^\top \mathcal{J} \left(\int_{kT}^{(k+1)T} \lambda_r(s)ds \right) = (1-a)T_r v(k)^\top \mathcal{J}\lambda_r(k)$$

and replace the expression above in the discretized speed equation to get

$$\lambda_r(k+1) = a\lambda_r(k) + (1-a)L_{sr}v(k) \tag{11.43}$$

$$\dot{q}_m(k+1) = \dot{q}_m(k) + \frac{(1-a)T_r}{D_m}\tau(k) - \frac{T}{D_m}\tau_L \tag{11.44}$$

$$\tau(k) = \frac{n_p L_{sr}}{L_r}v(k)^\top \mathcal{J}\lambda_r(k) \tag{11.45}$$

7.2   Analysis of discrete–time PBC

It is well known that the closed–loop system consisting of a continuous plant and the *ad hoc* discretization of a continuous-time nonlinear controller is not necessarily stable, even when the original (continuous–time) loop is stable. However, it is a reasonable conjecture that discretizing the PBC (11.24)–(11.25) with a *sufficiently small* sampling time will yield a stable closed loop. Let us consider, for simplicity, the case of constant desired rotor flux amplitude, i.e. $\dot{\beta}_* \equiv 0$. In this case, an *exact* discrete-time implementation of the PBC is obtained as

$$v(k) = \frac{1}{L_{sr}} \mathbf{e}^{\mathcal{J}\rho_d(k)} \left[\begin{array}{c} \beta_* \\ \frac{L_r}{n_p\beta_*}\tau_d(k) \end{array} \right] \tag{11.46}$$

$$\rho_d(k+1) = \rho_d(k) + \frac{TR_r}{n_p\beta_*^2}\tau_d(k)$$

$$\tau_d(k) = C(q)(\dot{q}_m(k) - \dot{q}_{m*}), \quad C(q) = -(K_P + \frac{K_I}{q-1}) \tag{11.47}$$

The stability of the system (11.23) in closed loop with (11.46)–(11.47) remains to be established.

Notice that the discretization above preserves the decoupling between flux and torque regulation in the controller structure. Due to discretization effects, the decoupling of torque and flux control may however be lost in the closed–loop system. In the next section we will present a modified FOC where the decoupling structure of the controller itself is not preserved, and it will be shown that in this way we can ensure GAS in the discrete–time case. This new controller will, of course, give decoupled control of flux and torque in closed–loop.

To motivate the new discrete-time algorithm, we will first highlight the difficulties in the stability analysis of (11.23), (11.46)–(11.47) introduced by the discretization. To this end, it is convenient to recall first that the stability analysis of the continuous–time case. It heavily relied on the cascaded structure of the error system, namely on the fact that $\tilde{\lambda}_r \to 0$ exponentially, which together with

$$\tau = (1 + \frac{1}{\beta_*^2}\lambda_{rd}^\top\tilde{\lambda}_r)\tau_d + \frac{n_p}{L_r}\lambda_{rd}^\top\mathcal{J}\tilde{\lambda}_r$$

allowed to complete the stability proof.

Unfortunately, this kind of analysis cannot be applied to the discrete–time controller (11.46)–(11.47). Essentially because, in contrast to (11.22), we cannot express $v(k)$ as a difference equation for the desired rotor flux, which is now defined as

$$\lambda_{rd}(k) = e^{\rho_d(k)\mathcal{J}} \begin{bmatrix} \beta_* \\ 0 \end{bmatrix} = \beta_* \begin{bmatrix} \cos(\rho_d(k)) \\ \sin(\rho_d(k)) \end{bmatrix}$$

Hence, we are unable to recover the asymptotically stable error equation for the rotor flux error.

7.3 A new discrete-time control algorithm

Several modifications to the PBC are required to obtain a globally stable discrete–time scheme. First, to ensure the desired behaviour for the rotor flux mentioned above, we propose to choose

$$v(k) = \frac{1}{L_{sr}} \left[\frac{1}{1-a}(\lambda_{rd}(k+1) - a\lambda_{rd}(k)) \right] \tag{11.48}$$

with the desired rotor flux defined as

$$\lambda_{rd}(k) = e^{\rho_d(k)\mathcal{J}} \begin{bmatrix} \beta_*(k) \\ 0 \end{bmatrix} = \beta_*(k) \begin{bmatrix} \cos(\rho_d(k)) \\ \sin(\rho_d(k)) \end{bmatrix} \tag{11.49}$$

As will be shown in the following analysis, it is essential for the stability analysis that $\beta_*(k)$ is a *time–varying* flux norm reference.

Inserting (11.48) in (11.43) leads to $\tilde{\lambda}_r(k+1) = a\tilde{\lambda}(k)$, and consequently $\tilde{\lambda}(k) \to 0$. Replacing this control law in (11.45) we get after some simple calculations

$$\tau(k) = \frac{n_p}{L_r(1-a)}\lambda_{rd}(k+1)^\top \mathcal{J}\lambda_{rd}(k) + \epsilon(k)$$

where

$$\epsilon(k) \triangleq \frac{n_p}{L_r(1-a)}(\lambda_{rd}(k+1) - a\lambda_{rd}(k))^\top \mathcal{J}\tilde{\lambda}_r(k)$$

is an exponentially decaying sequence. We see from the last equation that to recover the property of $\tau(k) \to \tau(k)^d$ we must ensure

$$\frac{n_p}{L_r(1-a)}\lambda_{rd}(k+1)^\top \mathcal{J}\lambda_{rd}(k) \equiv \tau_d(k) \tag{11.50}$$

To enforce this identity we first notice that $\lambda_{rd}(k)$ in (11.49) satisfies the difference equation

$$\lambda_{rd}(k+1) = \frac{\beta_*(k+1)}{\beta_*(k)} e^{Tb(k)\mathcal{J}}\lambda_{rd}(k)$$

with $Tb(k) \triangleq \rho_d(k+1) - \rho_d(k)$, which replaced in (11.50) yields

$$\frac{n_p}{L_r(1-a)} \lambda_{rd}(k+1)^\top \mathcal{J} \lambda_{rd}(k) = \frac{n_p}{L_r(1-a)} \beta_*(k) \beta_*(k+1) \sin(Tb(k))$$

Therefore, we choose

$$b(k) = \frac{1}{T} \arcsin \left[\frac{L_r(1-a)}{n_p \beta_*(k+1)} \beta_*(k) \tau_d(k) \right] \qquad (11.51)$$

To ensure that this equation has a solution we must choose $\beta_*(k)$ such that

$$\frac{L_r}{n_p \beta_*(k)} |\tau_d(k)| < \frac{\beta_*(k+1)}{1-a} \qquad (11.52)$$

holds $\forall k \in \mathbb{Z}_+$. Once we have proved that $\tau(k) \to \tau_d(k)$ we can follow *verbatim* the arguments of the continuous–time case for the stability analysis of the mechanical system. Stability limits for the PI gains can be computed by the insertion of the compensator (11.47) in the difference equation (11.44) and the use of Jury's stability criterion on the resulting polynomial in q.

We can now establish the following result:

Proposition 11.22 (Discrete–time controller from [215].) *Consider the motor model (11.23) with control inputs the (rotated) stator currents v and measurable output the rotor speed $\dot{q}_m$. Let $\dot{q}_{m*}$ denote the desired constant rotor speed, and assume load torque τ_L is also constant, though unknown. Define the discrete-time controller as*

$$v(k) = \frac{1}{L_{sr}} e^{\mathcal{J}\rho_d(k)} \left[\begin{array}{c} \frac{\beta_*(k+1)}{1-a} \cos(Tb(k)) - \frac{a}{1-a}\beta_*(k) \\ \frac{L_r}{\beta_*(k)n_p} \tau_d(k) \end{array} \right] \qquad (11.53)$$

$$\rho_d(k+1) = \rho_d(k) + Tb(k)$$

$$\tau_d(k) = C(q)(\dot{q}_m(k) - \dot{q}_{m*}), \quad C(q) = -(K_P + \frac{K_I}{q-1}) \qquad (11.54)$$

with $b(k)$ as in (11.51), sampling period $T > 0$, $v(k) = v(kT)$, $\dot{q}_m(k) = \dot{q}_m(kT)$, $a \triangleq e^{-\frac{T}{T_r}}$. The PI gains are chosen such that

$$0 < K_I < K_P < \frac{2D_m}{(1-a)T_r} \qquad (11.55)$$

Let $\beta_*(k) > 0$, which represents the reference value for the rotor flux amplitude $\|\lambda_{rd}(k)\|$, be chosen so as to satisfy[10] (11.52).

---

[10]Notice that the upper bounds (11.55) and (11.52) become arbitrarily large as $T \to 0$. Hence, relaxing the restrictions on the choice of K_P, K_I and β_{*_k}.

Under these conditions, for all initial conditions of the motor, we have that all internal signals are bounded and

(i) $\qquad \lim_{k \to \infty} |\dot{q}_m(k) - \dot{q}_{m*}| = 0$

(ii) $\qquad \lim_{k \to \infty} |\,\|\lambda_r(k)\| - \beta_*(k)| = 0$

$\square$

7.4 Discussion of discrete–time controller

We have introduced in the first component of (11.53) a term $\cos(b(k))$. Since $b(k)$ depends, via (11.51), on $\tau_d(k)$ we see that this term *couples* the flux and torque regulation objectives. It will be shown in the experiments below that the inclusion of this term enhances the rotor flux tracking performance.

As pointed out above, the introduction of a time–varying reference for the flux norm is essential to guarantee global stability via a suitable choice of this reference, i.e., (11.52). If the flux is kept constant we have to tradeoff between the sampling period and the reference speed.

From the definition of the control (11.48) we see that $v(k)$ is bounded, and in particular it satisfies the bound

$$\|v(k)\| \leq \frac{1}{1-a}(\beta_*(k+1) + a\beta_*(k))$$

Since the bound depends only on the rotor time constant and the desired rotor flux it might be used to study saturation effects.

We have considered here only the problem of speed regulation with constant load torques. It is clear from the proof that the extension to speed tracking or to *position* control can be solved simply by choosing a suitable compensator in (11.54). Also, we can easily treat the case of *unknown varying* load torque, e.g., a linearly parameterized function $\tau_L = \phi(\dot{q}_m, q_m)^\top \theta$, $\theta \in \mathbb{R}^q$, with a straightforward modification to (11.44) and use of classical adaptive control techniques.

7.5 Experimental results

A Experimental setup

The experimental system consists of a squirrel–cage induction motor with the rated parameters given in Table 11.4, a three–phase inverter, an anti–aliasing filter module and a dSPACE DS1102 DSP–board (see also Section **10**.6 for the description of a similar setup) in a host PC.

Parameter	Notation	Value	Unit
Nominal power	P	750	W
Number of pole pairs	n_p	2	
Power factor	$\cos\phi$	0.75	
Voltage	U	220	V
Nominal current	i_{ab}^N	3.8	A
Nominal speed	$\dot{q}_m^N$	1420	rpm
Rotor resistance	R_r	1.4	Ω
Rotor inductance	L_r	150	mH
Mutual inductance	L_{sr}	150	mH
Nominal torque	τ^N	7.5	Nm
Nominal flux	β^N	0.5	Wb

Table 11.4: Motor parameters.

The measurements of the stator voltages and currents are made using Hall–effect sensors. These sensors have a good accuracy (the linearity error is about 1%) and isolate the acquisition system electrically from the inverter. This reduces the measurement noise. Aliasing errors due to the high frequency modulation of the inverter are avoided by the use of four Bessel low–pass filters of order five. To obtain accurate matching of the filters, we used switched capacitor filters.

The rotor position is measured by a high–resolution optical incremental encoder with 5000 lines per revolution. This resolution is increased by a factor of 4 with a quadruple counter on the DSP-card. The motor is driven by a pulse–width modulated inverter with MOSFET bridges and current feedback loops. Only two of the phases are current controlled, and the third stator voltage is calculated to obtain a symmetric three–phase feeding.

For logging of signals and on–line tuning of parameters, the programs TRACE and COCKPIT were used to communicate with the DSP–board. The controller was implemented using multirate computation, with the calculations of the PWM running at a sampling rate of 0.55 μs (1.81 MHz), with a slower computation of reference currents and speed control at a rate of T in the interval [34, 1] ms (29.4 Hz–1 kHz).

Speed was estimated from position using the backward difference approximation proposed in [158]

$$\dot{\theta} \approx \frac{\theta(k) - \theta(k-1)}{T}$$

This backward difference approximation of speed was smoothened using a discrete–time implementation of the filter

$$\frac{1}{\frac{1}{\omega_0^2}p^2 + \frac{2\zeta}{\omega_0}p + 1}, \quad \omega_0 = 1000, \; \zeta = 0.7$$

In order to protect the equipment we also included a ±4 A current saturation.

To execute the program directly on the DSP-card, we used the Real-Time option in SIMULINK. The specific parameter values of our system as well as the integration method, in this case Runge-Kutta 3, could be specified directly by the use of this graphical interface. We used two different sampling times in the control calculation, a slow one $T \in [1, 34]$ ms for the determination of (11.53)–(11.54), and a fast sampling time $T \in [600, 700]$ μs for the rotation to the stator fixed reference frame. It should be pointed out that it is essential for a good performance of the algorithm to carry-out the latter calculation at a high sampling frequency.

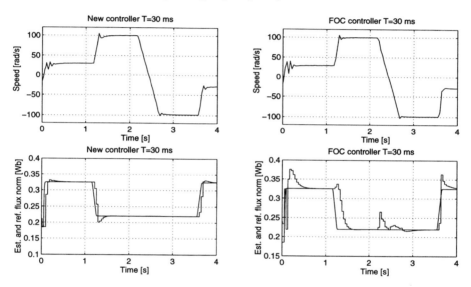

Figure 11.18: Improvement of flux tracking with new discrete controller.

B Experimental results

In the experiments it was aimed at the following performance evaluations:

- To compare the new algorithm with the discretized FOC for different sampling periods.

- To show the enhanced flux tracking performance of this novel discrete approach.

- To illustrate the robustness to variations in the motor parameters and the tuning gains.

• To evaluate the performance of the algorithm at low speed conditions.

For the sake of brevity we show here only the plots showing the enhanced flux tracking performance. The reader can consult [262] for more experimental results.

Interestingly enough, in these tasks the new controller systematically performed better than the standard FOC scheme, a feature that appeared more evident for slow sampling times. To illustrate this fact we carried-out an experiment where, to avoid current saturation, for a speed increase of 70 rad/s at $t = 1.2$ s the desired flux must be decreased from 0.33 Wb to 0.22 Wb. Fig. 11.18 shows the responses of speed and the estimated rotor flux norm for a sampling period of $T = 30$ ms. The latter was obtained with a standard open–loop estimator from stator currents as in [30], see also (10.91) and (10.92) in Section **10**.5.

8   Conclusions and further research

The current–fed model of the induction motor was given in Section 1 and the underlying assumption of this model explained. In Section 2 the standard direct and indirect FOC schemes were presented. An observer-based feedback–linearizing controller was also presented, and some of the drawbacks of this method were derived theoretically, and illustrated by simulations and experiments.

In Section 3 we proved that the PBC for the voltage–fed machine reduces to indirect FOC. This gives a *passivity interpretation* to FOC, whose importance can hardly be overestimated, since it provides a deep system–theoretic foundation to this popular strategy and paves the way for subsequent analysis. In Section 5 we established *exponential* stability (with a quadratic Lyapunov function) of PBC and robustness when $\hat{R}_r \neq R_r$ of FOC.

Simple off-line rules for PI gain tuning (with guaranteed stability) were presented in Section 6. Our contention, which is validated with simulations and experiments, is that with this rules we can improve the transient performance. The performance enhancement is quantitatively measured with an indicator of robustness of the stability with respect to uncertainty in the rotor resistance, namely the largest allowable estimation error under which global stability is preserved. A very simple algorithm that evaluates these ranges for each PI setting has been presented.

Finally, a globally stable *discrete–time* version of PBC was given in Section 7 and experimental result, showing enhanced flux tracking, were given in Section 7.5.

In spite of the remarkable stability robustness properties of PBC which we have presented here, there is an obvious interest of considering adaptive controllers that estimate R_r to improve performance. Experimental evidence that substantiates this claim may be found in [131]. Providing a satisfactory solution to this fundamental problem is the main driving force of the field now. Important steps towards its solution were given in [170]. Other results may be found in [1, 55, 171, 276]. Also, it is interesting to see whether the performance limitation imposed by rotor dynamics can be overcomed.

Chapter 12

Feedback interconnected systems: Robots with AC drives

Throughout the book we have stressed the fact that PBC is compatible with one of the important viewpoints of systems theory that complicated systems are best thought of as being *interconnections* of simpler subsystems, each one of them being characterized by its dissipation properties. This aggregation procedure has three important implications. First, it is consistent with the dominating approaches for modeling and simulation based on some kind of network representation and energy flow. Second, it help us to think in terms of the structure of the system and to realize that sometimes the pattern of the interconnections is more important than the detailed behaviour of the components. Finally, it is indeed a design–oriented methodology which allows us to isolate the "free subsystems" – sensors and actuators.

In Chapters **9** and **10** we have already shown how, via a decomposition of the system dynamics into its electrical and mechanical parts, we can exploit this feature of PBC to design practically useful controllers for electrical machines. We considered in those chapters the case where the dynamics of the mechanical subsystem is essentially linear. Even though this crude model is suitable for a vast array of problems, there are many modern applications that require the incorporation of a more detailed mathematical model of the mechanical load to meet the performance requirements. A typical example, that we consider in this chapter[1], is the problem of motion control of robot manipulators actuated by AC drives. In this case a linear model cannot capture the behaviour of the robot when it is moving fast and we have to look at the complete nonlinear coupled dynamics.

The approach that we develop in this chapter is applicable, not just to electromechanical systems, but to a very large class of feedback interconnected systems. For this reason we consider first a more general feedback interconnection problem, and then

---

[1]The material reported in this chapter is based on work done in collaboration with Elena Panteley and Paulo Aquino.

derive as a particular case the example of robots with AC drives. Namely, we assume the forward subsystem is an underactuated EL system and that, in the absence of the latter, a stabilizing controller is known for the feedback subsystem. (In the case of robots with AC drives the subsystems are the motors and the robot, respectively. The assumption of a known controller for the robot amounts to the standard assumption of neglecting the motor dynamics.) This scenario, which appears in many practical applications with the forward subsystem representing the actuator dynamics, leads naturally to the classical cascaded (nested–loop) control scheme that we have encountered already in previous chapters. We are then interested in establishing conditions under which we can design a passivity–based inner–loop controller for the EL system such that global tracking is achieved. These are expressed in terms of actuator–sensor couplings, the "strength" of the subsystems interconnection, and the requirement of linear dependence on the unmeasurable variables. Interestingly enough, this analysis does not invoke the standard time–scale separation assumptions prevalent in cascaded schemes, but uses instead some "growth" conditions on the interconnections.

The inner–loop controller is designed following the developments of Chapter **9**, hence we will review it only briefly. Further details may also be found in [217]. The main obstacle that we must overcome when the feedback (mechanical) subsystem is not LTI is that we cannot simply use an approximate differentiation filter (as done in Chapter **10**) to avoid the need for the derivative of the torque reference. To solve this problem we add a *nonlinear observer*.

After presenting our general result, we then use it in the design of an output feedback global position tracking controller for robot manipulators actuated by AC drives. Similarly to the material in Chapter **9**, the result applies to a fairly large class of AC drives, which includes as particular cases induction, synchronous and stepper motors. Instrumental for the observer design is the utilization of a new robot controller which is linear in the link velocities. We also present simulation results which compare our controller with the one reported in [107], which was derived using backstepping ideas.

1   Introduction

1.1   Cascaded systems

We are in this chapter interested in the problem of controlling feedback interconnected systems of the form depicted in Fig. 12.1. We assume that both subsystems, Σ_e and Σ_m, are nonlinear with u the control input vector, y_e, y_m the measurable outputs and τ an (unmeasurable) coupling signal. The control objective is to make y_m asymptotically track a desired (time–varying) reference y_{m*} with internal stability.

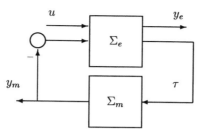

Figure 12.1: Feedback interconnected system.

This type of configuration appears in many practical applications. For instance, in Chapter **9** where we consider the electrical machine, the subsystem Σ_e contains the electrical dynamics, and Σ_m the mechanical part of the motor. In this case τ is a force (or torque) of electrical origin. Another situation when this scheme arises is when Σ_e represents the actuator dynamics and Σ_m the plant to be controlled. In these instances it is reasonable to assume that, if τ were a manipulated (control) variable, then we dispose of a suitable controller for the subsystem Σ_m, say $\mathcal{C}_{ol}$. This scenario, which is adopted in Chapter **9** is also followed here, and it leads naturally to the cascaded (*i.q.*, nested–loop) controller configuration of Fig. 12.2, where $\mathcal{C}_{il}$ is an inner–loop controller to be designed such that τ tracks τ_d "sufficiently fast" to avoid upsetting the stability of the outer loop. As already pointed out in Chapter **9**, this task is complicated by the dependence of Σ_e on y_m and the need to estimate $\dot{\tau}_d$.

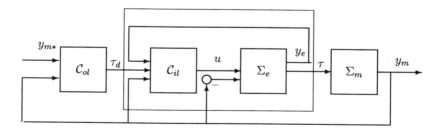

Figure 12.2: Cascaded (nested–loop) control configuration.

Cascaded controls are typically designed for linear systems invoking time-scale separation assumptions. That is, if the inner–loop is designed to have a large bandwidth, then it essentially behaves as a static gain for the outer loop, and stability bounds can then be estimated using, for instance, singular perturbation techniques [136]. This reasoning is most suitable for the case when Σ_e represents the actuator dynamics, which is typically faster than the plant itself. Besides the additional difficulty due to the dependence of Σ_e on y_m, the extension of this technique to the nonlinear case does not seem obvious. Furthermore, the time–scale separation assumption becomes questionable in some modern technological applications with increasing performance requirements.

In pure cascade systems, where y_m is not fed back into Σ_e, stabilizability of Σ_e is combined with a growth condition on the interconnection to insure stabilizability of the composition $\Sigma_m \circ \Sigma_e$, (see *e.g.* [127, 138, 142, 174] for a summary of the latest developments). It is therefore expected that in our case, besides the growth condition on the interconnection, we will require a "stronger" form of stabilizability (namely, uniform in y_m) for the subsystem Σ_e.

Conditions for stability of the cascaded scheme can be easily derived under fairly general conditions on the subsystems. For instance, assume Σ_m is a (sufficiently smooth) state–space system, linear in the input τ, which is exponentially stabilized by a (possibly nonlinear) controller $\mathcal{C}_{ol}$, whose output is linearly bounded by y_m. Further, assume that $\mathcal{C}_{il}$ is such that –for all y_m– we have the tracking performance

$$|\tau - \tau_d| \leq \varepsilon_t + \varepsilon_t |\tau_d| \tag{12.1}$$

where ε_t is used to denote exponentially decaying terms. Then, converse Lyapunov theorems can be invoked to prove suitable convergence properties of the overall scheme. There are (at least) two difficulties with this approach. First, the dependence of Σ_e on y_m makes the definition of an inner–loop controller $\mathcal{C}_{il}$ that ensures (12.1) for all y_m very difficult, unless some coupling conditions between the two subsystems are satisfied. These conditions are expressed in terms of some restrictions on the functional dependence of τ on y_m. Second, to ensure the desired tracking of the mappings $\tau_d \mapsto \tau$ and $y_{m*} \mapsto y_m$, the controllers $\mathcal{C}_{il}$ and $\mathcal{C}_{ol}$ will usually require the knowledge of higher order derivatives of τ_d and y_m, respectively. When both are combined this will translate, through the dynamics of Σ_m, into the need of having τ available for measurement. It will be shown in the following sections that this difficulty can be removed by the use of an observer, provided some linearity assumptions on $\mathcal{C}_{ol}$ and $\mathcal{C}_{il}$ are imposed.

In this chapter we illustrate how the ideas discussed above can be applied to the case when Σ_e is an underactuated EL system with partial state measurements. The inner–loop controller will then be a PBC as the one discussed in Chapter **9**. We will then establish conditions on the actuator–sensor coupling in Σ_e, the "strength" of the subsystems interconnection, and the nature of the dependence on the unmeasurable variables (*i.e.*, linearity), to ensure global tracking for the mapping $y_{m*} \mapsto y_m$. In essence, these three conditions are required to stabilize Σ_e (uniformly in y_m), to ensure a growth requirement for the subsystems coupling (similar to the one required in [227]), and for the design of the observer, respectively.

1.2   Robots with AC drives

An important corollary of our main result is the design of a global position tracking controller for rigid robot manipulators actuated by AC drives when the only variables available for measurement are the link positions and velocities, and the currents of the stator windings. Control of robot manipulators with AC drives is an interesting research topic, both from application oriented and theoretical viewpoints[2]. The problem was first postulated for the case of induction motors in [53] where a locally asymptotically stable scheme was presented. The scheme presented here extends the existing results in several directions:

1. We prove global asymptotic stability of the closed loop, which is strictly stronger than the local result of [53] or the ultimate boundedness condition of [107]. In [105], which heavily borrows from [209], asymptotic stability is also established.

2. We establish this result for the (fairly large) class of AC machines treated in Chapter **9**, which includes induction, permanent magnet synchronous and stepper motors as particular cases.

3. In [53] some of the terms coupling the robot and the motor dynamics are cancelled by the control to obtain a cascade connection of two subsystems, instead of the feedback configuration of Fig. 12.2. An important drawback of this approach is, of course, the lack of robustness of the nonlinearity cancelation which is clearly exhibited in [132]. The solution presented here does not suffer from this drawback.

4. From the viewpoint of computational complexity, our controller is several orders of magnitude simpler than the schemes given in [105, 107], see Section 6. It should be mentioned, however, that in [105] the more challenging adaptive control problem is studied.

5. As usual in backstepping–based schemes, the controllers in [105, 107] contain higher order terms which essentially act as high gains [89]. The latter will not only degrade performance in the presence of noise, but also create serious numerical problems in a practical implementation, as discussed in Section 6.

Instrumental for our development is the utilization of the robot controller of Proposition 4.8 in Chapter **4** (proposed in [22]), which is linear in the link velocities. This feature is essential for the design of an observer.

---

[2]We refer the reader to [53] and [107] for further motivation on the problem.

2   General problem formulation

We consider feedback interconnected systems of the form depicted in Fig.12.1, where Σ_m is described by the state equation

$$\Sigma_m : \begin{cases} \dot{y}_{m1} &= y_{m2} \\ \dot{y}_{m2} &= f_m(y_m) + g_m(y_m)\tau \end{cases} \qquad (12.2)$$

with the full state $y_m = [y_{m1}^\top, y_{m2}^\top]^\top \in \mathbb{R}^{2n_m}$ measurable, and $\tau \in \mathbb{R}^{n_m}$ the unmeasurable coupling signal. We will assume that $f_m(y_m)$ is locally Lipschitz in y_m, and that $g_m(y_m)$ is bounded, that is,

$$\sup_{y_m} \|g_m(y_m)\| \le \alpha < \infty \qquad (12.3)$$

Consistent with our cascaded control approach, we will make the following assumption:

A 12.1 For any given bounded reference signal $y_{m*} \stackrel{\triangle}{=} [y_{m1*}^\top, \ \dot{y}_{m1*}^\top]^\top \in \mathbb{R}^{2n_m}$, with known bounded derivative, we know a globally *exponentially* stabilizing static state feedback controller for (12.2), which is linear in y_{m2}, and linearly bounded in y_m. That is, we know τ_d of the form

$$\begin{aligned} \tau_d(y_m) &\stackrel{\triangle}{=} K_1(y_{m1}) + K_2(y_{m1})y_{m2} & (12.4) \\ \|K_1(y_{m1})\| &\le \alpha\|y_{m1}\| + \alpha \\ \|K_2(y_{m1})\| &\le \alpha \end{aligned}$$

such that, if $\tau \equiv \tau_d$, then $y_m = y_{m*}$ is a globally exponentially stable equilibrium of (12.2).

The subsystem Σ_e is an EL system with generalized coordinates $q_e \in \mathbb{R}^{n_e}$ and Lagrangian $\mathcal{L}_e(q_e, \dot{q}_e, y_{m1})$. Notice the dependence of $\mathcal{L}_e(q_e, \dot{q}_e, y_{m1})$ on y_{m1}, this will establish the coupling between the two subsystems. The behaviour of Σ_e is modeled by the EL equations of motion studied in Chapter **2**, that is

$$\frac{d}{dt}\Big[\frac{\partial \mathcal{L}_e}{\partial \dot{q}_e}(q_e, \dot{q}_e, y_{m1})\Big] - \frac{\partial \mathcal{L}_e}{\partial q_e}(q_e, \dot{q}_e, y_{m1}) = Q_e$$

where $Q_e \in \mathbb{R}^{n_e}$ are the external (dissipative and control) forces. We will assume that

$$\begin{aligned} \mathcal{L}_e(q_e, \dot{q}_e, y_{m1}) &= \frac{1}{2}\dot{q}_e^\top D_e(y_{m1})\dot{q}_e \\ Q_e &= -R_e\dot{q}_e + M_e u \end{aligned}$$

with $\infty > d_{Me} \geq D_e(y_{m1}) = D_e(y_{m1})^\top \geq d_{me} > 0$, $R_e = R_e^\top \geq 0$ and[3] $M_e^\top = [I_{n_e/2}, 0]$. We will further assume that $D_e(y_{m1})$ is globally Lipschitz in y_{m1}, consequently

$$\sup_{y_{m1}} \|W_i(y_{m1})\| \leq \alpha < \infty \tag{12.5}$$

for $i \in \bar{n}_m \triangleq \{1, \cdots, n_m\}$, with $(\cdot)_i$ the i–th component of a vector, and

$$W_i(y_{m1}) \triangleq \frac{\partial D_e}{\partial (y_{m1})_i}(y_{m1}), \ i \in \bar{n}_m$$

In summary, Σ_e is described by

$$\Sigma_e : \begin{cases} D_e(y_{m1})\ddot{q}_e + \frac{d}{dt}\left(D_e(y_{m1})\right)\dot{q}_e + R_e\dot{q}_e &= M_e u \\ y_e &= M_e^\top \dot{q}_e \end{cases} \tag{12.6}$$

Notice that, following [197], we have considered the natural outputs available for measurement. For our further developments we find it convenient to define $\dot{q}_e \triangleq [y_e^\top, \dot{q}_r^\top]^\top$.

Finally, we will assume that τ, *i.e.*, the input to the subsystem Σ_m, is given by

$$\tau = \frac{\partial \mathcal{L}_e}{\partial y_{m1}}(q_e, \dot{q}_e, y_{m1})$$

whose components can be alternatively written as

$$\tau_i = \tfrac{1}{2}\dot{q}_e^\top W_i(y_{m1})\dot{q}_e, \ i \in \bar{n}_m \tag{12.7}$$

The problem we consider in this chapter is formulated as follows

Definition 12.1 (Global tracking problem.) *Given the feedback system of Fig.12.1 described by (12.2), (12.6), (12.7) with Assumption* **A12.1**. *Find conditions on the generalized inertia matrix, $D_e(y_{m1})$, and the dissipation matrix, R_e, of Σ_e which will ensure the existence of an inner–loop PBC of the form $\mathcal{C}_{il} : (y_e, \ y_m, \ \tau_d) \mapsto u$ such that*

$$\lim_{t \to \infty} \|y_m - y_{m*}\| = 0$$

with internal stability.

Remark 12.2 It will become clear in the sequel that the results of this chapter are, *mutatis mutandis*, applicable to a broader class of systems. For instance, $\mathcal{L}_e(q_e, \dot{q}_e, y_{m1})$ may contain potential energy terms, and the dissipation structure need not be linear as assumed here. Also, the controller $\mathcal{C}_{ol}$ may be, in general, a dynamic output feedback.

---

[3]We concentrate here in the more interesting case of underactuated systems, the problem is considerably simpler in the fully actuated case, *i.e.*, when $M_e = I_{n_e}$. See also Section 3.2.

Remark 12.3 Notice that we have considered an EL model for Σ_e that does not depend on the generalized coordinates q_e, only on its derivatives. That is, these coordinates are cyclic (also called ignorable) [97]. Again, this is done for the sake of clarity of presentation. Furthermore, this is the case of the electrical machines considered in Section 5.

Remark 12.4 It is important to remark that, due to the dependence on y_{m1}, the subsystem Σ_e (12.6) does *not* define a passive operator $u \mapsto y_e$.

3   Assumptions

Before presenting the solution to the global tracking problem we discuss here the assumptions needed for its solvability.The inner–loop PBC that we present here is largely inspired by the PBC derived in Chapter **9** for the generalized AC machine. In this chapter we want to extend this result to a larger class of EL systems Σ_e with the additional complication that the feedback subsystem Σ_m is not LTI anymore. It is useful then to revisit the design proposed for AC machines, particularly the assumptions required for the realizability of the controller. This analysis will help us identify the class of EL systems for which our cascaded scheme will work. We will first concentrate in the torque control problem, and then consider the extension to position control. Besides the invertibility requirement, in speed/position control applications we also have the difficulty that the calculation of the PBC needs $\dot{\tau}_d$, which in its turn implies knowledge of acceleration. While this obstacle was removed in Section **10.3** with a linear filter for the single machine case, a nonlinear observer will be needed to handle the case of nonlinear Σ_m. We recall that the inner–loop must be stabilized uniformly in y_m, hence some severe restrictions on $D_e(y_{m1})$ are expected. In particular, we have seen in the previous chapters that in order to be able to explicitly solve the controller equations we need an invertibility assumption on the subsystem Σ_e. To give a system–theoretic flavor to this assumption we will postulate it in terms of the solvability of a tracking subproblem. Then, we enunciate some additional assumptions which are essentially related with the linearity required for the solution of the observer problem.

3.1   Realizability of the controller

Motivated by the controller for electrical machines we propose an inner–loop PBC $\mathcal{C}_{il}$ of the form

$$M_e u = \dot{\lambda}_d + R_e D_e^{-1}(y_{m1})\lambda_d + \begin{bmatrix} K_1(y_m) & 0 \\ 0 & 0 \end{bmatrix} D_e^{-1}(y_{m1})\tilde{\lambda} \tag{12.8}$$

$$(\tau_d)_i = \frac{1}{2}\lambda_d^\top C_i(y_{m1})\lambda_d, \; i \in \bar{n}_m \tag{12.9}$$

where

$$C_i(y_{m1}) \triangleq D_e^{-1}(y_{m1})W_i(y_{m1})D_e^{-1}(y_{m1}), \ i \in \bar{n}_m \tag{12.10}$$

and $K_1(y_m)$ is a damping injection gain to be defined later.

To address the realizability issue of this controller we first notice that, due to the structure of M_e, the first $n_e/2$ equations of the PBC (12.8) can always be solved for u once λ_d and $\dot\lambda_d$ are given. Hence we concentrate exclusively on the last $n_e/2$ equations, which we write as

$$\dot\lambda_{rd} = A(y_{m1})\lambda_{rd} + B(y_{m1})\lambda_{sd}$$

where we have introduced the partition $\lambda_d \triangleq [\lambda_{sd}^\top, \lambda_{rd}^\top]^\top$, with $\lambda_{sd}, \ \lambda_{rd} \in \mathbb{R}^{n_e/2}$, and defined

$$\begin{bmatrix} \times & \times \\ B(y_{m1}) & A(y_{m1}) \end{bmatrix} \triangleq -R_e D_e^{-1}(y_{m1}) \tag{12.11}$$

with $\times$ denoting some matrices (dependent on y_{m1}). The realizability problem will be solved if we can find a function $\lambda_{sd} = g(y_{m1}, \lambda_{rd}, \tau_d)$ such that the solutions of

$$\dot\lambda_{rd} = A(y_{m1})\lambda_{rd} + B(y_{m1})g(y_{m1}, \lambda_{rd}, \tau_d)$$

satisfy the constraint (12.9).

Motivated by this observation, we will now define an auxiliary problem, whose solvability implies the realizability of the inner–loop PBC.

A 12.2 Consider the linear time–varying system[4]

$$\dot\lambda_{rd} = A(t)\lambda_{rd} + B(t)\lambda_{sd} \tag{12.12}$$

with $A(t), B(t)$ defined by (12.11). Then, for arbitrary (possibly unbounded) $y_{m1}(t)$, $\tau_d(t)$, there exists a state feedback of the form

$$\lambda_{sd} = K_3(\lambda_{rd}, y_{m1}) + K_4(\lambda_{rd}, y_{m1})\tau_d \tag{12.13}$$

such that the closed loop system (12.12), (12.13) satisfies

$$\|\lambda_d\| \le \alpha + \alpha\|\tau_d\| \tag{12.14}$$

and

$$\lim_{t\to\infty} \left| \frac{1}{2}\lambda_d^\top C_i(t)\lambda_d - (\tau_d)_i \right| = 0$$

with exponential rate of convergence, where $C_i(t)$ is given by (12.10).

---

[4]Notice that we are treating $A(y_{m1})$, $B(y_{m1})$ as functions of time, λ_{sd} as an input and λ_{rd} as the state.

Remark 12.5 The following remarks are in order:

• We require λ_{sd} to be affine in τ_d, as indicated in (12.13), to be able to implement the observer in the next stage.

• The requirement in (12.14) is needed for the proof of boundedness of λ_d, which as we discussed in Section **9**.5, is a requirement to ensure that the implication $(\tilde{\lambda} \to 0 \Rightarrow \tau \to \tau_d)$ holds.

• Notice that the "clamping" condition (12.9) must only be satisfied asymptotically.

3.2 Other assumptions

We need three additional assumptions. First, that the non–actuated coordinates of Σ_e are suitably damped, that is

A 12.3 R_e is of the form

$$R_e \triangleq \left[\begin{array}{cc} R_s & 0 \\ 0 & R_r \end{array} \right]$$

with $R_r > 0$.

Second, that the subsystems coupling is "weak" with respect to the unmeasurable signals. Specifically, we need to ensure that τ does not contain terms which are quadratic in[5] $\dot{q}_r$. This is ensured by

A 12.4 $D_e(y_{m1})$ is such that

$$W_i(y_{m1}) \triangleq \left[\begin{array}{cc} \times & \times \\ \times & 0_{n_e/2 \times n_e/2} \end{array} \right]$$

The final assumption concerns the order of Σ_m, which is restricted to avoid complex tensor notation, and hence simplify the presentation. We will in Section 5 see how to handle the case of robots with AC drives, where this assumption is not verified.

A 12.5 The subsystem Σ_m is of order two, that is $n_m = 1$.

Remark 12.6 Assumptions **A12.2**–**A12.4** are imposed by the fact that Σ_e is under–actuated. They are not needed in the fully actuated case when $M_e = I_{n_e}$.

---

[5]Recall that we have defined $\dot{q}_e \triangleq [y_e^\top, \ \dot{q}_r^\top]^\top$.

4   Problem solution

We are now in position to present the main result of this chapter.

Theorem 12.7 *Given the feedback system of Fig. 12.2 described by (12.2), (12.6), (12.7). Assume that an outer loop controller $\mathcal{C}_{ol} : (y_{m*}, y_m) \mapsto \tau_d$, which satisfies the conditions of Assumption* **A12.1***, is known. Under these conditions, there exists an inner–loop PBC, $\mathcal{C}_{il} : (y_e, y_m, \tau_d) \mapsto u$ such that*

$$\lim_{t \to \infty} \|y_m - y_{m*}\| = 0$$

with internal stability provided **A12.2**–**A12.5** *are satisfied.*   □

4.1   Proof of Theorem 12.7

The proof consists of two major steps, the design of the inner–loop controller, and the stability analysis of the overall cascaded configuration. For the design of $\mathcal{C}_{il}$, we take–off from the PBC (12.8), (12.9). This control law is, unfortunately, *non–implementable* because it requires the measurement of $\dot{y}_m$. Thus, we propose an implementable scheme that uses an observer. It is at this point that we use the assumptions of linearity of $\mathcal{C}_{ol}$ on y_{m2} stated in **A12.1**, and linearity of $\mathcal{C}_{il}$ with respect to τ_d as expressed in (12.13). The design of the observer is the main technical contribution of the chapter.

A   Inner–loop

A.1 Non–implementable PBC

The inner–loop PBC, implicitly defined by (12.8), (12.9), can (in principle) be explicitly solved invoking **A12.2**. More precisely, the assumption insures that, from measurements of y_{m1} and τ_d, we can solve on–line (12.12), (12.13) to get λ_d. $\dot{\lambda}_{sd}$ follows from differentiation of (12.13) and knowledge of $\dot{y}_{m1}$ and $\dot{\tau}_d$, and can in its turn be replaced in (12.8), which can be solved for u.

Let us now prove the convergence of $\tilde{\lambda} \to 0$. To this end, we derive the error system

$$\dot{\tilde{\lambda}} + R_e D_e^{-1}(y_{m1})\tilde{\lambda} + \begin{bmatrix} K_1(y_m) & 0 \\ 0 & 0 \end{bmatrix} D_e^{-1}(y_{m1})\tilde{\lambda} = 0$$

and consider the desired energy function[6] $\mathcal{L}_{ed} = \frac{1}{2}\tilde{\lambda}^T D_e^{-1}(y_{m1})\tilde{\lambda}$, whose derivative

---

[6]This is the same function as the one used in Section **9**.4, but it is now written in terms of λ_d instead of $\dot{q}_{ed}$.

along trajectories of the error system gives

$$\dot{\mathcal{L}}_{ed} = -\tilde{\lambda}^T D_e^{-1}(y_{m1}) R_{es}(y_m) D_e^{-1}(y_{m1}) \tilde{\lambda}$$

where

$$R_{es}(y_m) \triangleq R_e + \frac{1}{2} \frac{d}{dt} D_e(y_{m1}) + \begin{bmatrix} K_1(y_m) & 0 \\ 0 & 0 \end{bmatrix}$$

See (9.18) in Chapter **9**. Notice that, in view of **A12.5**, $\frac{d}{dt} D_e(y_{m1}) = y_{m2} W_1(y_{m1})$, and from **A12.4** we have that $(W_1)_{22}(y_{m1}) = 0$, where $(\cdot)_{ij}$ denotes the (i,j)-th submatrix of a block matrix. This, together with **A12.3**, ensures that when

$$K_1(y_m) = K_1^\top(y_m) > \sup_{y_m} \left\{ \frac{y_{m2}^2}{4} (W_1)_{12} R_r^{-1}(W_1)_{12}^\top - \frac{1}{2}(W_1)_{11} y_{m2} \right\} \quad (12.15)$$

we will have $\inf_{y_m} \underline{\lambda}(R_{es}) \geq \alpha > 0$, where $\underline{\lambda}(\cdot)$ is the minimum eigenvalue. This proves that $\dot{\mathcal{L}}_{ed} \leq -\alpha \mathcal{L}_{ed}$, hence $\tilde{\lambda} \to 0$ exponentially fast.

Writing down (12.7) in terms of the errors $\tilde{\lambda}$ (as we did in Section **9**.5), invoking the convergence proof above and the bound (12.5) we get the bounds

$$\begin{aligned} \|\tau - \tau_d\| &\leq \varepsilon_t + \varepsilon_t \|\lambda_d\| \\ &\leq \varepsilon_t + \varepsilon_t \|\tau_d\| \end{aligned} \quad (12.16)$$

where ε_t are some exponentially decaying functions and we have used (12.14) to get the last bound.

Notice that **A12.2** requires the controller equation (12.9) to be satisfied only asymptotically, but as will be shown later this will not affect the stability proof.

A.2 Observer design

Unfortunately, this controller cannot be implemented in the cascade scheme of Fig. 12.2 because, as discussed above, u requires the knowledge of $\dot{\lambda}_{sd}$ which demands $\dot{\tau}_d$. This, in its turn, given that τ_d is a function of y_m (12.4), would require $\dot{y}_m$. In Section **10**.3 we proved that, in the case when Σ_m is linear, we can use a linear filter to overcome this problem. This solution is not feasible in the nonlinear case, for which we introduce a nonlinear observer and use the assumptions of linearity on y_{m2} and τ_d.

We will see now how we can construct an observer that will estimate $\dot{\tau}_d$ indirectly. To this end, let us denote with u_N the (non–implementable) control that uses $\dot{\lambda}_d$. From (12.8) we see that u depends linearly on $\dot{\lambda}_d$. The following chain of calculations allows us to write u_N as

$$\begin{aligned} u_N &= f_1(y_e, y_m, \tau_d) + f_2(y_e, y_m, \tau_d)\dot{\lambda}_d \\ &= f_3(y_e, y_m, \tau_d) + f_4(y_e, y_m, \tau_d)\dot{\tau}_d \\ &= f_5(y_e, y_m, \tau_d) + f_6(y_e, y_m, \tau_d)\dot{y}_{m2} \\ &= f_7(y_e, y_m, \tau_d) + f_8(y_e, y_m, \tau_d)\tau \end{aligned}$$

where f_i, $i = 1, \cdots, 8$ are some suitably defined functions and we have used (12.13), (12.4) and (12.2) to get the second, third and fourth equations, respectively. Finally, we observe that **A12.4** ensures that τ in (12.7) is linear in the unmeasurable part of the state of Σ_e, *i.e.*, $\dot{q}_r$, consequently

$$u_N = f_9(y_e, y_m, \tau_d) + f_o(y_e, y_m, \tau_d)\dot{q}_r \qquad (12.17)$$

We propose now the *control law*

$$u = f_9(y_e, y_m, \tau_d) + f_o(y_e, y_m, \tau_d)z_r \qquad (12.18)$$

where $z \triangleq [z_s^\top, z_r^\top]^\top \in \mathbb{R}^{n_e}$ will be an estimate of $\dot{q}_e$ generated as

$$D_e(y_{m1})\dot{z} + \frac{1}{2}\frac{d}{dt}\left(D_e(y_{m1})\right)z + R_{es}(y_m)z = M_e v - L(y_e, y_m, \tau_d, z)$$

with

$$v \triangleq u + K_1(y_m)y_e$$

and $L(y_e, y_m, \tau_d, z)$ an output injection to be defined below. The observer is motivated from the fact that the system equations, after the damping injection, are of the form (see (9.17) in Section **9.4**)

$$D_e(y_{m1})\ddot{q}_e + \frac{1}{2}\frac{d}{dt}\left(D_e(y_{m1})\right)\dot{q}_e + R_{es}(y_m)\dot{q}_e = M_e v$$

Hence the error equation results in

$$D_e(y_{m1})\dot{\tilde{z}} + \frac{1}{2}\frac{d}{dt}\left(D_e(y_{m1})\right)\tilde{z} + R_{es}(y_m)\tilde{z} = L(y_e, y_m, \tau_d, z)$$

where $\tilde{z} \triangleq \dot{q}_e - z$. Consider now the quadratic function $V_o \triangleq \frac{1}{2}\tilde{z}^\top D_e(y_{m1})\tilde{z}$, whose derivative satisfies

$$\dot{V}_o \leq -\alpha\|\tilde{z}\| + \tilde{z}^\top L(y_e, y_m, \tau_d, z)$$

Now, (12.18) can be written as $u = u_N - f_o(y_e, y_m, \tau_d)\tilde{z}_r$, yielding the error equation for Σ_e

$$D_e(y_{m1})\ddot{\tilde{q}}_e + \frac{1}{2}\frac{d}{dt}\left(D_e(y_{m1})\right)\dot{\tilde{q}}_e + R_{es}(y_m)\dot{\tilde{q}}_e = -M_e f_o(y_e, y_m, \tau_d)\tilde{z}_r$$

Thus the derivative of the desired energy function $\mathcal{L}_{ed}$ takes now the form

$$\dot{\mathcal{L}}_{ed} \leq -\alpha\|\dot{\tilde{q}}_e\|^2 - (y_e - \dot{q}_{ed})^\top f_o(y_e, y_m, \tau_d)\tilde{z}_r$$

The calculations above motivate the following choice for the *output injection*

$$L(y_e, y_m, \tau_d, z) \triangleq \bar{M}_e f_o^\top(y_e, y_m, \tau_d)(y_e - \dot{q}_{ed})$$

with $\bar{M}_e^\top = [0, I_{n_e/2}]$, which exactly cancels the cross term and leads to

$$\dot{V}_o + \dot{\mathcal{L}}_{ed} \leq -\alpha(\|\dot{\tilde{q}}_e\|^2 + \|\tilde{z}\|^2)$$

From here we conclude that $\dot{\tilde{q}}_e \to 0$ (and also $\tilde{z} \to 0$) exponentially fast.

B   Outer loop

Once we have established convergence of the inner–loop we turn our attention to the outer loop. First, adding and subtracting τ_d, we rewrite (12.2) in terms of $\tilde{y}_m \triangleq y_m - y_{m*}$ as

$$\dot{\tilde{y}}_m = F_m(\tilde{y}_m) + \begin{bmatrix} 0 \\ g_m(y_{m1}) \end{bmatrix} (\tau - \tau_d) \tag{12.19}$$

From **A12.1** and the converse Lyapunov theorem 11.1 of Chapter 2 in [139] we have that $F_m(\tilde{y}_m)$, – the closed loop vector field obtained by setting $\tau \equiv \tau_d$, satisfies

$$\frac{\partial^\top V_m(\tilde{y}_m)}{\partial \tilde{y}_m} F_m(t, \tilde{y}_m) \leq -\alpha \|\tilde{y}_m\|^2 \tag{12.20}$$

for some Lyapunov function $V_m(\tilde{y}_m)$ which furthermore verifies

$$\left\| \frac{\partial V_m}{\partial \tilde{y}_m} \right\| \leq \alpha \|\tilde{y}_m\| \tag{12.21}$$

Combining these inequalities and using the various assumptions we get the bounds

$$\begin{aligned} \dot{V}_m &\leq -\alpha \|\tilde{y}_m\|^2 + \alpha \|\tilde{y}_m\| \|\tau - \tau_d\| \\ &\leq -\alpha \|\tilde{y}_m\|^2 + \varepsilon_t \|\tilde{y}_m\| (1 + \|\tau_d\|) \\ &\leq -\alpha \|\tilde{y}_m\|^2 + \varepsilon_t \|\tilde{y}_m\| (1 + \|\tilde{y}_m\|) \end{aligned} \tag{12.22}$$

where we have used (12.19)–(12.21) and (12.3) to get the first bound, (12.16) for the second bound and (12.4) (and boundedness of y_{m*}) for the last one. Asymptotic convergence of $\tilde{y}_m$ to zero follows immediately from the last inequality.

■

Remark 12.8 The steps of the proof given above may be summarized as follows. First, we use the assumptions on actuator–sensor coupling, damping of the non–actuated dynamics, and decoupling between Σ_e and Σ_m (Assumptions **A12.2, A12.3,** and **A12.4,** respectively) to design an inner–loop controller which ensures that τ converges exponentially to τ_d (uniformly in y_m). This controller requires $\dot{y}_{m2}$ hence its not implementable. At this point we invoke the assumptions on linearity of τ_d and y_{m2} for the design of an observer. The proof is completed using **A12.1** on exponential stabilizability of Σ_m and a converse Lyapunov theorem.

Remark 12.9 The arguments used for the proof of stability of the outer loop are similar to the ones used in [227] to study stabilizability of cascaded systems of the form

$$\dot{x} = f(x, 0) + g(x, \varepsilon_t) \varepsilon_t$$

where $\dot{x} = f(x, 0)$ is globally exponentially stable. Notice however that the key "growth" assumption in [227], namely

$$\|g(x, \varepsilon_t)\| \leq \gamma(\varepsilon_t) \|x\|$$

with $\gamma(\cdot)$ a class $\mathcal{K}$ function, does not hold in our case, since –(as seen from the derivations detailed in (12.22))– we dispose only of the weaker condition

$$\|g(x, \varepsilon_t)\| \leq \varepsilon_t \|x\| + \varepsilon_t$$

It should be pointed out that similar line of reasoning was used before in [232].

5 Application to robots with AC drives

In this section we apply our general Theorem 12.7 to the problem of motion control of robot manipulators with AC drives.

5.1 Model

We consider rigid robot manipulators where each joint is independently actuated by an AC motor. In this case the Lagrangian of the whole system satisfies the decomposition property of Proposition 2.10, that is

$$\mathcal{L}(q, \dot{q}) = \mathcal{L}_e(q_e, \dot{q}_e, q_m) + \mathcal{L}_m(q_m, \dot{q}_m)$$

where we have partitioned $q \triangleq [q_e^\top, q_m^\top]^\top$, then the system can be represented as in Fig. 12.1 where both subsystems are passive. As discussed in Chapter **2**, robots are EL system whose Lagrangian is of the form

$$\mathcal{L}_m(\dot{q}_m, q_m) = \frac{1}{2} \dot{q}_m^\top D_m(q_m) \dot{q}_m - V_m(q_m)$$

with $q_m = y_{m1} \in \mathbb{R}^{n_m}$ the joint positions, $\infty > d_M \geq D_m(q_m) = D_m^\top(q_m) \geq d_m > 0$ the inertia matrix, and $V_m(q_m)$ the potential energy.

The dynamics of an n_m–degrees of freedom rigid robot given in Chapter **2** can be written in the state space form (12.2) with

$$\begin{aligned} f_m(y_m) &\triangleq -D_m^{-1}(y_{m1})[C_m(y_{m1}, y_{m2})y_{m2} + G_m(y_{m1})] \\ g_m(y_m) &\triangleq D_m^{-1}(y_{m1}) \end{aligned} \tag{12.23}$$

See Chapter **2** for the definition of the terms and some relevant properties of the model. We simply recall here that the Coriolis matrix satisfies

$$\|C_m(y_{m1}, y_{m2})\| \leq c_M \|y_{m2}\| \tag{12.24}$$

Also, the local Lipschitz condition on $f_m(y_m)$, and (12.3) are satisfied. Notice that we have assumed that the link positions and velocities are available for measurement.

In Section **9**.4 the AC motors were classified into underactuated and fully actuated, that is, machines where the voltages can be applied only to stator windings (*e.g.*, induction motor), or to both stator and rotor windings (*e.g.*, synchronous motor with field windings). The latter class also includes machines, like the PM synchronous, PM stepper and variable reluctance motors, where the generalized coordinates consist only of stator variables which are directly actuated by the stator voltages. Control of underactuated machines is the most challenging problem, hence we will restrict our attention here to this class. The extension to the fully actuated machines is straightforward. We therefore consider Σ_e to be a block diagonal operator consisting of n_m subsystems of the form (12.6) with $n_{ei} = 4$, that is

$$\Sigma_{ei} : \left\{ \begin{array}{rcl} D_{ei}((y_{m1})_i)\ddot{q}_{ei} + \frac{d}{dt}\left(D_{ei}((y_{m1})_i)\right)\dot{q}_{ei} + R_{ei}\dot{q}_{ei} &=& M_e u_i \\ y_{ei} &=& M_e^\top \dot{q}_{ei} \end{array} \right. \tag{12.25}$$

where $i \in \bar{n}_m$, $\dot{q}_{ei} = [y_{ei}^\top, \dot{q}_{ri}^\top]^\top \in \mathbb{R}^4$ are the stator and rotor currents, and the inductance and resistance matrices are given as

$$D_{ei}((y_{m1})_i) = \left[\begin{array}{cc} L_{si}I_2 & L_{sri}e^{\mathcal{J}(y_{m1})_i} \\ L_{sri}e^{-\mathcal{J}(y_{m1})_i} & L_{ri}I_2 \end{array} \right] > 0, \quad R_{ei} = \left[\begin{array}{cc} R_{si}I_2 & 0 \\ 0 & R_{ri}I_2 \end{array} \right] > 0 \tag{12.26}$$

with

$$\mathcal{J} = \left[\begin{array}{cc} 0 & -1 \\ 1 & 0 \end{array} \right], \quad e^{\mathcal{J}(y_{m1})_i} = \left[\begin{array}{cc} \cos((y_{m1})_i) & -\sin((y_{m1})_i) \\ \sin((y_{m1})_i) & \cos((y_{m1})_i) \end{array} \right]$$

The overall inductance and resistance matrices of Σ_e are defined as

$$D_e(y_{m1}) \triangleq \text{block diag}\{D_{ei}((y_{m1})_i)\}, \quad R_e \triangleq \text{block diag}\{R_{ei}\}$$

respectively. Notice that $n_e = 4n_m$ and that the rotor currents $\dot{q}_{ri}$ are not measurable. See Chapter **10** for the definition of all the terms above.

Since $n_m \neq 1$, **A12.5** does not hold and we cannot apply directly Theorem 12.7. However, we notice that since $D_{ei}((y_{m1})_i)$ depends only on component i of y_{m1}, the components of the torque vector are given, for all $i \in \bar{n}_m$, by

$$\tau_i = \frac{1}{2}\dot{q}_{ei}^\top W_i((y_{m1})_i)\dot{q}_{ei}$$

with

$$W_i((y_{m1})_i) = \frac{\partial D_{ei}}{\partial(y_{m1})_i}((y_{m1})_i) \tag{12.27}$$

This nice block diagonal structure simplifies the first part of the design of $\mathcal{C}_{il}$, allowing us to approach the problem as n_m independent tasks, and easily adapt Theorem 12.7.

5.2   Global tracking controller

Before stating our result we find convenient to repeat here Proposition 4.8 of Section 4.2.2, which pertains to the outer–loop controller $\mathcal{C}_{ol}$ for the robot subsystem. This controller was reported in [22].

Proposition 12.10 *Let*

$$\tau_d = D_m(y_{m1})\ddot{y}_{m1*} + C_m(y_{m1}, \dot{y}_{m1*})y_{m2r} + G_m(y_{m1}) - K_d\dot{\tilde{y}}_{m1} - K_p\tilde{y}_{m1} \quad (12.28)$$

where $y_{m1*}(t)$ is a bounded, three times continuously differentiable reference, and

$$y_{m2r} = y_{m2} - \frac{c_0\tilde{y}_{m1}}{1 + \|\tilde{y}_{m1}\|} \quad (12.29)$$

where the gains $c_0 > 0$, $K_d = K_d^\top > 0$ and $K_p = K_p^\top > 0$ are chosen such that

$$c_0 < \min\left\{\frac{\underline{\lambda}(K_d)}{3\overline{\lambda}(D_m) + 2c_M}, \frac{4\underline{\lambda}(K_p)}{\overline{\lambda}(K_d) + \underline{\lambda}(K_d)}, \frac{2\sqrt{\underline{\lambda}(D_m)\underline{\lambda}(K_p)}}{\overline{\lambda}(D_m)}\right\} \quad (12.30)$$

with c_M as in (12.24). Under these conditions, every solution of (12.2), (12.23) determined by the control law $\tau = \tau_d$ satisfies

$$\|\tilde{y}_m(t)\|^2 \le \alpha e^{-\alpha t}\|\tilde{y}_m(0)\|^2 \text{ for all } t \ge 0$$

$\square$

We can now state the main result of this section:

Proposition 12.11 *Consider the rigid robot dynamics (12.2), (12.23) with AC motor drives (12.25)–(12.27), where we assume that the link positions y_{m1}, velocities y_{m2}, and stator currents y_e of the motors are available for measurement. Then there exists compensators $\mathcal{C}_{il}$, $\mathcal{C}_{ol}$ of the cascaded control configuration shown in Fig. 12.2, which ensure global asymptotic tracking for all (bounded, three times continuously differentiable) position references y_{m1*}.* $\square$

A   Proof of Proposition 12.11

Due to the block diagonal structure of the system mentioned above, the first part of the proof –until the definition of the observer–, consists of verifying the conditions of Theorem 12.7. For the sake of clarity of presentation we will divide again the derivations into outer and inner–loop controller, and observer design.

A.1 Outer loop control

In this the following we will prove that the robot controller of [22] satisfies assumption **A12.1**. It is actually shown in [22] that

$$V_m(\tilde{y}_m) \; \triangleq \; \frac{1}{2}(\tilde{y}_{m2} + c_1\tilde{y}_{m1})^{\top} D_m(\tilde{y}_{m2} + c_1\tilde{y}_{m1}) + \frac{1}{2}\tilde{y}_{m1}^{\top} K_p \tilde{y}_{m1}$$

with $c_1 > 0$ verifying

$$c_0 \left(1 + \sqrt{\frac{2V_m(0)}{\lambda(K_p)}}\right)^{-1} \le c_1 < c_0$$

qualifies as a Lyapunov function, which clearly satisfies (12.20) and (12.21).

Besides exponential stabilizability, **A12.1** imposes the crucial requirements that τ_d must be linear in y_{m2} and linearly bounded by y_m. This can be easily verified from (12.28).[7]

A.2 Inner–loop control

The inner–loop controller exactly coincides with the PBC for induction machines presented in **10.3**, hence we go through it very briefly. From (12.26) we immediately see that, for each subsystem Σ_{ei}, **A12.3** and **A12.4** are satisfied. We will now prove that Assumption **A12.2** is also satisfied. With some lengthy, but straightforward, calculations we can show that in this case the matrices A_i and $B_i((y_{m1})_i)$ of (12.11) are given by

$$A_i \triangleq -\frac{R_{ri}L_{si}}{\sigma_i}I_2, \quad B_i((y_{m1})_i) \triangleq \frac{R_{ri}L_{sri}}{\sigma_i}e^{-\mathcal{J}(y_{m1})_i}$$

where $\sigma_i \triangleq L_{si}L_{ri} - L_{sri}^2$. A state feedback that satisfies the conditions of Assumption **A12.2** is given in (10.29) of Section **10.3**. It can be written in terms of the fluxes as

$$\lambda_{sdi} = \left(\frac{\sigma_i\tau_{di} + \beta_{i*}^2 L_{si}}{L_{sri}\beta_{i*}^2}\right) e^{\mathcal{J}(y_{m1})_i}\lambda_{rdi}$$

with $\beta_{i*} > 0$ a design parameter, (that defines the desired value for $\|\lambda_r\|$). Actually, in this case we can solve (12.12) and obtain an explicit expression for λ_{rdi} as

$$\lambda_{rdi} \; = \; e^{\rho_{di}\mathcal{J}}\begin{bmatrix} \beta_{i*} \\ 0 \end{bmatrix} \tag{12.31}$$

where

$$\dot{\rho}_{di} \; = \; \frac{R_{ri}}{\beta_{i*}^2}\tau_{di}, \quad \rho_{di}(0) = 0 \tag{12.32}$$

---

[7]In this respect it is worth pointing out that, even though there are other controllers that exponentially stabilize the robot, *e.g.*, [229], to the best of our knowledge, only the one given above satisfies the additional linearity requirements of Assumption **A12.1**.

It is easy to check that with the definitions above we have

$$\frac{1}{2}\lambda_{di}^{\top}C_i(t)\lambda_{di} = (\tau_d)_i \text{ for all } t \geq 0$$

which is clearly stronger than the asymptotic property required by Assumption **A12.2**.

The ideal control of Chapter **10**.3, which in this case is non–implementable, can be summarized as

$$u_{Ni} = v_{Ni} - (K_1)_i((y_{m1})_i)\dot{q}_{si}$$
$$v_{Ni} = L_{si}\ddot{q}_{sdi} + L_{sri}e^{\mathcal{J}(y_{m1})_i}\ddot{q}_{rdi} + L_{sri}\mathcal{J}e^{\mathcal{J}(y_{m1})_i}(y_{m2})_i\dot{q}_{rdi} + R_{si}\dot{q}_{sdi} + (K_1)_i\dot{q}_{sdi}$$

where

$$(K_1)_i((y_{m1})_i) = \frac{L_{sri}^2}{4b_i}\dot{y}_{m1i}^2 I_2, \quad 0 < b_i < \min\{R_{si}, R_{ri}\} \tag{12.33}$$

has been chosen to satisfy the damping injection condition (12.15), and the desired currents are defined as

$$\dot{q}_{edi} = \begin{bmatrix} \dot{q}_{sdi} \\ \dot{q}_{rdi} \end{bmatrix} = \begin{bmatrix} \frac{1}{L_{sri}}e^{\mathcal{J}(y_{m1})_i}[I_2 + \frac{\tau_{di}L_{ri}}{\beta_{i*}^2}\mathcal{J}]\lambda_{rdi} \\ -\frac{\tau_{di}}{\beta_{i*}^2}\mathcal{J}\lambda_{rdi} \end{bmatrix} \tag{12.34}$$

The description of $\mathcal{C}_{il}$ is completed with (12.31) and (12.32) and the definition of τ_{di} in (12.28).

A.3 Observer design

As in the proof of Theorem 12.7, the control law above is not implementable because of the dependence of v_{Ni} on $\ddot{q}_{edi}$. This dependence translates, via (12.34), into a dependence on $\dot{\tau}_{di}$, which in its turn implies –due to (12.28)– knowledge of the joint acceleration $\ddot{y}_{m1}$. To overcome this problem, we must implement an observer. To this end, we follow the derivations of Section 4 and write $u_{Ni} = f_{3i}(y_e, y_m, \tau_d) + f_{4i}(y_e, y_m, \tau_d)\dot{\tau}_{di}$. Up to this point, the block diagonal structure is still preserved. Now, looking at (12.28) we see that, due to the presence of the Coriolis matrix (taking K_d diagonal), each component $\dot{\tau}_{di}$ depends on the whole vector $\dot{y}_{m2}$. For the subsequent substitutions we have to introduce, in an obvious manner, the vector notation $u_N = f_3(y_e, y_m, \tau_d) + f_4(y_e, y_m, \tau_d)\dot{\tau}_d$, to finally obtain, as in Section 4, $u_N = f_9(y_e, y_m, \tau_d) + f_o(y_e, y_m, \tau_d)\dot{q}_r$, with $q_r \triangleq [q_{r1}^{\top}, q_{r2}^{\top}, \cdots, q_{rn_m}^{\top}]^{\top}$. The actual control law is

$$u = f_{10}(y_e, y_m, \tau_d) + f_o(y_e, y_m, \tau_d)z_r$$

where, exploiting the block diagonal structure of Σ_e, we propose the observer

$$\mathcal{D}(y_m)\dot{z} + \left(\frac{d}{dt}(\mathcal{D}(y_m)) + \mathcal{R}\right)z = \mathcal{M}_e v - L$$

with

$$\mathcal{D}(y_m) \triangleq \text{block diag}\{D_{ei}((y_{m1})_i)\}$$

$$\mathcal{R} \triangleq \text{block diag}\left\{R_{ei} + \begin{bmatrix} (K_1)_i((y_{m1})_i) & 0 \\ 0 & 0 \end{bmatrix}\right\}$$

$$\mathcal{M}_e \triangleq \text{block diag}\{\bar{M}_e\}$$

and the output injection

$$L(y_e, y_m, \tau_d, \dot{\tilde{q}}_s) = \mathcal{M}_e f_0^\top (y_e, y_m, \tau_d)\dot{\tilde{q}}_s$$

Considering the quadratic function for the whole system

$$V = \frac{1}{2}\tilde{z}^\top \mathcal{D}(y_m)\tilde{z} + \frac{1}{2}\dot{\tilde{q}}_e^\top \mathcal{D}(y_m)\dot{\tilde{q}}_e$$

we get exponential convergence of signals $\tilde{z}$ and $\dot{\tilde{q}}_e$ as in the proof of Theorem 12.7.

∎

Remark 12.12 (Controller structure and removal of observer.) There are two major drawbacks of the scheme presented above. First, even though the block diagonal nature of Σ_e somehow propagates through the controller calculations, in the observer stage this decentralized structure is unfortunately lost due to the motor cross–couplings that appear in the evaluation of (12.17). Second, the inclusion of the observer considerably complicates the controller structure, increasing its (dynamical) order and introducing –via $f_o(y_e, y_m, \tau_d)$– higher order terms. In order to remove the need of the observer we have recently derived a robot controller independent of joint velocity [218]. More precisely, we can show that

$$\tau = D_m(y_{m1})\ddot{y}_{m1*} + C_m(y_{m1}, y_{m2*})y_{m2*} + G_m(y_{m1}) - K_d(\theta - y_{m2*}) - K_p\tilde{y}_{m1}$$

with

$$\dot{\theta} = -\left(\alpha_1 + \alpha_2\|\theta - y_{m2}\| + \alpha_3\|\tilde{y}_{m1}\|\right)(\theta - y_{m2}) + \ddot{y}_{m1*}$$

in closed loop with (12.2), (12.23) ensures global uniform asymptotic stability (provided the gains are suitably chosen). Since the observer is now obviated we obtain a simpler decentralized scheme.

Remark 12.13 (Global tracking with FOC.) An interesting corollary of Proposition 12.11 is that for current–fed induction machines (where the stator currents y_e are now the control inputs) the standard indirect field–oriented control scheme, see Section 11.2, without any observer, can be combined with the robot controller of [22] to yield a global tracking controller.

6   Simulation results

To illustrate the performance of the controller in Proposition 12.11, and compare it with a backstepping–based scheme, we have performed in SIMNON simulations of the two–link robot with induction motors presented in Subsection X-B of [107]. The dynamic model of the robot and induction motors, as well as the initial states, are taken from that paper. It is worth pointing out that in [107] a time–varying flux reference β_{i*} (ψ_{di} in [107]) is needed to avoid singularities in the control calculations. This (somewhat artificial) provision is not needed for our controller which is globally defined. Consequently, we have used $\beta_{i*} = 2$ Wb which is the average value (in the time interval of interest) of the reference given in [107]. Our controller contains only four tuning parameters: The coefficients b_i of the damping injection gains (12.33), the proportional K_p and derivative K_d gains of the robot controller (12.28), and the gain c_0 (12.29). Following the suggestion of Section **10**.6 we set $b_i = 0.5R_{ri}$, where R_{ri} are the rotor resistances. After some straightforward tuning, we set up

$$K_p = 1000I_2, \quad K_d = 20I_2$$

and chose $c_0 = 0.1$ which satisfies (12.30). The resulting position tracking errors, motor torques, stator voltages and currents are depicted in Figs. 12.3–12.6. To facilitate the comparison, the plots are given in the same scale as the corresponding figures in [107].

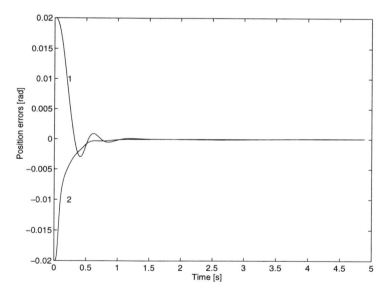

Figure 12.3: Position errors $(\tilde{y}_m)_i$, $i = 1, 2$.

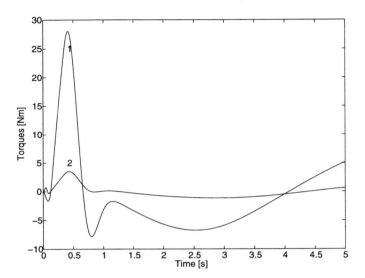

Figure 12.4: Motor torques τ_i, $i = 1, 2$.

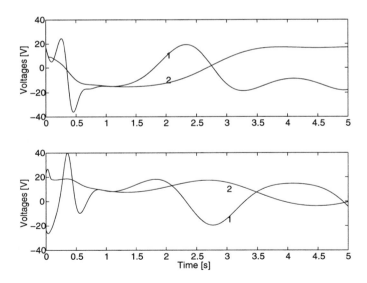

Figure 12.5: Stator voltages u for the two motors..

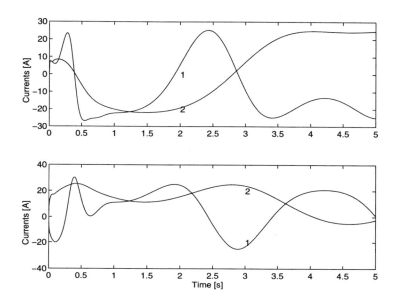

Figure 12.6: Stator currents $\dot{q}_s$ for the two motors.

The following remarks are in order:

1. The position errors of our scheme converge to zero in approximately 0.5 s, with a negligible overshoot in the first error (denoted by 1 in the figures). This should be contrasted with Fig. 7 of [107] where the errors converge to a neighbourhood of zero after 1 s and exhibits afterwards a sustained oscillation. We should recall that the analysis in [107] predicts only ultimate boundedness of the errors. The simulations show that inside the residual set the behaviour is oscillatory.

2. Our simulations proved to be highly insensitive to the parameter c_0. However, as thoroughly discussed in Section **10**.6, the gain of the nonlinear damping term b_i is critical for large speeds. Our experimental experience has shown that this term amplifies the noise and induces voltage saturations. This difficulty can be alleviated by use of integral action in stator currents as explained in Section **10**.3. This provision can also be incorporated to the theoretical analysis presented here.

3. To compare the computational complexity of the proposed passivity–based controller (PBC) with the backstepping–based controller (BBC) of [107] and the observer–less controller OL-PBC of Remark 12.12 we provide in Table 12.1 the number of operations needed by each algorithm (per joint).

Method	Numerical operation		
	Additions and subtractions	Multiplications and divisions	Total
PBC	114	271	385
BBC	138	318	456
OL-PBC	77	162	239

Table 12.1: Comparison of computational requirements.

4. In spite of the slow convergence rate of the position errors in [107], there is a highly oscillatory initial transient in torque, voltages and currents in the simulations of that paper. This seem to stem from the flux observation transient. The transient is partially alleviated by the particular (slowly increasing) choice of position reference, it gets worse for other more realistic tasks. Our scheme does not require flux observers and, as seen from the figures, the responses are very smooth.

5. In trying to reproduce the simulations of [107] we encountered some difficulties. Not only is the number of code lines two to three times larger than for our controller, but a high sensitivity to numerical errors was also experienced. We attribute these difficulties to the high gains introduced by the higher order terms inherited from the backstepping design. This problem is still present in a later work of the authors [105] where the control signals contain terms of order seven. Since these terms are of the same nature as our damping injection, we expect similar noise problems to appear in a practical setting. This seems to be the case in the experimental results of Section X-A in [107].

7 Concluding remarks

In this chapter we have studied the problem of output feedback global tracking of an EL system in feedback interconnection with a general nonlinear system. This configuration often appears in practical applications. In some instances it is possible to introduce a control action that removes the feedback path from the interconnection, leading to a cascade system. From a practical viewpoint, this does not seem to be advisable since cancelation of nonlinearities might induce serious robustness problems. See [132] for a conclusive example.

For cascaded systems it is by now well known that stabilizability of the driving system combined with a growth condition on the interconnection insures stabilizability of the composition. In our feedback interconnected case, besides the growth condition on the interconnection, we require a "stronger" form of stabilizability, namely that it must be uniform in the feedback signal. Relying on these assumptions we have investigated the feasibility of a cascaded control configuration which is particularly well suited for EL systems with "block diagonal" kinetic energy functions. For these systems, we have a natural decomposition into feedback interconnected passive subsystems. Furthermore, in the controller design we can easily use energy shaping plus damping injection ideas to design a PBC. Since we are interested in obtaining output feedback controllers we introduced a nonlinear observer. The design of the latter is the main technical contribution of the chapter.

One of the main motivations of our research was the problem of global tracking of robots with AC drives. A solution, based on backstepping principles, may be found in [107] (see also [105] for some important extensions). We believe these results, though interesting from the theoretical viewpoint, are of limited practical interest because of their high computational complexity and reliance on higher order terms which act as high gains. The first problem is a fundamental issue which, unfortunately, is not fully recognized in the recent literature on nonlinear control. It pertains, not just to the availability of fast and cheap "number crunchers", but also to the numerical sensitivity of the calculations and its impact on the tuning procedure, see [132]. The deleterious effects of injecting high gains into the control loop are well documented.

A fundamental building block for the cascade systems stabilization theory is the proof of [252] that global asymptotic stabilizability implies global input-to-state stabilizability with respect to input disturbances. Unfortunately, as shown in [88], such a result does not hold for output disturbances, which makes the problem of stabilization of feedback interconnected systems more challenging. We hope the material presented in this chapter will motivate the researchers to look into this practically important problem.

Chapter 13

Other applications and current research

The objective of this final chapter is twofold. First, we point out to other applications of PBC that went beyond the scope of this book. Second, we collect specific problems in PBC of EL systems on which we are currently working. They have not yet been fully resolved, and thus put forth further avenues of study.

The application of passivity (or the more general concept of dissipativity) for control of *large flexible structures* has a very long history dating at least as far back as the work of Opdenacker and Jonckheere in 1985 [201]. It is well known that for this class of systems, in the absence of gyroscopic effects, the structure transfer matrix is symmetric. Furthermore, the use of collocated sensor and actuators makes the structure dissipative. The practical significance of this feature is that, if placed in feedback with a passive controller, the closed loop will be robustly stable to dissipative unmodeled system modes. This property is particularly important in mechanical vibration problems since, in spite of their nice structural properties, their accurate representation requires many (ideally infinitely many) eigenmodes. Two new problems arise at this point, a convenient model reduction technique that preserves the dissipation structure of the system. Then, a systematic design methodology that ensures the resulting controller is passive. In [201] the first problem is addressed *via* LQG balancing techniques, while in [102] the authors determine conditions ensuring that an LTI system with an LQG compensator defines a dissipative loop. See also the recent book [121] for a review of some more recent developments in the linear case, and some extensions to treat nonlinear multibody flexible space structures.

1 Other applications

In the applications of PBC to robot manipulators described in this book we restricted ourselves to unconstrained robots, that is without interaction with their environment. It can easily be shown that the passivity property of the robot is preserved even in the later case. It has therefore been possible to extend many of the results reported here for practically important case of force and impedance control. See [161] for an exhaustive list of references. Another robotics area where passivity has played a prominent role is in teleoperation, where the fundamental work [6] opened the road for many additional extensions.

In the applications above, it is the ability of passivity to handle infinite dimensional systems, that is suitably exploited. It is quite natural then that the concept has also found its way on vibration control of civil engineering structures with active dampers, see [234] for a recent interesting survey. In [234] a very original application of structural controller with spacewise distributed actuators is also reported.

Some applications to *chemical engineering* processes of PBC have been reported in [241], where the theory presented here for systems in EL form, is applied for systems in extended Hamiltonian form 6.3. A new line of research has been opened in [290] where an elegant mathematical framework has been developed for PBC of thermodynamical systems.

The Department of Engineering Cybernetics of the Norwegian University of Science and Technology has also been very actively involved in applications of PBC. In particular the research on *marine vessels* is remarkable, see for instance [85] which contains some *full scale* experimental results of a passivity-based observer which successfully filters out the noise from the position measurements while attenuating perturbations due to environmental disturbances. We invite the reader to see also [86] where different *Lagrangian* models of marine vessels are included.

We should also mention the work carried out on controlled Lagrangians at Princeton University and CalTech by Nahomi Leonhard and Jerrold Marsden, respectively. Applications of this theory to underwater vehicles may be found in the URL http://www.cds.caltech.edu/~marsden/. This work, although much more theoretically–oriented, is intimately related to PBC as developed in this book.

The realm of application of PBC is clearly rich and diverse, and it is by no means limited to EL systems. The list of references above is, of course, not exhaustive, being necessarily constrained to the authors' knowledge, background and preferences.

2 Current research

In our research on PBC we currently address two main topics, the extension to other engineering areas of application, and the assessment of the *performance* of PBC. In the first research area we have given particular attention to power electronics and power systems applications, which we briefly describe below. To tackle the problem of power systems stabilization we were confronted with the question of passivation of systems with constant forcing inputs. The state of the art of this more fundamental research is also discussed below. The routes we have taken to study performance issues are described at the end of the chapter.

2.1 Power electronics

The research on power electronics is a natural continuation of the work reported here for DC–to–DC converters to tackle other types of devices. Motivated by some recent work of Stanković and co–workers [256], we have been looking at the application of PBC to the practically important *series–resonant* DC–to–DC converters. The dynamics of this system, although described by a simple third order model with well defined static nonlinearities, exhibits a very rich behaviour that cannot be globally characterized. To get a better understanding of the system structure we have recently carried–out some research on existence and attractivity of periodic orbits in [15]. In this work we also provide estimation of the signals amplitude as we move away from the resonant case. One of the main stumbling blocks for the development of PBCs (or for that matter any other kind of model–based controllers) for series–resonant DC–to–DC converters is the mathematical difficulty of dealing with strong discontinuities in the systems model. To overcome this problem, [256] uses a so–called phasor dynamic model, which consists of a set of smooth differential equations that describe the dynamic behaviour of the Fourier coefficients of the first harmonic approximation. We have been recently studying the utilization of this model for PBC. This has given rise to the question of evaluation of the domain of validity of the model, as well as some issues concerning the "inversion" of the averaging process (used to derive the model) to obtain an explicit realization of the PBC.

Another power electronic application that we have studied recently is the AC–to–DC converter, and more specifically the three–phase voltage sourced reversible rectifiers. The outcome of all our research on power electronics is reported in the Ph D thesis [71].

Our PBC methodology has been applied with some success in [93] to the problem of nonlinear adaptive control of static condensers. These devices, which belong to the class of flexible AC transmission systems, has gained wide popularity in the power systems community for voltage stabilization and control of reactive power.

2.2   Power systems

In the area of *power systems* we are currently interested in the problem of suppression of low frequency oscillations. These oscillations appear in strongly interconnected networks because of changes in load and topology, and they may cause loss of synchronism and generator tripping. This problem is usually studied in the power systems literature adopting a sinusoidal quasi–steady state approximation that ignores the dynamics of the network circuit elements. Further, the interaction between the generator and the network is characterized via terminal voltage phasor and complex power pairs. This sinusoidal characterization may prove inadequate in some studies basically due to the proliferation of switching controls in modern power systems.

A unifying framework to study this broad problem from the perspective of passivity has recently been advanced in [257] using linear models. As discussed above there is a clear motivation to try to extend this study to the nonlinear case. As a first step towards this end, we concentrate our attention in [213] on the design of power systems stabilizers with the generator exciter being the actuation point. The generator to be controlled is described by a standard EL model, with three forcing terms: The mechanical torque coming from the turbine, the terminal voltage of the network and the field voltage, which is our control variable. In view of the significant differences between the mechanical and the electrical time scales, the first signal can be treated as a constant disturbance. The terminal voltage may be viewed as the output of an operator, –defined by the remaining part of the network–, which is in feedback interconnection with the generator. Our basic assumption is that the network is always absorbing energy from the generator, whence the interconnection subsystem (as viewed from the generator) is passive. The control objective is then to close a loop around the field voltage so as to *passivate* the generator system. As a preliminary result, we characterize, in terms of a simple linear matrix inequality, a class of linear state–feedback controllers which achieve this objective.

Related research was recently reported in [18] where a PBC, in the form of $L_g V$ or Jurdjević–Quinn control [237], was utilized to enlarge the domain of attraction for the particular case of a single–generator with infinite bus connection.

2.3   Generation of storage functions for forced EL systems

The passivation objective described above is done with respect to the standard error dynamics storage function. That is, we take as usual $\mathcal{H}_d = \tilde{q}^\top D(q)\tilde{q}$, with $\tilde{q}$ some error signal. Although in some other applications, like the ones reported in this book, this choice proved suitable[1], in the power systems stabilization problem the "energy–shaping ability" of the controller is seriously crippled with the use of this storage

---

[1]Although, frankly speaking, difficult to rationalize from any physical viewpoint, this choice turned out to be mathematically convenient.

function. This stems from the fact that the main instability source in the problem is the existence of an external disturbance, the mechanical load, which although constant, shifts the equilibrium away from the origin. This problem triggered in [177] our interest towards the closely related fundamental question of generation of Lyapunov functions for extended Hamiltonian systems (of the form **6**.3) with external disturbances.

Motivated by the energy balance equation of passive systems

$$\dot{\mathcal{H}} = \bar{u}^\top y - x^\top \mathcal{R} x$$

where $\mathcal{H}, \mathcal{R}, \bar{u}, x$ are the total energy of the system, the dissipation matrix, the constant external disturbance, and the systems state, respectively, it seemed natural to postulate

$$\mathcal{H}(x) - \bar{u} \int_0^t y(\tau)d\tau$$

as a candidate Lyapunov function. Notice that the term $\bar{u}^\top \int_0^t y(\tau)d\tau$ is precisely the energy externally supplied to the system. Hence the new function is exactly the difference between the energy of the system and the supplied energy.

To check whether this function can be used as a Lyapunov function, the first basic question is, of course, if we can write $\bar{u}^\top \int_0^t y(\tau)d\tau$ as a function of the state $x(t)$. Unfortunately, this is not always possible. However, we identify some cases for which we can construct a suitable new "output". Interestingly enough, for linear systems the resulting Lyapunov function is the incremental energy, thus our derivations provide a physical explanation of it. An easily verifiable necessary and sufficient condition for the applicability of the technique in the general nonlinear case is also given.

This new storage function has recently been utilized as a basis for the generation of a new family of PBC for DC–to–DC converters in [71].

2.4 Performance

As we have shown throughout this book, the issue of stabilization of a large class of EL systems is essentially settled. The next natural step is to assess their performance. This is a fundamental question for which an answer is needed to convince the practitioners of the advantages of our theoretical developments.

The first performance indicator to be considered is *disturbance attenuation*, —the Achilles's heel of high–gain designs. If we consider the class of disturbances with finite energy, this translates into the evaluation of the $\mathcal{L}_2$-induced norm of the closed-loop operator mapping the disturbances to the output of interest. Clearly, this indicator may be very conservative, particularly for nonlinear systems, where the possibility of frequency-mixing, stymies the utilization of LTI filters to discriminate the more viable disturbances. In spite of this potential limitation, we carried out in [233] a comparison, from a disturbance attenuation perspective, of PBC and feedback linearizing control of a boost DC-to-DC converter and a rigid robot manipulator. For the former we proved that for both controllers there exists a *lower bound* to the achievable attenuation level which is independent of the design parameters. Furthermore, for the PBC we obtained an *upper bound* for the disturbance attenuation, which is ensured provided we sacrifice the convergence rate. It came as a rather nice surprise that this study gave us some clues on the basic tradeoff between robust stability and convergence for PBC. For the case of rigid robots we showed that both approaches yield *arbitrarily good* disturbance attenuation *without* compromising the convergence rate. See Proposition 4.4.

Performance analysis has typically been recast in the literature in terms or robustness of *stability* with respect to unmodeled effects.[2] PBC are indeed robustly stable, because of their inverse optimality properties and the "strong" form of stability that we can typically establish for these designs, e.g., GAS for mechanical systems, and even exponential stability for electrical machines. But this does not mean that they will give good performance, for instance fast and smooth transient responses. We have seen already in Sections 7.4 and 8.5 that the inability to add damping in some underactuated systems puts a hard bound on the achievable bandwidth. Also, the convergence rate of the PBCs developed for electrical machines is limited by the time constant of the unactuated electrical subsystem.[3]

In the authors' opinion the main motivation to consider nonlinear models of physical systems (at least the class considered in this book) is that they faithfully capture the actual behaviour of the system. The formulation of performance evaluation in terms of unmodeled effects seems then quite contrived. This is particularly distressing when the class of unmodeled effects is reduced to *linear* dynamics. This brings us back again into the role of modeling in control systems design. To elaborate on this point let us consider the problem of friction in mechanical systems studied in Chapter **5**. Given the dissipative nature of this phenomenon, it can certainly be claimed that PBC will be robustly stable with respect to friction. But friction, if it is not explicitly taken into account in the controller design, will make the system behave below par. Therefore, the question of performance for this problem has to be addressed through a careful study of models that capture the frictions effect, and not by "robustifying" the design.

We close this chapter with some scattered thoughts on the question of performance:

- **Adaptation.** There is no need to elaborate on the interest of adaptation since, after optimal control, this is the most performance–oriented technique in control theory. It is also unquestionably needed in many applications. For instance,

---

[2] We underscore the word "stability" because some far-fetched claims have been made regarding robust *performance*, particularly in the $\mathcal{H}_\infty$–control literature.

[3] It is fair to say that, as pointed out in Section **11.4** the same bound applies, this time to the observer convergence rate, for feedback linearizing schemes.

PBC of robot manipulators, which is certainly robustly stable to uncertainty in the payload, because the variations cannot destroy the physical property of positivity of the inertia matrix, or the existence of minima in the potential well. However, adding adaptation features to PBC we can actually identify the parameters of the inertia matrix and relocate the minimum of the potential energy at their desired value.

- **Analysis tools.** It is clear that the tools we use for stability analysis are not suited to evaluate performance, let alone design highly-performant controllers. Lyapunov analysis gives us, at best, estimates on convergence rates. Input–output analysis provides us additionally with estimates on the gains of the operators mapping the various signals. These estimates are very hard to obtain, and are usually very conservative.

- **Gain scheduling via flatness.** It has been mentioned already that flatness characterizes a class of systems for which trajectory planning is trivial. Many physical systems are flat –for that matter, all the ones we considered in this book– with, furthermore, a physically meaningful flat output. Now, in PBC some controller parameters depend on the reference signals. For instance, the gain δ of the PBC for flexible joint robots studied in Section 3.2.4.B.1 depends on the link reference q_{p1*}, which turns out to be a flat output for that system. In its current form the virtual robot simply places itself at the final position and then pulls up the robot, hence it is not surprising that the performance is not very satisfactory as witnessed by the simulations presented there. A more sensible approach would be to plan a smooth trajectory for the virtual robot, this translates into the definition of a time–varying gain δ which depends on the flat output. Current investigations are under way in this direction.

- **Performance as optimal energy transfer.** It is useful to view control as a mean to regulate the energy transfer in a physical system. For instance, we have shown in Section 6.2 that the action of the switch in the boost converter is simply to modify the dissipation structure to permit the transfer of magnetic energy in the inductance to electric energy in the capacitor. The same analogy can be made for mechanical systems, where PBC adds dashpots and springs to enforce a certain pattern of energy transfer. Adopting this perspective, it is reasonable to expect that a sensible "performance theory" should be closely linked to energy considerations. This point of view is adopted in the interesting paper [16] where it is claimed that the workless forces, which are disregarded in standard PBC, can be actually "shaped". We should point out that the additional terms needed towards this end are similar to the ones introduced in backstepping design. This has been illustrated in the example of the levitated ball in Chapter **8**.

We would like to finish this book with a wise warning from Michael Faraday, whose remarkable scientific and personal life sets forth an unvaluable example for the new generations:

> *"By adherence to a favorite theory, many errors have at times been introduced into general science which have required much labor for their removal... To guard against this requires a large proportion of mental humility, submission and independence."*

M. Faraday.

Appendix A

Dissipativity and passivity

Dissipativity is a fundamental property of physical systems closely related with the rather intuitive phenomena of loss or dissipation of energy. It was introduced by Willems as a generalization of the well–know property of passivity in the seminal paper [288]. Typical examples of dissipative systems are electrical circuits, in which part of the electrical and magnetic energy is dissipated as heat in the resistors. A similar role is played by friction in mechanical systems. To mathematically define the property of dissipativity we must introduce two functions: the supply rate, that is the rate at which energy flows into the system; and the storage function, which measures the amount of energy that is stored inside the system. These functions are related via the dissipation inequality, which states that along the time trajectories of a dissipative system the supply rate is not less than its increase in storage. This expresses the fact that a dissipative system cannot store more energy than is supplied to it from the outside, with the difference being the dissipated energy.

In its more general formulation the notion of dissipativity does not require the definition of inputs and outputs of the dynamical system. For the purposes of this book we find, however, useful to make this distinction, and call them $u \in \mathbb{R}^m$ and $y \in \mathbb{R}^m$, respectively. We will further restrict ourselves to a particular class of dissipative systems, namely *passive* systems, for which the supply rate function is simply $u^\top y$.

In the remaining of the appendix we recall some basic properties and classical results of passive systems. Most of the technical lemmas, theorems and propositions are borrowed from [69, 272]. For another recent text covering more modern concepts such as *feedback passivity* see [237]. Before proceeding with the precise mathematical definitions let us illustrate the basic concept with a simple example.

1   Circuit example

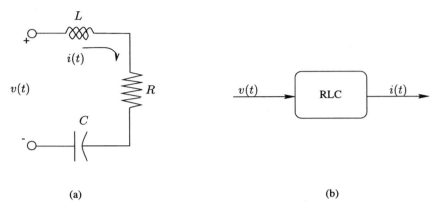

(a)   (b)

Figure A.1: RLC network

Consider the LTI RLC–circuit of Fig. A.1. The dynamic behaviour of the circuit can be easily obtained applying Kirchoff's law as

$$v = Ri + \frac{1}{C}\int_0^t i(\tau)d\tau + L\frac{di}{dt}$$

Now, multiplying by i we get

$$iv = Ri^2 + \frac{1}{C}i\int_0^t i(\tau)d\tau + Li\frac{di}{dt}$$

or equivalently

$$\frac{d}{dt}\left(\underbrace{\frac{1}{2C}(\int_0^t i(\tau)d\tau)^2}_{\mathcal{V}} + \underbrace{\frac{L}{2}i^2}_{\mathcal{T}}\right) = vi - Ri^2 \tag{A.1}$$

where we used the functions $\mathcal{V}$ and $\mathcal{T}$ to denote the electric energy stored in the capacitor and the magnetic energy of the inductance, respectively. Integrating (A.1) from 0 to t we get the energy balance equation

$$\underbrace{\mathcal{H}(t)}_{\text{available}} = \underbrace{\mathcal{H}(0)}_{\text{initial}} + \underbrace{\int_0^t v(\tau)i(\tau)d\tau}_{\text{supplied}} - \underbrace{\int_0^t Ri^2(\tau)d\tau}_{\text{dissipated}}$$

where we have introduced $\mathcal{H} \triangleq \mathcal{V} + \mathcal{T}$ –the total energy of the circuit. In this example the supply rate function vi is the power delivered from the external source to the network, while the storage function $\mathcal{H}$ is the total energy of the system. According to the definition above the RLC circuit is clearly dissipative.

If in the RLC–circuit above we adopt the convention of viewing v as input and i as output –although there is no particular preference for either one of them– we have that the system $\Sigma : u \mapsto y$ is passive.

2 $\mathcal{L}_2$ and $\mathcal{L}_{2e}$ spaces

We consider the set Ξ of all measurable real–valued n–dimensional functions of time $f(t) : \mathbb{R}_+ \rightarrow \mathbb{R}^n$. We define the set

$$\mathcal{L}_2 \triangleq \{x \in \Xi \mid \|f\|_2^2 \triangleq \int_0^\infty \|f(t)\|^2 dt < \infty\}$$

with $\|\cdot\|$ the standard Euclidean norm. This set forms a normed vector space over the field of real numbers with norm $\|\cdot\|_2$. We introduce now the extended space $\mathcal{L}_{2e}$ as

$$\mathcal{L}_{2e} \triangleq \{x \in \Xi \mid \|f\|_{2T}^2 \triangleq \int_0^T \|f(t)\|^2 dt < \infty, \ \forall T\}$$

Clearly $\mathcal{L}_2 \subset \mathcal{L}_{2e}$, because the extended space contains signals whose $\mathcal{L}_2$ norm may grow to infinity, but only at infinity.

We also define the inner product and the truncated inner product of two functions u and y as

$$\langle u \mid y \rangle \triangleq \int_0^\infty u(t)^\top y(t) dt$$

$$\langle u \mid y \rangle_T \triangleq \int_0^T u(t)^\top y(t) dt$$

3 Passivity and finite–gain stability

Even though the concepts of passivity and input–output stability are developed independently of the definition of the system state, see e.g. [69], for the sake of simplicity we will restrict ourselves here to systems of the form

$$\Sigma : \begin{cases} \dot{x} = f(x, u), \ x(0) = x_0 \in \mathbb{R}^n \\ y = h(x, u) \end{cases} \tag{A.2}$$

with state $x \in \mathbb{R}^n$, input $u \in \mathbb{R}^m$ and output $y \in \mathbb{R}^m$. In this way, (A.2) defines a causal dynamic operator $\Sigma : \mathcal{L}_{2e} \rightarrow \mathcal{L}_{2e} : u \mapsto y$.

We have the following definitions:

Definition A.1 (Dissipativity.) Σ *is dissipative with respect to the supply* $w(u, y)$:
$\mathbb{R}^m \times \mathbb{R}^m \to \mathbb{R}$ *if and only if there exists a storage function* $\mathcal{H} : \mathbb{R}^n \to \mathbb{R}_{\geq 0}$, *such that*

$$\mathcal{H}(x(T)) \leq \mathcal{H}(x(0)) + \int_0^T w(u(t), y(t)) dt \qquad (A.3)$$

for all u, *all* $T \geq 0$ *and all* $x_0 \in \mathbb{R}^n$.

Definition A.2 (Passivity.) Σ *is passive if it is dissipative with supply rate* $w(u, y) = u^\top y$. *It is input strictly passive (ISP) if it is dissipative with supply rate* $w(u, y) = u^\top y - \delta_i \|u\|^2$, *where* $\delta_i > 0$. *Finally,* Σ *is output strictly passive (OSP) if it is dissipative with supply rate* $w(u, y) = u^\top y - \delta_o \|y\|^2$, *where* $\delta_o > 0$.

Definition A.3 ($\mathcal{L}_2$ stability.) Σ *is said to be* $\mathcal{L}_2$ *stable*[1] *if there exists a positive constant* γ *such that for every initial condition* x_0, *there exists a finite constant* $\beta(x_0)$ *such that*
$$\|y\|_{2T} \leq \gamma \|u\|_{2T} + \beta(x_0).$$

The following corollary is obvious from the definitions.

Corollary A.4 ($\mathcal{L}_2$–stability and dissipativity.) *A state space system* Σ *is* $\mathcal{L}_2$–*stable if it is dissipative with supply rate* $w(u, y) = \frac{1}{2}\gamma^2 \|u\|^2 - \|y\|^2$, *for some* $\gamma > 0$.

The proposition below follows immediately from the definitions and an argument of completion of the squares.

Proposition A.5 (OSP $\Rightarrow \mathcal{L}_2$–stability.) *If* $\Sigma : u \mapsto y$ *is OSP then it is* $\mathcal{L}_2$–*stable.*
$\square$

Proof. The proof follows straight forward observing that OSP implies the existence of $\delta_o > 0$ and $\beta \in \mathbb{R}$ such that

$$\delta_o \|y\|_{2T}^2 \leq \langle u \mid y \rangle_T - \beta$$

Hence

$$\delta_o \|y\|_{2T}^2 \leq \langle u \mid y \rangle_T - \beta + \frac{1}{2} \left\| \frac{1}{\sqrt{\delta_o}} u - \sqrt{\delta_o} y \right\|_{2T}^2$$

also holds, and therefore

$$\frac{\delta_o}{2} \|y\|_{2T}^2 \leq \frac{1}{2\delta_o} \|u\|_{2T}^2 - \beta.$$

$\blacksquare$

---

[1] This type of stability is sometimes referred as *strong* (or stability with finite gain) to distinguish it from the strictly weaker property of $\Sigma : \mathcal{L}_2 \to \mathcal{L}_2$.

4 Feedback systems

We present in this section some well known results about the feedback system depicted in Fig. A.2. Each of the subsystems Σ_i, $i = 1, 2$ is a state–space system of the form (A.2). We further assume that the interconnection is well–posed, that is the operator $u \mapsto y$, with $u \triangleq (u_1, u_2)$ and $y \triangleq (y_1, y_2)$, is causal and maps $\mathcal{L}_{2e}$ signals into $\mathcal{L}_{2e}$ signals.

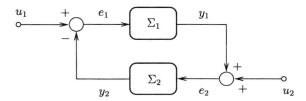

Figure A.2: Feedback interconnection of passive systems

We start with the basic property of invariance, under feedback interconnection, of passivity.

Proposition A.6 (Invariance of passivity.) *Consider the* input-output *system depicted in Fig. A.2. If Σ_1 and Σ_2 are both passive then $\Sigma : u \mapsto y$ is also passive. If furthermore they are OSP then $\Sigma : u \mapsto y$ is also OSP.* $\square$

The main theorem concerning stability of the feedback system is given below.

Theorem A.7 (Passivity theorem) *Consider the* input-output *system depicted in Fig. A.2. Suppose there exists constants $\delta_{i1}, \delta_{o1}, \delta_{i2}, \delta_{o2}, \beta_1, \beta_2$ such that*

$$\langle e_1 \mid y_1 \rangle_T \geq \delta_{i1} \|e_1\|_{2T}^2 + \delta_{o1} \|y_1\|_{2T}^2 + \beta_1$$

and

$$\langle e_2 \mid y_2 \rangle_T \geq \delta_{i2} \|e_2\|_{2T}^2 + \delta_{o1} \|y_2\|_{2T}^2 + \beta_2$$

for all $e_1, e_2 \in \mathcal{L}_{2e}$ and all $T \geq 0$. Then, $\Sigma : u \mapsto y$ is $\mathcal{L}_2$–stable provided

$$\delta_{i1} + \delta_{o2} > 0, \quad \delta_{o1} + \delta_{i2} > 0$$

$\square$

Several interesting criteria can be obtained as special cases of this theorem. For instance, that input (or output) strict passivity of the subsystems ensures $\mathcal{L}_2$–stability. Notice though that the theorem does not require that *both* operators be passive, since the excess of passivity of one of them can compensate for the lack of it on the other one.

5   Internal stability and passivity

It is rather clear that input-output stable systems are also internally stable, i.e. stable in the sense of Lyapunov, if some observability properties are satisfied [46, 103]. To formalize this relationship we need the following definition:

Definition A.8 (Zero-state observability and detectability) *A state-space system* $\dot{x} = f(x)$, $x \in \mathbb{R}^n$ *is* zero-state observable *from the output* $y = h(x)$, *if for all initial conditions* $x(0) \in \mathbb{R}^n$ *we have* $(y(t) \equiv 0 \Rightarrow x(t) \equiv 0)$. *It is* zero-state detectable *if* $(y(t) \equiv 0 \Rightarrow \lim_{t\to\infty} x(t) = 0)$.

To simplify the presentation we will specialize in this section to systems *affine* in u, that is

$$\Sigma_a \ : \ \begin{cases} \dot{x} = f(x) + g(x)u, \ x(0) = x_0 \in \mathbb{R}^n \\ y = h(x) \end{cases}$$

with $g(x)$ and $n \times m$ matrix.

The main result of this section is due to Hill and Moylan [103]. We present here a version given in [272].

Proposition A.9 (Proposition 3.2.3 of [272]) *Suppose the system* Σ_a *is OSP with positive semidefinite storage function* $\mathcal{H} \geq 0$.

(i) *If* Σ_a *is zero-state observable then* $\mathcal{H}(x) > 0$ *for all* $x \neq 0$.

(ii) *If* $\mathcal{H}(x) > 0$ *for all* $x \neq 0$, $\mathcal{H}(0) = 0$ *and* Σ_a *is zero-state detectable, then* $x = 0$ *is a locally asymptotically stable equilibrium of* $\dot{x} = f(x)$. *Furthermore, if* $\mathcal{H}$ *is radially unbounded, the stability is global.* □

For the feedback system of Fig. A.2 we have the following state–space version of the passivity theorem.

Proposition A.10 (Proposition 3.2.5 of [272])

(i) *Suppose* Σ_1 *and* Σ_2 *are passive with storage functions which have strict local minima in* $x_1 = \bar{x}_1$ *and* $x_2 = \bar{x}_2$, *respectively. Then* $(\bar{x}_1, \bar{x}_2)$ *is a stable equilibrium of the feedback system, without external inputs, i.e.,* $u_1 = u_2 = 0$.

(ii) *Suppose that* Σ_1 *and* Σ_2 *are output strictly passive and zero-state detectable, and the corresponding storage functions are proper, have a global and unique minimum in* $x_1 = 0$, *respectively* $x_2 = 0$. *Then* $(0, 0)$ *is a globally asymptotically stable equilibrium of the feedback system without external inputs.*

□

6 The Kalman–Yakubovich–Popov lemma

Lemma A.11 (The Kalman–Yakubovich–Popov lemma.) *Consider a stable LTI system with minimal state space representation*

$$\dot{x} = Ax + Bu, \quad x(0) = x_0$$
$$y = Cx$$

where $x \in \mathbb{R}^n$ and $u, y \in \mathbb{R}^m$, and the corresponding transfer matrix $H(s) = C(sI - A)^{-1}B$.

The following statements are equivalent

(i) *$H(s)$ is positive real, that is, all poles are on the open left–hand plane and those on the $j\omega$ axis are simple with Hermitian positive definite residues, and $\forall \omega \in [0, \infty)$ (which is not a pole of $H(s)$) we have*

$$H(j\omega) + H^\top(-j\omega) \geq 0, \ \forall \ \omega \in \mathbb{R};$$

(ii) *There exists matrices $P = P^\top > 0$ and $Q = Q^\top \geq 0$ such that*

$$A^\top P + PA = -Q$$
$$PB = C^\top \tag{A.4}$$

(iii) *The operator $H : u \mapsto y$ is passive with storage function $V(x) = \frac{1}{2}x^\top Px$.*

Also, the statements below are equivalent:

(i') *$H(s)$ is strictly positive real, meaning that $H(s - \epsilon)$ is positive real for some $\epsilon > 0$, that is, all poles are on the closed left–hand plane*

$$H(j\omega) + H^\top(-j\omega) > 0, \ \forall \ \omega \in \mathbb{R};$$

and

$$\lim_{\omega \to \infty} \omega^2[H(j\omega) + H^\top(-j\omega)] > 0;$$

(ii') *There exists matrices $P = P^\top > 0$ and $Q = Q^\top > 0$ such that (A.4) holds;*

(iii') *The operator $H : u \mapsto y$ is passive and furthermore, for all $t \geq 0$, the following identity holds*

$$< u|y >_t = \frac{1}{2}\|x^\top Qx\|_{2t}^2 + V(x(t)) - V(x(0))$$

□

A proof of this lemma can be found in [127] and the references therein. We close this appendix with the following basic result.

Lemma A.12 ([69, 212]) *Let $y = G(p)u$, where $G(p)$ is an $n \times m$ strictly proper, exponentially stable transfer function and $p = \frac{d}{dt}$. Then $u \in \mathcal{L}_2^n$ implies that $y \in \mathcal{L}_2^n \cap \mathcal{L}_\infty^n$, $\dot{y} \in \mathcal{L}_2^n$, $y(t)$ is continuous, and $y(t) \to 0$ as $t \to \infty$. If in addition, $u(t) \to 0$ as $t \to \infty$, then $\dot{y}(t) \to 0$.*   □

Appendix B

Derivation of the Euler-Lagrange equations

In this appendix the EL equations of motion are derived from an integral principle. This fundamental equations are used throughout the text in this book for the formulation of the dynamic equations of motion of the physical systems we deal with.

The first section of the appendix covers the "Lagrangian formulation", while the last one gives some connections with the "Hamiltonian formulation".

caveat A complete derivation of the EL equations of motion, in the generality needed in this book (i.e., covering mechanical, electrical and electromechanical systems), would take us to far away from our main objective of control design. The material is, therefore, not meant to be self–contained, and is presented in a form that highlights the main ideas, rather than mathematical rigor. Detailed derivations of the EL equations may be found in [179], [285] and [97].

1 Generalized coordinates and velocities

The configuration of a physical system is generally described by a set of quantities called coordinates. For a single mass particle in space, the coordinates needed to describe the configuration could be a three dimensional vector of quantities describing the position of the particle relative to some reference point in a coordinate system (e.g. $x-$, $y-$, and $z-$coordinates in a Cartesian coordinate system).

From a dynamic point of view, a physical system can be considered as consisting of several particles, with interconnections between particles, giving constraints on the behavior of the system, and relations between the coordinates which would be independent without the interconnections.

Since a physical system is often an isolated part of a much larger system, the surroundings will also impose constraints on the behavior. In a more general setting, a physical system can be considered as consisting of a set of subsystems, with interconnections giving internal constraints and possibly dependence between coordinates, and additional constraints given by the environment.

The choice of coordinates for a physical system with many interconnected subsystems is often somewhat arbitrary, but in general a subset of the total set of coordinates can be associated with each of the subsystems.

For systems in static equilibrium, the configuration coordinates describe the system completely, but if the system is *dynamic*, an extra set of dynamic variables, which gives information about how the configuration of the system is changing, is needed to describe the system. The first derivatives of the coordinates (the velocities) can be chosen as this extra set of dynamic variables, or another set of dynamic variables, called the momenta, can alternatively be chosen. The momenta and the velocities are said to be associated variables[1].

When considering a system as a set of interconnected subsystems, it may be possible that the constraints of the system allows for a reduction of the number of variables used for describing the system, because the constraints give relations between the various variables which are not independent. This fact motivates us to distinguish two different types of constraints, *holonomic* and *non-holonomic*.

Holonomic constraints are expressed as relations between coordinates, or relations on differential form (relations between velocities) which can be integrated to yield relations between coordinates. Holonomic constraints are expressible in the form

$$f_j(q_1, \ldots, q_n; t) = 0, \quad j = 1, \ldots, m \tag{B.1}$$

where q_i, $i = 1, \ldots, n$ are the coordinates of the system.

The m relations above can then be used to reduce the number of coordinates to $(n - m)$.

For a system which has only holonomic constraints, it is possible to select a set of independent coordinates such that the constraint equations are no longer needed. This means that if there are n coordinates and m holonomic constraints, a set of $(n - m)$ independent coordinates can be derived. These coordinates together with their associated dynamic variables (velocities or momenta), constitute a set of $2(n-m)$ variables, which describe the dynamic motion of the system uniquely. The minimum number of $N = (n-m)$ that can be found is the *degrees of freedom* of the system, and when a system is described by a set of coordinates which eliminate the constraints, these coordinates are generally called the *generalized coordinates* of the system, with associated generalized dynamic variables.

---

[1] The coordinates could also have a set of associated variables, called forces.

Constraints which can not be the written on the form (B.1), as relations between the coordinates, are called non-holonomic. An example is a constraint of the form (differential equation)

$$f(\dot{q}_1, \ldots, \dot{q}_n; t) = 0$$

which can not be integrated to the form (B.1). This is the case for electric machines with commutators.

Another type of non-holonomic constraints can be written on the form

$$f(q_1, \ldots, q_n; t) \leq 0$$

which can be used to describe the motion of gas particles within a container, where the walls limit the motion of the particles.

For physical systems with non-holonomic constraints, it is not possible to find a set of N generalized coordinates, with N the number of degrees of the system. The number of coordinates must be equal to the degrees of freedom plus the number of non-holonomic constraints. This means that there will coordinates which are not independent. In this text, the derivation of the EL equation will be based on the use of generalized coordinates, an it will not apply to non-holonomic systems. There are other methods for the derivation of the dynamic equations of motion for non-holonomic systems, but this topic will not be considered here, and the interested reader should consult [191].

Another case that is sometimes encountered in the modeling of physical systems, is the use of *quasi coordinates*. It is possible to have many different sets of *true coordinates* for instance $q_1, \ldots, q_N$ and $q'_1, \ldots, q'_N$, but with this type of coordinates, it is always possible to find linear relations between the different sets, expressed as

$$q_i = \sum_{j=1}^{N} a_{ij} q'_j, \quad i = 1, \ldots, N$$

If it is not possible to find such relations between the different sets of coordinates, but instead relations of the form

$$\dot{q}_i = \sum_{j=1}^{N} b_{ij} \dot{q}'_j, \quad i = 1, \ldots, N$$

with[2] ($b_{ij} \neq \partial q'_i / \partial q_j$) or ($(\partial b_{ij} / \partial q_k) \neq \partial b_{ik} / \partial q_j$)) can be found, then the coordinates $q'_1, \ldots, q'_N$ are called *quasi coordinates*. An example of this is the relation between rotor currents and terminal currents in a commutator machine, or currents in an

---

[2]The given relations between the coefficients ensure that the differential equations are not integrable.

AC machine, which has been rotated from their natural frame, where they can be integrated to physical charges, to another frame of reference. In both these cases, integration of the currents will result in quasi coordinates without any physical meaning.

The main problem with quasi coordinates in the modeling of electromechanical systems, is that sometimes the coordinates are not explicitly used in the derivation of the model, only the velocities. Thus, a set of velocities which are not the true velocities can be used in the modeling procedure by mistake, but this is never discovered, since there is no need for evaluating the coordinates.

In the following, the use of quasi coordinates in the derivation of the EL equations will not be considered. In the systems considered in this book, the generalized velocities will be chosen based on physical interpretation, and they can be integrated to give physically interpretable true coordinates.

To summarize the above, the dynamic motion of a physical holonomic system with N degrees of freedom can be completely described by a set of generalized independent coordinates $q_1(t), \ldots, q_N(t)$ describing the configuration of the system as a function of time, and a set of N dynamic variables, given either as generalized velocities $\dot{q}_1(t), \ldots, \dot{q}_N(t)$, or generalized momenta $p_1(t), \ldots, p_N(t)$. Thus, the *state*[3] of a dynamic system can be presented in a $2N$ dimensional space. The coordinates $q_1(t), \ldots, q_N(t)$ evolve in the *configuration space* of the system, and the space formed of the coordinates $q_1(t), \ldots, q_N(t), p_1(t), \ldots, p_N(t)$ is called the $2N$-dimensional *phase space* of the system.

Once a choice of independent generalized coordinates has been made a state function, that characterizes the system, must be defined. Although other state functions can be chosen, (e.g., the Hamiltonian), we will select here the Lagrangian, that we denote $\mathcal{L}(q, \dot{q}, t)$. This choice is motivated by several reasons: first, the fact that the resulting equations of motion for the electrical portion of our systems will be identical to those obtained from Kirchhoff's laws, which is an appealing features for electrical engineers. Second, as we have seen in Chapter **2**, the EL formalism is more suitable to reveal the workless forces, a fundamental step in our PBC synthesis.

We should mention that the problem of selecting the proper set of independent variables in a dynamical system always presents some difficulty. For instance, the Lagrangian used in classical mechanics is defined as the difference between the kinetic and the potential energy, that is

$$\mathcal{L}(q, \dot{q}, t) = \mathcal{T}(q, \dot{q}, t) - \mathcal{V}(q, t)$$

It turns out that to treat electromechanical systems this definition is not sufficiently general. To handle these cases we should use instead the kinetic *co-energy*. Although these functions coincide for linear systems, they will differ in general.

---

[3]The state of a *static* system can be completely described by its N generalized coordinates, since there are no dynamic changes.

2   Hamilton's principle

The path of a dynamic system will evolve in the configuration space from one configuration $q(t_1) = [q_1(t_1), \dots, q_N(t_1)]^\top$ at time $t = t_1$, to a new configuration $q(t_2) = [q_1(t_2), \dots, q_N(t_2)]^\top$ at time $t = t_2$. The equations of motion governing this change of configuration can be found from different principles.

They can be derived from basic physical relations like force laws, which form a set of differential principles, concerning incremental changes in the system.

An alternative way is the use of a *variational method*, based on integral principles. These principles relates to the gross motion of the system.

Hamilton's principle is considered to be one of the most important integral principles. It can be derived from d'Alembert's principle and the principle of virtual work, but it is a more general principle than the former, and proves to be significant for more than just mechanical systems.

Hamilton's principle states that the *actual dynamic path of a system described by a state function $\mathcal{L}(q, \dot{q}, t)$, from time t_1 to t_2 is such that the line integral*

$$I = \int_{t_1}^{t_2} \mathcal{L}(q, \dot{q}, t) dt \tag{B.2}$$

is an extremum for this path.

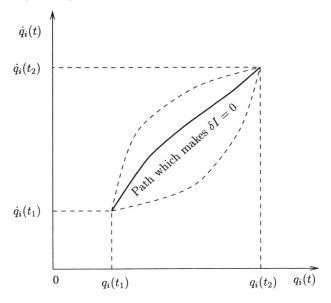

Figure B.1: The path of motion according to Hamilton's principle.

Thus, with Hamilton's principle we have that the derivation of the equations of motion has been reduced to an optimization problem. The main mathematical tool to solve this problem is the calculus of variations, which may be viewed as an extension of ordinary calculus, which is used to find extrema of functions, to the case when we are interested in the extrema of functionals, (which are functions of functions, for instance the integral (B.2) above).

3 From Hamilton's principle to the EL equations

Using standard techniques of calculus of variations, it can be shown that for the dynamic path to be and extremum, the first variation of the line integral I must be equal to zero, that is

$$\delta I = \delta \int_{t_1}^{t_2} \mathcal{L}(q, \dot{q}, t) dt = 0$$

subject to the end constraints

$$\delta q(t_1) = 0 \qquad \delta q(t_2) = 0$$

where δ means a *time-independent variation*, analogous to the differential used in ordinary differential calculus.[4]

Evaluating this variation, and setting it equal to zero, yields

$$\delta I = \int_{t_1}^{t_2} \sum_{i=1}^{N} \left\{ \frac{\partial \mathcal{L}}{\partial q_i} - \frac{d}{dt} \frac{\partial \mathcal{L}}{\partial \dot{q}_i} \right\} \delta q_i dt = 0, \; i = 1, \ldots, N \qquad \text{(B.3)}$$

Now, the extremum $\delta I = 0$ must hold for all variations δq_i of any coordinate q_i. Furthermore, since the N coordinates q_i are independent, the only way one can satisfy (B.3) is by setting the term in brackets equal to zero for all i, yielding

$$\frac{\partial \mathcal{L}}{\partial q_i} - \frac{d}{dt} \left(\frac{\partial \mathcal{L}}{\partial \dot{q}_i} \right) = 0, \; i = 1, \ldots, N \qquad \text{(B.4)}$$

Equations (B.4) are known as the (first) differential equations of EL[5] or in short, "EL equations". This equations yield a set of ordinary differential equations that describe the dynamics of any *conservative* system with independent coordinates.

---

[4]For readers unfamiliar with calculus of variations, it is useful to think of the problem of finding an extremum as one of finding the points of zero slope of I, which are obtained by setting some differential of I equal to zero.

[5]Euler obtained this equation heuristically in 1736, and Lagrange obtained it (incorrectly) in 1755. The correct derivation was done by P. du Bois-Reymond in 1879.

4   EL equations for non-conservative systems

If there are non-conservative forces acting on the system, the derivation of the EL equations above can sometimes be carried through with a Lagrangian which has been modified to take the nonconservative forces into account. We consider in the book two types of non–conservative forces: controls u and external disturbances Q_ζ, which are independent of the generalized coordinates and velocities, and dissipation. The former are easily incorporated in a nonconservative potential function

$$\mathcal{V}_{NC}(q, t) = -(u + Q_\zeta)^\top q$$

while the latter are added in a nonconservative kinetic potential, which is the Rayleigh dissipation function $\mathcal{F}(\dot{q})$.

A nonconservative Lagrangian can now be formulated as

$$\mathcal{L}_{NC}(q, \dot{q}, t) \;=\; \mathcal{T}(q, \dot{q}) + \int_0^t \mathcal{F}(\dot{q}) dt - (\mathcal{V}(q) + \mathcal{V}_{NC}(q, t))$$

and the equations of motion can be derived from Hamilton's principle with this new Lagrangian, giving

$$\frac{d}{dt}\left[\frac{\partial \mathcal{L}_{NC}}{\partial \dot{q}}\right] - \frac{\partial \mathcal{L}_{NC}}{\partial q} = 0$$

which upon evaluation yields

$$\frac{d}{dt}\left[\frac{\partial \mathcal{L}}{\partial \dot{q}}\right] - \frac{\partial \mathcal{L}}{\partial q} = u + Q_\zeta - \frac{\partial \mathcal{F}}{\partial \dot{q}}$$

Throughout the book we will refer to this set of differential equations as *EL equations*.

5   List of generalized variables

For the systems considered in this book, the generalized coordinates and velocities, with associated forces and momenta, will be chosen as shown in Table B.1.

6   Hamiltonian formulation

In the Hamiltonian formulation we seek to describe the motion in terms of first order differential equations. This formulation is viewed here, for simplicity, as a *consequence* of the EL equations and a simple change of variables. However, it is well–known that a direct Hamiltonian formulation is possible for many systems, and that there are

Generalized variables:	Electro-magnetic part	Mechanical part
Generalized coordinates, q_i	electric charges	mechanical displacements
Generalized velocities, $\dot{q}_i$	electric currents	mechanical velocities
Generalized forces, $\mathcal{Q}_i$	negative electric voltages	mechanical forces
Generalized momenta, p_i, λ_i	flux linkages	mechanical momenta

Table B.1: Definition of variables.

even systems which do not easily admit the EL formulation but admit a Hamiltonian one.

We first observe that the EL equations (B.5) provides the time derivative of $\frac{\partial \mathcal{L}_{NC}}{\partial \dot{q}}$, leading to a second order differential equation in $\dot{q}$. With the goal of obtaining first order differential equations, define the conjugate variables p (called the generalized conjugate momenta)

$$p \triangleq \frac{\partial \mathcal{L}_{NC}(q, \dot{q}, t)}{\partial \dot{q}} \tag{B.5}$$

Assume that this relation allows for a global, unique solution for $\dot{q}$ in terms of the remaining quantities $\dot{q} = g(q, p, t)$. (This is true under some rather weak conditions of strict convexity and "quadratic boundedness" of $\mathcal{L}_{NC}$ with respect to $\dot{q}$.) Let us now define a useful scalar quantity

$$\mathcal{G}(q, \dot{q}, t) \triangleq \frac{\partial \mathcal{L}_{NC}^{\mathsf{T}}(q, \dot{q}, t)}{\partial \dot{q}} \dot{q} - \mathcal{L}_{NC}(q, \dot{q}, t)$$

Define now the Hamiltonian as

$$\mathcal{H}(q, p, t) \triangleq \mathcal{G}(q, g(q, p, t), t)$$

This is, of course, the well known Legendre transformation. Notice that, by definition $\mathcal{G}$ and $\mathcal{H}$ are numerically equal, but are expressed with *different* arguments.

We now proceed from the second order differential equations of (B.5) to first order state space form. The development is quite standard; we summarize it for completeness and to stress the role of the Rayleigh dissipation function. From the definitions of $g, \mathcal{G}$ and $\mathcal{H}$ and some lengthy, but straightforward derivations, we get

$$\begin{bmatrix} \dot{q} \\ \dot{p} \end{bmatrix} = \begin{bmatrix} 0 & I \\ -I & 0 \end{bmatrix} \begin{bmatrix} \frac{\partial \mathcal{H}(q,p,t)}{\partial q} \\ \frac{\partial \mathcal{H}(q,p,t)}{\partial p} \end{bmatrix} - \begin{bmatrix} 0 \\ \frac{\partial \mathcal{F}(q,\dot{q},t)}{\partial \dot{q}} \end{bmatrix}$$

where we choose to maintain $\mathcal{F}$ in terms of $\dot{q}$ because it typically relates more naturally to velocities than to momenta.

Before closing this section let us evaluate the rate of change of the Hamiltonian along solution trajectories, this gives

$$\frac{d}{dt}\left(\mathcal{H}(q,p,t)\right) = -\dot{q}^\top \frac{\partial \mathcal{F}(q,\dot{q},t)}{\partial \dot{q}} - \frac{d}{dt}\mathcal{L}_{NC}(q,\dot{q},t)$$

which is our usual power balance equation.

Appendix C

Background material

The definitions below concern the minima of a vector function. Let $f : \mathbb{R}^n \mapsto \mathbb{R}$ be a smooth function then we define the following [175].

Definition C.1 (Critical Point) *A point $x^* \in \mathbb{R}^n$ is called* critical point *of $f(x)$ if and only if $\frac{\partial f}{\partial x}(x^*) = 0$*

Definition C.2 (Minimum) *A point $x^* \in \mathbb{R}^n$ is a* local minimum *of $f(x)$ if there is a neighbourhood B_δ of x^* with $0 < \delta < \infty$ such that $f(x) \geq f(x^*)$ for all $x \in B_\delta$.*

Definition C.3 (Absolute or global minimum) *A point $x^* \in \mathbb{R}^n$ is an* absolute or global minimum *of $f(x)$ if $f(x) \geq f(x^*)$ for all $x \in \mathbb{R}^n$.*

Definition C.4 (Unique minimum) *A point $x^* \in \mathbb{R}^n$ is an* unique minimum *of $f(x)$ if there are no other local minima of $f(x)$ in $\mathbb{R}^n$.*

Definition C.5 (Strict minimum) *A point $x^* \in \mathbb{R}^n$ is a* strict local minimum *of $f(x)$ if there exists a neighbourhood B_δ of x^* with $0 < \delta \leq \infty$ such that $f(x) > f(x^*)$ for all $x \in B_\delta$.*

Theorem C.6 [Mean value] *Assume that $f : \mathbb{R}^n \mapsto \mathbb{R}$ is differentiable at each point x of an open set $S \in \mathbb{R}^n$. Let x and y be two points of S such that the line segment $L(x, y) \subset S$. Then there exists a point z of $L(x, y)$ such that*

$$f(y) - f(x) = \left(\frac{\partial f}{\partial x}(z) \right) (y - x)$$

$\square$

Lemma C.7 [175] *Let $f(x) : \mathbb{R}^n \mapsto \mathbb{R}$ and $B_\sigma \subseteq \mathbb{R}^n$, such that*

(i) $f(0) = 0$

(ii) $\frac{\partial f}{\partial x}(0) = 0$

(iii) $\frac{\partial^2 f}{\partial x^2} > I_n \varepsilon > 0$, $\varepsilon > 0$, for all $x \in B_\sigma$

then $f(x)$ has a unique strict minimum at the origin, locally in B_σ. If $B_\sigma = \mathbb{R}^n$ then the minimum is global and unique.   □

The lemma C.7 above appears too strong in some cases. Notice that a necessary condition for (iii) is that $f(x) = \mathcal{O}(\|x\|^2)$ in the ball B_σ, specially when globality is to be assured it is desirable to find milder conditions. The lemma below establishes weaker sufficient conditions for a function $f(x)$ to have a global minimum at the origin.

Lemma C.8 [162]. *Let* $f(x) : \mathbb{R}^n \to \mathbb{R}$ *be a* $\mathcal{C}^1$ *function. Assume,*

1. $f(x) > 0$, for all $x \in \mathbb{R}^n$, $x \neq 0$ and $f(0) = 0$

2. $\left\| \frac{\partial f}{\partial x}(x) \right\| > 0$, for all $x \neq 0 \in \mathbb{R}^n$

Then the function $f(x)$ *is globally positive definite with an unique and global minimum at* $x = 0$.   □

Condition 1 implies that $f(x)$ is positive definite with 0, a strict global minimum. Nevertheless, it is important to remark that this condition alone does not imply the uniqueness of the minimum. Condition 2 implies that 0 is the only critical point, hence that 0 is also an unique minimum of $f(x)$.

Appendix D

Proofs

1 Proofs for the PI^{2}D controller

1.1 Properties of the storage $\mathcal{H}_3(\tilde{q}, \dot{q}, \vartheta)$

A   Positive definiteness of $\mathcal{H}_3(\tilde{q}, \dot{q}, \vartheta)$

Let us partition $\mathcal{H}_3$ as $\mathcal{H}_3 = W_1 + W_2 + W_3 + W_4$ where

$$W_1 = \frac{1}{8}\dot{q}^\top D\dot{q} + \frac{1}{8}\tilde{q}^\top K'_p\tilde{q} + \varepsilon\tilde{q}^\top D\dot{q}, \tag{D.1}$$

$$W_2 = \frac{1}{8}\tilde{q}^\top K'_p\tilde{q} + U_g + \tilde{q}^\top K'_p(q_d - \delta) + \frac{1}{2}(q_d - \delta)^\top K'_p(q_d - \delta) + c_1, \tag{D.2}$$

$$W_3 = \frac{1}{8}\dot{q}^\top D\dot{q} + \frac{1}{4}\vartheta^\top K_d B^{-1}\vartheta - \varepsilon\vartheta^\top D\dot{q}, \tag{D.3}$$

$$W_4 = \frac{1}{4}\tilde{q}^\top K'_p\tilde{q} + \frac{1}{4}\vartheta^\top K_d B^{-1}\vartheta + \frac{1}{4}\dot{q}^\top D\dot{q}. \tag{D.4}$$

Under the conditions of proposition 3.21, W_2 is positive definite [125]. W_1 is positive definite if

$$\frac{1}{2}\sqrt{\frac{k'_{p_m}}{d_M}} > \varepsilon$$

while

$$2\sqrt{\frac{k_{d_m}}{b_M d_M}} > \varepsilon \tag{D.5}$$

insures W_3 to be positive definite. Thus, V is positive definite for ε sufficiently small.

B Time derivative of $\mathcal{H}_3(\tilde{q}, \dot{q}, \vartheta)$

Using the properties **P2.1 - P2.2** and after some straightforward calculations one obtains that the time derivative of $\mathcal{H}_3$ along the trajectories of (3.64), (3.65) is bounded by

$$
\dot{\mathcal{H}}_3 \leq -\gamma_1 \|\tilde{q}\|^2 - \gamma_2 \|\dot{q}\|^2 - \gamma_3 \|\vartheta\|^2 - \varepsilon \left[\frac{b_m d_m}{4} - d_M - k_c \|\vartheta\| - k_c \|\tilde{q}\| \right] \|\dot{q}\|^2 + z^{\top} \zeta
$$

$$
- \frac{\varepsilon}{2} \left\{ \left[\begin{array}{c} \|\tilde{q}\| \\ \|\vartheta\| \end{array} \right]^{\top} Q_1 \left[\begin{array}{c} \|\tilde{q}\| \\ \|\vartheta\| \end{array} \right] - \left[\begin{array}{c} \|\vartheta\| \\ \|\dot{q}\| \end{array} \right]^{\top} Q_2 \left[\begin{array}{c} \|\vartheta\| \\ \|\dot{q}\| \end{array} \right] \right\} - \left[\frac{k_{d_m} a_m}{4 b_M} - \varepsilon k_{d_m} \right] \|\vartheta\|^2
$$

$$\tag{D.6}$$

where we defined

$$
Q_1 \triangleq \left[\begin{array}{cc} k'_{p_m} - 2k_g & -k'_{p_M} - k_{d_m} - k_g \\ -k'_{p_M} - k_{d_m} - k_g & \frac{k_{d_m} a_m}{2 \varepsilon b_M} \end{array} \right], \quad Q_2 \triangleq \left[\begin{array}{cc} \frac{k_{d_m} a_m}{2 \varepsilon b_M} & -a_M d_M \\ -a_M d_M & b_m d_m \end{array} \right].
$$

and the constants

$$
\gamma_1 \triangleq \frac{\varepsilon b_m d_m}{2}, \quad \gamma_2 \triangleq \frac{\varepsilon k'_{p_m}}{2}, \quad \gamma_3 \triangleq \frac{k_{d_m} a_m}{4 b_M}. \tag{D.7}
$$

We derive now sufficient conditions for $\dot{\mathcal{H}}_3$ to be negative semidefinite in $(\tilde{q}, \dot{q}, \vartheta)$ with $\zeta \equiv 0$. If

$$
\frac{(k'_{p_m} - 2k_g) k_{d_m} a_m}{2 b_M \left[k'_{p_M} + k_{d_M} + k_g \right]^2} > \varepsilon \tag{D.8}
$$

we have $Q_1 > 0$. In a similar way, (for all $\frac{b_m}{b_M} < \infty$), we have that $Q_2 > 0$ if

$$
\frac{k_{d_m} a_m d_m}{2 [a_m d_M]^2} > \varepsilon. \tag{D.9}
$$

The third right hand term of (D.6) is negative if[1]

$$
\frac{1}{2 k_c} \left[\frac{1}{4} b_m d_m - d_M \right] > \|x\| \tag{D.10}
$$

where the left hand side is positive due to (3.76). Finally, the last term in (D.6) is negative if

$$
\frac{k_{d_m} a_m}{4 b_M k_{d_M}} > \varepsilon \tag{D.11}
$$

while (D.8), (D.9) and (D.11) are satisfied for ε sufficiently small. Therefore, $\dot{\mathcal{H}}_3$ is locally negative semidefinite with $\zeta \equiv 0$.

---

[1] Observe that we give this condition in terms of the original state x, instead of x'. This in order to derive the domain of attraction (and prove the semiglobal stability claim that requires ε arbitrarily small) in the coordinates x.

1.2 Lyapunov stability of the PI²D

A Domain of attraction

To define the *domain of attraction* we will first find some positive constants α_1, α_2 such that

$$\alpha_1\|x\|^2 \leq V(x) \leq \alpha_2\|x\|^2. \tag{D.12}$$

Notice that

$$V \geq W_4 \geq \frac{1}{4}\left[k'_{p_m}\|\tilde{q}\|^2 + \frac{k_{d_m}}{b_M}\|\vartheta\|^2\right] + \frac{1}{4}\left[d_m\|\dot{q}\|^2 + \frac{2\varepsilon}{k_{i_M}}\|z\|^2\right].$$

To obtain the lower bound in terms of x we need the following inequality

$$\left[1 - \frac{k_{i_M}}{\varepsilon}\right]\|\tilde{v}\|^2 + \left[\frac{k_{i_m}^2}{\varepsilon^2} - \frac{k_{i_M}}{\varepsilon}\right]\|\tilde{q}\|^2 \leq \|z\|^2$$

which leads to

$$V \geq \frac{1}{4}\left\{\left[k'_{p_m} + 2\frac{k_{i_m}^2}{\varepsilon k_{i_M}} - 2\right]\|\tilde{q}\|^2 + d_m\|\dot{q}\|^2\right\} + \frac{1}{4}\left\{\frac{k_{d_m}}{b_M}\|\vartheta\|^2 + 2\left[\frac{\varepsilon}{k_{i_M}} - 1\right]\|\tilde{v}\|^2\right\}$$

so we define α_1 as

$$\alpha_1 \triangleq \frac{1}{4}\min\left\{k'_{p_m} + 2\left[\frac{k_{i_m}^2}{\varepsilon k_{i_M}} - 1\right], \frac{k_{d_m}}{b_M}, d_m, 2\left[\frac{\varepsilon}{k_{i_M}} - 1\right]\right\}.$$

In a similar manner, an upperbound on V is

$$V \leq \left\{\frac{1}{4}\left[k'_{p_m} + 2k_g\right] + \frac{1}{2}\left[\frac{\varepsilon}{2}d_M + k'_{p_M}\right]\right\}\|\tilde{q}\|^2 + \left[(\varepsilon + \frac{1}{2})d_M\right]\|\dot{q}\|^2$$

$$+ \frac{1}{2}\left[\varepsilon d_M + \frac{k_{d_m}}{b_M}\right]\|\vartheta\|^2 + \frac{\varepsilon}{2k_{i_m}}\|z\|^2.$$

Now using $K_p \triangleq K'_p + \frac{1}{\varepsilon}K_i$ we have

$$\|z\|^2 \leq \left[1 + \frac{k_{i_M}}{\varepsilon}\right]\left[\|\tilde{v}\|^2 + \frac{k_{i_M}}{\varepsilon}\|\tilde{q}\|^2\right] \leq 2[\|\tilde{q}\|^2 + \|\tilde{v}\|^2]$$

so we define

$$\alpha_2 \triangleq \max\left\{\left[(\varepsilon + \frac{1}{2})d_M\right], \frac{1}{2}\left[\varepsilon d_M + \frac{k_{d_M}}{b_m}\right], \frac{1}{4}\left[k'_{p_m} + 2k_g\right] + \frac{1}{2}\left[\frac{\varepsilon}{2}d_M + k'_{p_M}\right] + \frac{\varepsilon}{k_{i_m}}\right\}.$$

From (D.12) and (D.10) we conclude that the domain of attraction contains the set

$$\|x\| \leq c_3 \triangleq \frac{1}{2k_c}\left[\frac{1}{2}b_m d_m - d_M\right]\sqrt{\frac{\alpha_1}{\alpha_2}}. \tag{D.13}$$

2   Proof of positive definiteness of $f(\tilde{q}_p)$ defined in (3.43)

We will establish the proof that

$$f(\tilde{q}_p) \triangleq \sum_{i=1}^{n} \left\{ k_{3i} \int_0^{\tilde{q}_{p_i}} \mathrm{sat}(x_i) dx_i \right\} + V_p(\tilde{q}_p + q_{p*}) - V_p(q_{p*}) - \tilde{q}_p^{\top} \frac{\partial V_p}{\partial q_p}(q_{p*})$$

is positive definite and radially unbounded by verifying the conditions of Lemma C.8 so as to define a $k_{3_i}^{\min}$ that ensures this to be the case.

Condition 1.
To prove that $f(\tilde{q}_{p_i}) > 0$ for all $\tilde{q}_p \neq 0 \in \mathbb{R}^n$ we shall prove first that there exists $\varepsilon > 0$ such that

$$\sum_{i=1}^{n} k_{3_i} \int_0^{\tilde{q}_{p_i}} \mathrm{sat}(x) dx \geq \min_i \{k_{3i}\} \frac{\mathrm{sat}(\varepsilon)}{2\varepsilon} \|\tilde{q}_p\|^2 \quad \forall \ \|\tilde{q}_p\| < \varepsilon \qquad (\mathrm{D}.14)$$

$$\sum_{i=1}^{n} k_{3_i} \int_0^{\tilde{q}_{p_i}} \mathrm{sat}(x) dx \geq \min_i \{k_{3i}\} \frac{\mathrm{sat}(\varepsilon)}{2} \|\tilde{q}_p\| \quad \forall \ \|\tilde{q}_p\| \geq \varepsilon. \qquad (\mathrm{D}.15)$$

Let ε be a constant that satisfies inequalities (3.29) and (3.30). For the sake of clarity we consider two cases separately:

Case 1: $\|\tilde{q}_p\| < \varepsilon$

Notice that in this case we have that $|\tilde{q}_{p_i}| < \varepsilon, \ \forall \ i \in \underline{n}$, then using **P3.1** and (3.29) we get

$$\sum_{i=1}^{n} k_{3_i} \int_0^{\tilde{q}_{p_i}} \mathrm{sat}(x) dx \geq \sum_{i=1}^{n} k_{3_i} \frac{\mathrm{sat}(\varepsilon)}{2\varepsilon} \tilde{q}_{p_i}^2 \geq \min_i \{k_{3i}\} \frac{\mathrm{sat}(\varepsilon)}{2\varepsilon} \|\tilde{q}_p\|^2. \qquad (\mathrm{D}.16)$$

Case 2: $\|\tilde{q}_p\| \geq \varepsilon$

Within this case we shall consider three different cases:

case a: $|\tilde{q}_{p_i}| < \varepsilon \ \forall i \in \underline{n}$
First notice that

$$\sum_{i=1}^{n} k_{3_i} \frac{\mathrm{sat}(\varepsilon)}{2\varepsilon} \tilde{q}_{p_i}^2 \geq \min_i \{k_{3i}\} \frac{\mathrm{sat}(\varepsilon)}{2\varepsilon} \|\tilde{q}_p\|^2 \geq \min_i \{k_{3i}\} \frac{\mathrm{sat}(\varepsilon)}{2} \|\tilde{q}_p\|.$$

Using (3.29) and **P3.1**, (D.15) follows.

case b: $|\tilde{q}_{p_i}| \geq \varepsilon \;\; \forall i \in \underline{n}$

From **P3.1** and (3.30) notice that

$$\sum_{i=1}^{n} k_{3_i} \int_0^{\tilde{q}_{p_i}} \mathrm{sat}(x)dx \geq \sum_{i=1}^{n} k_{3_i} \frac{\mathrm{sat}(\varepsilon)}{2}|\tilde{q}_{p_i}| \geq \min_i \{k_{3i}\} \frac{\mathrm{sat}(\varepsilon)}{2} \sum_{i=1}^{n} |\tilde{q}_{p_i}|, \tag{D.17}$$

then (D.15) easily follows observing that $\|\tilde{q}_p\| \leq \sum_{i=1}^{n} |\tilde{q}_{p_i}|$.

case c: $|\tilde{q}_{p_i}| \geq \varepsilon, \;\; |\tilde{q}_{p_j}| < \varepsilon \;\; \forall i, j \in \underline{n}, \; i \neq j$

Without loss of generality we can take $i \leq n/2$ and $1 \leq j < n/2$, then a simple analysis along the lines of cases a and b, shows that (D.15) holds as well in this case.

Now we prove that, for all $\varepsilon > 0$ there exists constants $\beta_1(\varepsilon), \beta_2(\varepsilon) \in \mathbb{R}$ such that

$$\mathcal{V}_p(q_p) - \mathcal{V}_p(q_{p*}) - \tilde{q}_p^\top \frac{\partial \mathcal{V}_p(q_p)}{\partial q_p}(q_{p*}) \geq \begin{cases} \beta_1 \|\tilde{q}_p\|^2 & \forall \; \|\tilde{q}_p\| < \varepsilon \\ \beta_2 \|\tilde{q}_p\| & \forall \; \|\tilde{q}_p\| \geq \varepsilon. \end{cases} \tag{D.18}$$

On one hand, notice that using Lemma C.7 it follows from (2.18) that

$$\mathcal{V}_p(q_p) - \mathcal{V}_p(q_{p*}) - \tilde{q}_p^\top \frac{\partial \mathcal{V}_p(q_p)}{\partial q_p}(q_{p*}) \geq -\frac{k_g}{2}\|\tilde{q}_p\|^2, \tag{D.19}$$

on the other hand, invoking the Mean Value Theorem we have that $\exists\; \xi \in \mathbb{R}^n$ such that

$$\mathcal{V}_p(q_{p*}) - \mathcal{V}_p(q_p) = \left(\frac{\partial \mathcal{V}_p(q_p)}{\partial q_p}(\xi)\right)(q_{p*} - q_p) \leq k_v\|q_{p*} - q_p\|,$$

then using (2.19) we can write

$$\mathcal{V}_p(q_p) - \mathcal{V}_p(q_{p*}) - \frac{\partial \mathcal{V}_p(q_p)}{\partial q_p}(q_{p*})^\top \tilde{q}_p \geq -2k_v\|\tilde{q}_p\|. \tag{D.20}$$

Since (D.19) and (D.20) hold for all $\tilde{q}_p \in \mathbb{R}^n$, then (D.18) holds with $\beta_1 = -\frac{k_g}{2}$ and $\beta_2 = -k_v$. We then conclude from (D.14), (D.15), and (D.18) that

$$f(\tilde{q}_p) \geq \begin{cases} \left(\min_i \{k_{3i}\} \frac{\mathrm{sat}(\varepsilon)}{2\varepsilon} - \frac{k_g}{2}\right)\|\tilde{q}_p\|^2 & \forall \; \|\tilde{q}_p\| < \varepsilon \\ \left(\min_i \{k_{3i}\} \frac{\mathrm{sat}(\varepsilon)}{2} - 2k_v\right)\|\tilde{q}_p\| & \forall \; \|\tilde{q}_p\| \geq \varepsilon. \end{cases}$$

From here it's easy to see that condition 1 is satisfied provided

$$\min_i \{k_{3i}\} \geq k_{3_i}^{\min} > \max\left\{\frac{\varepsilon k_g}{\mathrm{sat}(\varepsilon)}, \frac{4k_v}{\mathrm{sat}(\varepsilon)}\right\} \tag{D.21}$$

holds with ε as in **P3.2**, k_g and k_v defined by (2.19) and (2.18) respectively.

Condition 2.

Taking the partial derivatives of $f(\tilde{q})$ we get

$$\frac{\partial f}{\partial \tilde{q}_p}(\tilde{q}_p) = K_3 \begin{bmatrix} \text{sat}(\tilde{q}_{p_1}) \\ \text{sat}(\tilde{q}_{p_2}) \\ \vdots \\ \text{sat}(\tilde{q}_{p_n}) \end{bmatrix} + \frac{\partial \mathcal{V}_p}{\partial \tilde{q}_p}(\tilde{q}_p) - \frac{\partial \mathcal{V}_p}{\partial q_p}(q_{p*}).$$

Now, taking the norm and using the triangle inequality we get

$$\left\| \frac{\partial f}{\partial \tilde{q}_p}(\tilde{q}_p) \right\| \geq \left\| K_3 \begin{bmatrix} \text{sat}(\tilde{q}_{p_1}) \\ \text{sat}(\tilde{q}_{p_2}) \\ \vdots \\ \text{sat}(\tilde{q}_{p_n}) \end{bmatrix} \right\| - \left\| \frac{\partial \mathcal{V}_p}{\partial \tilde{q}_p}(\tilde{q}_p) - \frac{\partial \mathcal{V}_p}{\partial q_p}(q_{p*}) \right\|. \tag{D.22}$$

On one hand, from (2.18), (2.19) and using the Mean Value Theorem we have that for all $\varepsilon > 0$

$$-\left\| \frac{\partial \mathcal{V}_p}{\partial \tilde{q}_p}(\tilde{q}_p) - \frac{\partial \mathcal{V}_p}{\partial q_p}(q_{p*}) \right\| \geq \begin{cases} -k_g \|\tilde{q}_p\| & \text{if } \|\tilde{q}_p\| < \varepsilon \\ -2k_v & \text{if } \|\tilde{q}_p\| \geq \varepsilon \end{cases}$$

and on the other hand since K_3 is diagonal and using **P3.2**, we obtain

$$\left\| K_3 \begin{bmatrix} \text{sat}(\tilde{q}_{p_1}) \\ \text{sat}(\tilde{q}_{p_2}) \\ \vdots \\ \text{sat}(\tilde{q}_{p_n}) \end{bmatrix} \right\| \geq \begin{cases} \min_i \{k_{3i}\} \, \frac{\text{sat}(\varepsilon)}{\varepsilon} \|\tilde{q}_p\| & \text{if } \|\tilde{q}_p\| < \varepsilon \\ \min_i \{k_{3i}\} \, \text{sat}(\varepsilon) & \text{if } \|\tilde{q}_p\| \geq \varepsilon. \end{cases}$$

Thus, we are able to write

$$\left\| \frac{\partial f}{\partial \tilde{q}_p}(\tilde{q}_p) \right\| \geq \begin{cases} \left(\min_i \{k_{3i}\} \, \frac{\text{sat}(\varepsilon)}{\varepsilon} - k_g \right) \|\tilde{q}_p\| & \text{if } \|\tilde{q}_p\| < \varepsilon \\ (\min_i \{k_{3i}\} \, \text{sat}(\varepsilon) - 2k_v) & \text{if } \|\tilde{q}_p\| \geq \varepsilon \end{cases}$$

which happens to hold provided (D.21) is satisfied.

In the case of $\text{sat}(x) = \tanh(x)$ we have that **P3.2** is true with $\varepsilon \triangleq \frac{4k_v}{k_g}$. Substitution of this ε in (D.21) implies (3.41).

3   The BP transformation

3.1   Proof of Proposition 9.20

For the proof of Proposition 9.20, the following lemma is needed.

Lemma D.1

$$\frac{dD_e(q_m)}{dq_m} \neq 0 \;\Rightarrow\; U + U^\top = 0 \tag{D.23}$$

$\square$

Proof. Note that the differential equation (9.31) has the unique solution $D_e(q_m) = \mathbf{e}^{\,Uq_m} D_e(0)\,\mathbf{e}^{\,-Uq_m}$ [114].

$$
\begin{aligned}
D_e(q_m) = D_e^\top(q_m) \;\Rightarrow\;& \mathbf{e}^{\,Uq_m} D_e(0)\,\mathbf{e}^{\,-Uq_m} = \mathbf{e}^{\,-U^\top q_m} D_e(0)\,\mathbf{e}^{\,U^\top q_m} \\
\Rightarrow\;& \mathbf{e}^{\,U^\top q_m}\,\mathbf{e}^{\,Uq_m} D_e(0) = D_e(0)\,\mathbf{e}^{\,U^\top q_m}\,\mathbf{e}^{\,Uq_m} \\
\Rightarrow\;& \mathbf{e}^{\,U^\top q_m}\,\mathbf{e}^{\,Uq_m} = I \\
\Rightarrow\;& U^\top + U = 0
\end{aligned}
$$

The third implication follows from the fact that U and $D_e(q_m)$ do not commute, unless $\frac{dD_e(q_m)}{dq_m} = 0$, see (9.31), and this implies that $f(U) = \mathbf{e}^{\,U^\top q_m}\,\mathbf{e}^{\,Uq_m}$ and $D_e(q_m)$ cannot commute, unless $f(U) \equiv I$.

$\blacksquare$

Proof of Proposition 9.20
From (9.30) it follows that

$$
\begin{aligned}
\dot{q}_e &= \mathbf{e}^{\,Uq_m} P_1^{-1} \dot{z}_e \\
\ddot{q}_e &= \mathbf{e}^{\,Uq_m} P_1^{-1} \ddot{z}_e + U\,\mathbf{e}^{\,Uq_m} P_1^{-1} \dot{q}_m \dot{z}_e
\end{aligned}
$$

Inserting these two equations into (9.6), and multiplying from the left by $\mathbf{e}^{\,-Uq_m}$ results in

$$
\begin{aligned}
&\mathbf{e}^{\,-Uq_m} D_e(q_m)\,\mathbf{e}^{\,Uq_m} P_1^{-1} \ddot{z}_e + \mathbf{e}^{\,-Uq_m} D_e(q_m) U\,\mathbf{e}^{\,Uq_m} P_1^{-1} \dot{q}_m \dot{z}_e \\
&+ \mathbf{e}^{\,-Uq_m} W_1(q_m)\dot{q}_m\,\mathbf{e}^{\,Uq_m} P_1^{-1} \dot{z}_e + \mathbf{e}^{\,-Uq_m} W_2(q_m)\dot{q}_m \\
&+ \mathbf{e}^{\,-Uq_m} R_e\,\mathbf{e}^{\,Uq_m} P_1^{-1} \dot{z}_e = \mathbf{e}^{\,-Uq_m} M_e u
\end{aligned} \tag{D.24}
$$

From $D_e(q_m) = \mathbf{e}^{\,Uq_m} D_e(0)\,\mathbf{e}^{\,-Uq_m}$, and since (9.32) implies that $\mathbf{e}^{\,Uq_m} R_e = R_e\,\mathbf{e}^{\,Uq_m}$ [147], notice that

$$
\begin{aligned}
\mathbf{e}^{\,-Uq_m} D_e(q_m)\,\mathbf{e}^{\,Uq_m} &= D_e(0) \tag{D.25} \\
\mathbf{e}^{\,-Uq_m} D_e(q_m) U\,\mathbf{e}^{\,Uq_m} &= D_e(0) U \\
\mathbf{e}^{\,-Uq_m} W_1(q_m)\,\mathbf{e}^{\,Uq_m} &\overset{(9.31)}{=} \mathbf{e}^{\,-Uq_m} \big[U\,\mathbf{e}^{\,Uq_m} D_e(0)\,\mathbf{e}^{\,-Uq_m} \\
&\qquad - \mathbf{e}^{\,Uq_m} D_e(0) U\,\mathbf{e}^{\,-Uq_m} \big]\,\mathbf{e}^{\,Uq_m} \\
&= U D_e(0) - D_e(0) U \tag{D.26} \\
\mathbf{e}^{\,-Uq_m} R_e\,\mathbf{e}^{\,Uq_m} &= R_e \tag{D.27}
\end{aligned}
$$

In addition, it follows from (9.33) that $W_2(q_m) = \mathbf{e}^{\ U q_m} W_2(0)$. This implies that $\mathbf{e}^{\ -U q_m} W_2(q_m) = W_2(0)$, which is constant with respect to q_m. Finally, inserting (D.25–D.27) into (D.24), gives

$$D_e(0)P_1^{-1}\ddot{z}_e + U D_e(0)P_1^{-1}\dot{q}_m\dot{z}_e + W_2(0)\dot{q}_m + R_e P_1^{-1}\dot{z}_e = \mathbf{e}^{\ -U q_m} M_e u$$

For the transformed mechanical system Σ_m, it follows that

$$\dot{q}_e^{\top} W_1(q_m)\dot{q}_e = \dot{z}_e^{\top} P_1^{-\top} \mathbf{e}^{\ U^{\top} q_m} W_1(q_m) \mathbf{e}^{\ U q_m} P_1^{-1}\dot{z}_e$$
$$\overset{(D.23)}{=} \dot{z}_e^{\top} P_1^{-\top} \mathbf{e}^{\ -U q_m} W_1(q_m) \mathbf{e}^{\ U q_m} P_1^{-1}\dot{z}_e$$
$$\overset{(D.26)}{=} \dot{z}_e^{\top} P_1^{-\top} [U D_e(0) - D_e(0)U] P_1^{-1}\dot{z}_e$$
$$= 2\dot{z}_e^{\top} P_1^{-\top} U D_e(0) P_1^{-1}\dot{z}_e$$
$$= 2\dot{z}_e^{\top} P_1^{-\top} D_e(0) U^{\top} P_1^{-1}\dot{z}_e$$

$$W_2^{\top}(q_m)\dot{q}_e = W_2^{\top}(q_m) \mathbf{e}^{\ U q_m} P_1^{-1}\dot{z}_e$$
$$= W_2^{\top}(0) P_1^{-1}\dot{z}_e$$

This completes the proof.   ∎

3.2   A Lemma on the BP Transformation

Lemma D.2 *Unless $U \equiv 0$, the velocities $\dot{z} = [\dot{z}_e^{\top}, \dot{q}_m]^{\top}$ introduced by the BP transformation cannot be derived from a transformation $z = Z(q)$ of the generalized coordinates $q = [q_e^{\top}, q_m]^{\top}$.*   □

Proof.   The transformation from the generalized electrical velocities $\dot{q}_e$ and the generalized mechanical velocity $\dot{q}_m$ to $\dot{z} = [\dot{z}_e^{\top}, \dot{q}_m]^{\top}$ is

$$\begin{bmatrix} \dot{z}_e \\ \dot{q}_m \end{bmatrix} = \begin{bmatrix} P_1 \mathbf{e}^{\ -U q_m} & 0 \\ 0 & 1 \end{bmatrix} \begin{bmatrix} \dot{q}_e \\ \dot{q}_m \end{bmatrix}$$

If $z = Z(q)$, then

$$\frac{\partial Z}{\partial q} = \begin{bmatrix} P_1 \mathbf{e}^{\ -U q_m} & 0 \\ 0 & 1 \end{bmatrix}$$

since $\dot{z} = \frac{\partial Z}{\partial q}\dot{q}$. From this, it can be seen that z_e must be of the form

$$z_e = Z_e(q) = P_1 \mathbf{e}^{\ -U q_m} q_e + c, \ c \in \mathbb{R}^{n_e}$$

Taking the total time derivate gives

$$\dot{z}_e = \frac{\partial Z_e}{\partial q}\dot{q} = -P_1 \mathbf{e}^{\ -U q_m} U \dot{q}_m q_e + P_1 \mathbf{e}^{\ -U q_m} \dot{q}_e$$

from which it follows that since the BP transformation (see (9.30)) is defined as $\dot{z}_e = P_1 \, \mathbf{e} \, ^{-Uq_m} \dot{q}_e$, it must be true that

$$-P_1 \, \mathbf{e} \, ^{-Uq_m} U \dot{q}_m q_e \;=\; 0, \; \forall q_e \in \mathbb{R}^{n_e}$$

For this to hold, U must be the zero matrix, since $P_1 \, \mathbf{e} \, ^{-Uq_m}$ is nonsingular, and consequently for $U \neq 0$ there is no transformation $z_e = Z_e(q)$ such that $\dot{z}_e = \frac{\partial Z_e}{\partial q} \dot{q}$. ■

4 Proof of Eqs. (10.41) and (10.77)

4.1 A theorem on positivity of a block matrix

For use in the following proofs, a theorem on positivity of a block matrix is needed. The results is given in terms of the block elements on the diagonal of the matrix and their corresponding Schur complements. A proof of this theorem can be found in [141].

Theorem D.3 *An arbitrarily partitioned Hermitian[2] matrix of the form*

$$Q \;=\; \begin{bmatrix} Q_{11} & Q_{12} \\ Q_{12}^H & Q_{22} \end{bmatrix}$$

is positive definite $(Q > 0)$ *if and only if either*

$$\begin{cases} Q_{11} > 0 \\ Q_{22} - Q_{12}^H Q_{11}^{-1} Q_{12} > 0 \end{cases}$$

or

$$\begin{cases} Q_{22} > 0 \\ Q_{11} - Q_{12} Q_{22}^{-1} Q_{12}^H > 0 \end{cases}$$

□

For a necessary and sufficient condition on the matrix to be positive semidefinite when one of the block matrices Q_{11} or Q_{22} is positive definite (and hence invertible), the requirement to the Schur complements can be relaxed from *greater than zero*, to *greater than or equal to zero*. For necessary and sufficient conditions of positive semidefiniteness in the case where none of the block matrices are invertible, see [141].

4.2 Proof of Eq. (10.77)

Proof. It must be shown that

$$\mathcal{M} \;=\; \begin{bmatrix} \mathcal{R} + \mathcal{K}(\dot{q}_d) & -\frac{1}{2}\mathcal{S}(q_m, \dot{q}_d) \\ -\frac{1}{2}\mathcal{S}^\top(q_m, \dot{q}_d) & R_e \end{bmatrix} \geq \delta I_9 > 0$$

---

[2]Superscript H is used to denote the conjugate transpose of a complex matrix.

with $\mathcal{R} = \text{diag}\{R_e, R_m\}$, $\mathcal{K}(\dot{q}_d) = \text{diag}\{K_1(\dot{q}_{md})I_2, 0, 0, K_2(\dot{q}_d)\}$ and

$$
\mathcal{S}(q_m, \dot{q}_d) = \begin{bmatrix} 0_{2\times 2} & n_p L_{sr} \mathcal{J} e^{\mathcal{J} n_p q_m} \dot{q}_{md} \\ 0_{1\times 2} & 0_{1\times 2} \\ 0_{2\times 2} & -n_p L_{sr} \dot{q}_{sd}^{\top} \mathcal{J} e^{\mathcal{J} n_p q_m} \end{bmatrix} \in \mathbb{R}^{5\times 4}
$$

For the use of the theorem in Section 4.1 it must be checked if there exists a $\delta > 0$ such that $\mathcal{M} - \delta I_9 \geq 0$ with the given definition of $\mathcal{K}(\dot{q}_d)$.

Since $R_e \geq \min\{R_s, R_r\}I_2$, under the assumption that $0 < \delta < \min\{R_s, R_r\}$, which ensures invertibility of $R_e - \delta I_4$, the theorem in the previous section can be used, and it must only be checked if there exists a δ within these limits such that

$$
\mathcal{R} + \mathcal{K} - \frac{1}{4}\mathcal{S}\left\{R_e - \delta I_4\right\}^{-1}\mathcal{S}^{\top} - \delta I_5 \geq 0
$$

Writing out this expression, it follows that

$$
\begin{bmatrix} \{R_s + K_1(\dot{q}_{md}) - \delta\}I_2 & 0_{2\times 2} & 0_{2\times 1} \\ 0_{2\times 2} & \{R_r - \delta\}I_2 & 0_{2\times 1} \\ 0_{1\times 2} & 0_{1\times 2} & R_m + K_2(\dot{q}_d) - \delta \end{bmatrix}
$$
$$
- \frac{1}{4}\begin{bmatrix} 0_{2\times 2} & n_p L_{sr} \mathcal{J} e^{\mathcal{J} n_p q_m} \dot{q}_{md} \\ 0_{2\times 2} & 0_{2\times 2} \\ 0_{1\times 2} & -n_p L_{sr} \dot{q}_{sd}^{\top} \mathcal{J} e^{\mathcal{J} n_p q_m} \end{bmatrix}\begin{bmatrix} \frac{1}{R_s - \delta}I_2 & 0_{2\times 2} \\ 0_{2\times 2} & \frac{1}{R_r - \delta}I_2 \end{bmatrix}
$$
$$
\times \begin{bmatrix} 0_{2\times 2} & 0_{2\times 2} & 0_{2\times 1} \\ -n_p L_{sr} \mathcal{J} e^{-\mathcal{J} n_p q_m} \dot{q}_{md} & 0_{2\times 2} & n_p L_{sr} \mathcal{J} e^{-\mathcal{J} n_p q_m} \dot{q}_{sd} \end{bmatrix}
$$

$$
= \begin{bmatrix} \{R_s + K_1(\dot{q}_{md}) - \delta\}I_2 & 0_{2\times 2} & 0_{2\times 1} \\ 0_{2\times 2} & \{R_r - \delta\}I_2 & 0_{2\times 1} \\ 0_{1\times 2} & 0_{1\times 2} & R_m + K_2(\dot{q}_d) - \delta \end{bmatrix}
$$
$$
- \frac{1}{4}\begin{bmatrix} 0_{2\times 2} & \frac{n_p L_{sr}}{R_r - \delta} \mathcal{J} e^{\mathcal{J} n_p q_m} \dot{q}_{md} \\ 0_{2\times 2} & 0_{2\times 2} \\ 0_{1\times 2} & -\frac{n_p L_{sr}}{R_r - \delta} \dot{q}_{sd}^{\top} \mathcal{J} e^{\mathcal{J} n_p q_m} \end{bmatrix}
$$
$$
\times \begin{bmatrix} 0_{2\times 2} & 0_{2\times 2} & 0_{2\times 1} \\ -n_p L_{sr} \mathcal{J} e^{-\mathcal{J} n_p q_m} \dot{q}_{md} & 0_{2\times 2} & n_p L_{sr} \mathcal{J} e^{-\mathcal{J} n_p q_m} \dot{q}_{sd} \end{bmatrix}
$$

$$
= \begin{bmatrix} \{R_s + K_1(\dot{q}_{md}) - \delta\}I_2 & 0_{2\times 2} & 0_{2\times 1} \\ 0_{2\times 2} & \{R_r - \delta\}I_2 & 0_{2\times 1} \\ 0_{1\times 2} & 0_{1\times 2} & R_m + K_2(\dot{q}_d) - \delta \end{bmatrix}
$$
$$
- \frac{1}{4}\begin{bmatrix} \frac{n_p^2 L_{sr}^2}{R_r - \delta} \dot{q}_{md}^2 I_2 & 0_{2\times 2} & -\frac{n_p^2 L_{sr}^2}{R_r - \delta} \dot{q}_{md} \dot{q}_{sd} \\ 0_{2\times 2} & 0_{2\times 2} & 0_{2\times 1} \\ -\frac{n_p^2 L_{sr}^2}{R_r - \delta} \dot{q}_{md} \dot{q}_{sd}^{\top} & 0_{1\times 2} & \frac{n_p^2 L_{sr}^2}{R_r - \delta} \dot{q}_{sd}^{\top} \dot{q}_{sd} \end{bmatrix}
$$

and it must be required that the matrix below be positive semidefinite.

$$
\begin{bmatrix}
\left\{ R_s + K_1(\dot{q}_{md}) - \frac{n_p^2 L_{sr}^2}{4(R_r - \delta)}\dot{q}_{md}^2 - \delta \right\} I_2 & 0_{2\times 2} & \frac{n_p^2 L_{sr}^2}{4(R_r - \delta)}\dot{q}_{md}\dot{q}_{sd} \\
0_{2\times 2} & \left\{ R_r - \delta \right\} I_2 & 0_{2\times 1} \\
\frac{n_p^2 L_{sr}^2}{4(R_r - \delta)}\dot{q}_{sd}^\top \dot{q}_{md} & 0_{1\times 2} & R_m + K_2(\dot{q}_d) - \frac{n_p^2 L_{sr}^2}{4(R_r - \delta)}\dot{q}_{sd}^\top \dot{q}_{sd} - \delta
\end{bmatrix}
$$

Under the assumption that the 4×4 upper left submatrix is positive definite and hence invertible, only the positive semidefiniteness of its Schur complement must be checked according to the theorem.

The upper 4×4 matrix is invertible if and only if

$$
R_s + K_1(\dot{q}_{md}) - \frac{n_p^2 L_{sr}^2}{4(R_r - \delta)}\dot{q}_{md}^2 \; > \; \delta
$$

This condition can be satisfied by choosing the gain as

$$
K_1(\dot{q}_{md}) \; \triangleq \; \frac{n_p^2 L_{sr}^2}{4\epsilon_1}\dot{q}_{md}^2 + k_1, \; 0 < \epsilon_1 < R_r, \; k_1 \geq 0 \tag{D.28}
$$

For each choice of ϵ_1, there will be a corresponding $0 < \delta < \min\{R_s, R_r\}$ such that the requirement above is satisfied, but as $\epsilon_1 \to R_r$, $\delta \to 0$.

Calculation of the Schur complement for the upper 4×4 matrix, results in

$$
R_m + K_2(\dot{q}_d) - \frac{n_p^2 L_{sr}^2}{4(R_r - \delta)}\dot{q}_{sd}^\top \dot{q}_{sd}
$$

$$
- \frac{n_p^4 L_{sr}^4}{\left[R_s + K_1(\dot{q}_{md}) - \frac{n_p^2 L_{sr}^2}{4(R_r - \delta)}\dot{q}_{md}^2 - \delta \right] 16(R_r - \delta)^2}\dot{q}_{sd}^\top \dot{q}_{sd}\dot{q}_{md}^2 \geq \delta \tag{D.29}
$$

Now, since

$$
R_s + K_1(\dot{q}_{md}) - \frac{n_p^2 L_{sr}^2}{4(R_r - \delta)}\dot{q}_{md}^2 - \delta =
$$

$$
\frac{4\epsilon_1 (R_s + k_1 - \delta)(R_r - \delta) + n_p^2 L_{sr}^2 \dot{q}_{md}^2 (R_r - \delta - \epsilon_1)}{4\epsilon_1 (R_r - \delta)}
$$

(D.29) can be rewritten as

$$
R_m + K_2(\dot{q}_d)
$$

$$
- \frac{n_p^2 L_{sr}^2}{4}\left[\frac{4\epsilon_1 (R_s + k_1 - \delta) + n_p^2 L_{sr}^2 \dot{q}_{md}^2}{4\epsilon_1 (R_s + k_1 - \delta)(R_r - \delta) + n_p^2 L_{sr}^2 \dot{q}_{md}^2 (R_r - \delta - \epsilon_1)} \right]\dot{q}_{sd}^\top \dot{q}_{sd} \geq \delta
$$

Choosing $K_2(\dot{q}_d)$ as

$$K_2(\dot{q}_d) = \frac{n_p^2 L_{sr}^2}{4\epsilon_1} \dot{q}_{sd}^\top \dot{q}_{sd}$$

gives the requirement

$$R_m + \frac{n_p^2 L_{sr}^2}{4} \left[\frac{1}{\epsilon_1} - \frac{4\epsilon_1(R_s + k_1 - \delta) + n_p^2 L_{sr}^2 \dot{q}_{md}^2}{4\epsilon_1(R_s + k_1 - \delta)(R_r - \delta) + n_p^2 L_{sr}^2 \dot{q}_{md}^2(R_r - \delta - \epsilon_1)} \right] \dot{q}_{sd}^\top \dot{q}_{sd} \geq \delta$$

A rearrangement of the terms finally gives

$$R_m + \frac{n_p^2 L_{sr}^2}{4} \left[\frac{4\epsilon_1(R_s + k_1 - \delta)(R_r - \delta - \epsilon_1) + n_p^2 L_{sr}^2 \dot{q}_{md}^2(R_r - \delta - 2\epsilon_1)}{4\epsilon_1^2(R_s + k_1 - \delta)(R_r - \delta) + n_p^2 L_{sr}^2 \epsilon_1 \dot{q}_{md}^2(R_r - \delta - \epsilon_1)} \right] \dot{q}_{sd}^\top \dot{q}_{sd} \geq \delta$$

From this equation it can be seen that there exists a $0 < \delta \leq \min\{R_s, R_r, R_m\}$ such that the above inequality is satisfied, at least for any $0 < \epsilon_1 < \frac{1}{2}R_r$. As ϵ_1 approaches its upper limit, δ goes to zero.

This bound on ϵ_1 becomes the restricting bound. However, δ goes to zero with the mechanical damping R_m, even if ϵ_1 can be chosen independent of R_m. This dependence on R_m can be avoided by adding a constant $k_2 > 0$ to $K_2(\dot{q}_d)$, giving

$$K_2(\dot{q}_d) \triangleq \frac{n_p^2 L_{sr}^2}{4\epsilon_1} \dot{q}_{sd}^\top \dot{q}_{sd} + k_2, \ 0 < \epsilon_1 < \frac{1}{2}R_r, \ k_2 > 0$$

∎

4.3   Proof of Eq. (10.41)

Proof. For the proof of Eq. (10.41), it must be shown that

$$[R(q_m, \dot{q}_m) + \mathcal{K}(\dot{q}_m)]_{\text{es}} - \delta I_4 \ \geq \ 0$$

with

$$[R(q_m, \dot{q}_m) + \mathcal{K}(\dot{q}_m)]_{\text{es}} = \begin{bmatrix} R_s I_2 + K_1(\dot{q}_m) I_2 & \frac{1}{2} n_p L_{sr} \mathcal{J} e^{\mathcal{J} n_p q_m} \dot{q}_m \\ -\frac{1}{2} n_p L_{sr} \mathcal{J} e^{-\mathcal{J} n_p q_m} \dot{q}_m & R_r I_2 \end{bmatrix}$$

for some $\delta > 0$, with the given choice of the nonlinear gain $K_1(\dot{q}_m)$.

Using the results from Section 4.1, the fact that $R_r > 0$ and calculating the Schur complement of the lower 2×2 matrix, gives the requirement

$$R_s + K_1(\dot{q}_m) - \frac{n_p^2 L_{sr}^2}{4(R_r - \delta)} \dot{q}_m^2 \ \geq \ \delta$$

Using the results from the derivation of (D.28), it follows that the requirement is fulfilled for some $0 < \delta \leq \min\{R_s, R_r\}$ if

$$K_1(\dot{q}_m) \triangleq \frac{n_p^2 L_{sr}^2}{4\epsilon} \dot{q}_m^2 + k_1, \ 0 < \epsilon < R_r, \ k_1 \geq 0$$

∎

5   Derivation of Eqs. (10.55) and (10.56)

In this section it will be shown how (10.55) and (10.56) are derived for a general torque reference τ_*.

5.1   Derivation of Eq. (10.55)

The starting point is (10.54)

$$P_{\text{loss}} = u^\top \dot{q}_s - \tau \dot{q}_m$$

The control u is first eliminated from the above expressions by using (10.1), which can be rewritten as

$$u = L_s \ddot{q}_s + L_{sr}\, \mathbf{e}^{\,\mathcal{J} n_p q_m} \ddot{q}_r + n_p L_{sr} \dot{q}_m \mathcal{J}\, \mathbf{e}^{\,\mathcal{J} n_p q_m} \dot{q}_r + R_s \dot{q}_s$$

$$0 = L_r \ddot{q}_r + L_{sr}\, \mathbf{e}^{\,-\mathcal{J} n_p q_m} \ddot{q}_s - n_p L_{sr} \dot{q}_m \mathcal{J}\, \mathbf{e}^{\,-\mathcal{J} n_p q_m} \dot{q}_s + R_r \dot{q}_r$$

The derivative of the rotor currents, $\ddot{q}_r$, can be eliminated from the stator equation by using the last of the equations above. This results in

$$
\begin{aligned}
u = {} & \left(L_s - \frac{L_{sr}^2}{L_r} \right) \ddot{q}_s + R_s \dot{q}_s + n_p \frac{L_{sr}^2}{L_r} \dot{q}_m \mathcal{J} \dot{q}_s \\
& + \left(-\frac{L_{sr} R_r}{L_r} I_2 + n_p L_{sr} \dot{q}_m \mathcal{J} \right) \mathbf{e}^{\,\mathcal{J} n_p q_m} \dot{q}_r
\end{aligned}
$$

Substitution of the expression above together with $\tau = n_p L_{sr} \dot{q}_s^\top \mathcal{J}\, \mathbf{e}^{\,\mathcal{J} n_p q_m} \dot{q}_r$ in P_{loss}, and use of the fact that $z^\top \mathcal{J} z = 0, \forall z \in \mathbb{R}^2$ (skew-symmetry) gives

$$
\begin{aligned}
P_{\text{loss}} = {} & \dot{q}_s^\top u - \tau \dot{q}_m \\
= {} & \left(L_s - \frac{L_{sr}^2}{L_r} \right) \dot{q}_s^\top \ddot{q}_s + R_s \dot{q}_s^\top \dot{q}_s + n_p \frac{L_{sr}^2}{L_r} \dot{q}_m \dot{q}_s^\top \mathcal{J} \dot{q}_s - \frac{L_{sr} R_r}{L_r} \dot{q}_s^\top\, \mathbf{e}^{\,\mathcal{J} n_p q_m} \dot{q}_r \\
& + n_p L_{sr} \dot{q}_m \dot{q}_s^\top \mathcal{J}\, \mathbf{e}^{\,\mathcal{J} n_p q_m} \dot{q}_r - n_p L_{sr} \dot{q}_m \dot{q}_s^\top\, \mathbf{e}^{\,\mathcal{J} n_p q_m} \dot{q}_r \\
= {} & \left(L_s - \frac{L_{sr}^2}{L_r} \right) \dot{q}_s^\top \ddot{q}_s + R_s \dot{q}_s^\top \dot{q}_s - \frac{L_{sr} R_r}{L_r} \dot{q}_s^\top\, \mathbf{e}^{\,\mathcal{J} n_p q_m} \dot{q}_r \\
= {} & \left(L_s - \frac{L_{sr}^2}{L_r} \right) \dot{q}_s^\top \ddot{q}_s + R_s \dot{q}_s^\top \dot{q}_s - \frac{L_{sr} R_r}{L_r^2} \dot{q}_s^\top \left(\mathbf{e}^{\,\mathcal{J} n_p q_m} \lambda_r - L_{sr} \dot{q}_s \right) \\
= {} & \left(L_s - \frac{L_{sr}^2}{L_r} \right) \dot{q}_s^\top \ddot{q}_s + \left(R_s + R_r \frac{L_{sr}^2}{L_r^2} \right) \dot{q}_s^\top \dot{q}_s - \frac{L_{sr} R_r}{L_r^2} \dot{q}_s^\top\, \mathbf{e}^{\,\mathcal{J} n_p q_m} \lambda_r \quad \text{(D.30)}
\end{aligned}
$$

which is identical to (10.55).

5.2 Derivation of Eq. (10.56)

Under the assumption of perfect control, i.e. that the stator current tracking error has converged to zero, the desired functions for $\dot{q}_s$ and $\ddot{q}_s$ defined in (10.29) can be substituted into (10.55). For convenience they are rewritten here as

$$
\begin{aligned}
\dot{q}_s \equiv \dot{q}_{sd} &= \frac{1}{L_{sr}} \left[\left(1 + \frac{L_r \dot{\beta}_*}{R_r \beta_*} \right) I_2 + \frac{L_r}{n_p \beta_*^2} \tau_* \mathcal{J} \right] e^{\mathcal{J} n_p q_m} \lambda_{rd} \\
&\triangleq [C I_2 + D \mathcal{J}] e^{\mathcal{J} n_p q_m} \lambda_{rd} \\
\ddot{q}_s \equiv \ddot{q}_{sd} &= \frac{1}{L_{sr}} \left\{ \left[\frac{R_r L_r \ddot{\beta}_* \beta_* - R_r L_r \dot{\beta}_*^2}{R_r^2 \beta_*^2} I_2 + \left(-\frac{2 L_r \dot{\beta}_* \beta_*}{n_p \beta_*^4} \tau_* + \frac{L_r}{n_p \beta_*^2} \dot{\tau}_* \right) \mathcal{J} \right] \right. \\
&\quad \left. + \frac{n_p \dot{q}_m}{L_{sr}} \left[-\frac{L_r}{n_p \beta_*^2} \tau_* I_2 + \left(1 + \frac{L_r \dot{\beta}_*}{R_r \beta_*} \right) \mathcal{J} \right] \right\} e^{\mathcal{J} n_p q_m} \lambda_{rd} \\
&\quad + \frac{1}{L_{sr}} \left[\left(1 + \frac{L_r \dot{\beta}_*}{R_r \beta_*} \right) I_2 + \frac{L_r}{n_p \beta_*^2} \tau_* \mathcal{J} \right] e^{\mathcal{J} n_p q_m} \left(\frac{\dot{\beta}_*}{\beta_*} I_2 + \frac{R_r}{n_p \beta_*^2} \tau_* \mathcal{J} \right) \lambda_{rd} \\
&= \frac{1}{L_{sr}} \left[\left(\frac{L_r \ddot{\beta}_*}{R_r \beta_*} - \frac{L_r \dot{q}_m}{\beta_*^2} \tau_* + \frac{\dot{\beta}_*}{\beta_*} - \frac{L_r R_r}{n_p^2 \beta_*^4} \tau_*^2 \right) I_2 \right. \\
&\quad \left. + \left(\frac{L_r}{n_p \beta_*^2} \dot{\tau}_* + n_p \dot{q}_m + \frac{n_p L_r \dot{q}_m \dot{\beta}_*}{R_r \beta_*} + \frac{R_r}{n_p \beta_*^2} \tau_* \right) \mathcal{J} \right] e^{\mathcal{J} n_p q_m} \lambda_{rd} \\
&\triangleq [A I_2 + B \mathcal{J}] e^{\mathcal{J} n_p q_m} \lambda_{rd}
\end{aligned}
$$

where the constants A, B, C and D have been introduced to simplify later calculations. In the above calculations $\dot{\lambda}_{rd}$ from (10.31) was used.

The above expressions substituted into (D.30) results in

$$
\begin{aligned}
P_{\text{loss}} &= \left(L_s - \frac{L_{sr}^2}{L_r} \right) \lambda_{rd}^\top (C I_2 - D \mathcal{J}) (A I_2 + B \mathcal{J}) \lambda_{rd} \\
&\quad + \left(R_s + R_r \frac{L_{sr}^2}{L_r^2} \right) \lambda_{rd}^\top (C I_2 - D \mathcal{J}) (C I_2 + D \mathcal{J}) \lambda_{rd} \\
&\quad - \frac{R_r L_{sr}}{L_r^2} \lambda_{rd}^\top (C I_2 - D \mathcal{J}) \lambda_{rd} \\
&= \left(L_s - \frac{L_{sr}^2}{L_r} \right) (AC + BD) \lambda_{rd}^\top \lambda_{rd} \\
&\quad + \left(R_s + R_r \frac{L_{sr}^2}{L_r^2} \right) (C^2 + D^2) \lambda_{rd}^\top \lambda_{rd} - \frac{R_r L_{sr}}{L_r^2} C \lambda_{rd}^\top \lambda_{rd}
\end{aligned}
$$

where the skew-symmetry of $\mathcal{J}$ has been used in the last transition, and the constants

AC, BD, C^2 and D^2 are given as

$$AC = \frac{1}{L_{sr}^2}\left[\frac{L_r\ddot{\beta}_*}{R_r\beta_*} - \frac{L_r\dot{q}_m}{\beta_*^2}\tau_* + \frac{\dot{\beta}_*}{\beta_*} - \frac{L_rR_r}{n_p^2\beta_*^4}\tau_*^2\right]\left[1 + \frac{L_r\dot{\beta}_*}{R_r\beta_*}\right]$$

$$= \frac{1}{L_{sr}^2}\left[\frac{L_r\ddot{\beta}_*}{R_r\beta_*} - \frac{L_r\dot{q}_m}{\beta_*^2}\tau_* + \frac{\dot{\beta}_*}{\beta_*} - \frac{L_rR_r}{n_p^2\beta_*^4}\tau_*^2\right.$$

$$\left. + \frac{L_r^2\ddot{\beta}_*\dot{\beta}_*}{R_r^2\beta_*^2} - \frac{L_r^2\dot{q}_m\dot{\beta}_*}{R_r\beta_*^3}\tau_* + \frac{L_r\dot{\beta}_*^2}{R_r\beta_*^2} - \frac{L_r^2\dot{\beta}_*}{n_p^2\beta_*^5}\tau_*^2\right]$$

$$BD = \frac{1}{L_{sr}^2}\left[\frac{L_r}{n_p\beta_*^2}\dot{\tau}_* + n_p\dot{q}_m + \frac{n_pL_r\dot{q}_m\dot{\beta}_*}{R_r\beta_*} + \frac{R_r}{n_p\beta_*^2}\tau_*\right]\frac{L_r}{n_p\beta_*^2}\tau_*$$

$$= \frac{1}{L_{sr}^2}\left[\frac{L_r^2}{n_p^2\beta_*^4}\tau_*\dot{\tau}_* + \frac{L_r\dot{q}_m}{\beta_*^2}\tau_* + \frac{L_r^2\dot{q}_m\dot{\beta}_*}{R_r\beta_*^3}\tau_* + \frac{L_rR_r}{n_p^2\beta_*^4}\tau_*^2\right]$$

$$C^2 = \frac{1}{L_{sr}^2}\left[1 + \frac{2L_r\dot{\beta}_*}{R_r\beta_*} + \frac{L_r^2\dot{\beta}_*^2}{R_r^2\beta_*^2}\right]$$

$$D^2 = \frac{1}{L_{sr}^2}\frac{L_r^2}{n_p^2\beta_*^4}\tau_*^2$$

Use of the above expressions results in

$$AC + BD = \frac{1}{L_{sr}^2}\left[\frac{L_r\ddot{\beta}_*}{R_r\beta_*} + \frac{\dot{\beta}_*}{\beta_*} + \frac{L_r^2\ddot{\beta}_*\dot{\beta}_*}{R_r^2\beta_*^2} + \frac{L_r\dot{\beta}_*^2}{R_r\beta_*^2} - \frac{L_r^2\dot{\beta}_*}{n_p^2\beta_*^5}\tau_*^2 + \frac{L_r^2}{n_p^2\beta_*^4}\tau_*\dot{\tau}_*\right]$$

$$C^2 + D^2 = \frac{1}{L_{sr}^2}\left[1 + \frac{2L_r\dot{\beta}_*}{R_r\beta_*} + \frac{L_r^2\dot{\beta}_*^2}{R_r^2\beta_*^2} + \frac{L_r^2}{n_p^2\beta_*^4}\tau_*^2\right]$$

By the use of the above results and the fact that $\lambda_{rd}^T\lambda_{rd} = \beta_*^2$, P_{loss} is finally found to be

$$
\begin{aligned}
P_{\text{loss}} &= \frac{L_s L_r}{L_{sr}^2 R_r}\ddot{\beta}_*\beta_* + \frac{L_s}{L_{sr}^2}\dot{\beta}_*\beta_* + \frac{L_r^2 L_s}{L_{sr}^2 R_r^2}\ddot{\beta}_*\dot{\beta}_* + \frac{L_r L_s}{L_{sr}^2 R_r}\dot{\beta}_*^2 - \frac{L_r^2 L_s \dot{\beta}_*}{n_p^2 L_{sr}^2 \beta_*^3}\tau_*^2 \\
&\quad + \frac{L_r^2 L_s}{n_p^2 L_{sr}^2 \beta_*^2}\tau_*\dot{\tau}_* - \frac{1}{R_r}\ddot{\beta}_*\beta_* - \frac{L_r}{R_r^2}\ddot{\beta}_*\dot{\beta}_* + \frac{L_r \dot{\beta}_*}{n_p^2 \beta_*^3}\tau_*^2 - \frac{L_r}{n_p^2 \beta_*^2}\tau_*\dot{\tau}_* \\
&\quad + \frac{R_s}{L_{sr}^2}\beta_*^2 + 2\frac{L_r R_s}{L_{sr}^2 R_r}\dot{\beta}_*\beta_* + \frac{L_r^2 R_s}{L_{sr}^2 R_r^2}\dot{\beta}_*^2 + \frac{L_r^2 R_s}{n_p^2 L_{sr}^2 \beta_*^2}\tau_*^2 + \frac{R_r}{n_p^2 \beta_*^2}\tau_*^2 \\
&= \frac{L_r^2 L_s - L_{sr}^2 L_r}{L_{sr}^2 R_r^2}\ddot{\beta}_*\dot{\beta}_* + \frac{L_s L_r - L_{sr}^2}{R_r}\ddot{\beta}_*\beta_* \\
&\quad + \frac{L_s R_r + 2 L_r R_s}{L_{sr}^2 R_r}\dot{\beta}_*\beta_* + \frac{R_r L_r L_s + L_r^2 R_s}{L_{sr}^2 R_r}\dot{\beta}_*^2 \\
&\quad + \frac{L_r L_{sr}^2 - L_r^2 L_s}{n_p^2 L_{sr}^2}\tau_*^2\frac{\dot{\beta}_*}{\beta_*^3} + \frac{R_s}{L_{sr}^2}\beta_*^2 \\
&\quad + \left[\frac{L_r^2 L_s - L_r L_{sr}^2}{n_p^2 L_{sr}^2}\dot{\tau}_*\tau_* + \frac{L_r^2 R_s + R_r L_{sr}^2}{n_p^2 L_{sr}^2}\tau_*^2\right]\frac{1}{\beta_*^2}
\end{aligned}
$$

which is identical to (10.56) when $\tau_* = \tau_d$.

6   Boundedness of all signals for indirect FOC

6.1   Proof of Proposition 11.10

Starting from (11.25) we have

$$
\begin{aligned}
\dot{\tau}_d &= -\left(K_P + \frac{K_I}{p}\right)\ddot{q}_m \\
&= -\left(K_P + \frac{K_I}{p}\right)\left(v^\top \mathcal{J}\lambda_{rd} + v^\top \mathcal{J}\tilde{\lambda} - \tau_L\right) \\
&= -\left(K_P + \frac{K_I}{p}\right)\left(\tau_d - \tau_L - \left[\begin{array}{cc} -\frac{\tau_d}{\beta_*} & \beta_* \end{array}\right]e^{-\mathcal{J}\rho_d}\tilde{\lambda}\right)
\end{aligned}
$$

The rest of the proof consist in expressing $e^{-\mathcal{J}\rho_d}\tilde{\lambda}$ as a bounded signal, which ensures a bounded nonlinear feedback since it is multiplied by τ_d.

A Boundedness of $\mathrm{e}^{-\mathcal{J}\rho_d}\tilde{\lambda}$

Since $\tilde{\lambda} \triangleq \lambda_r - \lambda_{rd}$ and $\tilde{R}_r \triangleq \hat{R}_r - R_r$

$$
\begin{aligned}
\dot{\tilde{\lambda}} &= \dot{\lambda}_r - \dot{\lambda}_{rd} \\
&= -R_r(\lambda_r - \lambda_{rd}) - \tilde{R}_r v + \tilde{R}_r \lambda_{rd} \\
&= -R_r\tilde{\lambda} + \tilde{R}_r(\lambda_{rd} - v)
\end{aligned}
$$

The term $\lambda_{rd} - v$ can be replaced by

$$
\lambda_{rd} - v = -\frac{1}{\hat{R}_r}\dot{\lambda}_{rd} = -\frac{1}{\beta_*^2}\tau_d\mathcal{J}\lambda_{rd}
$$

so that

$$
\dot{\tilde{\lambda}} = -R_r\tilde{\lambda} - \tilde{R}_r\frac{1}{\beta_*^2}\tau_d\mathcal{J}\lambda_{rd}
$$

$\mathcal{J}\lambda_{rd}$ is bounded since

$$
\mathcal{J}\lambda_{rd} = \mathcal{J}\beta_* \, \mathrm{e}^{\mathcal{J}\rho_d}\begin{bmatrix} 1 \\ 0 \end{bmatrix} = \beta_* \, \mathrm{e}^{\mathcal{J}\rho_d}\begin{bmatrix} 0 \\ 1 \end{bmatrix}
$$

so that the derivative of $\tilde{\lambda}$ becomes

$$
\dot{\tilde{\lambda}} = -R_r\tilde{\lambda} - \frac{\tilde{R}_r}{\beta_*^2}\tau_d\beta_* \, \mathrm{e}^{\mathcal{J}\rho_d}\begin{bmatrix} 0 \\ 1 \end{bmatrix}
$$

From this equation, $\tilde{\lambda}(t)$ follows from the convolution integral

$$
\begin{aligned}
\tilde{\lambda}(t) - \mathrm{e}^{-R_r t}\tilde{\lambda}(0) &= -\frac{1}{\beta_*}\int_0^t \mathrm{e}^{-R_r(t-s)}\tilde{R}_r\tau_d(s) \, \mathrm{e}^{\mathcal{J}\rho_d(s)}\begin{bmatrix} 0 \\ 1 \end{bmatrix}\mathrm{d}s \\
&= -\frac{\tilde{R}_r\beta_*}{\hat{R}_r}\int_0^t \mathrm{e}^{-R_r(t-s)}\dot{\rho}_d(s)\mathcal{J} \, \mathrm{e}^{\mathcal{J}\rho_d(s)}\mathrm{d}s\begin{bmatrix} 1 \\ 0 \end{bmatrix} \\
&= -\frac{\tilde{R}_r\beta_*}{\hat{R}_r}\int_0^t \mathrm{e}^{-R_r(t-s)}\frac{\mathrm{d}\mathrm{e}^{\mathcal{J}\rho_d(s)}}{\mathrm{d}s}\mathrm{d}s\begin{bmatrix} 1 \\ 0 \end{bmatrix} \\
&= -\frac{\tilde{R}_r\beta_*}{\hat{R}_r}\mathrm{e}^{-R_r t}\int_0^t \mathrm{e}^{R_r s}\frac{\mathrm{d}\mathrm{e}^{\mathcal{J}\rho_d(s)}}{\mathrm{d}s}\mathrm{d}s\begin{bmatrix} 1 \\ 0 \end{bmatrix}
\end{aligned}
$$

Integration by parts results in an expression for the integral

$$
\begin{aligned}
\int_0^t \mathrm{e}^{R_r s}\frac{\mathrm{d}\mathrm{e}^{\mathcal{J}\rho_d(s)}}{\mathrm{d}s}\mathrm{d}s &= \mathrm{e}^{R_r s}\, \mathrm{e}^{\mathcal{J}\rho_d(s)}\big|_0^t - \int_0^t \mathrm{e}^{\mathcal{J}\rho_d(s)}R_r\mathrm{e}^{R_r s}\mathrm{d}s \\
&= \mathrm{e}^{R_r t}\, \mathrm{e}^{\mathcal{J}\rho_d(t)} - \mathrm{e}^{\mathcal{J}\rho_d(0)} - R_r\int_0^t \mathrm{e}^{\mathcal{J}\rho_d(s)}\mathrm{e}^{R_r s}\mathrm{d}s
\end{aligned}
$$

So that, with initial conditions $\lambda_r(0) = 0$, $\lambda_{rd}(0) = [\beta_*, 0]^\top$,

$$\mathbf{e}^{-\mathcal{J}\rho_d(t)}\tilde{\lambda}(t) - \mathbf{e}^{-\mathcal{J}\rho_d(t)}\mathbf{e}^{-R_r t}\tilde{\lambda}(0)$$

$$= -\frac{\tilde{R}_r \beta_*}{\hat{R}_r}\mathbf{e}^{-R_r t}\,\mathbf{e}^{-\mathcal{J}\rho_d(t)}\left(\mathbf{e}^{R_r t}\,\mathbf{e}^{\mathcal{J}\rho_d(t)} - \mathbf{e}^{\mathcal{J}\rho_d(0)} - R_r\int_0^t \mathbf{e}^{\mathcal{J}\rho_d(s)}\mathbf{e}^{R_r s}\mathrm{d}s\right)\begin{bmatrix} 1 \\ 0 \end{bmatrix}$$

$$= -\frac{\tilde{R}_r \beta_*}{\hat{R}_r}\left(I_2 - \mathbf{e}^{-R_r t}\,\mathbf{e}^{-\mathcal{J}(\rho_d(t)-\rho_d(0))} - R_r\int_0^t \mathbf{e}^{-\mathcal{J}(\rho_d(t)-\rho_d(s))}\mathbf{e}^{-R_r(t-s)}\mathrm{d}s\right)\begin{bmatrix} 1 \\ 0 \end{bmatrix}$$

$$= -\left(I_2 - \mathbf{e}^{-R_r t I_2 - \mathcal{J}(\rho_d(t)-\rho_d(0))}\right)\frac{\tilde{R}_r \beta_*}{\hat{R}_r}\begin{bmatrix} 1 \\ 0 \end{bmatrix} + \frac{\tilde{R}_r \beta_*}{\hat{R}_r}R_r\int_0^t \mathbf{e}^{-R_r(t-s)}\begin{bmatrix} \cos z \\ -\sin z \end{bmatrix}\mathrm{d}s$$

where $z(t, s)$ is defined as

$$z(t, s) \stackrel{\triangle}{=} \rho_d(t) - \rho_d(s)$$

B Elimination of $\tilde{\lambda}$ from the expression for $\dot{\tau}_d$

For elimination of $\mathbf{e}^{-\mathcal{J}\rho_d}\tilde{\lambda}$ in the equation

$$\dot{\tau}_d = -\left(K_P + \frac{K_I}{p}\right)\left(\tau_d - \tau_L - \begin{bmatrix} -\frac{\tau_d}{\beta_*} & \beta_* \end{bmatrix}\mathbf{e}^{-\mathcal{J}\rho_d}\tilde{\lambda}\right)$$

the expression

$$\mathbf{e}^{-\mathcal{J}\rho_d}\tilde{\lambda} = \mathbf{e}^{-R_r t}\begin{bmatrix} f_1(t) \\ f_2(t) \end{bmatrix} - \frac{\tilde{R}_r \beta_*}{\hat{R}_r}\begin{bmatrix} 1 \\ 0 \end{bmatrix} + \frac{\tilde{R}_r \beta_*}{\hat{R}_r}R_r\int_0^t \mathbf{e}^{-R_r(t-s)}\begin{bmatrix} \cos z \\ -\sin z \end{bmatrix}\mathrm{d}s$$

can be used, where the functions $f_1, f_2 \in \mathcal{L}_\infty$ depend on the initial values $\tilde{\lambda}(0), \rho_d(0)$, so that

$$\begin{bmatrix} -\frac{\tau_d(t)}{\beta_*} & \beta_* \end{bmatrix}\mathbf{e}^{-\mathcal{J}\rho_d}\tilde{\lambda} = -\mathbf{e}^{-R_r t}f_1(t)\frac{\tau_d}{\beta_*} + \beta_*\mathbf{e}^{-R_r t}f_2(t) + \frac{\tilde{R}_r}{\hat{R}_r}\tau_d(t)$$

$$-\frac{\tilde{R}_r \beta_*}{\hat{R}_r}R_r\int_0^t \mathbf{e}^{-R_r(t-s)}\left(\frac{\tau_d(t)}{\beta_*}\cos z + \beta_*\sin z\right)\mathrm{d}s$$

which finally leads to

$$\dot{\tau}_d = -\left(K_P + \frac{K_I}{p}\right)\left(\tau_d - \tau_L + \mathbf{e}^{-R_r t}f_1(t)\frac{\tau_d}{\beta_*} - \beta_*\mathbf{e}^{-R_r t}f_2(t)\right.$$

$$\left. -\frac{\tilde{R}_r}{\hat{R}_r}\tau_d(t) + \frac{\tilde{R}_r \beta_*}{\hat{R}_r}R_r\int_0^t \mathbf{e}^{-R_r(t-s)}\left(\frac{\tau_d(t)}{\beta_*}\cos z + \beta_*\sin z\right)\mathrm{d}s\right)$$

By use of the differential operator $p \triangleq d/dt$, this can be written as

$$\left(p + \left(\frac{pK_P + K_I}{p}\right)\left(1 - \frac{\tilde{R}_r}{\hat{R}_r}\right)\right)\tau_d =$$
$$\left(K_P + \frac{K_I}{p}\right)\left(\tau_L - e^{-R_r t}f_1(t)\frac{\tau_d}{\beta_*} + \beta_* e^{-R_r t}f_2(t)\right.$$
$$\left. - \frac{\tilde{R}_r \beta_*}{\hat{R}_r}R_r \int_0^t e^{-R_r(t-s)}\left(\frac{\tau_d(t)}{\beta_*}\cos(z) + \beta_* \sin z\right)ds\right)$$

so that τ_d is the output of a LTI operator $G(p)$ with a nonlinear feedback $b(t)$

$$\tau_d(t) = -\frac{pK_P + K_I}{p^2 + (pK_P + K_I)\left(1 - \frac{\tilde{R}_r}{\hat{R}_r}\right)} \times$$
$$\left(\frac{\tilde{R}_r \beta_* R_r}{\hat{R}_r}\int_0^t e^{-R_r(t-s)}\left(\frac{\tau_d(t)}{\beta_*}\cos z + \beta_* \sin z\right)ds + e^{-R_r t}f_1(t)\frac{\tau_d}{\beta_*}\right)$$
$$+ \frac{pK_P + K_I}{p^2 + (pK_P + K_I)\left(1 - \frac{\tilde{R}_r}{\hat{R}_r}\right)}(\tau_L + \beta_* e^{-R_r t}f_2(t))$$

C Feedback gain calculation

The bounded signal $b(t)$ which is multiplied with τ_d is

$$b(t) = -\frac{\tilde{R}_r R_r}{\hat{R}_r}\int_0^t e^{-R_r(t-s)}\cos z ds - e^{-R_r t}f_1(t)\frac{1}{\beta_*}$$

and it follows that $b(t) = b_\infty(t) + b_1(t)$ with

$$\|b_\infty(t)\|_\infty = |\frac{\tilde{R}_r}{\hat{R}_r}|, \; b_1(t) \in \mathcal{L}_1$$

■

Bibliography

[1] T. Ahmed-Ali, F. Lamnabhi-Lagarrigue, and R. Ortega. A globally stable adaptive indirect field–oriented controller for current–fed induction motors. In *Proc. American Control Conference*, Philadelphia, PA, 1998.

[2] A. Ailon. Output controllers based on iterative schemes for set-point regulation of uncertain flexible-joint robot models. *Automatica*, 32(10):1455, 1996.

[3] A. Ailon and R. Ortega. An observer-based set-point controller for robot manipulators with flexible joints. *Syst. Contr. Letters*, 21:329–335, 1993.

[4] J. Amin, B. Friedland, and A. Harnoy. Implementation of a friction estimation and compensation technique. *IEEE Contr. Syst. Mag.*, 17:71–76, 1997.

[5] Y. Amran, F. Huliehlel, and S. Ben-Yaacov. Unified SPICE compatible average model of PWM converters. *IEEE Trans. Pow. Elec.*, 6:585–594, 1991.

[6] J. R. Anderson and M. Spong. Asymptotic stability for force reflecting teleoperators with time delay. *Int. J. Robotics Res.*, 11(2):135–149, 1992.

[7] S. Arimoto. A class of quasi-natural potentials and hyper-stable PID servo-loops for nonlinear robotic systems. *Trans. Soc. Instrument Contr. Engg.*, 30(9):1005–1012, 1994.

[8] S. Arimoto, S. Kawamura, and T. Naniwa. Proposal of friction/gravity–free robots and generalization of impedance control. In *Proc. 7th. Symp. Robot Contr.*, Nantes, France, 1997.

[9] S. Arimoto and F. Miyazaki. Stability and robustness of PD feedback control with gravity compensation for robot manipulator. In F. W. Paul and D. Yacef-Toumi, editors, *Robotics: Theory and Applications – DSC –*, volume 3, pages 67–72, 1986.

[10] S. Arimoto and T. Naniwa. A VSS analysis for robot dynamics under coulomb frictions and a proposal of inertia-only robots. In Proc. IEEE Workshop on Variable Structure Syst., Tokio, Japan, 1996.

515

[11] B. Armstrong-Hélouvry, P. Dupont, and C. Canudas de Wit. A survey of models, analysis tools and compensation methods for the control of machines with friction. *Automatica*, 30(7):1083–1138, 1994.

[12] V. Arnold. *Mathematical Methods of Classical Mechanics*. Springer-Verlag, New York, 2nd edition, 1989.

[13] K. J. Åström and B. Wittenmark. *Adaptive Control*. Addison–Wesley, 1995. ISBN 0-201-55866-1.

[14] N. Barabanov and R. Ortega. Necessary and sufficient conditions for passivity of the LuGre friction model. Technical report, Supélec, France, March 1998. Submitted to *IEEE Trans. Automat. Contr.*

[15] N. Barabanov and R. Ortega. Qualitative behaviour of series–resonant dc–to–dc converters. Technical report, LSS–Supelec, Gif s/Yvette, France, 1998.

[16] O. Bas, V. Davidkovich, A. M. Stankovich, and G. Tadmor. Passivity–based sensorless control of a smooth rotor permanent magnet synchronous motor. In *Proc. 36th. IEEE Conf. Decision Contr.*, San Diego, CA, 1997.

[17] S. Battiloti, L. Lanari, and R. Ortega. On the role of passivity and output injection in the output feedback stabilization problem: Application to robot control. *European J. of Contr.*, 3(2):92–103, 1997.

[18] A. Bazanella, A. Silva, and P. Kokotovic. Lyapunov design for excitation control of synchronous machines. In *Proc. 36th. IEEE Conf. Decision Contr.*, San Diego, CA, USA, 1997.

[19] H. Berghuis. *Model based robot control: From theory to practice*. PhD thesis, University of Twente, Enschede, The Netherlands, 1993.

[20] H. Berghuis and H. Nijmeijer. Global regulation of robots using only position measurements. *Syst. Contr. Letters*, 21:289–293, 1993.

[21] H. Berghuis and H. Nijmeijer. A passivity approach to controller-observer design for robots. *IEEE Trans. on Robotics Automat.*, 9(6):740–754, 1993.

[22] H. Berghuis, R. Ortega, and H. Nijmeijer. A robust adaptive controller for robot manipulators. *IEEE Trans. on Robotics Automat.*, 9:740–754, 1993.

[23] G. Besançon. Simple global output feedback tracking control for one-degree-of-freedom Euler-Lagrange systems. In *IFAC Conf. on Systems Structure and Control*, 1998.

[24] F. Blaschke. The principle of field orientation as applied to the new TRANSVEKTOR closed-loop control system for rotating field machines. *Siemens Review*, XXXIX(5):217–220, 1972.

[25] A. J. Blauch, M. Bodson, and J. Chiasson. High-speed parameter estimation of stepper motors. *IEEE Trans. Contr. Syst. Technol.*, 1(4):270–279, 1993.

[26] P. Bliman and M. Sorine. Easy-to-use realistic dry friction models for automatic control. In *Proc. European Cont. Conf.*, pages 3788–3794, Rome, Italy, 1995.

[27] A.M. Bloch, N.E. Leonard, and J.E. Marsden. Stabilization of mechanical systems using controlled Lagrangians. In *Proc. 36th. IEEE Conf. Decision Contr.*, pages 2356–2361, San Diego, CA, USA, 1997.

[28] M. Bodson. Emerging technologies in control engineering. *IEEE Contr. Syst. Mag.*, 14(6):10–12, 1994.

[29] M. Bodson and J. Chiasson. A systematic approach for selecting optimal flux references in induction motors. In *Proc. IEEE IAS Annual Meeting*, pages 531–537, Houston, TX, USA, 1992.

[30] M. Bodson, J. Chiasson, and R. Novotnak. High-performance induction motor control via input-output linearization. *IEEE Contr. Syst. Mag.*, 14(4):25–33, Aug. 1994.

[31] M. Bodson, J. Chiasson, and R. Novotnak. Nonlinear servo control of an induction motor with saturation. In *Proc. 33rd. IEEE Conf. Decision Contr.*, pages 1832–1837, Orlando, FL, 1994.

[32] M. Bodson, J. Chiasson, and R.T. Novotnak. Nonlinear speed observer for high-performance induction motor control. *IEEE Trans. on Ind. Electr.*, 42(4):337–343, 1995.

[33] M. Bodson, J.N. Chiasson, R.T. Novotnak, and R.B. Rekowski. High-performance nonlinear feedback control of a permanent magnet stepper motor. *IEEE Trans. Contr. Syst. Technol.*, 1(1):5–14, 1993.

[34] B. K. Bose. *Modern Power Electronics: Evolution, Technology and Applications.* IEEE Press, New York, 1992.

[35] B. K. Bose, editor. Special Issue on Power Electronics and Motion Control, *Proceedings of the IEEE*, volume 82, No. 8, Aug. 1994.

[36] B.K. Bose. *Power Electronics and AC Drives.* Prentice Hall, 1986. ISBN 0-13-686882-7.

[37] B.K. Bose. Power electronics and motion control - technology status and recent trends. *IEEE Trans. on Ind. Appl.*, 29(5):902–909, 1993.

[38] B.K. Bose. Expert systems, fuzzy logic, and neural network applications in power electronics and motion control. In Bose [35], pages 1303–1323.

[39] B. Brogliato and R. Lozano. Corrections to "Adaptive control of robot manipulators with flexible joints". *IEEE Trans. on Automat. Contr.*, 41(6):920–922, 1996.

[40] B. Brogliato, R. Ortega, and R. Lozano. Global tracking controllers for flexible-joint manipulators: A comparative study. *Automatica*, 31(7):941–956, 1995.

[41] R.T. Bupp and D.S. Bernstein. A benchmark problem for nonlinear control design: Problem statement, experimental testbed and passive nonlinear compensation. In *Proc. American Control Conference*, pages 4363–4367, Seattle, WA, 1995.

[42] I.V. Burkov. Asymptotic stabilization of nonlinear Lagrangian systems without measuring velocities. In *Proc. Internat. Sym. Active Control in Mechanical Engineering*, pages 37–41, Lyon, France, 1995.

[43] I.V. Burkov. Mechanical system stabilization via differential observer. In *IFAC Conference on System Structure and Control*, pages 532–535, Nantes, France, 1995.

[44] I.V. Burkov. Stabilization of mechanical systems via bounded control and without velocity measurement. In *Proc. 2nd Russian-Swedish Control Conf.*, pages 37 – 41, St. Petersburg, Russia, 1995.

[45] I.V. Burkov, A.A. Pervozvanski, and L.B. Freidovich. Algorithms of robust global stabilization of flexible manipulators. In P. Dauchez, editor, *Proc. World Automation Congress*, Montepellier, France, May 1996.

[46] C. Byrnes, A. Isidori, and J. C. Willems. Passivity, feedback equivalence, and the global stabilization of minimum phase nonlinear systems. *IEEE Trans. on Automat. Contr.*, 36(11):1228–1240, 1991.

[47] C.I. Byrnes and C.F. Martin. An integral-invariance principle for nonlinear systems. *IEEE Trans. on Automat. Contr.*, 40(6):983–994, 1995.

[48] A. Quintiliani C. Cecati and N. Rotondale. A low cost induction motor drive based on the passivity theory approach. In *IAS-Annual Meeting*, pages 581–586, San Diego, CA, Oct. 1996.

[49] L. Cai and G. Song. A smooth robust nonlinear controller for robot manipulators with joint stick-slip friction. In *Proc. IEEE Conf. Robotics Automat.*, pages 449–454, Atlanta, GA, 1993.

[50] C. Canudas. Correction to: "A new model for control of systems with friction". *IEEE Trans. on Automat. Contr.*, 1998. To appear.

[51] C. Canudas and P. Lischinsky. Adaptive friction compensation with partially known dynamic friction model. *Int. J. Adapt. Contr. Sign. Process.*, 11:65–80, 1991.

[52] C. Canudas de Wit, H. Olsson, K.J. Åström, and P. Lischinsky. A new model for control of systems with friction. *IEEE Trans. on Automat. Contr.*, 40(3):419–426, 1995.

[53] C. Canudas de Wit, R. Ortega, and S. Seleme. Robot motion control using induction motor drives. In *Proc. IEEE Conf. Robotics Automat.*, pages 533–538, Atlanta, GA, 1993.

[54] L. Cava, M. Picardi, and C. Ranieri. Application of the extended Kalman filter to parameter and state estimation of induction motors. *Int. J. Model. and Simul.*, 9(3):85–89, 1989.

[55] G. Chang, J. P. Hespanha, A. S. Morse, M. Netto, and R. Ortega. Supervisory field–oriented control of induction motors with uncertain rotor resistance. In *IFAC Workshop on Adaptive Control*, Glasgow, UK, Aug. 1998.

[56] A. Chelouah, E. Delaleau, P. Martin, and P. Rouchon. Differential flatness and control of IM. In *Proc. CESA'96 IMACS Multiconference*, pages 80–85, 1996.

[57] J. Chiasson. Control of electric drives. Lecture notes, Dept. Elec. Eng., University of Pittsburgh.

[58] J. Chiasson. Dynamic feedback linearization of the induction motor. *IEEE Trans. on Automat. Contr.*, 38(10):1588–1594, 1993.

[59] J.N. Chiasson. A new approach to dynamic feedback linearization control of induction motors. In *Proc. 34th. IEEE Conf. Decision Contr.*, pages 2173–2178, New Orleans, LA, 1995.

[60] R. Colbaugh and K. Glass. Adaptive regulation of rigid-link electrically-driven manipulators. In *Proc. IEEE Conf. Robotics Automat.*, pages 293–300, Nagoya, Japan, 1995.

[61] J.M. Coron, A. Teel, and L. Praly. Feedback stabilization of nonlinear systems: Sufficient and necessary conditions and Lyapunov Input-Output techniques. In A. Isidori, editor, *New trends in control*, pages 293–347. Springer-Verlag, New York, 1995.

[62] P. Crouch and A. J. van der Schaft. *Variational and Hamiltonian Control Systems*, volume 101 of *Lecture Notes in Control and Information Sciences*. Springer-Verlag, Berlin, 1987.

[63] S. Ćuk. *Modeling, Analysis and Design of Switching Converters.* PhD thesis, CALTECH, Pasadena, CA, 1976.

[64] D. Czarkowski and M. K. Kazimierczuk. Energy-conservation approach to modeling PWM DC-to-DC converters. *IEEE Trans. on Aero. and Elect. Syst.*, 29:1059–1063, 1993.

[65] D.M. Dawson, J. Hu, and T.C. Burg. *Nonlinear Control of Electric Machinery.* Marcel Dekker, 1998. ISBN 0-8247-0180-1.

[66] A. De Luca and L. Lanari. Robots with elastic joints are linearizable via dynamic feedback. In *Proc. 34th. IEEE Conf. Decision Contr.*, pages 3895–3897, New Orleans, LA, 1995.

[67] A. De Luca and G. Ulivi. Full linearization of induction motors via nonlinear state-feedback. In *Proc. 26th. IEEE Conf. Decision Contr.*, pages 1765–1770, Los Angeles, CA, 1987.

[68] P. de Wit, R. Ortega, and I. Mareels. Indirect field-oriented control of induction motors is robustly globally stable. *Automatica*, 32(10):1393–1402, 1996.

[69] C.A. Desoer and M. Vidyasagar. *Feedback Systems: Input-Output Properties.* Academic Press, New York, 1975.

[70] Editorial Board. Scanning the issue. *IEEE Trans. on Automat. Contr.*, 38(2):193, 1993.

[71] G. Escobar. *Nonlinear control of power electronic converters.* PhD thesis, LSS-Supelec, Gif s/Yvette, France, 1998. Under preparation.

[72] G. Escobar and R. Ortega. Output-feedback global stabilization of a nonlinear benchmark system using a saturated passivity-based controller. In *Proc. 36th. IEEE Conf. Decision Contr.*, pages 4340–4341, San Diego, CA, 1997. To appear in *IEEE Trans. Contr. Syst. Technol.*

[73] G. Escobar, R. Ortega, and M. Reyhanoglu. Regulation and tracking of the nonholonomic double integrator: A field-oriented control approach. *Automatica*, 34(1):125–131, 1998.

[74] G. Escobar, A. J. van der Schaft, and R. Ortega. A Hamiltonian viewpoint in the modeling of switching power converters. *Automatica*, 1998. To appear.

[75] G. Escobar, I. Zein, R. Ortega, H. Sira-Ramírez, and J. Vilain. An experimental comparison of several nonlinear controllers for power converters. In *Proc. IEEE Int. Symp. on Ind. Elec. (ISIE)*, University of Minho, Guimarães, Portugal, 1997.

[76] G. Espinosa-Pérez and R. Ortega. State observers are unnecessary for induction motor control. *Syst. Contr. Letters*, 23(5):315–323, 1994.

[77] G. Espinosa-Pérez and R. Ortega. An output feedback globally stable controller for induction motors. *IEEE Trans. on Automat. Contr.*, 40(1):138–143, 1995.

[78] G. Espinosa-Pérez, R. Ortega, G. Chang, and E. Mendes. On FOC of induction motors: Tuning of the PI gains for performance enhancement. In *Proc. 37th. IEEE Conf. Decision Contr.*, Tampa, FL, Dec. 1998.

[79] G. Espinosa-Pérez, R. Ortega, and P.J. Nicklasson. Torque and flux tracking of induction motors. *Int. J. Rob. and Nonl. Contr.*, 7:1–9, 1997.

[80] A.E. Fitzgerald, C. Kingsley Jr., and S.D. Umans. *Electric Machinery*. McGraw-Hill, 5th edition, 1992. ISBN 0-07-707708-3.

[81] M. Fliess. Generalized controller canonical forms for linear and nonlinear dynamics. *IEEE Trans. on Automat. Contr.*, 35:994–1001, 1990.

[82] M. Fliess, J. Lévine, P. Martin, and P. Rouchon. Sur les systèmes linéaires différentiellement plats. *C. R. de la Acad. de Sci. Paris. Serie I, Automatique*, 315:619–624, 1992.

[83] M. Fliess, J. Lévine, P. Martin, and P. Rouchon. Flatness and defect of non-linear systems: introductionary theory and examples. *Int. J. of Contr.*, 61(6):1327–1361, 1995.

[84] M. Fliess and H. Sira-Ramírez. Regimes glissants, structures variables linéaires et modules. *C. R. de la Acad. de Sci. Paris, Serie 1, Automatique* , pages 703–706, 1993.

[85] T. I. Fossen and J. P. Strand. Passive nonlinear observer design for ships using Lyapunov methods: Full-scale experiments with a supply vessel. *Automatica*, 1998. (to appear).

[86] T.I. Fossen. *Guidance and control of ocean vehicles*. John Wiley & Sons, 1994.

[87] G.F. Franklin, J.D. Powell, and M.L. Workman. *Digital Control of Dynamic Systems*. Addison–Wesley, 2nd edition, 1990. ISBN 0-201-51-884-8.

[88] R. Freeman. Global internal stabilizability does not imply global external stabilizability for small sensor disturbances. *IEEE Trans. on Automat. Contr.*, 40:2119–2122, 1995.

[89] R. Freeman and P. V. Kokotović. Design of 'softer' robust nonlinear control laws. *Automatica*, 29:1425–1437, 1993.

[90] M. Gäfvert. Comparison of two friction models. Master's thesis, Lund Inst. of Techn., 1996. ISRN LUTFD2/TFRT–5561–SE.

[91] F. Gantmacher. *Lectures in Analytical Mechanics*. Mir Publishers, Moscow, 1970.

[92] G.O. Garcia, J.C. Mendes Luís, R.M. Stephan, and E.H. Watanabe. An efficient controller for an adjustable speed induction motor drive. *IEEE Trans. on Ind. Electr.*, 41(5):533–539, 1994.

[93] D. Georges, E. Oyarbide, and S. Bacha. Passivity–based control for static condensers. In *Proc. 23th Int. Conf. Ind. Elec., Contr., and Instr.: IECON '97*, New Orleans, LA, USA, 1997.

[94] G. Georgiou, A. Chelouah, S. Monaco, and D. Normand-Cyrot. Nonlinear multirate adaptive control of a synchronous motor. In *Proc. 31st. IEEE Conf. Decision Contr.*, pages 3523–3528, Tucson, AZ, 1992.

[95] L.U. Gökdere. *Passivity-based Methods for Control of Induction Motors*. PhD thesis, University of Pittsburgh, USA, 1996.

[96] L.U. Gökdere and M.A. Simaan. A passivity-based method for induction motor control. *IEEE Trans. on Ind. Electr.*, 44(5):688–695, 1997.

[97] H. Goldstein. *Classical Mechanics*. Addison–Wesley, 2nd edition, 1980. ISBN 0-201-02969-3.

[98] M.B. Gorzałczany and T. Stefański. Fuzzy control and fuzzy neural network control of inverter-fed induction motor drive for electrical vehicle. In *Proc. 3rd. European Contr. Conf.*, pages 820–825, Rome, Italy, 1995.

[99] D. Haessig and B. Friedland. On the modelling and simulation of friction. *J. Dyn Syst Meas. Control Trans. ASME*, 113(3):354–362, 1991.

[100] K. Hasse. *Zur Dynamik drehzahlgeregelter Antriebe mit stromrichtergespeisten Asynchron-Kurzschluß-läfermaschinen*. PhD thesis, Technischen Hochschule Darmstadt, 1969.

[101] N. Hemati. Non-dimensionalization of the equations of motion for permanent-magnet machines. *Elec. Mach. Pow. Syst.*, 23(5):541–556, 1995.

[102] G. Hewer and C. Kenney. Dissipative LQG control systems. *IEEE Trans. on Automat. Contr.*, 34(8):866–870, 1989.

[103] D. Hill and P. Moylan. The stability of nonlinear dissipative systems. *IEEE Trans. on Automat. Contr.*, pages 708–711, 1976.

[104] J. Hu and D.M. Dawson. Adaptive control of induction motor systems despite rotor resistance uncertainty. In *Proc. American Control Conference*, pages 1397–1402, Seattle, WA, 1995.

[105] J. Hu, D.M. Dawson, and Y. Ou. A global adaptive link position tracking controller for robot manipulators driven by induction motors. In *Proc. 34th. IEEE Conf. Decision Contr.*, pages 33–38, New Orleans, LA, 1995.

[106] J. Hu, D.M. Dawson, and Y. Qian. Position tracking control of an induction motor via partial state feedback. *Automatica*, 31(7):989–1000, 1995.

[107] J. Hu, D.M. Dawson, and Y. Qian. Tracking control for robot manipulators driven by induction motors without flux measurement. *IEEE Trans. on Robotics Automat.*, 12(3):419–438, 1996.

[108] J. Hu, D.M. Dawson, and Z. Qu. Adaptive tracking control of an induction motor with robustness to parametric uncertainty. *IEE. Proc.-B. Electr. Power. Appl.*, 141(2):85–94, March 1994.

[109] The Institute of Marine Engineers. *Electric Propulsion: The Effective Solution, Part 1*, London, Oct. 1995. ISBN 0-907-206-65-4.

[110] A. Isidori. *Nonlinear Control Systems*. Springer-Verlag, 2nd edition, 1995. ISBN 3-540-19916-0.

[111] M. Janković, D. Fontaine, and P. Kokotović. TORA example: Cascaded and passivity control designs. *IEEE Trans. Contr. Syst. Technol.*, 4(3):292–297, 1996.

[112] P.L. Jansen and R.D. Lorentz. Transducerless position and velocity estimation in induction motors and salient ac machines. *IEEE Trans. on Ind. Electr.*, 31(2):240–247, 1995.

[113] E. Jonckheere. Lagrangian theory of large scale systems. In Proc. European Conf. on Circuit Th. and Design, pages 626–629, The Hague, The Netherlands, 1981.

[114] T. Kailath. *Linear Systems*. Prentice Hall, 1980. ISBN 0-13-536961-4.

[115] I. Kanellakopoulos and P.T. Krein. Integral-action nonlinear control of induction motors. In *Proc. 12th. IFAC World Congress*, volume 7, pages 251–254, Sydney, Australia, 1993.

[116] I. Kanellakopoulos, P.T. Krein, and F. Disilvestro. Nonlinear flux-observer-based control of induction motors. In *Proc. American Control Conference*, pages 1700–1704, 1992.

[117] J. G. Kassakian, M. Schlecht, and G. C. Verghese. *Principles of Power Electronics.* Addison–Wesley, 1991.

[118] M. K. Kazimierczuk and D. Czarkowski. Application of the principle of energy conservation to modeling the PWM converters. In *Proc. 2nd IEEE Conf. Cont. Appl.*, volume 1, pages 291–296, Vancouver, BC , 1993.

[119] M.P. Kaźmierkowski and A.B. Kasprowicz. Improved direct torque and flux vector control of PWM inverter-fed induction motor drives. *IEEE Trans. on Ind. Electr.*, 42(4):344–349, 1995.

[120] M.P. Kaźmierkowski and H. Tunia. *Automatic Control of Converter-Fed Drives*, volume 45 of *Studies in Electrical and Electronic Engineering.* Elsevier, 1994. ISBN 0-444-98660-X.

[121] A. Kelkar and S. Joshi. *Control of nonlinear multibody flexible space structures*, volume 221 of *Lecture Notes in Control and Information Sciences.* Springer-Verlag, Berlin, 1996.

[122] R. Kelly. Comments on: Adaptive PD control of robot manipulators. *IEEE Trans. on Robotics Automat.*, 9(1):117–119, 1993.

[123] R. Kelly. A simple set–point robot controller by using only position measurements. In *Proc. 12th. IFAC World Congress*, volume 6, pages 173–176, Sydney, Australia, 1993.

[124] R. Kelly, R. Carelli, and R. Ortega. Adaptive motion control design of robot manipulators: An Input-Output approach. *Int. J. of Contr.*, 49(12):2563–2581, 1989.

[125] R. Kelly, R. Ortega, A. Ailon, and A. Loria. Global regulation of flexible joints robots using approximate differentiation. *IEEE Trans. on Automat. Contr.*, 39(6):1222–1224, 1994.

[126] R. Kelly, V. Santibañez, and H. Berghuis. Global regulation for robot manipulators under actuator constraints. Technical report, CICESE, Ensenada, BC, Mexico., 1994.

[127] H.K. Khalil. *Nonlinear systems.* Prentice Hall, 2nd edition, 1996. ISBN 0-13-228024-8.

[128] F.M.H. Khater, R.D. Lorenz, D.W. Novotny, and K. Tang. Selection of flux level in field-oriented induction machine controllers with consideration of magnetic saturation effects. *IEEE Trans. on Ind. Appl.*, 23(2):276–282, 1987.

[129] D. Kim, I. Ha, and M. Ko. Control of induction motors via feedback linearization with input-output decoupling. *Int. J. of Contr.*, 51(4):863–883, 1990.

[130] G.-S. Kim, I.-J. Ha, and M.-S. Ko. Control of induction motors for both high dynamic performance and high power efficiency. *IEEE Trans. on Ind. Electr.*, 39(4):323–333, 1992.

[131] K.C. Kim. *Experiments on Nonlinear Control of Induction Motors.* PhD thesis, Université de Technologie de Compiègne, FRANCE, 1996.

[132] K.C. Kim, R. Ortega, A. Charara, and J.P. Vilain. Theoretical and experimental comparison of two nonlinear controllers for current-fed induction motors. *IEEE Trans. Contr. Syst. Technol.*, 5(3):1393–1402, 1997.

[133] D. E. Koditschek. Natural motion of robot arms. In *Proc. 23rd. IEEE Conf. Decision Contr.*, Las Vegas, NV, 1984.

[134] D. E. Koditschek. Application of a new Lyapunov function to global adaptive attitude tracking. In *Proc. 27th. IEEE Conf. Decision Contr.*, Austin, TX, 1988.

[135] D. E. Koditschek. Robot planning and control via potential functions. In Khatib O., Craig J. J. and Lozano-Pérez T., editors, *The Robotics review 1*, pages 349–367. The MIT Press, 1989.

[136] P. V. Kokotović, H. Khalil, and J. O'Reilly. *Singular perturbations in systems and control.* Academic Press, New York, 1986.

[137] P. V. Kokotović and H. J. Sussman. A positive real condition for global stabilization of nonlinear systems. *Syst. Contr. Letters*, 13(4):125–133, 1989.

[138] A. Kolesnikov and A. Gelfgat. *Design of Multicriterion Systems.* Electroatomizdat, Moscow, 1993. In Russian.

[139] N. N. Krasovskii. *Some problems of the theory of stability of motion.* Gos. Izd-vo Phyz. Mat. Lit., Moscow, 1959. In Russian.

[140] P.C. Krause. *Analysis of Electric Machinery.* McGraw-Hill, 1986.

[141] E. Kreindler and A. Jameson. Conditions for nonnegativeness of partitioned matrices. *IEEE Trans. on Automat. Contr.*, pages 147–148, 1972.

[142] M. Krstić, I. Kanellakopoulos, and P. V. Kokotović. *Nonlinear and Adaptive Control Design.* John Wiley & Sons, 1995. ISBN 0-471-12732-9.

[143] Z. Krzemiński. Nonlinear control of induction motor. In *Proc. 10th IFAC World Congress*, pages 349–354, Münich, Germany, 1987.

[144] M. LaCava, G. Paletta, and C. Piccardi. Stability analysis of PWM systems with PID regulators. *Int. J. of Contr.*, 39(5):987–1005, 1984.

[145] L. Lanari, P. Sicard, and J. T. Wen. Trajectory tracking of flexible joint robots:a passivity approach. In *Proc. 2nd. European Contr. Conf.*, pages 886–892, Groningen, The Netherlands, 1993.

[146] L. Lanari and J. T. Wen. Asymptotically stable set point control laws for flexible robots. *Syst. Contr. Letters*, 19:119–129, 1992.

[147] P. Lancaster and M. Tismenetsky. *The Theory of Matrices: With Applications.* Academic Press, 2nd edition, 1985.

[148] C. Lanczos. *The Variational Principles of Mechanics.* Dover, 4th edition, 1970.

[149] I. D. Landau. *Adaptive control: The model reference approach.* Marcel Dekker, New York, 1979.

[150] Y.S. Lee. A systematic and unified approach to modeling switches in switch-mode power supplies. *IEEE Trans. on Ind. Electr.*, 32:445–448, 1985.

[151] A. A. J. Lefeber. (Adaptive) Control of chaotic and robot systems via bounded feedback control. Master's thesis, University of Twente, Enschede, The Netherlands, 1996.

[152] W. Leonhard. 30 years space vectors, 20 years field orientation, 10 years digital signal processing with controlled AC-drives, a review (Part 1). *EPE Journal*, 1(1):13–20, 1991.

[153] W. Leonhard. *Control of Electrical Drives.* Springer-Verlag, 2nd edition, 1996. ISBN 3-540-59380-2.

[154] J. Levine, J. Lottin, and J. Ponsart. A nonlinear approach to the control of magnetic bearings. *IEEE Trans. Contr. Syst. Technol.*, 4(5), 1996.

[155] S. Lim, D. Dawson, J. Hu, and M.S. de Queiroz. An adaptive link position tracking controller for rigid link, flexible joint robots without velocity measurements. In *Proc. 33rd. IEEE Conf. Decision Contr.*, pages 351–357, Orlando, FL, 1994.

[156] T. Lipo. Synchronous reluctance machines - a viable alternative for AC drives. *Elec. Mach. Pow. Syst.*, 19:659–672, 1991.

[157] X. Liu, G. Verghese, J. Lang, and M. Önder. Generalizing the Blondel-Park transformation of electrical machines: Necessary and sufficient conditions. *IEEE Trans. on Circ. Syst.*, 36(8):1085–1067, 1989.

[158] R.D. Lorenz, T.A. Lipo, and D.W. Novotny. Motion control with induction motors. In Bose [35], pages 1215–1240.

[159] R.D. Lorenz and K.W. Van Patten. High-resolution velocity estimation for all-digital, ac servo drives. *IEEE Trans. on Ind. Appl.*, 27(4):701–705, 1991.

[160] A. Loria. Global tracking control of one degree of freedom Euler-Lagrange systems without velocity measurements. *European J. of Contr.*, 2(2), 1996.

[161] A. Loria. *On output feedback control of Euler-Lagrange systems*. PhD thesis, Université de Technologie de Compiègne, FRANCE, Oct. 1996.

[162] A. Loría, R. Kelly, R. Ortega, and V. Santibañez. On output feedback control of Euler-Lagrange systems under input constraints. *IEEE Trans. Automat. Contr.* , 1996.

[163] A. Loria and R. Ortega. On tracking control of rigid and flexible joints robots. *Appl. Math. and Comp. Sci., Special issue on Mathematical Methods in Robotics*, 5(2):101–113, 1995. Eds. K. Tchon and A. Gosiewsky.

[164] A. Loria, E. Panteley, and H. Nijmeijer. Control of the chaotic Duffing equation with uncertainty in all parameters. *IEEE Trans. Circs. Syst. I: Fund. Th. and Appl.*, 1997. To appear. Available upon request from the authors.

[165] A. Loria, E. Panteley, H. Nijmeijer, and T. I. Fossen. Robust adaptive control of passive systems with unknown disturbances. In *Proc. IFAC NOLCOS*, Enschede, The Netherlands, 1998. To appear.

[166] R. Lozano and B. Brogliato. Adaptive control of robot manipulators with flexible joints. *IEEE Trans. on Automat. Contr.*, 37(2):174–181, 1992.

[167] H. Ludvigsen, R. Ortega, P. Albertos, and O. Egeland. On hybrid control of nonlinear systems under slow sampling: Application to induction motors. In *Proc. IFAC NOLCOS*, Enschede, The Netherlands, 1998.

[168] R. Marino and S. Nicosia. On the feedback control of industrial robots with elastic joints: a singular perturbation approach. In *1st IFAC Symp. Robot Control*, pages 11–16, Barcelona, Spain, 1985.

[169] R. Marino, S. Peresada, and P. Tomei. Adaptive output feedback control of current-fed induction motors. In *Proc. 12th. IFAC World Congress*, volume 2, pages 451–454, Sydney, Australia, 1993.

[170] R. Marino, S. Peresada, and P. Tomei. Exponentially convergent rotor resistance estimation for induction motors. *IEEE Trans. on Ind. Electr.*, 42(5):508–515, 1995.

[171] R. Marino, S. Peresada, and P. Tomei. Global adaptive output feedback control of induction motors with uncertain rotor resistance. In *Proc. 35th. IEEE Conf. Decision Contr.*, pages 4701–4706, Kobe, Japan, Dec. 1996.

[172] R. Marino, S. Peresada, and P. Valigi. Adaptive input-output linearizing control of induction motors. *IEEE Trans. on Automat. Contr.*, 38(2):208–221, 1993.

[173] R. Marino and P. Tomei. Global adaptive output feedback control of nonlinear systems. Part I : Linear parameterization. *IEEE Trans. on Automat. Contr.*, 38:17–32, 1993.

[174] R. Marino and P. Tomei. *Nonlinear Control Design, Geometric, Adaptive and Robust.* Prentice Hall, 1995. ISBN 0-13-342635-1.

[175] J.E. Marsden and A. J. Tromba. *Vector Calculus.* W. H. Freeman and Company, New York, 3rd edition, 1988.

[176] P. Martin and P. Rouchon. Two remarks on induction motors. In *Proc. CESA'96 IMACS Multiconference*, pages 76–79, Lille, 1996.

[177] B. Maschke, R. Ortega, and A. J. van der Schaft. Energy–based Lyapunov functions for forced hamiltonian systems with dissipation. In *Proc. 37th. IEEE Conf. Decision Contr.*, Tampa, Florida, 1998. Submitted.

[178] B. M. Maschke, A. J. van der Schaft, and P. C. Breedveld. An intrinsic hamiltonian formulation of the dynamics of LC circuits. *IEEE Trans. on Circ. Syst.*, 42c(2):73–82, 1995.

[179] J. Meisel. *Principles of Electromechanical-Energy Conversion.* McGraw-Hill, 1966.

[180] D. Merkin. *Introduction to the theory of stability.* Texts in Applied Mathematics. Springer–Verlag, Berlin, 1997.

[181] R. D. Middlebrook. A continuous model of the tapped boost converter. In *Proc. IEEE Pow. Elec. Spec. Conf. (PESC)*, 1975.

[182] R. D. Middlebrook and S. Ćuk. A general unified approach to modelling switching–converter power stages. In *Proc. IEEE Pow. Elec. Spec. Conf. (PESC)*, pages 18–34, 1976.

[183] E. Milent. *Contribution à l'étude d'un actionneur asynchrone à contrôle vectoriel et ses possibilités d'utilisation dans des applications embarquées.* PhD thesis, Université de Technologie de Compiègne, FRANCE, 1992.

[184] T.J.E. Miller. *Brushless Permanent-Magnet and Reluctance Motor Drives.* Clarendon Press, Oxford, 1993. ISBN 0-19-859369-4.

[185] C. Moons and B. De Moor. Parameter identification of induction motor drives. *Automatica*, 31(8):1137–1147, 1995.

[186] R. Morici, C. Rossi, A. Tonielli, and S. Peresada. Adaptive output feedback control of current fed induction motor A rotating reference frame approach. In *Proc. 3rd. European Contr. Conf.*, pages 313–318, Rome, Italy, 1995.

[187] A. S. Morse. Overcoming the obstacle of relative dgree. *European J. of Contr.*, 2(29):29–35, 1996.

[188] P. Moylan and B. Anderson. Nonlinear regulator theory and an inverse optimal control problem. *IEEE Trans. on Automat. Contr.*, 18:460–465, 1973.

[189] R.M. Murray. Nonlinear control of mechanical systems: A Lagrangian perspective. In *Proc. IFAC NOLCOS*, pages 378–389, Tahoe City, CA, June 1995.

[190] A. Nabae, K. Otsuka, H. Uchino, and R. Kurosawa. An approach to flux control of induction motors operated with variable frequency power supply. *IEEE Trans. on Ind. Appl.*, 16(3):342–350, 1980.

[191] J.I. Neĭmark and N.A. Fufaev. *Dynamics of Nonholonomic Systems*. Vol. 33 of Translations of Mathematical Monographs. American Mathematical Society, Providence, Rhode Island, 1972.

[192] P.J. Nicklasson. *Passivity-Based Control of Electric Machines*. PhD thesis, Dept. of Eng. Cybernetics, NTNU, Trondheim, Norway, 1996.

[193] P.J. Nicklasson, R. Ortega, and G. Espinosa-Pérez. Passivity-based control of a class of Blondel-Park transformable electric machines. *IEEE Trans. on Automat. Contr.*, 42(5):629–647, 1997.

[194] S. Nicosia and P. Tomei. Robot control by using only joint position measurement. *IEEE Trans. on Automat. Contr.*, 35-9:1058–1061, 1990.

[195] S. Nicosia and P. Tomei. A method to design adaptive controllers for flexible joints robots. *J. of Robotic Systems*, 10(6):835–846, 1993.

[196] S. Nicosia and P. Tomei. A tracking controller for flexible joint robots using only link position feedback. *IEEE Trans. on Automat. Contr.*, 40(5):885–890, 1995.

[197] H. Nijmeijer and A. J. van der Schaft. *Nonlinear Dynamical Control Systems*. Springer-Verlag, New York, 1990.

[198] R. Nilsen and M.P. Kaźmierkowski. Reduced-order observer with parameter adaption for fast rotor flux estimation in induction machines. *IEE Proc.-D*, 136(1):35–43, 1989.

[199] H. Olsson. *Control systems with friction*. PhD thesis, Lund Inst. of Techn., 1996. ISRN LUTFD2/TFRT–1045–SE.

[200] H. Olsson, K. J. Åström, C. Canudas, M. Gäfvert, and P. Lischinky. Friction models and friction compensation. *European J. of Contr.*, 1998. To appear.

[201] Ph. Opdenacker and E. Jonckheere. LQG balancing and reduced LQG compensation of symmetric passive systems. *Int. J. of Contr.*, 41(1):73–109, 1985.

[202] R. Ortega. Stability analysis of adaptive systems: An approach based on controller structure optimality. *Opt. Cont. Appl. and Methods*, 9(1):87–91, 1988.

[203] R. Ortega. Applications of Input-output techniques to control problems. In *Proc. 1st. European Contr. Conf.*, pages 1307–1313, July 1991.

[204] R. Ortega, C. Canudas de Wit, and S. Seleme. Nonlinear control of induction motors: Torque tracking with unknown load disturbance. *IEEE Trans. on Automat. Contr.*, 38(11):1675–1679, Nov. 1993.

[205] R. Ortega and G. Espinosa-Pérez. A controller design methodology for systems with physical structures:Application to Induction Motors. In *Proceedings of the 30th IEEE Conference on Decision and Control*, pages 2345–2349, Brighton, England, 1991.

[206] R. Ortega and G. Espinosa-Pérez. Torque regulation of induction motors. *Automatica*, 29(3):621–633, 1993.

[207] R. Ortega, A. Loria, and R. Kelly. A semiglobally stable output feedback PI²D regulator for robot manipulators. *IEEE Trans. on Automat. Contr.*, 40(8):1432–1436, 1995.

[208] R. Ortega, A. Loria, R. Kelly, and L. Praly. On passivity-based output feedback global stabilization of Euler-Lagrange systems. *Int. J. Rob. Nonl. Contr., (Special issue on control of nonlinear mechanical systems)*, 5(4):313–325, 1995. H. Nijmeijer and A. J. van der Schaft Eds.

[209] R. Ortega, P.J. Nicklasson, and G. Espinosa-Pérez. On speed control of induction motors. *Automatica*, 32(3):455–460, 1996.

[210] R. Ortega, A. Rodriguez, and G. Espinosa. Adaptive stabilization of nonlinearizable systems under a matching condition. In *Proc. American Control Conference*, San Diego, CA, USA, 1990.

[211] R. Ortega and H. Sira-Ramírez. Lagrangian modeling and control of switch regulated dc-to-dc power converters. In S. Morse, editor, *Control Using Logic-Based Switchings*, volume 22 of *Lecture Notes in Control and Information Sciences*, pages 151–161. Springer-Verlag, 1996.

[212] R. Ortega and M. Spong. Adaptive motion control of rigid robots: a tutorial. *Automatica*, 25(6):877–888, 1989.

[213] R. Ortega, A. Stanković, and P. Stefanov. A passivation approach to power systems stabilization. In *Proc. IFAC NOLCOS*, Enschede, The Netherlands, 1998.

[214] R. Ortega and D. Taoutaou. Indirect field-oriented speed regulation for induction motors is globally stable. *IEEE Trans. on Ind. Electr.*, 43(2):340–341, 1996.

[215] R. Ortega and D. Taoutaou. On discrete-time control of current-fed induction motors. *Syst. Contr. Letters*, 28(3):123–128, 1996.

[216] B. Paden and R. Panja. Globally asymptotically stable PD+ controller for robot manipulators. *Int. J. of Contr.*, 47:1697–1712, 1988.

[217] E. Panteley and R. Ortega. Cascaded control of feedback interconnected nonlinear systems: Application to robots with AC drives. *Automatica*, 33(11):1935–1947, 1997.

[218] E. Panteley, R. Ortega, and P. Aquino. Cascaded control of feedback interconnected nonlinear systems:Application to robots with AC drives. In *Proc. 4th. European Contr. Conf.*, Brussels, Belgium, 1997.

[219] E. Panteley, R. Ortega, and M. Gäfvert. An adaptive friction compensator for global tracking of robot manipulators. *Syst. Contr. Letters*, 33(5):307–314, 1998.

[220] E. Polak. Stability and graphical analysis of first order pulse-width-modulated sampled-data regulator systems. *IEEE Trans. on Automat. Contr.*, 6:276–282, 1961.

[221] H. Pota and P. Moylan. Stability of locally dissipative interconnected systems. *IEEE Trans. on Automat. Contr.*, 38(2):308–312, 1993.

[222] G. K. Pozharitskii. On asymptotic stability of equilibria and stationary motions of mechanical systems with partial disspation. *Prikl. Mat. i Mekh.*, 25:657–667, 1961. Engl. transl. in *J. Appl. Math. Mech.*, vol. 25, 1962.

[223] M. Queiroz and D. Dawson. Nonlinear control of magnetic bearing: A backstepping approach. *IEEE Trans. Contr. Syst. Technol.*, 4(5), 1996.

[224] K. Rajashekara, A. Kawamura, and K. Matsuse. *Sensorless Control of AC Motor Drives*. IEEE Press, 1996. ISBN 0-471-30576-6.

[225] A. Rodriguez and R. Ortega. Adaptive stabilization of nonlinear systems: The non-feedback-linearizabole case. In *Proc. 11th. IFAC World Congress*, volume 4, pages 121–124, Tallinn, USSR, 1990.

[226] N. Rouche and J. Mawhin. *Ordinary differential equations II: Stability and periodical solutions.* Pitman publishing Ltd., London, 1980.

[227] A. Saberi, P. V. Kokotović, and H. J. Sussman. Global stabilization of partially linear systems. *SIAM J. Contr. and Optimization*, 28:1491–1503, 1990.

[228] A. Sabonović, N. Sabanović, and K. Ohnishi. Sliding modes in power converters and motion control systems. *Int. J. of Contr.*, 57(5):1237–1259, 1993.

[229] N. Sadegh and R. Horowitz. Stability and robustness analysis of a class of adaptive controllers for robot manipulators. *Int. J. Rob. Res.*, 9:74–92, 1990.

[230] R. Sanders, G. C. Verghese, and D. F. Cameron. Nonlinear control laws for switching power converters. In *Proc. 25th. IEEE Conf. Decision Contr.*, Athens, Greece, 1986.

[231] S.R. Sanders and G.C. Verghese. Lyapunov-based control for switched power converters. *IEEE Trans. on Pow. Elect.*, 7(1):17–24, 1992.

[232] S. Sastry and A. Isidori. Adaptive control of linearizable systems. *IEEE Trans. on Automat. Contr.*, 34:1123–1131, 1989.

[233] J. Scherpen and R. Ortega. Disturbance attenuation properties of nonlinear controllers for euler–lagrange systems. *Syst. Contr. Letters*, 29(6):300–308, 1997.

[234] K. Schlacher and A. Kugi. Robust design of nonlinear mechatronic systems, experimental and theoretical results. In Proc. COSY Work. on Control of Nonlinear and Uncertain Systems, London, 1998.

[235] F. Schütte. Digitale regelung eines asynchronservoantriebs mit dem signalprozessorboard DS1102 unter verwendung des SIMULINK-Real-Time-Interface. Master's thesis, Universität–Gesamthochschule Paderborn und dSPACE GmbH Paderborn, Germany, 1994. Katalog-Nr. 53.

[236] P. Seibert and R. Suárez. Global stabilization of cascade systems. *Syst. Contr. Letters*, 14(4), 1990.

[237] R. Sepulchre, M. Janković, and P. V. Kokotović. *Constructive nonlinear control.* Springer-Verlag, 1997.

[238] R. P. Severns and G. E. Bloom. *Modern DC-to-DC Switchmode Power Converter Circuits.* Van Nostrand Reinhold, New York, 1983.

[239] H. Sira-Ramírez. Dynamical pulse width modulation control of nonlinear systems. *Syst. Contr. Letters*, 18(2):223–231, 1992.

[240] H. Sira-Ramirez. A geometric approach to pulse-width modulated control in nonlinear dynamical systems. *IEEE Trans. on Automat. Contr.*, 34(2):184–187, 1989.

[241] H. Sira-Ramírez and M. Delgado. A lagrangian approach to modeling of DC-to-DC power converters. *IEEE Trans. Circs. Syst. I: Fund. Th. and Appl.*, 43(5):427–430, 1996.

[242] H. Sira-Ramírez and M. Ilic. Exact linearization in switch mode DC-to-DC power converters. *Int. J. of Contr.*, 50(2):511–524, 1989.

[243] H. Sira-Ramírez and M. Ilic-Spong. A geometric approach to the feedback control of switchmode DC-to-DC power supplies. *IEEE Trans. on Circ. Syst.*, 35(10), 1988.

[244] H. Sira-Ramírez and P. Lischinsky-Arenas. The differential algebraic approach in nonlinear dynamical compensator design for DC-to-DC power converters. *Int. J. of Contr.*, 54:111–134, 1991.

[245] H. Sira-Ramírez, R. Ortega, and M. García-Esteban. Adaptive passivity-based control of average DC-to-DC power converter models. *Int. J. Adapt. Contr. Sign. Process.*, 12(1):63–80, 1998.

[246] H. Sira-Ramírez, R. Pérez-Moreno, R. Ortega, and M. García-Esteban. Passivity-based controllers for the stabilization of DC-to-DC power converters. *Automatica*, 33(4):499–513, 1997.

[247] H. Sira-Ramírez and M. Rios-Bolívar. Sliding mode control of DC-to-DC power converters via extended linearization. *IEEE Trans. Circs. Syst. I: Fund. Th. and Appl.*, pages 652–661, 1994.

[248] H. Sira-Ramírez, M. Rios-Bolívar, and A. S. I. Zinober. Adaptive dynamical input-output linearization of DC–to–DC power converters: A backstepping approach. *Int. J. Rob. and Nonl. Contr.*, 7:279–296, 1997.

[249] H. Sira-Ramírez, R. Tarantino-Alvarado, and O. Llanes-Santiago. Adaptive feedback stabilization in PWM controlled dc-to-dc power supplies. *Int. J. of Contr.*, 57:599–625, 1993.

[250] R. A. Skoog and G. Blankenship. On the stability of pulse-width-modulated feedback systems. *IEEE Trans. on Automat. Contr.*, 13(5):532–538, 1968.

[251] J.J. Slotine and W. Li. Adaptive manipulator control: a case study. *IEEE Trans. on Automat. Contr.*, 33:995–1003, 1988.

[252] E. Sontag. Smooth stabilization implies coprime factorization. *IEEE Trans. on Automat. Contr.*, 34(4):435–443, 1989.

[253] M. Spong. Modeling and control of elastic joint robots. *ASME J. Dyn. Syst. Meas. Contr.*, 109:310–319, 1987.

[254] M. Spong, R. Ortega, and R. Kelly. Comments on "Adaptive manipulator control: a case study". *IEEE Trans. on Automat. Contr.*, 35:761–762, 1990.

[255] M. Spong and M. Vidyasagar. *Robot Dynamics and Control.* John Wiley & Sons, New York, 1989.

[256] A. Stanković, D. Perreault, and K. Sato. Analysis and experimentation with dissipative nonlinear controllers for series–resonant DC–to–DC converters. In *28th IEEE Power Electronics Specialists Conf.*, pages 679–685, St. Louis, Mo, USA, 1997.

[257] A. Stanković, P. Stefanov, G. Tadmor, and D. Sobajic. A dissipativity approach to supression of low frequency oscillations in power systems. Technical report, Northeastern University, 1997.

[258] H.C. Stanley. An analysis of the induction machine. *A.I.E.E. Transactions*, 57:751–755, 1938.

[259] C.P. Steinmetz. The alternating current induction machine. *A.I.E.E. Transactions*, 14:185–217, 1897.

[260] A. Săbanović and D.B. Izosimov. Application of sliding modes to induction motor control. *IEEE Trans. on Ind. Appl.*, 17(1):41–49, 1981.

[261] M. Takegaki and S. Arimoto. A new feedback method for dynamic control of manipulators. *ASME J. Dyn. Syst. Meas. Contr.*, 103:119–125, 1981.

[262] D. Taoutaou, R. Puerto, R. Ortega, and L. Loron. A new field–oriented discrete-time controller for current-fed induction motors. *Control Engg. Practice*, 5(2):209–217, 1997.

[263] D.G. Taylor. Nonlinear control of electric machines: An overview. *IEEE Contr. Syst. Mag.*, 14(6):41–51, Dec. 1994.

[264] J. Theocharis and V. Petridis. Neural network observer for induction motor control. *IEEE Contr. Syst. Mag.*, 14(2):26–37, 1994.

[265] P. Tomei. Adaptive PD control for robot manipulators,. *IEEE Trans. on Robotics Automat.*, 7(4):565–570, 1991.

[266] P. Tomei. A simple PD controller for robots with elastic joints. *IEEE Trans. on Automat. Contr.*, 36(10):1208–1213, 1991.

[267] M. Torres and R. Ortega. *Perspectives on Control*, chapter "Feedback linearization, integrator backstepping and passivity–based control: A comparative example". Springer-Verlag, 1998. D. Normand–Cyrot, ed.

[268] Y. Z. Tsypkin. *Relay Control Systems*. Cambridge University Press, Cambridge, UK, 1984.

[269] V. I. Utkin. *Sliding modes in control optimization*. Springer-Verlag, Berlin, Heidelberg, 1992.

[270] V.I. Utkin. *Sliding Modes and Their Applications in Variable Structure Systems*. MIR Editors, Moscow, 1978.

[271] V.I. Utkin. Sliding mode control design principles and applications to electric drives. *IEEE Trans. on Ind. Electr.*, 40(1):23–36, 1993.

[272] A. J. van der Schaft. $\mathcal{L}_2$-*Gain and passivity techniques in nonlinear control*. Number 218 in Lecture Notes in Control and Information Sciences. Springer-Verlag, Heidelberg, 1996.

[273] A. J. van der Schaft and B. M. Masche. The Hamiltonian formulation of energy conserving physical systems with external ports. *Int. J. of Electronics and Communications*, 49(5/6):362–371, 1995.

[274] T. van Raumer, J.M. Dion, L. Dugard, and J.L. Thomas. Applied nonlinear control of an induction motor using digital signal processing. *IEEE Trans. Contr. Syst. Technol.*, 2(4):327–335, 1994.

[275] P. Vas. *Vector Control of AC Machines*. Clarendon Press, Oxford, 1990. ISBN 0-19-859-370-8.

[276] P. Vedagarbha, D.M. Dawson, and T. Burg. Adaptive control for a class of induction motors via an on-line flux calculation method. In *Proc. IEEE Int. Conf. on Contr. Appl.*, pages 620–625, Dearborn, MI, 1996.

[277] P. Vedagarbha, D.M. Dawson, T. Burg, and Y. Ou. Velocity tracking/setpoint control of induction motors with improved efficiency. In *Proc. 13th. IFAC World Congress*, pages 37–42, San Francisco, CA, 1996.

[278] V. Venkataramanan, A. Sabanovic, and S. Cúk. Sliding mode control of DC–to–DC converters. In *Proc. Int. Conf. Ind. Elec., Contr., and Instr.: IECON '85*, pages 251–258, 1985.

[279] G.C. Verghese, J.H. Lang, and L.F. Casey. Analysis of instability in electrical machines. *IEEE Trans. on Ind. Appl.*, 22(5):853–864, 1986.

[280] G.C. Verghese and S.R. Sanders. Observers for flux estimation in induction machines. *IEEE Trans. on Ind. Electr.*, 35(1):85–94, 1988.

[281] V. Vorpérian. Simplified analysis of pwm converters using the PWM switch, Part I: Continuous conduction mode. *IEEE Trans. Aero. and Elec. Syst.*, 26, 1990.

[282] J. T. Wen and S. Murphy. PID control for robot manipulators. Technical Report 54, CIRSSE, 1990.

[283] J.T. Wen and S.D. Bayard. New class of control laws for robot manipulators, Part I. *Int. J. of Contr.*, 47(5):1288–1300, 1988.

[284] L.L. Whitcomb, A.A. Rizzi, and D.E. Koditscheck. Comparative experiments with a new adaptive controller for robot arms. In *Proc. IEEE Conf. Robotics Automat.*, pages 2–7, Sacramento, CA, 1991.

[285] D.C. White and H.H. Woodson. *Electromechanical Energy Conversion*. John Wiley & Sons, 1959.

[286] J. C. Willems. *The Analysis of Feedback Systems*. The MIT Press, 1971.

[287] J. C. Willems. Dissipative dynamical systems. Part I: General theory. *Arch. Rat. Mech. and Analysis*, 45(5), 1972.

[288] J.L. Willems. A system theory approach to unified electrical machine analysis. *Int. J. of Contr.*, 15(3):401–418, 1972.

[289] G. Winston and P. Gilbert. Adaptive compensation for an optical tracking telescope. *Automatica*, 10(2):125–131, 1974.

[290] B.E. Ydstie and A.A. Alonso. Process systems and passivity via the Clausius–Planck inequality. *Syst. Contr. Letters*, 30(5):253–264, 1997.

[291] D. Youla, L. Castriota, and H. Carlin. Bounded real scattering matrices and the foundations of linear passive networks. *IRE Tran. Circ. Theory.*, 4(1):102–124, 1959.

[292] D.C. Youla and J.J. Bongiorno. A Floquet theory of the general rotating machine. *IEEE Trans. on Circ. Syst.*, 27(1):15–19, 1980.

[293] I. Zein. Power converters laws: An experimental comparative study. Master's thesis, Heudiasyc, URA CNRS 817, Université de Technologie de Comiègne, 1996.

[294] F. Zhang, D. M. Dawson, M. S. de Queiroz, and W. Dixon. Global adaptive output feedback tracking control of robot manipulators. In *Proc. 36th. IEEE Conf. Decision Contr.*, pages 3634–3639, San Diego, USA, 1997.

[295] M. Zribi and J. Chiasson. Position control of a PM stepper motor by exact linearization. *IEEE Trans. on Automat. Contr.*, 36(5):620–625, 1991.

Index

Z